Power Systems

Electrical power has been the technological foundation of industrial societies for many years. Although the systems designed to provide and apply electrical energy have reached a high degree of maturity, unforeseen problems are constantly encountered, necessitating the design of more efficient and reliable systems based on novel technologies. The book series Power Systems is aimed at providing detailed, accurate and sound technical information about these new developments in electrical power engineering. It includes topics on power generation, storage and transmission as well as electrical machines. The monographs and advanced textbooks in this series address researchers, lecturers, industrial engineers and senior students in electrical engineering.

** Power Systems is indexed in Scopus**

More information about this series at http://www.springer.com/series/4622

Rik W. De Doncker • Duco W.J. Pulle
André Veltman

Advanced Electrical Drives

Analysis, Modeling, Control

Second Edition

Springer

Rik W. De Doncker
ISEA - RWTH Aachen University
Aachen, Germany

Duco W.J. Pulle
EMsynergy
Milperra, Sydney, NSW, Australia

André Veltman
Piak Electronic Design B.V.
Culemborg, The Netherlands

Additional material to this book can be downloaded from http://www.springer.com.

ISSN 1612-1287 ISSN 1860-4676 (electronic)
Power Systems
ISBN 978-3-030-48976-2 ISBN 978-3-030-48977-9 (eBook)
https://doi.org/10.1007/978-3-030-48977-9

This Springer imprint is published by the registered company Springer Nature Switzerland AG.
The registered company address is: Gewerbestrasse 11, 6330 Cham, Switzerland

In memory of Prof. André J. A. Vandenput and Prof. Robert D. Lorenz who have inspired the authors.

Foreword to the Second Edition

It is very appropriate that a new edition of this textbook is being published at such an exciting time in the history of power electronics, electric machines, and adjustable-speed drives, as well as the many applications that these critical technologies serve. The electric power grid in developed parts of the world is in the early stages of undergoing a major modernization that is unlike anything it has experienced since its initial installation a century ago. This reimagined grid is characterized by major infusions of renewable energy sources (e.g., solar photovoltaic and wind power) that heavily depend on power electronics, machines, and drives for their existence. At the same time, nearly every major mode of transportation on the land, sea, and air is in some stage of being transformed via electrification. While this has been most apparent to many in the growing numbers of electric vehicles (EVs) on the road, there are technology revolutions simultaneously underway in the electrification of both ships and aircraft that offer major improvements in their efficiency and performance while significantly reducing greenhouse gas emissions. This cannot happen soon enough in light of the seemingly endless stream of reports raising alarms about the dangerous rise of the CO_2 concentration in our atmosphere that is approaching tipping-point levels.

As these electrification opportunities blossom, so does power electronics technology, particularly in the form of maturing wide-bandgap (WBG) power semiconductor devices. The SiC- and GaN-based power devices that are now becoming commercially available mark the vanguard of an unfolding revolution in power electronics, much the same way that MOS-gated Si-based power semiconductors in the form of IGBTs and power MOSFETs revolutionized power electronics 30 years ago. There is little doubt that major expansions in the future WBG-based power device offerings combined with continuing advances in ruggedized power electronics and additive manufacturing will lead to future generations of power converters and machine drives that are much smaller, lighter, less expensive, and more reliable than those that are available today.

Against the exciting backdrop of abundant applications and impressive technology advances described above, this textbook offers students a highly organized and comprehensive treatment of the modeling and control of electric drives for a wide range of electric machine types. Consistent with the observations made in the foreword written 8 years ago by my dear late colleague, Prof. Robert Lorenz, I would like to highlight the conscious efforts made by the authors to unify the underlying physical principles, modeling approaches, and control concepts applied to the several classes of machine drives addressed in their book. The technical material contained in this book provides the serious student a well-defined path towards mastering the modeling and control of nearly all major classes of electric drives, together with invaluable insights into how much these machines and their drives actually share in common. The many examples and simulation based tutorials that are provided throughout the book play a critical role in helping students to visualize (literally) the nature of these principles and their interrelationships.

Realizing the full potential of future breakthroughs in electric drive technology in demanding applications will require large numbers of highly qualified engineers with the best possible education and training in this field to realize the exciting future vision outlined above. Aspiring students who master the compelling technical material presented in this textbook will have made a major step towards preparing themselves to leave their own personal mark on the fulfillment of that inspiring collective vision.

Madison, WI, USA Thomas M. Jahns

Foreword to the First Edition

The value of a textbook is largely determined by how well its structure supports the reader in mastering the depth and breadth of the intended subject. This textbook provides a structure that can achieve that goal for engineers seeking to master key technologies for a wide range of advanced electrical drives.

To achieve that goal, this work wisely places very significant, but common background material in the early chapters, where it introduces the core topologies of power converters and the key issues needed to understand and apply practical power electronic converters. It also lays a sound foundation for understanding the two fundamental approaches for current regulators: hysteresis control and model based control. By providing a sound and detailed background on power converters and current regulators, the rest of the text is able to focus on the advanced electrical drive concepts that are unique to the major classes of machines: DC, AC synchronous machines, AC induction machines, and switched reluctance machines.

Common structures are used to great advantage. To develop a common basis for modeling and control, AC synchronous and AC induction (asynchronous) machines are modeled using an ideal rotating transformer. Common modules are used to provide uniformity in the discussion between the various machine types and to be directly compatible with a simulation modeling environment. A similar structure is extensively used for the control modules that follow the machine modules.

The text's separation of machine modeling from drive control is very helpful. Machine modeling lays a foundation such that controls can logically sequence from classical to advanced drive methodologies. The inclusion of both surface and interior permanent magnet synchronous machines is particularly relevant since those machines are beginning to dominate many applications. The significant treatment of field weakening operation is also critical. The inclusion of limits such as maximum current, maximum flux, maximum torque per flux, and maximum torque per ampere makes the range of operation of the machine drives very transparent. The universal field-oriented control structure is aptly used to unify the subsequent presentation of indirect and direct field orientation control methods.

A very clear transition is made from predominately Lorentz force based machines to purely reluctance torque-based machines. The detailed modeling and evaluation of switched reluctance machines allow drives engineers to correctly model the inherently pulsating torque that each phase provides. The treatment of saturation and its effect on power conversion leads nicely into evaluation of drives with these properties. By including a rigorous discussion of classical hysteresis current control and multi-phase direct instantaneous torque control, the reader can appreciate the structure needed for high-performance control of torque in switched reluctance drives.

Throughout the text, extensive tutorials tie modules that codify key concepts in the theory, to their implementation in a simulation environment. This makes it possible for the reader to quickly explore details and develop confidence in their mastery of major concepts for advanced electrical drives.

By following the approach of this book, I believe that advanced drive engineers will be able to develop depth and breadth that is not normally easy to achieve.

Madison, WI, USA Robert D. Lorenz

Preface

Mastering the synergy of electromagnetics, control, power electronics, and mechanical concepts remains an intellectual challenge. Nevertheless, this barrier must be overcome by engineers and senior students who have a need or desire to comprehend the theoretical and practical aspects of modern electrical drives. In this context, the term *drive* represents a plethora of motion control systems that are currently present in the industry.

This book *Advanced Electrical Drives* builds on basic concepts outlined in the book *Fundamentals of Electrical Drives* (2nd Edition) by the same authors. Hence, it is prudent for the uninitiated reader to consider this material prior to tackling the more advanced material presented in this book. Others well versed in the basic concepts of electrical drives should be able to readily assimilate the material presented as every effort has been made to ensure that the material presented can be mastered without the need to continually switch between the books.

In our previous work, the unique concept of an *ideal rotating transformer* (IRTF), as developed by the authors, was introduced to facilitate the basic understanding of torque production in electrical machines. The application of the IRTF module to modern electrical machines as introduced in *Fundamentals of Electrical Drives* is fully explored in this volume. As such it allows the user to examine a range of unique dynamic and steady-state machine models which cover brushed DC, non-salient/salient synchronous, and induction machines.

In addition, this volume explains the *universal field-oriented* (UFO) concept which demonstrates the concepts of modern vector control and exemplifies the seamless transition between the so-called *stator flux* and *rotor flux* oriented control techniques. Together with IRTF, the UFO concept forms a powerful tool for the development of flux oriented machine models of all types of rotating field machines. These models form the basis of UFO vector control techniques which are covered extensively together with traditional drive concepts. In the last sections of this book, attention is given to the dynamic modeling of *switched reluctance* (SR) drives, where a comprehensive set of modeling tools and control techniques are presented.

Similar to the first edition and second edition of the book *Fundamentals of Electrical Drives*, the interactive learning process using *build and play* modules is continued. The simulation tool *PLECS*, which contains a tailored set of modules, is used as a virtual experiment that brings to life the circuit and generic models introduced in the text. This approach provides the reader with the opportunity to interactively explore, fully comprehend, and visualize the concepts presented in this text. A link via the Springer website provides the reader with access to the tutorials associated with this book.

The text *Advanced Electrical Drives* should appeal to the readers in industry and universities who have a desire or need to understand the intricacies of modern electrical drives without losing sight of the fundamental principles. The book brings together the concepts of IRTF and UFO which allow a comprehensive and insightful analysis of AC electrical drives in terms of modeling and control. Particular attention is also given to switched reluctance drives modeling methods and modern control techniques.

The tutorials linked to this book are simulation-driven, but the reader is reminded of the fact that our companion book *Applied Control of Electrical Drives* provides a comprehensive set of "hands-on" laboratory examples, which complement the material presented in this book and *Fundamentals of Electrical Drives*.

Aachen, Germany — Rik W. De Doncker
Sydney, NSW, Australia — Duco W. J. Pulle
Culemborg, The Netherlands — André Veltman

Acknowledgements

That this work has come to fruition stems from a deep belief that the material presented in this book will be of profound value to the educational institutions and the engineering community as a whole. In particular, the fast but accurate simulations that accompany the tutorials provide a new way of learning that is highly interactive, so that they may stimulate the creativity of students and experts alike by virtue of virtual experiments.

The content of this book reflects on the collective academic and industrial experience of the authors and co-workers. In this context, the inputs of students and research associates cannot be overestimated. The authors wish to acknowledge the staff at the Institute for Power Electronics and Electrical Drives (ISEA) of RWTH Aachen University. In particular, the authors would like to thank (in alphabetical order) Matthias Bösing, Christian Carstensen, Martin Hennen, Knut Kasper, Markus Kunter, Christoph Neuhaus, and Daniel van Treek for their contribution to the first edition, as well as Rolf Loewenherz, Stefan Quabeck, Iliya Ralev, and Annegret Klein-Hessling for their contributions to this second edition.

Contents

List of Figures

List of Tables

Chapter 1
Modern Electrical Drives: An Overview

1.1 Introduction

An electrical drive, as shown in Fig. 1.1 can be defined in terms of its ability to efficiently convert energy from an electrical *power source* to a *mechanical load.* The main purpose of the drive is to control a mechanical load or process. The direction of energy flow is generally from electrical to mechanical, i.e., motoring mode with power flow from the *power source* to the *mechanical load* via the *converter* and *machine* as shown in Fig. 1.1. However, the energy flow can in some cases be reversed, in which case the drive often is configured bi-directional to also allow energy flow from the *mechanical load* to the *power source*, i.e., generating mode. Modern electrical drives, as considered in this book, utilize power electronic devices to (digitally) control this power conversion process, a feature which is highlighted in Fig. 1.1 by the presence of the *modulator* and *controller* unit. Note that in some cases the *modulator* is simply removed in which case the power electronic devices in the *converter* are controlled directly via the *controller* module. In addition, the controller module shown in Fig. 1.1 must be able to communicate with higher level computer systems because drives are progressively networked. Communication links to high-level computer networks are required to support a range of functions, such as commissioning, initialization, diagnostics, and higher level process control. The embedded digital controller shown in Fig. 1.1 houses the high speed logic devices, processors, and electronic circuitry needed to accommodate the *sensor signals* derived from mechanical and electrical sensors. Furthermore, and most importantly, suitable control algorithms must be developed to facilitate the power conversion processes within the drive. From this perspective, drive technology can be considered as a relative "newcomer." This statement maybe put in perspective by considering that electrical machine development commenced approximately one hundred and 60 years ago. However, with the advent of new materials and new design tools, novel machine concepts such as linear machines, PM magnet, switched reluctance, and transversal flux machines, to name only a

R. W. De Doncker et al., *Advanced Electrical Drives*, Power Systems,
https://doi.org/10.1007/978-3-030-48977-9_1

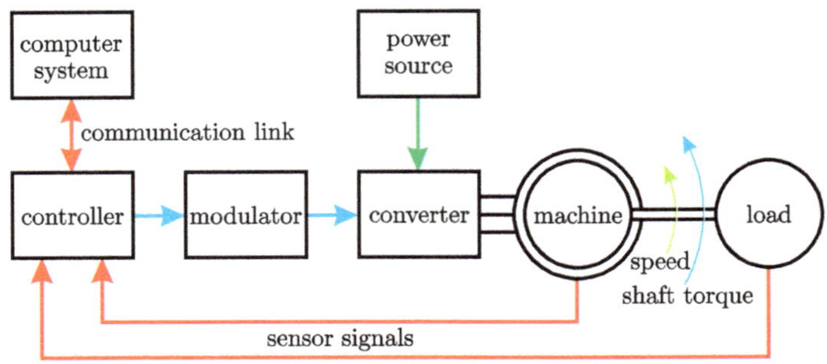

Fig. 1.1 Typical drive setup block diagram

few, have been developed over the past 30 years. Although the era of modern, silicon based power semiconductors started just about 60 years ago with the introduction of Silicon Controlled Rectifiers (SCRs), which are nowadays called thyristors, high power MOS-gated turn-off devices, which are essential to develop voltage source converters, became available only 35 years ago. Simultaneously, high speed digital signal processing devices have come to maturity, i.e. their computational speed increased from 5 MIPs to 100 MFLOPs, over the past 35 years. Furthermore, suitable control algorithms such as *field-oriented control* have been developed over the past 40 years. Ongoing drive development is fueled by the continuous emergence of new drive related drive products such as new processors, sensors, and most importantly new control algorithms aimed at, for example, the elimination of expensive position sensors in electrical drives. Such developments enhance drive robustness, improve reliability, and expand the use of electrical drives to other industrial applications, hitherto considered to be unfeasible. The power range associated with these industrial applications is impressive and typically ranges from a few milliwatts to hundreds of megawatt, which underlines the flexibility and broad application base of modern drive technology. Figure 1.2 shows two examples which are positioned at opposite ends of the drive power range.

1.2 Drive Technology Trends

This section aims to provide the reader with an overview of technology trends associated with key components of the electrical drive as shown in Fig. 1.1. Most importantly the objective in each of the ensuing subsections is to identify important developments and trends of key drive elements such as the machine, converter, and controller.

Fig. 1.2 Two examples of drive technology at opposite ends of the power range. Left, a very small (less than 1 W) drive. Right, a high power (in excess of 10 MW) example [17]

1.2.1 Electrical Machines

The primary electro-mechanical energy converter of the drive is the electrical machine, which must be controlled in accordance with the industrial processes in which the unit is deployed. Modern (rotating) electrical drives typically use one of the three electrical machine types shown in Fig. 1.3. These machine types are referred to as the *induction (asynchronous)*, *PM synchronous*, and *switched reluctance* machine and are shown in Fig. 1.3 consecutively from top to bottom and will be discussed extensively in this book. Both asynchronous and synchronous machine configurations depicted in Fig. 1.3 are shown with the typical three-phase winding, which is located in the stator slots of the machine. Note that other machine types, including the *brushed DC machine*, are also still in use. Of the three configurations shown in Fig. 1.3, the induction machine is most commonly found in industrial drives. This can be attributed to the inherent robustness of the machine itself and the presence of tried and proven drive components which form the basis of a reliable drive. Above all, the emergence of fast, low cost digital processors, and micro-controllers has been instrumental in achieving this market position, given that these controller units are able to accommodate well established control algorithms such as *field-oriented control*. The end result is a brushless and in many cases (position) sensorless induction machine drive with dynamic performance that outperforms that of the classical brushed DC machine.

An important performance parameter of the machine is the *power density*, i.e., the output power to weight ratio in (kW/kg). The power density over the past century has steadily increased from 0.02 kW/kg at the beginning of the twentieth century to 0.15 kW/kg by 1970, according to an "S" curve as typically found in maturing technologies. The expectation at that time was that further substantial increases in

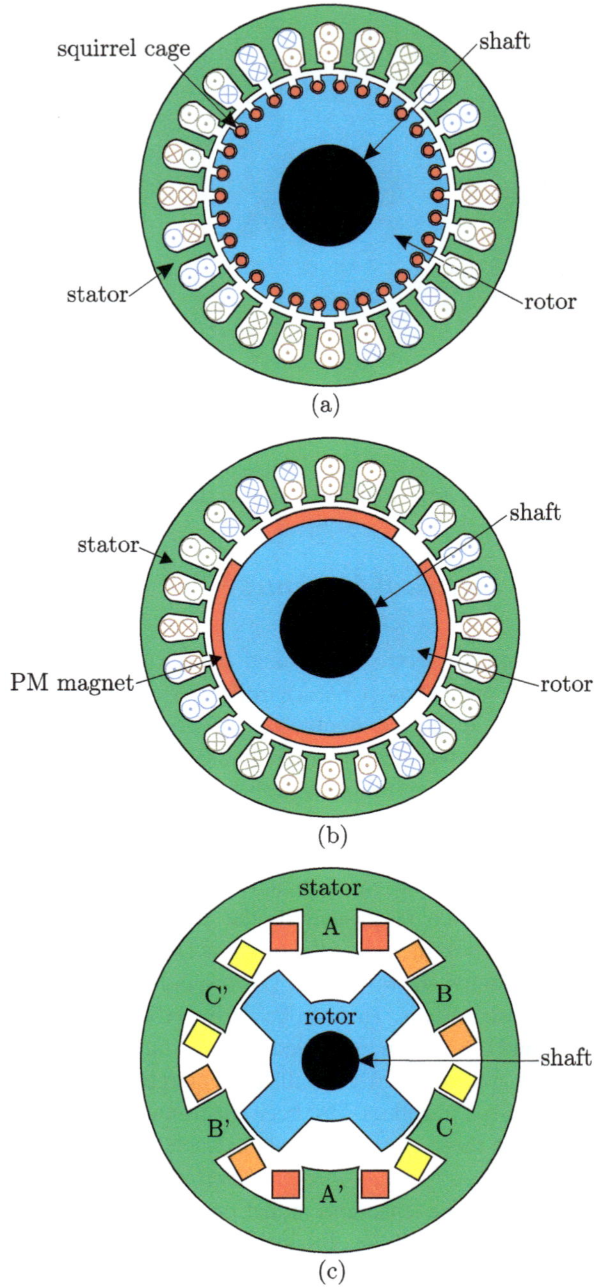

Fig. 1.3 Example of commonly used machine configurations. (**a**) Induction machine. (**b**) Synchronous machine. (**c**) Switched reluctance machine

power density were unlikely. Consequently, power density values were expected to level off around the 0.16 kW/kg value, given the need to keep operating temperatures within acceptable levels. The importance of this statement follows from Fig. 1.4,

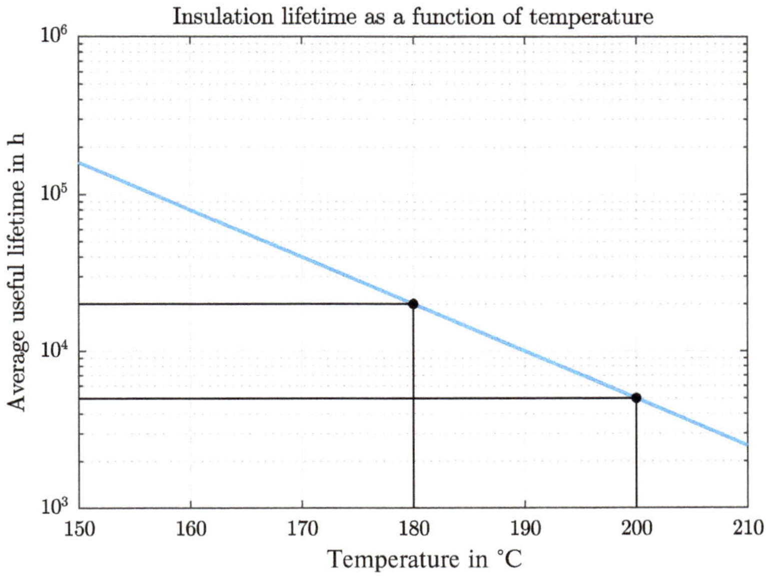

Fig. 1.4 Useful lifetime of insulation material versus operating temperature (class H)

which shows the relationship between operating temperature and lifetime of a class H insulation material. Hence, according to this figure, a lifetime of, for example, 5000 h can be expected provided operating temperatures are constrained to 200 °C whereby the latter is primarily dependent on the power density level.

Subsequently, improvements in power density could, according to thinking in the past century, only be realized by the development of more efficient magnetic materials, i.e., with lower eddy current and hysteresis losses, and/or improvements in insulation materials as to allow extended operation at higher temperatures, without compromising machine lifetime. Given these arguments, the reader may well be surprised to hear that machines with power densities between 1.2 and 3.5 kW/kg are now emerging despite the fact that the magnetic and insulation material have virtually remained unchanged. A set of factors have contributed to this substantial increase in power density namely:

- Improvements in bearing and gear technology have been primarily instrumental in achieving high power densities. In this context it is noted that output power is the product of shaft torque and shaft speed. Rated shaft torque is primarily determined by rotor volume and maximum flux density of 2 T, (which has not substantially changed over the past century). This implies that power density improvements have been achieved by substantially raising operating speeds from typically 3000 rpm (as commonly used around the turn of the last century) to 6000 rpm for traction machine and 16,000 rpm up to 25,000 rpm in hybrid and electrical vehicles. Machines with speeds reaching 100,000 rpm are now manufactured with inherent excellent cooling capabilities and power density values up to 3.5 kW/kg. Such machines operate under low torque conditions which implies that the design can be made light and compact.

- Improvements in quality control and automated production reduces losses and allows design with higher precision. A prime example is the use of copper injection techniques for manufacturing squirrel-cage rotors of induction machines [2].
- Availability of improved design tools (to be discussed in Sect. 1.3) for machines which allow the user to examine and fine-tune magnetic, thermal (to identify hot spots) and acoustical behavior. In addition, extensive simulation tool are now available which allow the user to examine dynamic operating cycles with different load and control scenarios.
- Improved cooling techniques to avoid hot spots, which leads to a better utilization of available insulation materials. Consequently, better thermal conductivity is achievable in machines.
- Reduced derating of converter connected machines. Traditionally a 15% derating figure was imposed to counter the temperature rise due to higher harmonics generated by the converter. Modern converter are able to operate with higher (than used in the previous century) operating frequencies without compromising cost and efficiency of the converter, as will become apparent in the ensuing subsection.

1.2.2 Power Converters

The development of power electronic converter has over the past 35 years been substantial as may be gauged from the volumetric power density parameter, i.e., the ratio of converter apparent power to volume $\mathrm{kVA/m^3}$. For industrial air-cooled *ac* to *ac* converters the volumetric power density has increased from $30\,\mathrm{kVA/m^3}$ (at the end of the last century) to present day values up to $500\,\mathrm{kVA/m^3}$. These substantial improvements to the volumetric power density can be attributed to a number of factors namely:

- Availability of power semiconductors able to switch more efficiently, i.e., with lower power dissipation in comparison with values found in devices present at the turn of the century.
- Improvements to heat-sink technology which means that smaller modules can be developed.
- Use of topologies and control techniques to minimize power device losses, such as, for example, soft-switching techniques.
- Use of better design tools as will become apparent in Sect. 1.3.
- Availability of compact high-performance digital processors with extensive I/O capabilities that can be readily interfaced with the equally compact power electronic drive circuitry required to control the switching devices.

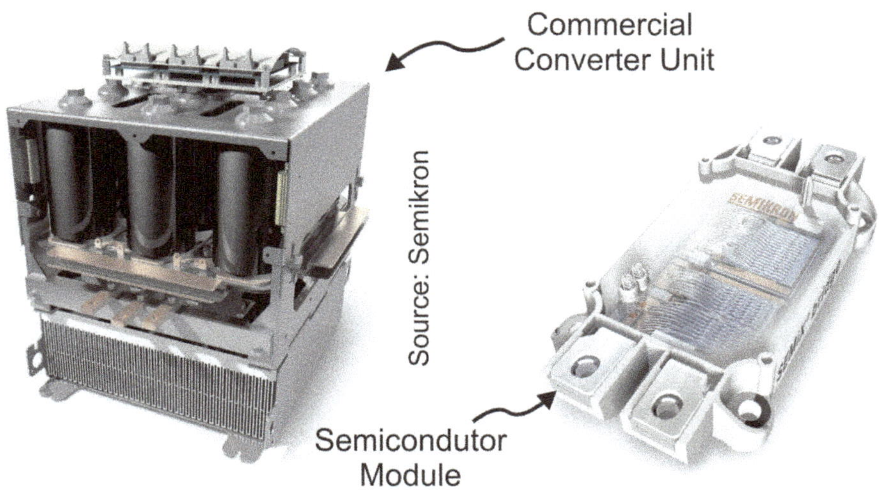

Fig. 1.5 Example of commercial converter unit and integrated semiconductor module which can be used to build an inverter [16]

- Design and manufacturing improvements in passive devices most notably in capacitors which play a key role in terms of overall voltage source converter sizing and costs.

The culmination of the improvements indicated above is exemplified by the availability of building blocks as shown in Fig. 1.5, to construct a complete inverter and "off-the-shelf" commercial inverters, as shown in Fig. 1.6, which can be readily interfaced to electrical machines. Note that the term *inverter* refers to a dc to ac converter, as shown in Fig. 1.1.

In this section, the emphasis has been predominantly placed on improvements in volumetric power density. However, improvements in converter technology leading to the ability to operate at much higher electrical fundamental frequencies have also been instrumental in realizing high speed drives.

The primary design constraint on the volumetric power density of the converter is thermal, i.e., the need to limit operating temperatures and guarantee sufficient thermal cycles of the semiconductor devices and corresponding packages. This implies that the volumetric power density is to a large extend governed by the specific losses of the devices in use, method of cooling, and drive operating conditions. In electric and hybrid vehicles high power density values for machine and converter are essential. An example of such as drive, as shown in Fig. 1.7, utilizes a liquid cooled DC to AC converter with a volumetric power density of 6000 kVA/m^3 and a 55 kW switched reluctance machine with a power density of approximately 1.2 kW/kg.

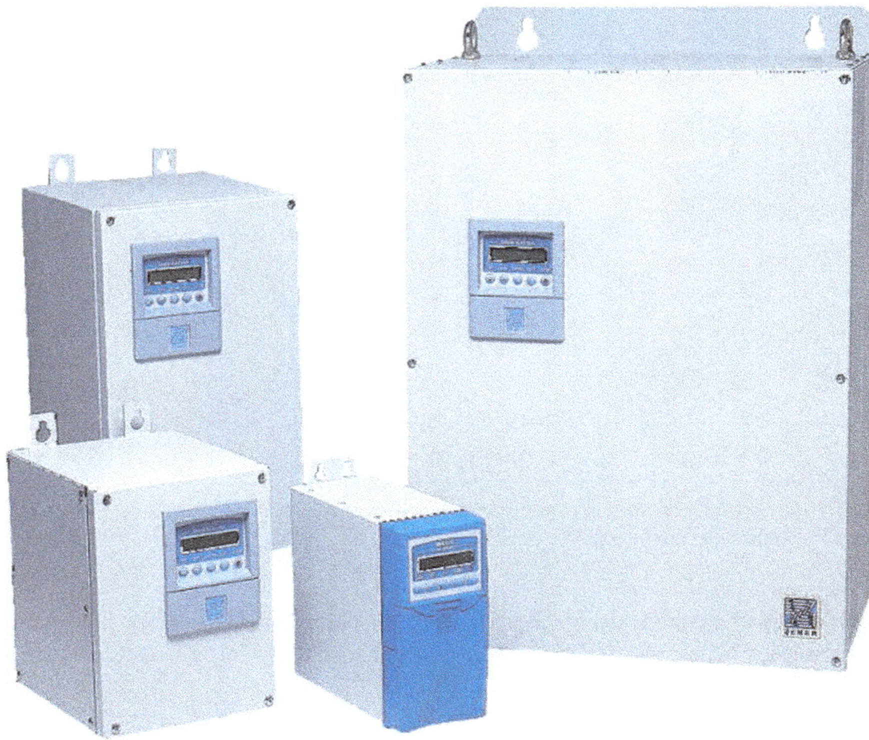

Fig. 1.6 Example of modern inverter technology with standard communication interfaces [25]

1.2.3 Embedded Control and Communication Links

The *controller* and *modulator* (if used), as shown in Fig. 1.1, are part of an embedded system which is interfaced with the switching device drivers, and sensors (voltage and/or current and position/speed measurements). In addition, these specialized computer systems in the form of digital signal processors or micro-controllers are specifically tailored for electrical drive applications. As such they are provided with extensive interface (digital/analog inputs and outputs) capabilities including networking capabilities for communication with other higher level computer systems. Both fixed and floating point processors are available with able computing power to accommodate the real time processing requirements of the drive. An example of a DSP unit as given in Fig. 1.8 clearly shows the multiple I/O *command* signals and other interface capabilities as required for an electrical drive.

The advancements in drive development could not have been realized without innovations over the past 40 years in the field of control algorithms, software tools, and hardware related to the controller. A brief overview of key innovations related to the controller can be summarized as follows:

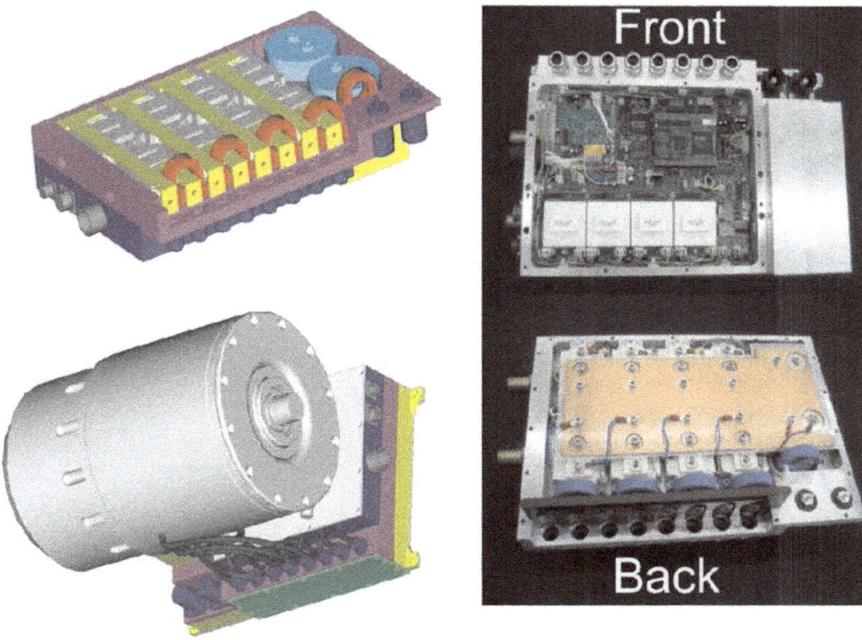

Fig. 1.7 Example of an electrical vehicle propulsion unit which utilizes a liquid cooled AC–DC converter and 55 kW switched reluctance machine [6, 8]

- Development of *Field-oriented control* (FOC) algorithms for AC machines. This control technique has led to the complete decoupling of the flux linkage and current. As a result, control of an AC machine is akin with the brushed DC machine, in terms of dynamic performance [5, 10].
- Synchronized space vector, pulse width modulation (PWM) techniques have been instrumental in achieving better DC bus voltage utilization and improvements to the output frequency spectrum of three-phase converters [23].
- Development of *Direct Torque control* (DTC) algorithms for AC and switched reluctance machines. Direct torque control for AC machines simplifies the overall drive technology given that the controller is directly interface with the drive circuitry of the switches, i.e., the modulator can be omitted. Direct torque control for switched reluctance drives (which do not utilize a modulator) empowers these highly nonlinear machines with servo drive performance capabilities [9].
- Development of control algorithms which make position and speed sensors superfluous [11, 20]. In addition, control techniques have also been used to minimize the number of electrical (voltage/current) sensors needed. Furthermore, power devices are now available with integrated current sensors which enhances overall cost effectiveness.

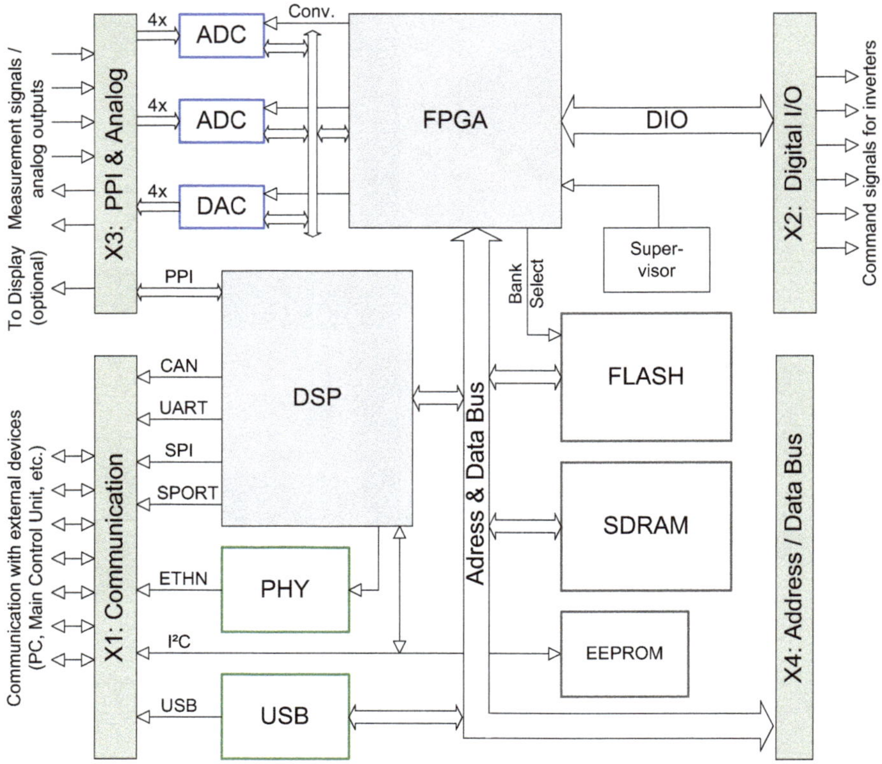

(a) DSP control board architecture

(b) DSP board example

Fig. 1.8 (**a**) Example of a DSP control board architecture, which demonstrate the flexibility available in terms of interfacing to other peripherals [1], (**b**) a Texas Instruments LAUNCHPAD board of the C2000 family DSP [19]

- The availability of high-performance (fast) fixed point as well as floating point digital signal processors (DSP) and micro-controllers (μC) have simplified the implementation of real time complex control algorithms as mentioned above.

- Availability of a range of programming tools for these DSP and μC units, which utilize high-level programming languages, such as C++, and graphical programming tools, such as MATLAB/SIMULINK [22], ACTIVATE [3], and EMBED, serve to shorten overall drive development times and enhance drive application flexibility.

These innovations related to embedded control have had a profound influence on drive development. Notably, the distinction between low and high-performance drives has become less pronounced because the same control platform can be used. Changes between the two are a matter of introducing different control algorithms and introducing or omitting electrical and/or position sensors. This implies that drives can be readily adapted to accommodate different industrial processes. Furthermore, changes to the power level of a drive can be carried through by using different converter packages. The net result is that plug-and play concepts are increasingly used by drive manufacturers where the various components shown in Fig. 1.1 can be integrated and interchanged to suit a diverse range of drive applications. In all cases diverse standardized communication links are available to the user so that drives can be readily integrated in larger automated systems.

1.3 Drive Design Methodology

In this section, attention is given to drive design methodology, i.e., the processes used to develop and evaluate electrical drives. A thorough appreciation of these processes is important as they demonstrate the need for specialized software tools and above all a clear understanding of all the key drive components as well as the nature of the industrial process to be controlled.

Typically, a closed loop iterative design process is used which encompasses the complete drive as may be observed from the example given in Fig. 1.9. The example shown in Fig. 1.9 considers the design optimization process of a switched reluctance drive, but the approach is equally valid for other drives.

A brief discussion of the process shown in Fig. 1.9 may be initiated by considering the type of *application* for which the drive is intended. In this context, knowledge of the industrial process to be controlled is essential. For example, drive development for an industrial process in the form of a compressor or fan is far less demanding than an off-shore drilling rig where dynamic load fluctuations must be controlled. Once the application is well defined, *design goals and constraints* can be established to form the basis for the *initial design of the machine and converter*. At this stage the type of machine and drive control algorithm to be used are also formulated, which may have a large impact on the time needed to complete the design process. For example, switched reluctance machines are highly nonlinear by nature and the design of the machine is tied to the converter and envisaged control techniques. Use of "off-the-shelf" rotating field AC machines, on the other hand, simplifies the design process as may be expected. In the latter

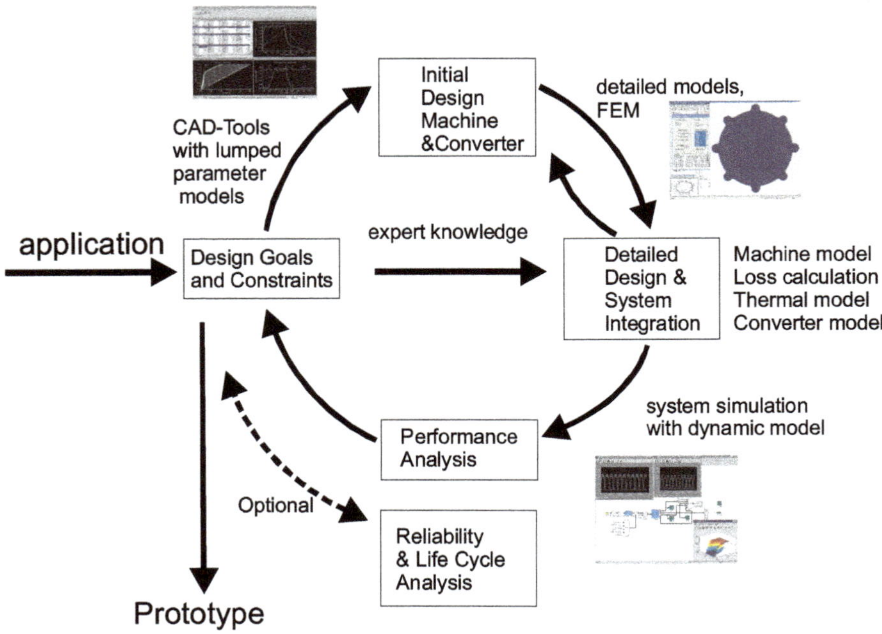

Fig. 1.9 Typical design methodology used for electrical drives [3, 7, 18, 22]

case, which is typical for AC machines, manufacturer data is usually available and can be used to generate a well-defined set of parameters. For switched reluctance machines, a more extensive set of characteristics must be defined using computer aided design (*CAD-Tools*). Once the *initial design machine and converter* (see Fig. 1.9) has been established, the *detailed design and system integration* phase can be entered. This may be done in several layers. The first is to develop a global simulation model of the drive using ideal components. Software tools such as, for example, MATLAB/SIMULINK [22], PLECS [14] or ACTIVATE/COMPOSE [3] can be used to build dynamic simulation models of the complete drive. Such models utilize preplanned control algorithms which, for example, provide precise torque control. Usually additional control structures are added at this stage to empower the drive with the ability to control the intended industrial process. Subsequent *system simulation* of the model provides the transient and steady-state data needed to undertake a provisional *performance analysis*. On the basis of this information, further refinements to the model are undertaken which may involve the *initial design of the machine and converter* and *detailed design and system integration* elements shown in Fig. 1.9.

The level of design refinement is heavily dependent on the nature of the drive application. For example, applications involving aircraft and electric vehicles need very high power density levels, which can be achieved by undertaking more extensive computer modeling of the machine and converter. Such studies call for the use of sophisticated two- or three-dimensional *finite element* (FEM) simulation

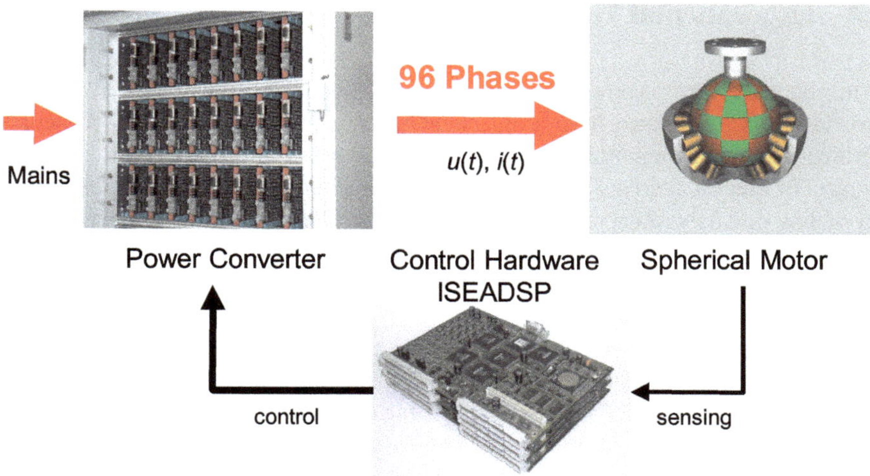

Fig. 1.10 Ninety six-phase spherical machine with corresponding converter and control platform with utilizes six DSP's [13]

which can evaluate the magnetic behavior of the machine. In addition software tools exist to examine the thermal behavior of the machine and converter. Acoustical noise and vibration modes of the machine can also be considered well before the prototyping stage is reached.

A prime example of a drive, for which the design process described above has been applied is given in Fig. 1.10. The machine shown is a 96-phase spherical permanent magnet machine that is connected to a power converter [13]. Each of the 96-phase converter modules was independently current controlled [12] by a dedicated control unit operating under field-oriented control. A global control algorithm was used to identify the appropriate phase currents to control the rotor with μm precision at high torque density level.

After completion of the design stage, a drive prototype was built and evaluated. Parameter adjustments and model improvements were needed. However, overall experimental performance was generally consistent with the results obtained during the design phase. Projects of this kind demonstrate the need for drive engineers to have an in depth understanding of wide ranging technologies linked to electrical machines, mechanical industrial processes, power electronics, passive and active electronic circuitry, control hardware, software design tools, communication, and control algorithms. Mastering and extending the boundaries of these technologies is a formidable challenge to the next generation of drive engineers.

1.4 Experimental Verification

In this book the focus is on simulation and control synthesis of electrical drives in the broadest sense. However, from a didactical and research perspective experimental verification remains important. It is for that reason that a companion book "Applied Control of Electrical drives" [15] was written. This book provides a comprehensive set of low cost laboratory examples that demonstrate many of the concepts to be discussed in this book. Notwithstanding the above, it is prudent to provide the reader with two examples which show the tool-chain associated with this experimental verification process.
The first example as given in Fig. 1.11 makes use of PLECS [14] for simulation and this program has the ability to generate C-Code that then compiles using Visual DSP, where it is embedded in the C-code of the test-bench.

Using the software "AixcontrolCenter," the compiled code is subsequently uploaded to the test-bench and the machine under consideration can be tested. The test-bench shown above consists of a DSP System and Power Rack and can accommodate a range of machines. Notably this test-bench includes asynchronous machines (ASM), switched reluctance machines (SRM), permanent magnet synchronous machines (PMSM), and DC machines (DCM). Online debugging of the test setup is possible using the software "AixScope," enabling: online read, write, and display internal variables of the DSP. Subsequent offline analysis can then be done using MATLAB.

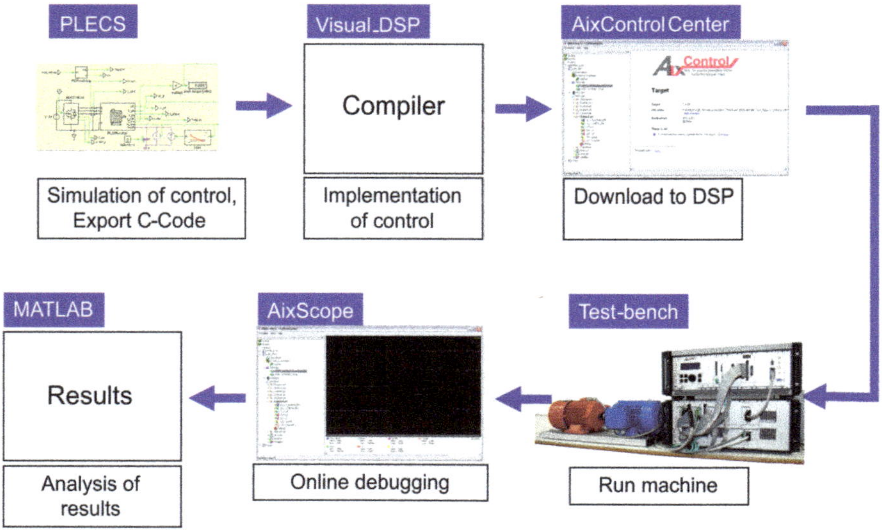

Fig. 1.11 Example 1: tool-chain for experimental verification using PLECS and AixControl [1, 4, 14, 22]

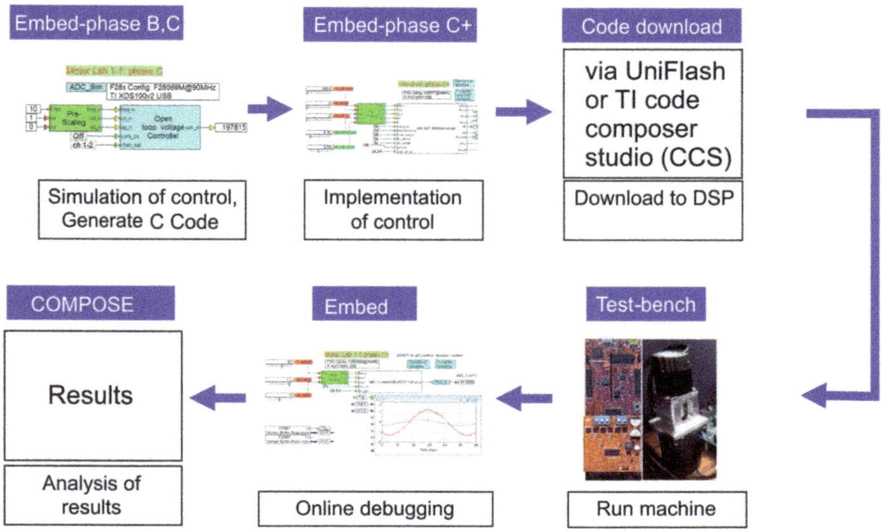

Fig. 1.12 Example 2: tool-chain for experimental verification using EMBED, COMPOSE, and Texas Instruments hardware [3, 15, 21]

The second tool-chain experimental verification example, shown in Fig. 1.12 and is the setup used in the companion book "Applied Control of Electrical drives." This tool-chain starts with the development of fixed point, embedded control phases B and C. During "phase B" a drive control example, developed for example with ACTIVATE, is converted to to a fixed point format using EMBED [3]. Then this controller is evaluated using a model of the machine under consideration. In development stage "phase C," the hardware interface modules, such as the PWM and ADC modules are added, after which automatic C-code compilation is done. At no stage is C-code literacy required, hence the user can concentrate on implementing the drive controller. The generated C-Code file is used during "phase C+," where the user can setup the input and output variables, scopes, sliders, etc. When this phase is started, the program either downloads the C-code directly to the DSP or the user can download to flash memory using for example UNIFLASH or CODE COMPOSER STUDIO (CCS). The test-bench hardware consists of a low cost Texas Instruments 28,069 MCU based LaunchpadXL board, attached to a boostpack [21]. The latter module, which is the power electronic converter can be connected to suitable motors that can accommodate a 24 V DC bus supply. This setup can be used with a DC machine (as shown in our book Fundamentals of Electrical drives, 2nd edition [24]), an induction machine or permanent magnet machine. During "phase C+," the user can control the drive and evaluate performance or debug online using real time plots. Offline analysis can, for example, be undertaken with the aid of COMPOSE. Furthermore, a comprehensive set of laboratory examples is included for sensorless control of single/three-phase induction and PM machines, where use of the InstaSpin algorithm [20] is made.

References

1. AixControl GmbH (2010). http://www.aixcontrol.de
2. Alcoa, Inc (2010). http://www.alcoa.com
3. Altair Engineering, Inc (2018). https://www.altair.com/
4. Analog Devices, Inc (2010). http://www.analog.com/en/embedded-processing-dsp/blackfin/vdsp-bf-sh-ts/processors/product.html
5. Blaschke F (1972) The principle of field orientation as applied to the new transvektor closed-loop control system for rotating-field machines. Siemens Rev 39(5):217–219
6. Carstensen CE, Inderka RB, Netzer Y, Doncker RWWD (2002) Implementation of a 75 kW switched reluctance drive for electric vehicles. In: 19th International electric vehicle symposium EVS19
7. CEDRAT Group (2010). http://www.cedrat.com/en/software-solutions/flux.html
8. De Doncker R (2006) Modern electrical drives: design and future trends. In: Proceedings of the international power electronics and motion control conference, IPEMC2006, Beijing, vol 1, pp 1–8. https://doi.org/10.1109/IPEMC.2006.4777944
9. Depenbrock M (1988) Direct self-control (DSC) of inverter fed induction machine. IEEE Trans Power Electron 3:4
10. Hasse K (1969) Zur dynamik drehzahlgeregelter antriebe mit stromrichtergespeisten asynchron-kurzschlussläufermaschinen. PhD Thesis, TH Darmstadt
11. Holtz J (1993) Speed estimation and sensorless control of AC drives. In: IEEE industrial electronics conference (IECON'1993), vol 2, pp 649–654. https://doi.org/10.1109/IECON.1993.339003
12. Kahlen K, Doncker RD (2000) Current regulators for multi-phase permanent magnet spherical machines. In: Conference record of the 2000 IEEE industry applications conference. Thirty-Fifth IAS annual meeting and world conference on industrial applications of electrical energy (Cat. No. 00CH37129). IEEE, Piscataway. https://doi.org/10.1109/ias.2000.882153
13. Kahlen K, Voss I, Priebe C, De Doncker R (2004) Torque control of a spherical machine with variable pole pitch. IEEE Trans Pow Electr 19(6):1628–1634. https://doi.org/10.1109/TPEL.2004.836623
14. Plexim GmbH (2018). https://www.plexim.com/plecs
15. Pulle DWJ, Darnell P, Veltman A (2015) Applied control of electrical drives. Springer International Publishing, Berlin. https://doi.org/10.1007/978-3-319-20043-9
16. Semikron International GmbH (2010). http://www.semikron.com
17. Siemens AG (2010). http://support.automation.siemens.com/
18. Speed Laboratory (2010). http://www.speedlab.co.uk/software.html
19. Texas Instruments Incorporated (2018). http://www.ti.com/tool/LAUNCHXL-F28379D
20. Texas Instruments Incorporated (2018). http://www.ti.com/ww/en/mcu/instaspin/#
21. Texas Instruments Incorporated (2018). http://processors.wiki.ti.com/index.php/C2000_LaunchPad#LAUNCHXL-F28069M
22. The MathWorks, Inc (2010). http://www.mathworks.com/
23. van der Broeck H, Skudelny HC, Stanke G (1988) Analysis and realization of a pulsewidth modulator based on voltage space vectors. IEEE Trans Ind Appl 24(1):142–150. https://doi.org/10.1109/28.87265
24. Veltman A, Pulle DW, Doncker RWD (2016) Fundamentals of electrical drives. Springer International Publishing, Berlin. https://doi.org/10.1007/978-3-319-29409-4
25. Zener Electric Pty Ltd (2010). http://www.zener.com.au

Chapter 2
Modulation for Power Electronic Converters

2.1 Introduction

At present, voltage source converters are mostly used in electrical drives. These converters utilize capacitors in the DC-link to temporarily store electrical energy. Switching the power electronic devices allows the DC voltage to be modulated which can result in a variable voltage and frequency waveform. The purpose of the modulator is to generate the required switching signals for these switching devices on the basis of user defined inputs. For this purpose, the voltage–time integral was introduced [3], which in turn is tied to the average voltage per sample $U(t_k)$ according to

$$U\left(t_k\right)=\frac{1}{T_{\mathrm{s}}}\int_{t_k}^{t_k+T_{\mathrm{s}}}u\left(t\right)\,\mathrm{d}t, \tag{2.1}$$

where T_{s} is a given sample interval and $u(t)$ represents the instantaneous voltage across a single-phase of a load. The introduction of the variable T_{s} assumes the use of a fixed sampling frequency which is normally judicially chosen higher than the fundamental frequency range required to control electrical machines. The upper sampling frequency limit is constrained by the need to limit the switching losses of the converter semiconductor devices.

The ability to control the converter devices in such a manner as to provide the load with a user defined mean reference voltage per sample $U^{*}(t_k)$ is instrumental to control current accurately. This statement can be made plausible by considering the incremental flux linkage for one sample interval of a load in the form of a coil with inductance L and resistance R which may be written as

$$\Delta\psi\left(t_k\right)=\int_{t_k}^{t_k+T_{\mathrm{s}}}\left(u\left(t\right)-Ri\left(t\right)\right)\,\mathrm{d}t. \tag{2.2}$$

R. W. De Doncker et al., *Advanced Electrical Drives*, Power Systems,
https://doi.org/10.1007/978-3-030-48977-9_2

The corresponding incremental change of load current (over a sample interval T_s) may be written as

$$\Delta i\,(t_k) = \frac{\Delta\psi\,(t_k)}{L} \tag{2.3}$$

in the event that magnetic saturation effects may be ignored. This expression can, with the aid of Eq. (2.2), be expressed as

$$\Delta i\,(t_k) = \frac{1}{L}\int_{t_k}^{t_k+T_s} u\,(t)\;\mathrm{d}t - \frac{R}{L}\int_{t_k}^{t_k+T_s} i\,(t)\;\mathrm{d}t \tag{2.4}$$

which may be reduced to

$$\Delta i\,(t_k) \cong \frac{U\,(t_k)\;T_s}{L} \tag{2.5}$$

when the time constant $\tau = L/R$ of the load is deemed to be relatively large, i.e., at least by a factor of ten, compared to T_s, as is normally the case for electrical machines. Central to the issue of controlling the incremental current is therefore, according to Eq. (2.5), the ability of the modulator to realize (within the constraint of this unit) the condition

$$U\,(t_k) = U^*\,(t_k) \tag{2.6}$$

for each sampling instance. Note that Eq. (2.6) simply states that the switching states of the converter must be controlled by the modulator to ensure that the average voltage (per sample) equals the user defined average reference value to ensure that the actual and reference incremental current change (per sample interval) are equal.

How this may be achieved will be outlined in subsequent sections for various converter topologies using an approach taken by Svensson [1]. In effect, this approach considers how the average voltage per sample $U(t_k)$ varies as function of the converter switch on/off time within a sample interval. Once this relation is known for the converter under consideration, the function in question is compared with the user defined reference value to determine the converter switch state within each sample. Initially, a single-phase half-bridge converter, as discussed in [3], will be considered followed by an analysis of a single-phase full-bridge converter and a three-phase converter. In the context of modulation for three-phase converters, the so-called *space vector modulation* [2] will also be considered, together with the need to impose a modulator strategy that can handle the finite switch on/off times of practical converter switches. In this chapter, a set of *build and play* tutorials are outlined, which will allow the reader to become better acquainted with the subject matter presented in this chapter.

2.2 Single-Phase Half-Bridge Converter

The so-called *half-bridge* converter configuration as shown in Fig. 2.1 consists of two switches which must be controlled by the modulator. The converter in question is connected to a single-phase symbolic load element Z (see Fig. 2.1).

For drive applications, as considered in this book, the load element is typically a phase of an electrical machine which may be represented by a load impedance in the form of an inductance L and resistance R circuit connected in series, together with a so-called back-EMF voltage source e. The two *ideal* switches are controlled by two logic signals Sw_{t}, Sw_{b}. Logic 1 corresponds to a *closed*, i.e., "on," switch state and logic 0 to an *open* switch state. There are therefore four possible switch combinations possible as may be observed from Table 2.1. Of the four states shown in Table 2.1, the *shoot-through mode* must be avoided in voltage source converters to prevent short-circuiting of the supply. The *idle* mode is normally used to disable the converter. Observation of Table 2.1 demonstrates that the two active switching states are complementary and can therefore be represented by a single logic switching state Sw with $Sw_{\mathrm{t}} = Sw$ and $Sw_{\mathrm{b}} = \overline{Sw}$.

An example of a typical output voltage waveform, which appears in a half-bridge converter, is given in Fig. 2.2 for the sampling interval $t_{\mathrm{k}-1} \ldots t_{\mathrm{k}+1}$, where the switching function Sw is assumed to be zero at $t = t_{\mathrm{k}-1}$ and changed to its

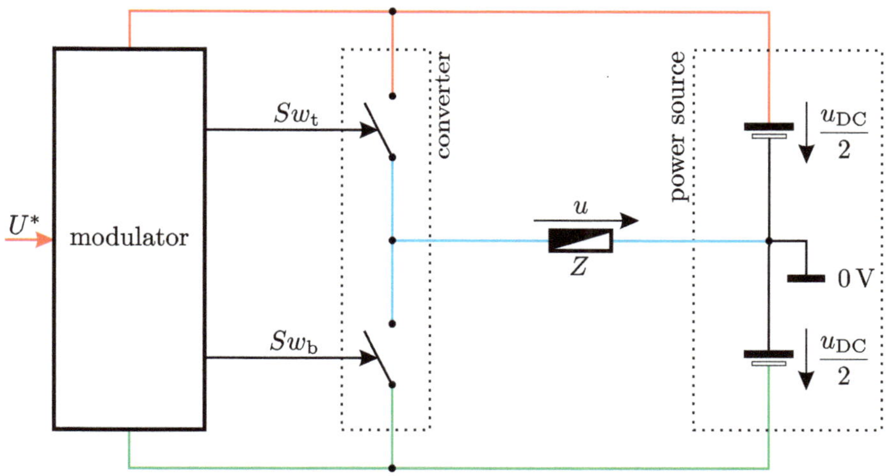

Fig. 2.1 Two switch *half-bridge* converter with power source and modulator

Table 2.1 Half-bridge switching states

Sw_{t}	Sw_{b}	Voltage u	Comment
0	0	–	Idle mode
0	1	$-\frac{u_{\mathrm{DC}}}{2}$	Active mode
1	0	$\frac{u_{\mathrm{DC}}}{2}$	Active mode
1	1	–	Shoot-through mode

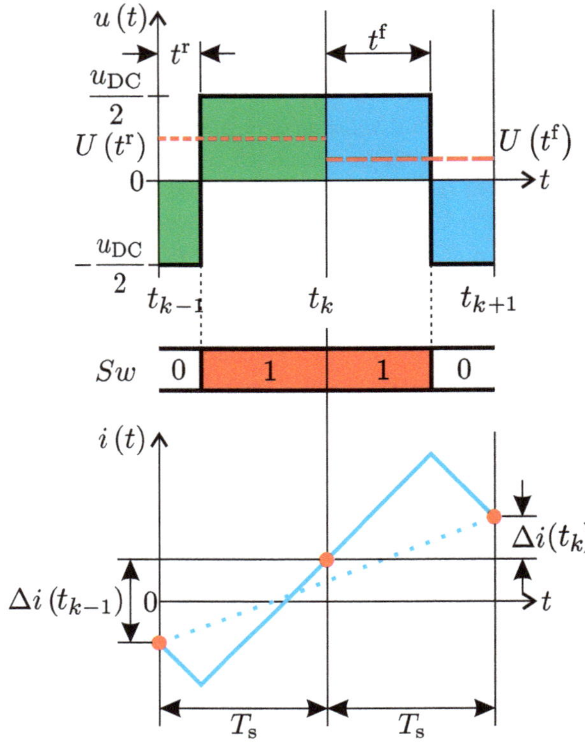

Fig. 2.2 Variation of the average voltage per sample, with half-bridge converter

logic state *1* at $t = t_{k-1} + t^r$. Subsequently, Sw is set to zero during the next sample interval at $t = t_k + t^f$, with t^r the time interval after which Sw rises and t^f the interval after which Sw falls. This switching sequence is repeated every two samples where t^r and t^f may be varied within the limits of the interval $[t_{k-1}, t_k]$ and $[t_k, t_{k+1}]$, respectively. This modulation strategy is known as *edged* PWM, because the rising and falling edge of the load voltage waveform are varied as function of t^r and t^f. The latter waveform is also shown in Fig. 2.2 for two samples of operation. They constitute one period $T_{PWM} = 2\,T_s = 1/f_{PWM}$ of modulator operation.

The corresponding average voltage functions $U(t^r)$ and $U(t^f)$ can be found using Eq. (2.1), which leads to equation set (2.7)

$$U(t^r) = \frac{u_{DC}}{2}\left(1 - \frac{2t^r}{T_s}\right); \quad 0 \le t^r \le T_s \tag{2.7a}$$

$$U(t^f) = \frac{u_{DC}}{2}\left(\frac{2t^f}{T_s} - 1\right); \quad 0 \le t^f \le T_s. \tag{2.7b}$$

where u_{DC} represents the DC bus voltage of the converter as shown in Fig. 2.1. Also shown in Fig. 2.2 is a typical load current trajectory in the event that the element Z is represented by an ideal (zero resistance) coil. The gradient of the current waveform

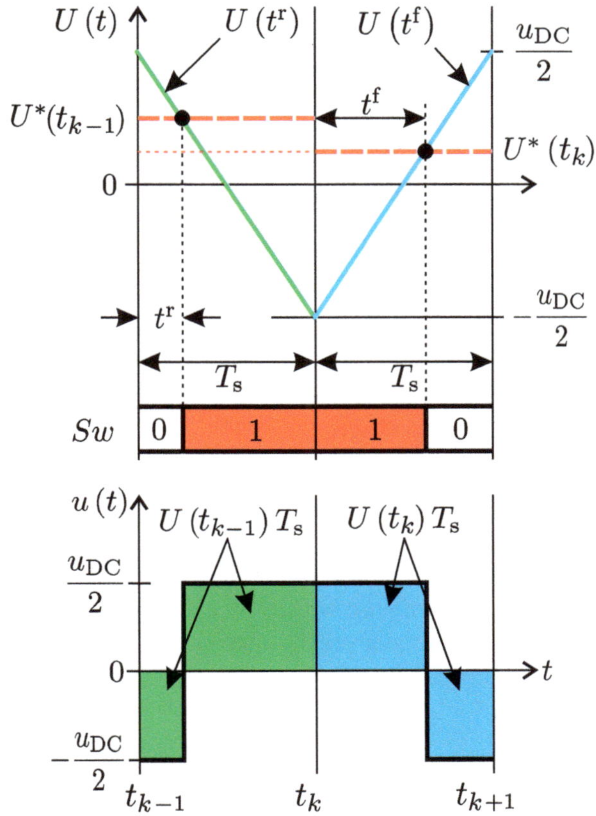

Fig. 2.3 Double edged PWM strategy, with half-bridge converter

is in this case dictated by the ratio $u(t)/L$, while the incremental current change per sample $\Delta i(t_k)$ (which is also shown in Fig. 2.2 for both sample intervals) is determined by the product $U\,T_s$ (see expression (2.5)).

The average voltage functions according to equation set (2.7) are shown in Fig. 2.3 together with the user defined reference values for both sample intervals. The required switching function Sw may be found by comparing the defined reference average voltage values $U^*(t_{k-1})$, $U^*(t_k)$ with a (triangular) sawtooth function (green and blue straight lines in top graph of Fig. 2.3), which leads to the values t^r and t^f needed to meet the condition specified by Eq. (2.6). The latter may also be confirmed by observing the *green* and *blue* areas shown in the load voltage $u(t)$ waveform of Figs. 2.2 and 2.3. These areas represent the actual voltage–time products $U(t_{k-1})\,T_s$ and $U(t_k)\,T_s$, respectively, as generated by the converter. Clearly identifiable in Fig. 2.3 are the switch Sw states over one period of modulator operation, where the *red* colored interval corresponds to logic level 1.

A generic implementation of a *double edged* PWM strategy, as given in Fig. 2.4, shows two A/D modules which are tied to the reference average voltage value (from the controller) and the measured DC bus voltage u_{DC} value (from the

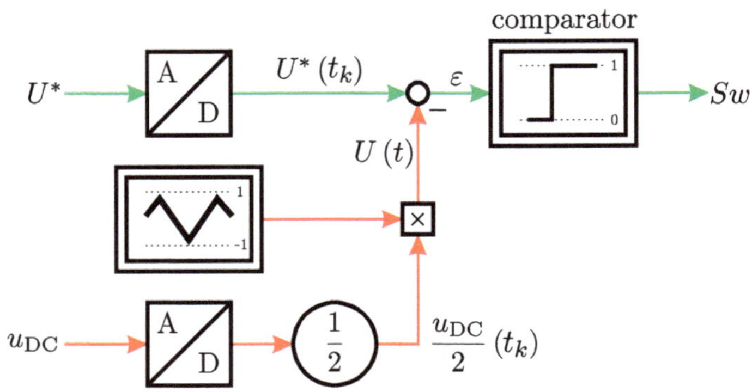

Fig. 2.4 Generic model of double edged PWM based half-bridge modulator

converter module). The sampled DC voltage is multiplied by a gain $1/2$ to provide the maximum sampled average voltage value $u_{DC}/2$. This value is multiplied by a *triangular* function which results in the function $U(t)$ defined by expression equation set (2.7). A summation module compares the two average voltage values U and U^*. Its output ε is used by a so-called comparator module with a transfer function of the form given by equation set (2.8).

$$\text{if } \varepsilon > 0 \quad \text{comparator output} = 1 \tag{2.8a}$$

$$\text{if } \varepsilon \leq 0 \quad \text{comparator output} = 0 \tag{2.8b}$$

The output of the comparator, known as switching function Sw, drives the two converter switches as discussed previously. Note that use of the measured DC bus voltage in the generic structure is beneficial because this provides the modulator with the capability to maintain the average voltage reference value, even when bus voltage variations occur.

Prior to introducing the full-bridge converter, it is helpful to introduce two modulation parameters which are applicable to sinusoidally varying average voltage reference signals that are commonly used in drive application. The first, known as the *amplitude modulation ratio* m_A, is defined as

$$m_A = \frac{\hat{U}^*}{u_{DC}/2}, \tag{2.9}$$

where $\hat{U}^*$ represents the peak value of the reference average phase voltage. The second modulation parameter in use is the *frequency modulation ratio* m_f which is defined as

$$m_f = \frac{f_{PWM}}{f^*} \tag{2.10}$$

where f_{PWM} represents the frequency of the triangular waveform shown in Fig. 2.3, which is half the sample frequency $f_s = {}^1/_{T_s}$. The frequency of a sinusoidally varying average voltage reference signal is defined as f^*. In tutorial 2.6.1, located at the end of this chapter, a simulation example is presented to underline the concepts presented in this section.

2.3 Single-Phase Full-Bridge Converter

The full-bridge, otherwise known as H-bridge converter, can be constructed by two half-bridge converters as shown in Fig. 2.5. A single-phase load impedance Z (as defined in the previous section) is again assumed, which in this case is represented by a virtual two-phase equivalent load, with a per phase impedance of ${}^Z/_2$ and phase currents $i_1 = i$, $i_2 = -i$. Use of a virtual two-phase load is particularly instructive for the development of a modulator structure given that this approach can be readily extended to three-phase systems. The virtual two-phase center point voltage with respect to the zero volt node, here the mid-point of the DC-link, is defined as u_0.

The key to determining the modulator strategy for this converter centers on Eq. (2.1) which, with the aid of Fig. 2.5, may be written as

$$U(t_k) = \underbrace{\frac{1}{T_s}\int_{t_k}^{t_k+T_s} u_1(t)\,\mathrm{d}t}_{U_1(t_k)} - \underbrace{\frac{1}{T_s}\int_{t_k}^{t_k+T_s} u_2(t)\,\mathrm{d}t}_{U_2(t_k)}\,. \tag{2.11}$$

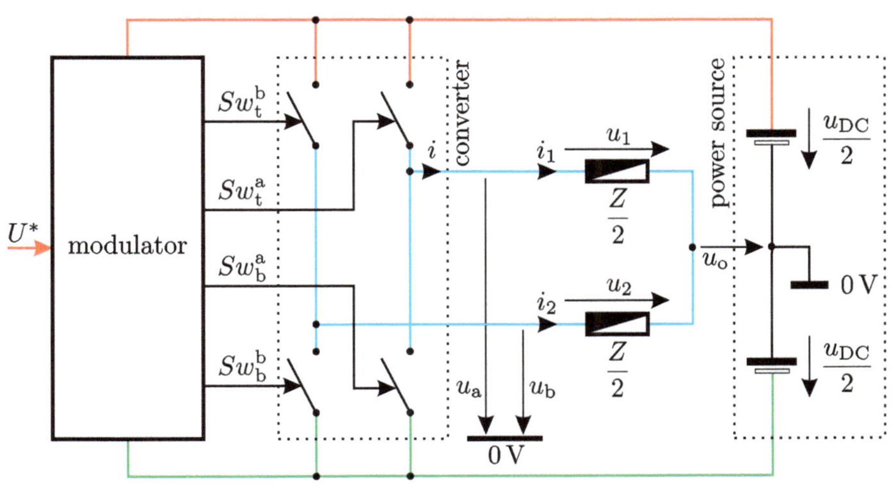

Fig. 2.5 H-bridge converter

In expression (2.11), the terms $U_1(t_k)$ and $U_2(t_k)$ are introduced which represent the average (per sample) voltages for both virtual phases. These may in turn be written as

$$U_1\,(t_k) = \underbrace{\frac{1}{T_\mathrm{s}}\int_{t_k}^{t_k+T_\mathrm{s}} u_\mathrm{a}\,(t)\;\mathrm{d}t}_{U_\mathrm{a}(t_k)} - \underbrace{\frac{1}{T_\mathrm{s}}\int_{t_k}^{t_k+T_\mathrm{s}} u_0(t)\,\mathrm{d}t}_{U_0(t_k)} \tag{2.12a}$$

$$U_2\,(t_k) = \underbrace{\frac{1}{T_\mathrm{s}}\int_{t_k}^{t_k+T_\mathrm{s}} u_\mathrm{b}\,(t)\;\mathrm{d}t}_{U_\mathrm{b}(t_k)} - \underbrace{\frac{1}{T_\mathrm{s}}\int_{t_k}^{t_k+T_\mathrm{s}} u_0(t)\,\mathrm{d}t}_{U_0(t_k)}, \tag{2.12b}$$

where $U_\mathrm{a}(t_k)$ and $U_\mathrm{b}(t_k)$ represent the half-bridge average voltage values. The required half-bridge average voltage references $U_\mathrm{a}^*(t_k)$, $U_\mathrm{b}^*(t_k)$ can, with the aid of equation set (2.12), be written in terms of the user defined average voltage references $U_1^*(t_k)$, $U_2^*(t_k)$ as

$$U_\mathrm{a}^*\,(t_k) = U_1^*\,(t_k) + U_0^*\,(t_k) \tag{2.13a}$$

$$U_\mathrm{b}^*\,(t_k) = U_2^*\,(t_k) + U_0^*\,(t_k)\,. \tag{2.13b}$$

In reality, a single-phase load exists, as mentioned above, to which a specified average voltage (per sample) value must be applied and this value is, according to Eq. (2.11), given by

$$U^*\,(t_k) = U_1^*\,(t_k) - U_2^*\,(t_k) \tag{2.14}$$

which, with the aid of equation set (2.13), may also be written as

$$U^*\,(t_k) = U_\mathrm{a}^*\,(t_k) - U_\mathrm{b}^*\,(t_k)\,. \tag{2.15}$$

Equation (2.15) shows that the virtual average voltage value $U_0^*(t_k)$ may be chosen freely (given that it is not present in this equation). This can also be observed from Fig. 2.6a where within two consecutive sample intervals the virtual average voltage value $U_0^*(t_k)$ is arbitrarily selected at values greater and less than zero, respectively, for the same average voltage reference value, i.e., $U^*(t_{k-1}) = U^*(t_k)$. An observation of Fig. 2.6a shows that the output pulses correspond to the required average voltage–time reference values $U^*(t_{k-1})\,T_\mathrm{s}$ and $U(t_k)\,T_\mathrm{s}$, which in this example were chosen equal. However, the relative position of the output pulses within the sampling interval is defined by the virtual average voltage values $U_0^*(t_{k-1})$ and $U_0^*(t_k)$. As mentioned before, in this example, the values are arbitrarily chosen to be positive and negative, respectively. Also shown in Fig. 2.6 are the half-bridge voltages u_a and u_b which toggle between $\pm\, u_\mathrm{DC}/2$.

From a practical perspective, it is prudent to choose the average voltage converter references $U_\mathrm{a}^*(t_k)$ and $U_\mathrm{b}^*(t_k)$, during each sample interval in such a manner that

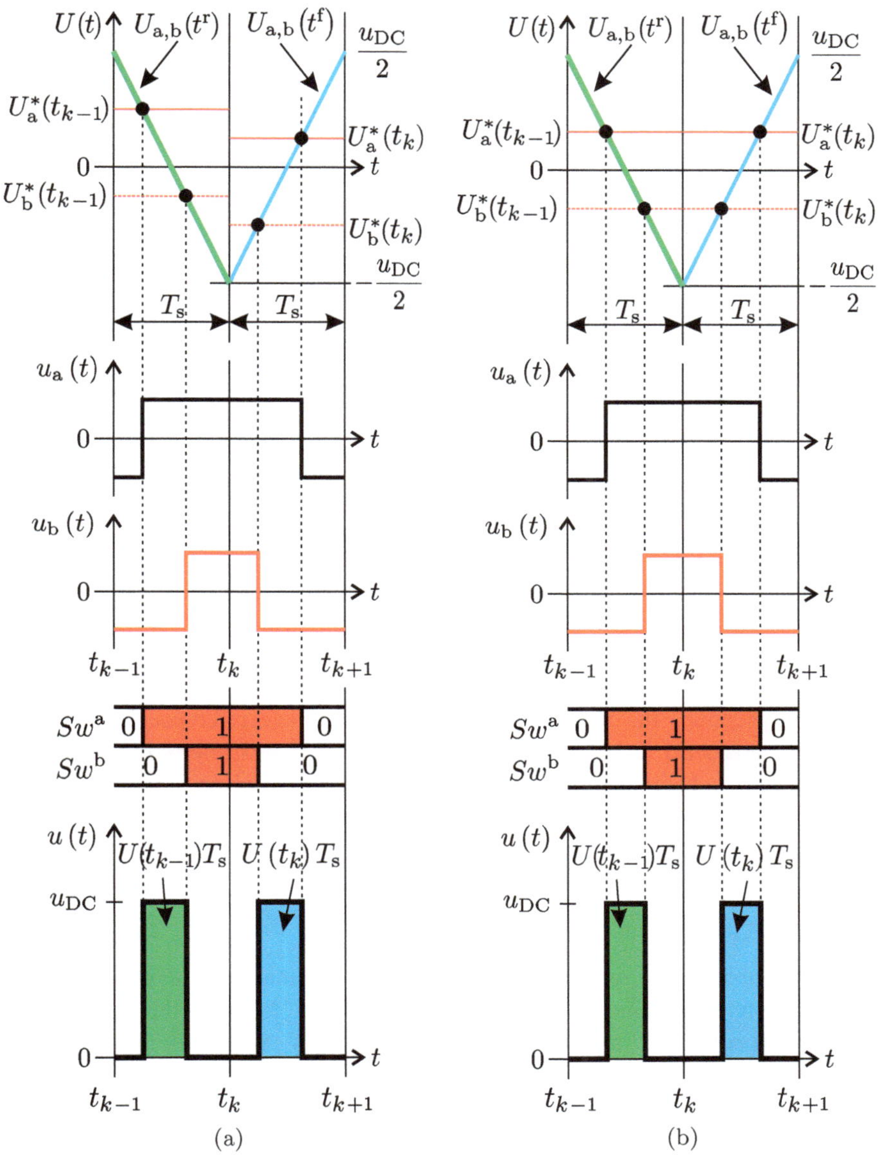

Fig. 2.6 Full-bridge converter with asymmetrically sampled PWM. (**a**) PMW, without pulse centering unit. (**b**) PWM, with pulse centering unit

they are symmetrically oriented with respect to the horizontal time axis. If this condition is maintained, the largest possible value of $U^* = u_{DC}$ can be realized by the modulator/converter. Symmetrical orientation of the references $U_a^*(t_k)$ and $U_b^*(t_k)$ can be achieved by imposing the condition

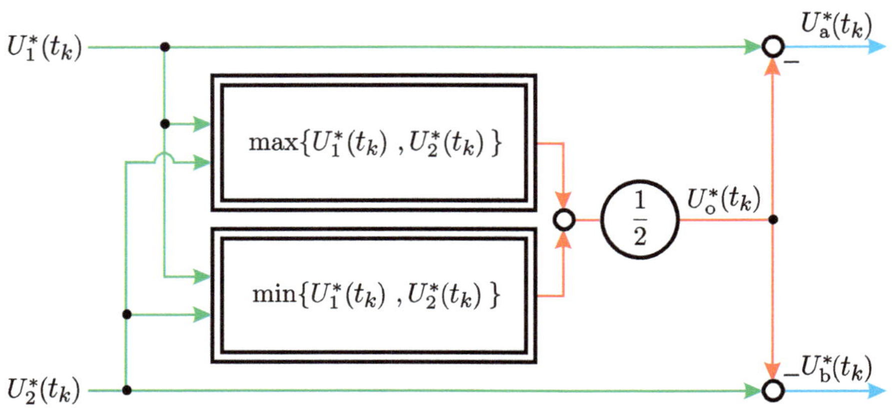

Fig. 2.7 Pulse centering module

$$\max\left\{U_{\mathrm{a}}^{*}(t_k), U_{\mathrm{b}}^{*}(t_k)\right\} + \min\left\{U_{\mathrm{a}}^{*}(t_k), U_{\mathrm{b}}^{*}(t_k)\right\} = 0. \tag{2.16}$$

With the aid of Eq. (2.13), this expression may also be written as

$$\max\left\{U_1^{*}(t_k), U_2^{*}(t_k)\right\} + \min\left\{U_1^{*}(t_k), U_2^{*}(t_k)\right\} + 2\,U_0^{*}(t_k) = 0 \tag{2.17}$$

which fully defines the required virtual average voltage reference level $U_0^*(t_k)$ needed to satisfy Eq. (2.16). The control structure, which is referred to as a *pulse centering unit* and corresponds to Eqs. (2.13) and (2.17), is given in Fig. 2.7.

Note that for the full-bridge modulator considered here, expression (2.16) may be written as

$$U_{\mathrm{a}}^{*}(t_k) + U_{\mathrm{b}}^{*}(t_k) = 0 \tag{2.18}$$

which symmetrizes the references with respect to the time axis. Input to the pulse centering reference value are the variables U_1^* and U_2^*, while the input to the modulator is equal to U^* which, according to Eq. (2.14), is equal to the difference of said variables. Correspondingly, one of the two reference values can be chosen arbitrarily and a convenient choice is as follows:

$$U_1^{*}(t_k) = U^{*}(t_k) \tag{2.19a}$$

$$U_2^{*}(t_k) = 0. \tag{2.19b}$$

The outputs from the pulse centering module are therefore of the form

$$U_{\mathrm{a}}^{*}(t_k) = \frac{1}{2}U^{*}(t_k) \tag{2.20a}$$

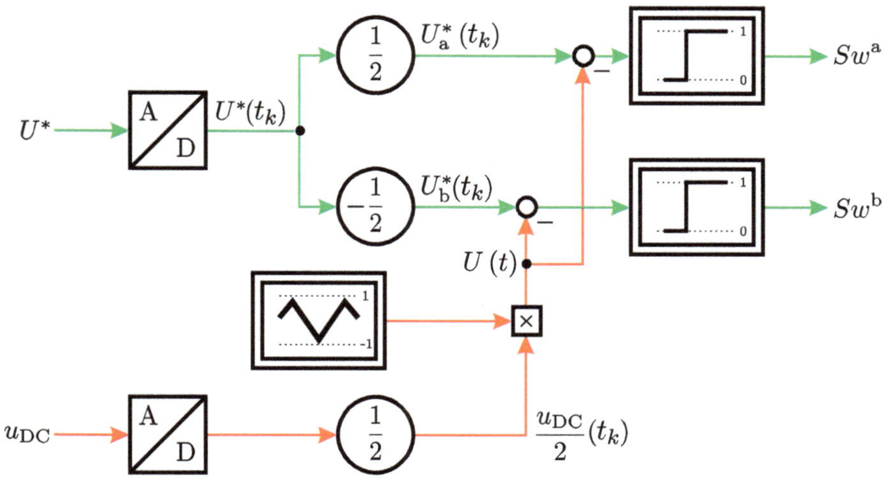

Fig. 2.8 Generic model of double edged PWM based modulator for full-bridge converter

$$U_b^*(t_k) = -\frac{1}{2}U^*(t_k). \tag{2.20b}$$

Figure 2.6b shows the impact of choosing the average voltage value U_0 according to Eq. (2.17). Observation of the load voltage waveform demonstrates that it has now been centered with respect to the middle of the sample interval. Consequently, increasing the value of the reference average voltage value U^* will increase the area underneath the voltage pulse, but its position relative to the sampling interval remains unchanged. Without the use of this pulse centering unit an increase in the reference U^* will lead to a situation where one of the half-bridge references will exceed the maximum value. For example, reference U_a^* in Fig. 2.6a will exceed the absolute maximum value before reference U_b^*. With pulse centering, both half-bridge average voltage reference values will remain centered. Hence, they will reach their maximum value simultaneously.

A generic representation of the modulator for the H-bridge converter topology is given in Fig. 2.8 and contains two comparators which provide the logic signals Sw^a and Sw^b that are in turn used to control the switches as discussed in the previous section. The tutorial given in Sect. 2.6.3 shows the generic H-bridge modulator with a full-bridge converter connected to an inductive load.

2.4 Three-Phase Converter

The three-phase converter topology, as shown in Fig. 2.9, consists of a symmetric balanced star connected load which is connected to three half-bridge converters as discussed in Sect. 2.2. The converter structure is similar to the virtual two-

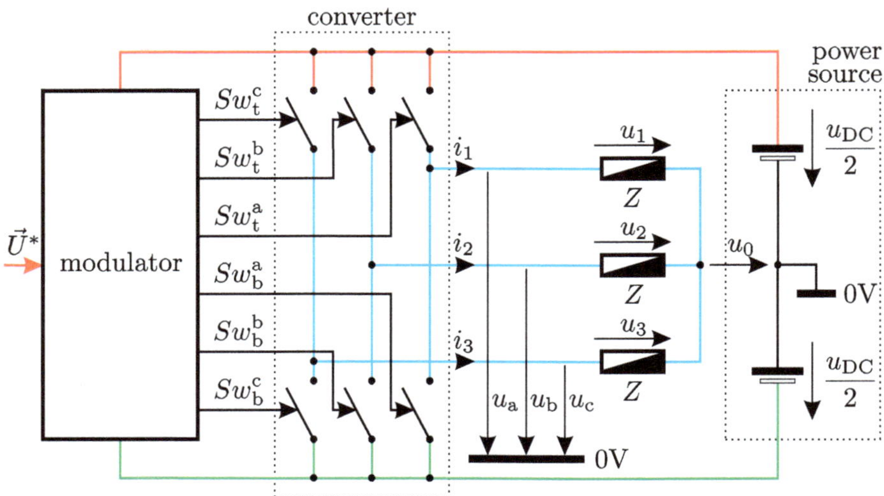

Fig. 2.9 Three-phase converter

phase structure introduced in the previous section. Accordingly, the mathematical handling required to obtain the average voltage references for the three half-bridges is very similar as will become apparent shortly. The aim is to determine (for a given sampling interval) a switching strategy for the six switches of the converter, which ensures that the average load phase voltage values $U_1(t_k)$, $U_2(t_k)$, $U_3(t_k)$ correspond with the three reference values $U_1^*(t_k)$, $U_2^*(t_k)$, and $U_3^*(t_k)$. The latter are in turn linked to the (amplitude invariant) space vector average voltage reference value $\vec{U}^*$ as given in Eq. (2.21), in its sampled form

$$\vec{U}^*(t_k) = \frac{2}{3}\left(U_1^*(t_k) + U_2^*(t_k)\,\mathrm{e}^{\mathrm{j}\gamma} + U_3^*(t_k)\,\mathrm{e}^{2\mathrm{j}\gamma}\right), \tag{2.21}$$

where $\gamma = 2\pi/3$. Furthermore, a modulator generic structure is to be developed to produce the required switching signals for the converter on the basis of the user defined average voltage reference vector $\vec{U}^*$.

The average voltage (per sample) values for the three-phase load can, with the aid of Eq. (2.1) and Fig. 2.9, be written as

$$U_1^*(t_k) = \underbrace{\frac{1}{T_s}\int_{t_k}^{t_k+T_s} u_a(t)\,\mathrm{d}t}_{U_a(t_k)} - \underbrace{\frac{1}{T_s}\int_{t_k}^{t_k+T_s} u_o(t)\,\mathrm{d}t}_{U_o(t_k)} \tag{2.22a}$$

$$U_2^*(t_k) = \underbrace{\frac{1}{T_s}\int_{t_k}^{t_k+T_s} u_b(t)\,\mathrm{d}t}_{U_b(t_k)} - \underbrace{\frac{1}{T_s}\int_{t_k}^{t_k+T_s} u_o(t)\,\mathrm{d}t}_{U_o(t_k)} \tag{2.22b}$$

$$U_3^*(t_k) = \underbrace{\frac{1}{T_s}\int_{t_k}^{t_k+T_s} u_c(t)\,dt}_{U_c(t_k)} - \underbrace{\frac{1}{T_s}\int_{t_k}^{t_k+T_s} u_o(t)\,dt}_{U_o(t_k)}, \tag{2.22c}$$

where $U_a(t_k)$, $U_b(t_k)$, and $U_c(t_k)$ represent the three half-bridge average voltages values. The required half-bridge average voltage references $U_a^*(t_k)$, $U_b^*(t_k)$, and $U_c^*(t_k)$ can, with the aid of equation set (2.22), be written in terms of the user defined average voltage references $U_1^*(t_k)$, $U_2^*(t_k)$, and $U_3^*(t_k)$ as

$$U_a^*(t_k) = U_1^*(t_k) + U_o^*(t_k) \tag{2.23a}$$

$$U_b^*(t_k) = U_2^*(t_k) + U_o^*(t_k) \tag{2.23b}$$

$$U_c^*(t_k) = U_3^*(t_k) + U_o^*(t_k) \tag{2.23c}$$

which contains a zero sequence average voltage (per sample) value $U_o^*(t_k)$ that can be defined in terms of the user defined average voltage values according to the approach set out in the previous section. In practice, it is helpful to rewrite equation set (2.23) in a space vector format. Using equation set (2.23) and Eq. (2.21) gives

$$\begin{aligned}\vec{U}^*(t_k) =& \frac{2}{3}\left(U_a^*(t_k) + U_b^*(t_k)\,e^{j\gamma} + U_c^*(t_k)\,e^{2j\gamma}\right) \\ & - \frac{2}{3}U_o(t_k)\underbrace{\left(1 + e^{j\gamma} + e^{2j\gamma}\right)}_{0}.\end{aligned} \tag{2.24}$$

The second term of expression (2.24) contains a vector sum of value zero, together with the zero sequence average voltage value $U_o^*(t_k)$, which implies that the latter can be chosen freely. It is noted that the approach discussed here is almost identical to that used in the previous section. The choice of the *zero sequence average voltage* value is therefore of no concern with respect to the choice of the average voltage half-bridge references. However, it is prudent to choose the value of $U_o(t_k)$ in such a manner that the maximum and minimum value of the half-bridge reference values are symmetrical with respect to the time axis. The reader is reminded of the fact that the same approach was purposefully introduced for the full-bridge modulator, which led to condition 2.16. This condition can simply be adapted to accommodate three instead of two variables allowing this expression to be written as

$$\begin{aligned}&\max\left\{U_a^*(t_k), U_b^*(t_k), U_c^*(t_k)\right\} \\ &+ \min\left\{U_a^*(t_k), U_b^*(t_k), U_c^*(t_k)\right\} = 0.\end{aligned} \tag{2.25}$$

With the aid of equation set (2.23), this expression may also be written as

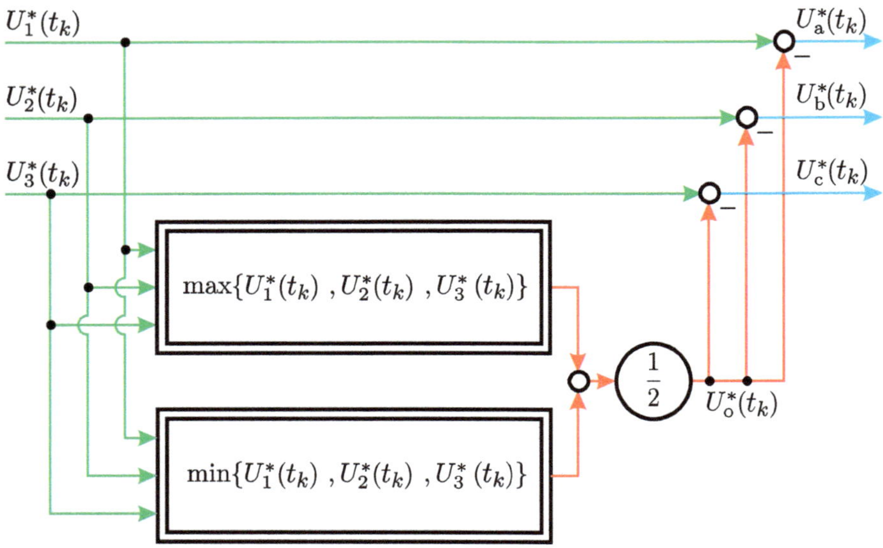

Fig. 2.10 Three-phase pulse centering module

$$\begin{aligned}&\max\left\{U_1^*(t_k), U_2^*(t_k), U_3^*(t_k)\right\}\\&+\min\left\{U_1^*(t_k), U_2^*(t_k), U_3^*(t_k)\right\}+2U_o^*(t_k) \quad =0.\end{aligned} \tag{2.26}$$

Further development of Eq. (2.26) gives

$$\begin{aligned}U_o^*(t_k)=-\frac{1}{2}\Big[&\max\left\{U_1^*(t_k), U_2^*(t_k), U_3^*(t_k)\right\}\\&+\min\left\{U_1^*(t_k), U_2^*(t_k), U_3^*(t_k)\right\}\Big]\end{aligned} \tag{2.27}$$

which fully defines the zero sequence average voltage value $U_o^*(t_k)$ for each sampling interval. The pulse centering unit according to Fig. 2.7 is readily modified to a three output/three input structure, as given in Fig. 2.10 that complies with Eq. (2.27).

An example which demonstrates the omission and use of the pulse centering unit is given in Fig. 2.11 for the case where a reference vector $\vec{U}^*(t_i)$ of constant amplitude $1/2\ u_{DC}$ is rotated by $\pi/6$ rad during a time interval T_s. The angle $\rho^*(t_i)$ between the reference vector and the real axis of a stationary complex plane is chosen to be zero at time t_{k-1}, i.e., $\rho^*(t_{k-1}) = 0$. The corresponding average load voltage reference values $U_{1,2,3}^*$ for each sampling interval may be found via a standard amplitude invariant vector to three-phase conversion, which gives

$$U_1^*(t_i)=\Re\left\{\vec{U}^*(t_i)\right\} \tag{2.28a}$$

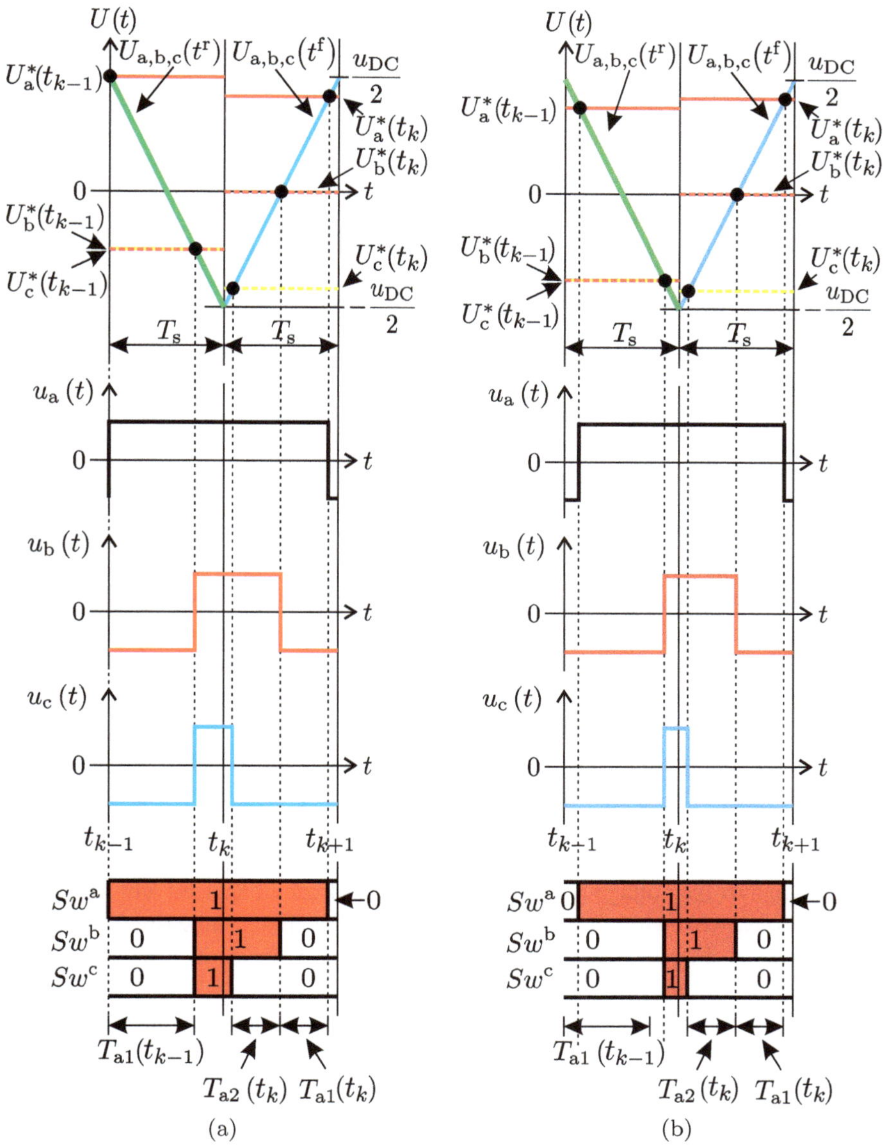

Fig. 2.11 Asymmetrically sampled PWM, three-phase converter, use of pulse centering unit. (**a**) PWM, without pulse centering unit. (**b**) PWM, with pulse centering unit

$$U_2^*(t_i) = \Re\left\{\vec{U}^*(t_i)\,\mathrm{e}^{-\mathrm{j}\gamma}\right\} \tag{2.28b}$$

$$U_3^*(t_i) = \Re\left\{\vec{U}^*(t_i)\,\mathrm{e}^{-\mathrm{j}2\gamma}\right\}. \tag{2.28c}$$

Table 2.2 Reference vector: $\vec{U}^*(t_{k-1}) = 1/2\, u_{DC}$, $\vec{U}^*(t_k) = 1/2\, u_{DC} e^{j\pi/6}$

Phase reference	Time t_{k-1}	Time t_k
U_1^*	$1/2\, u_{DC}$	$\sqrt{3}/4\, u_{DC}$
U_2^*	$-1/4\, u_{DC}$	0
U_3^*	$-1/4\, u_{DC}$	$-\sqrt{3}/4\, u_{DC}$
U_o^*	$-1/8\, u_{DC}$	0

Application of equation set (2.28) to the chosen reference vector for the two samples leads to the reference phase values given in Table 2.2, which also gives the value of U_o^* as calculated using Eq. (2.27). Note from Table 2.2 that the choice of reference space vector amplitude $|\vec{U}^*| = 1/2\, u_{DC}$ is such that the phase variable $U_1^*(t_k)$ is at the highest possible value of $1/2\, u_{DC}$. This implies that the largest orbit of the reference vector that can occur *without* pulse centering may be represented by a circle with radius $1/2\, u_{DC}$. The required half-bridge average voltage values $U_{a,b,c}^*$ may be found with the aid of the pulse centering module given in Fig. 2.10 in which case the zero sequence average voltage reference value U_o^* is calculated using Eq. (2.27).

In the first diagram, Fig. 2.11a, no pulse centering is used, i.e., U_o^* has been set to zero. An observation of the example given in Fig. 2.11b shows that the pulse centering unit symmetrizes the maximum and minimum average voltage references $U_{a,b,c}^*$ with respect to the time axis. The use of the pulse centering unit has, as may be observed from Fig. 2.11b, lowered the $U_1^*(t_k)$ value to $3/8\, u_{DC}$ which implies that the average voltage reference vector amplitude may be further increased before reaching the supply voltage limits $\pm u_{DC}/2$ of the converter. It will be shown in the next subsection that the introduction of pulse centering allows the user to increase the reference space vector amplitude from $|\vec{U}^*| = 1/2\, u_{DC}$ to $|\vec{U}^*| = \sqrt{1/3}\, u_{DC}$, which is an increase of approximately 15%. For the second sample $t_k \ldots t_{k+1}$, the half-bridge average voltage reference values were already balanced with respect to the time axis, which implies that the waveforms for this sample are identical to those shown in Fig. 2.11a for interval $t_k \ldots t_{k+1}$. Also shown in Fig. 2.11 are the half-bridge voltages u_a, u_b, u_c which toggle between $\pm\, u_{DC}/2$, with respect to the zero bus voltage node. In addition, the three comparator output switch states Sw^a, Sw^b, Sw^c are shown in Fig. 2.11.

The generic diagram for the three-phase converter as given in Fig. 2.12 is similar to the modulator concept described for the single-phase full-bridge converter. However, in this case the pulse centering module is used to generate the average voltage reference value for the three half-bridges on the basis of the user defined space vector average voltage reference.

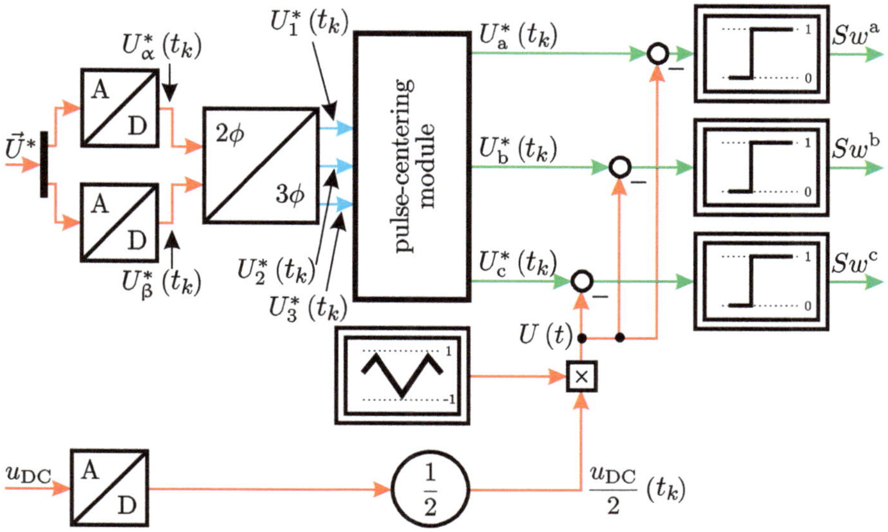

Fig. 2.12 Generic model of double edged PWM based modulator for three-phase converter

2.4.1 Space Vector Modulation

It is also helpful to consider the above modulator scheme from a space vector perspective. The so-called *space vector modulation*, introduced by van der Broeck et al. [2], is directly based on the use of Eq. (2.21). Using a pulse centering unit is important to maximize the linear operating region of the converter. Hence, further discussion will focus on Fig. 2.11, which also shows the switch states that correspond with this example. These switch states and corresponding half-bridge converter outputs can, with the aid of Eq. (2.21), be interpreted as a set of voltage space vectors. The switch states shown in Fig. 2.11 represent a subset of the eight possible converter switching combinations $\{Sw^a, Sw^b, Sw^c\}$ which gives

$$\vec{u}_{\{Sw^a,Sw^b,Sw^c\}} = \frac{2}{3}u_{DC}\left\{Sw^a + Sw^b\,e^{j\gamma} + Sw^c\,e^{j2\gamma}\right\}. \tag{2.29}$$

Evaluation of Eq. (2.29) shows that there are six active voltage vectors of magnitude $2/3u_{DC}$ that are displaced by $\pi/3$ rad as indicated in Fig. 2.13. In addition, two *zero* vectors are present in this figure corresponding to converter switch combinations {000} and {111}. The process by which the converter meets the condition specified by Eq. (2.30) may therefore be interpreted in terms of determining the two active voltage vectors adjacent to the reference average voltage vector $\vec{U}^*$ and determining the time interval for which they must be active during the sampling time T_s. In addition, the duration of the zero vectors must be determined for each sampling interval.

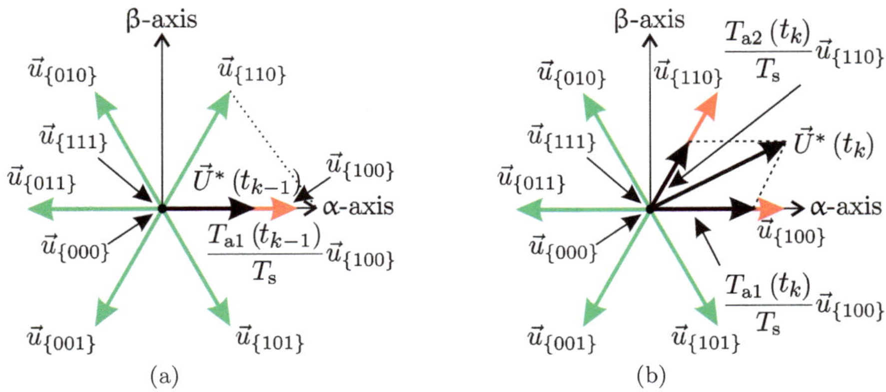

Fig. 2.13 Space vector representation of Fig. 2.11. (**a**) Space vectors, time $t_{k-1} \rightarrow t_k$. (**b**) Space vectors, time $t_k \rightarrow t_{k+1}$

$$\vec{U}(t_i) = \vec{U}^*(t_i) \tag{2.30}$$

A mathematical formulation of this strategy is of the form

$$\vec{U}^*(t_i) = \left(\frac{T_{a1}(t_i)}{T_s}\right)\vec{u}_{\bar{s}} + \left(\frac{T_{a2}(t_i)}{T_s}\right)\vec{u}_{\bar{s}}\, e^{j\frac{\pi}{3}}, \tag{2.31}$$

where $\vec{u}_{\bar{s}}$ is, according to Eq. (2.29), the active voltage vector which is clockwise adjacent to the voltage reference vector $\vec{U}^*$. In addition, the total duration of the time during which the active vectors can be deployed must satisfy the condition given in Eq. (2.32).

$$\frac{T_{a1}(t_i)}{T_s} + \frac{T_{a2}(t_i)}{T_s} \leq 1 \tag{2.32}$$

The equivalent space vector representation for the example discussed in the previous section with the aid of Fig. 2.11 is shown in Fig. 2.13. Included in both sub-figures are the eight possible converter voltage space vectors, together with the average voltage reference vectors for both samples t_{k-1} and t_k. The active voltage converter vector $\vec{u}_{\bar{s}}$, which must be used with Eq. (2.31), is in this example $\vec{u}_{\{100\}}$. Figure 2.13 shows that the PWM approach outlined earlier in this section determines the appropriate duration times for the active vectors as to satisfy equation set (2.31) and (2.32). Furthermore, the presence of the pulse centering unit ensures that the combined time interval in which active vectors are used is centered with respect to the *center* of the sampling interval. For the first sample $t_{k-1} \ldots t_k$, the average voltage reference vector coincides with the active average voltage converter vector $\vec{U}_{\{100\}}$. In this case, the modulation strategy determines the required time needed to activate said vector during the sample interval. For the second sample interval, two active vectors adjacent to the average voltage reference vector are activated for a time interval sufficient to ensure that the condition given by Eq. (2.30) is met.

It is emphasized that the modulation technique discussed above determines, on the basis of the reference voltage, the duration of the active vectors used within each sample. Furthermore, the pulse centering unit is responsible for centering the combined active vector time interval within the sample. An often quoted alternative to the strategy discussed above is the so-called *space vector modulation* [2] which is particularly suited to digital implementation as this approach calculates directly for a given sample interval (t_{i}) the duration of the active vectors adjacent to the specified average voltage reference vector $\vec{U}^{*}(t_{\mathrm{i}})$ with the aid of equation set (2.33), namely

$$\frac{T_{\mathrm{a2}}\left(t_{\mathrm{i}}\right)}{T_{\mathrm{s}}}=\frac{2}{\sqrt{3}}\Im\left\{\frac{\vec{U}^{*}\left(t_{\mathrm{i}}\right)}{\vec{u}_{\overline{\mathrm{s}}}}\right\} \tag{2.33a}$$

$$\frac{T_{\mathrm{a1}}\left(t_{\mathrm{i}}\right)}{T_{\mathrm{s}}}=\Re\left\{\frac{\vec{U}^{*}\left(t_{\mathrm{i}}\right)}{\vec{u}_{\overline{\mathrm{s}}}}\right\}-\frac{1}{2}\frac{T_{\mathrm{a2}}\left(t_{\mathrm{i}}\right)}{T_{\mathrm{s}}}. \tag{2.33b}$$

These expressions can be found using Eqs. (2.31) and (2.32) and give the user access to the required duration time of the active vectors, without having to implement a generic modular structure as given by, for example, Fig. 2.12.

The centering of the combined active time interval can be carried out by ensuring that zero vectors (if needed) are activated for an identical time instance at the beginning and end of each sample interval. The choice of zero vectors, which switch the converter in the so-called freewheeling states, at any particular instance is decided on the basis of a minimum number of switch actions needed to reach the nearest zero vector from the last active vector in use. In terms of modulation, the overall result achieved with space vector modulation is identical in its execution to the pulse width modulation strategy as undertaken with the aid of the generic modulator structure given by Fig. 2.12.

Finally, it is considered important to determine the maximum reference average voltage amplitude $\{U^{*}(t_{\mathrm{i}})\}^{\max}$ which may be used without imposing any constraints on the corresponding phase angle $\rho^{*}(t_{\mathrm{i}})$. Substitution of the active vector $\vec{u}_{\overline{\mathrm{s}}}$ (see Eq. (2.29)) into equation set (2.33) leads, after some mathematical manipulation, to

$$\left(\frac{T_{\mathrm{a1}}\left(t_{\mathrm{i}}\right)}{T_{\mathrm{s}}}\right)^{2}+\frac{T_{\mathrm{a1}}\left(t_{\mathrm{i}}\right)T_{\mathrm{a2}}\left(t_{\mathrm{i}}\right)}{T_{\mathrm{s}}^{2}}+\left(\frac{T_{\mathrm{a2}}\left(t_{\mathrm{i}}\right)}{T_{\mathrm{s}}}\right)^{2}=\left(\frac{U^{*}\left(t_{\mathrm{i}}\right)}{\frac{2}{3}u_{\mathrm{DC}}}\right)^{2} \tag{2.34}$$

in which the active vector $\vec{u}_{\overline{\mathrm{s}}}$ is arbitrarily chosen as $\vec{U}_{\{100\}}$. Equation (2.34) represents an ellipse in the $T_{\mathrm{a1}}/T_{\mathrm{s}}$, $T_{\mathrm{a2}}/T_{\mathrm{s}}$ plane and its size is determined by the variable $\nu = U^{*}(t_{\mathrm{i}})/\tfrac{2}{3}u_{\mathrm{DC}}$ as may be observed from Fig. 2.14.

Figure 2.14 shows how the duty cycle of two adjacent active vectors can be selected to achieve a given voltage reference value U^{*} (amplitude only) as expressed by the variable ν. Also shown in this diagram is a straight line which represents

Fig. 2.14 Active vector duty cycles for $\nu = \frac{3}{4}$ and $\nu = \sqrt{\frac{3}{4}}$

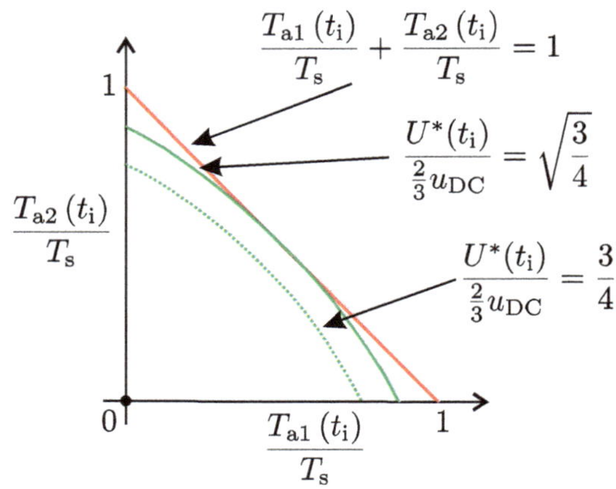

the sum of the two duty cycles which must be equal to one. Indeed, the time over which both vectors are activated cannot exceed the sample interval T_s. Observation of Fig. 2.14 shows two curves with values of $\nu = 3/4$ and $\nu = \sqrt{3/4}$ corresponding to a voltage reference amplitude of $U^* = 1/2\, u_{DC}$ and $U^* = 1/\sqrt{3}\, u_{DC}$, respectively. The latter value is the largest reference value which can be produced given that it corresponds with the largest ellipse (shown partly in the first quadrant) that can be used, given the constraints imposed by the linear function and two axes shown in Fig. 2.14. For comparison purposes part of a second ellipse (dotted curve) has also been drawn in Fig. 2.14. This ellipse corresponds to the largest ellipse with reference voltage $1/2\, u_{DC}$ that can be used without the presence of a pulse centering module while maintaining the phase references within the supply limits $\pm u_{DC}/2$. This leads to an important observation, namely that centering the active vectors within a sample either by use of a pulse centering PWM unit or by calculation using the space vector modulation approach allows the user to extend the linear operating range by a factor of $2/\sqrt{3} \simeq 15\%$, without encountering the supply level limits of the converter. Note that the ability to extend the linear operation of the modulator/converter by approximately 15% is particularly advantageous as it allows the converter to be operated with higher voltages and consequently lower currents. The latter implies that the kVA-rating of the converter can be increased by approximately 15% as a result of using pulse centering.

It is instructive to consider the circuit orbit which coincides with the highest allowable reference value, as discussed above, in a diagram together with the converter active and zero vectors. Such a diagram is given in Fig. 2.15 and shows that the circular orbit is the largest which can be located within a hexagon that is constructed from the six active vectors with length $2/3\, u_{DC}$. Observation of Fig. 2.15 shows the presence of a right-angled triangle of which the hypotenuse is represented by the active vector $\vec{u}_{\{100\}}$ and one of the legs which represents the radius $u_{DC}/\sqrt{3}$ of the circle. The triangle in question is a 30°–60°–90° acute triangle, of which the sides are in the ratio $1 : \sqrt{3} : 2$. This also shows that the largest circle which can

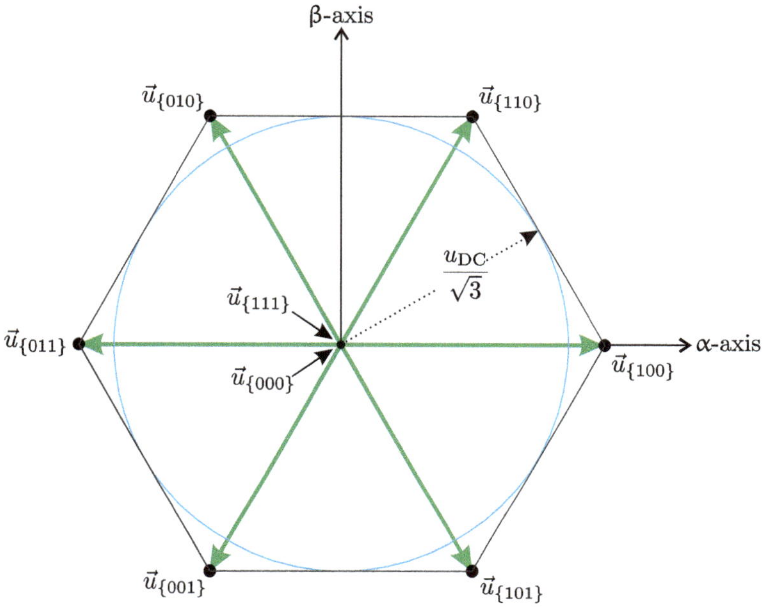

Fig. 2.15 Maximum average voltage vector orbit and available converter vectors

be placed inside the hexagon must have a radius that is $\sqrt{3}/2$ smaller than the active vector amplitude.

Prior to considering the effect of dead-time on the converter, it is instructive to examine a scenario where the reference voltage vector amplitude is increased beyond the maximum value of $u_{DC}/\sqrt{3}$. Under such conditions, the half-bridge reference values will exceed the DC supply limits during part of the circular orbit undertaken by the reference vector. This implies that the converter output voltage(s) may be held to the supply value for a period of time which is in excess of the sampling time. Consequently, the converter switching frequency will no longer be equal to half the sampling frequency. This mode of operation is referred to as *overmodulation* and will ultimately lead (with increasing U^*) to a situation where the converter waveforms are rectangular. In vector terms, this operation is characterized by switching from one active vector to the next, i.e., along the hexagon boundary shown in Fig. 2.15. This mode of operation is known as *six-step* operation and can be readily observed in the phase-voltage waveforms by increasing the reference voltage amplitude beyond the value $u_{DC}/\sqrt{3}$ in the tutorial given in Sect. 2.6.5.

2.5 Dead-Time Effects

In the half-bridge converter configuration shown in this chapter, ideal switches are used, which implies that they may be turned on or off instantaniously. Consequently the top- and bottom- signals are chosen complementary (top-ON, bottom-OFF and visa versa), as shown in Fig. 2.3. In reality, semiconductor based switches and diodes are used which require a finite turn-on and turn-off time. This implies that a so-called *dead-time* Δt_D must be imposed, which is the time when both switch signals are set to zero, to allow the switch turn-on/-off process to be completed. Failure to adhere to this policy can invoke the *shoot-through* mode as identified in Table 2.1 with usually disastrous consequences for the devices in question. The half-bridge configuration shown in Fig. 2.16 is representative of the converter topologies discussed in this section. Shown in Fig. 2.16 are two IGBT switches with corresponding diodes which are taken to be non-ideal, i.e., these require a finite turn-on/off time taken to be equal to Δt_D.

Consequently, the question arises how, for example, the generic modulator structure according to Fig. 2.4 may be modified to accommodate a *dead-time* equal to at least Δt_D. This technique can subsequently be equally applied to the full-bridge and three-phase converter. The envisaged approach is to introduce two comparators (to control the switches individually) and adding an average voltage offset $\pm U_D/2$ to the reference voltage value U^* as shown in Fig. 2.17. The value of U_D is defined as

$$U_D = u_{DC} \frac{\Delta t_D}{T_s} \tag{2.35}$$

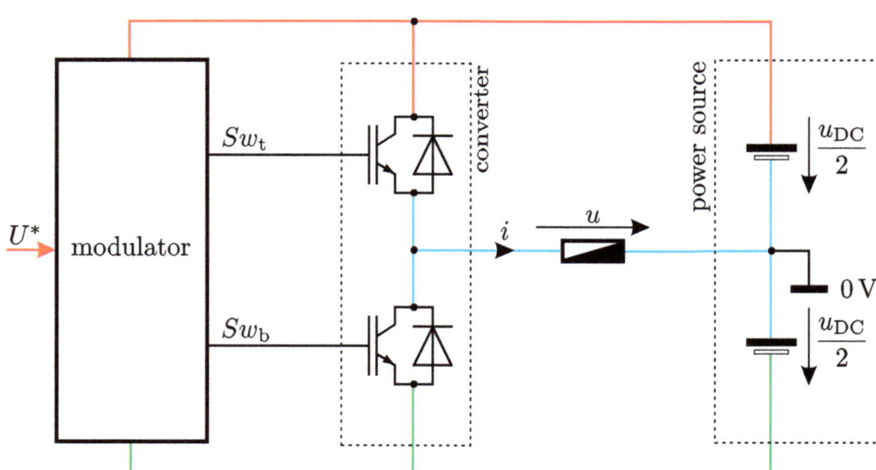

Fig. 2.16 IGBT based *half-bridge* converter with power source and modulator

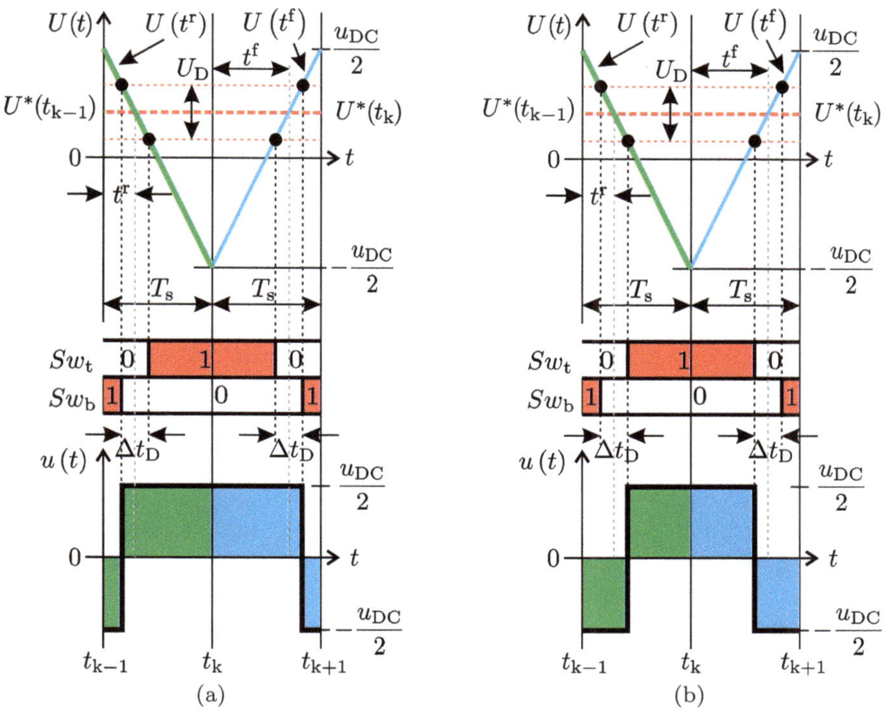

Fig. 2.17 Double edged PWM strategy, with half-bridge converter and dead-time effects. (**a**) Load current $i < 0$. (**b**) Load current $i > 0$

which may be found by considering the absolute gradient of the function $U(t)$ which is equal to u_{DC}/T_s. The value of $\pm U_D/2$ and corresponding dead-time value shown in Fig. 2.17 are purposely chosen large, in comparison to the sample time T_s for didactic reasons.

The consequence of introducing converter dead-time is that the modulator will not be able to set the control signals in such a manner that the delivered average voltage–time area to the load (per sample interval) is exactly equal to the reference average voltage–time area (see Eq. (2.6)). The reason for this effect may be illustrated with the aid of Fig. 2.17, which shows two examples where the reference average voltage level is purposely kept constant. In the first example given in Fig. 2.17a, the load current is assumed to be negative and the load is connected to the negative DC rail voltage as long as the bottom switch is closed. However, when the switch signal for this device is set to zero the turn-off process will start and the current will be commutated immediately to the top diode. This implies that the load is connected to the top supply rail during the dead-time period. After this dead-time interval the top switch is activated which means that the load remains connected to the top rail by virtue of either negative current through the top diode or positive current through the (conducting) top switch. In this case, the average voltage (per sample) across the load is larger than the required reference average voltage level as shown by Eq. (2.36a). In the second example, shown in Fig. 2.17b, a positive load current is assumed and events as described for the previous interval also occur with the difference that the bottom diode will remain conducting during the dead-time interval when the bottom switch turns off. Likewise, the bottom diode will also conduct during the dead-time interval as soon as the top switch turns off. Figure 2.17b demonstrates that the average voltage across the load will be less than the required reference average voltage value. The impact of the dead-time effect on the average load voltage may be observed with the aid of equation set 2.36.

$$U = U^* + \frac{U_D}{2} \quad (i < 0) \tag{2.36a}$$

$$U = U^* - \frac{U_D}{2} \quad (i > 0) \tag{2.36b}$$

Observation of equation set (2.36) and Eq. (2.35) shows that, in each switching interval, the error on the average load voltage with respect to the reference average load voltage depends on the polarity of the load current as well as the ratio between dead-time and sample time. The dead-time is a semiconductor device dependent variable (typically set to a constant, safe value), whereas the sample time may be chosen by the user. If the sample frequency $f_s = 1/T_s$ is increased, the ratio $\Delta t_D/T_s$ increases and consequently a larger average voltage error $U_D/2$ is introduced. In the tutorial section at the end of this chapter a simulation is shown (see Sect. 2.6.2) which demonstrates the issues set out in this subsection. Dead-time effects are not, for didactic reasons, included in the modulation strategies to be discussed in the remaining part of this book. However, dead-time effects must be accommodated in a practical power electronic converter by adjusting the PWM switching intervals based on the sign of the load currents as shown by equation set 2.36.

2.6 Tutorials

2.6.1 Tutorial 1: Half-Bridge Converter with Pulse width Modulation

A half-bridge converter structure, as given in Fig. 2.1, is implemented in this tutorial. The ideal switches are replaced by a combination of ideal semiconductors and diodes as shown in Fig. 2.16. The supply voltage is $u_{DC} = 300$ V. The load impedance Z is reduced to an ideal inductance of 100 mH. A modulator structure, as shown in Fig. 2.4, is implemented with a sampling period of $T_s = 1$ ms. Furthermore, a sinusoidal reference average voltage waveform U^* with an amplitude of 80 V and frequency of 50 Hz is used. The aim is to implement a simulation model and examine the sampled reference average voltage, voltage and

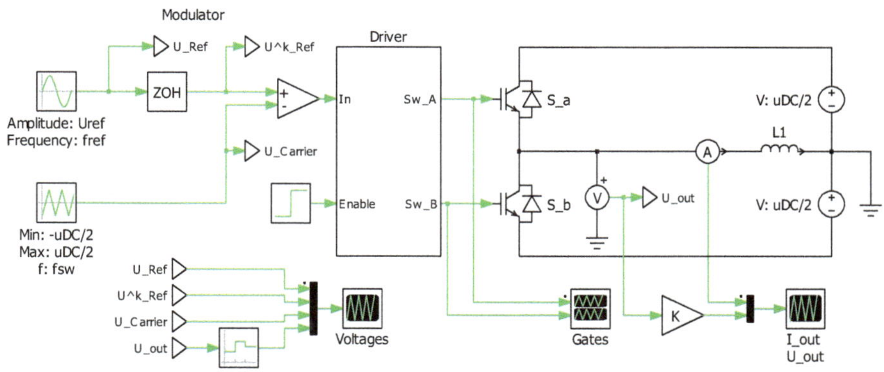

Fig. 2.18 Simulation of half-bridge converter with PWM in PLECS simulation tool

current waveforms of the load, as well as the logic switch signals in this model. A time interval of 40 ms is simulated.

An example of a simulation model which satisfies the needs for this tutorial is given in Fig. 2.18. The converter topology which consists of two ideal IGBTs and ideal diodes can be seen in this figure. The modulator structure is consistent with the generic structure shown in Fig. 2.4. An *enable* module is used together with a logic interface unit to activate the converter from time mark $t = 1$ ms onwards. Such a logic interface is usually implemented to ensure that the power electronics circuit is enabled prior to the controller to reduce the risk of accidental activation of both switches simultaneously during the power-up phase of operation.

The voltage across the load inductance and corresponding load current are provided in Fig. 2.19. The load current waveform consists of piecewise linear segments comprising two different slopes (up and down) for each switching interval. The change of average current of each switching period depends on the load average voltage–time product $U(t_k)\,T_{\mathrm{s}}$ (for a given load inductance value L) as shown earlier in Fig. 2.2. If the modulator operates correctly, the load average voltage $U_{\mathrm{conv}}(t_k)$ must be equal to the sampled reference average voltage $U^*(t_k)$. To observe the average, a periodic average module is employed in the simulation which calculates the average voltage per sample time on the basis of the instantaneous input value, which is in this case the voltage across the load. Note that the output voltage of the periodic average module corresponds to the calculated average voltage value of the previous sample.

The two complimentary switch signals given in the *Gate Signals* scope are directly tied to the comparator signal Sw. The ideal switches which are controlled by the signals Sw_{t}, Sw_{b} (logic one corresponds to a switch on-state) connect the converter center node to the appropriate supply rail. The modular switch signal Sw is in turn determined by the average voltage function $U(t)$ and the sampled reference average voltage $U^*(t_k)$ shown in the *Voltages* scope.

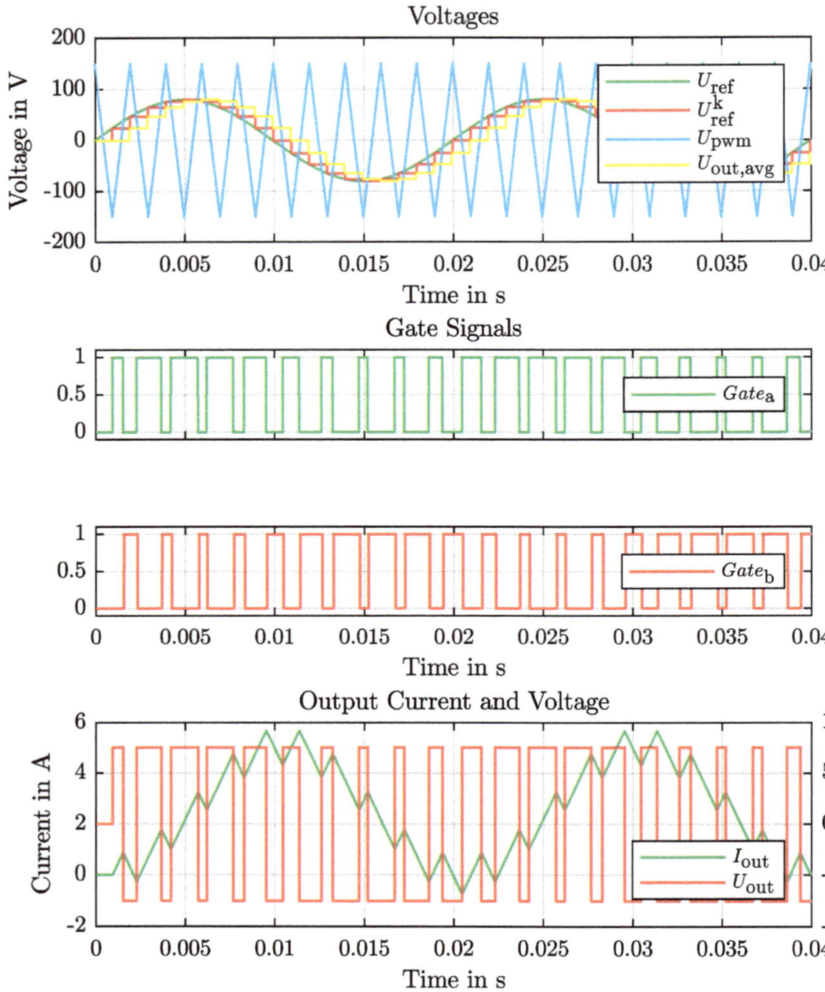

Fig. 2.19 Simulation results for half-bridge converter with PWM

2.6.2 Tutorial 2: Half-Bridge Converter with PWM and Dead-Time Effects

This tutorial considers the effects of dead-time as discussed in Sect. 2.5. For this purpose the tutorial exercise as discussed above should be modified to accommodate an unrealistically high dead-time of $\Delta t_{\mathrm{D}} = 100\,\mu\mathrm{s}$, which, according to Eq. (2.35), corresponds to a dead-time voltage value U_{D} of 30 V, given the supply voltage and sample time value introduced in the previous section. Note that the dead-time value has been purposely chosen high to better illustrate the effect of dead-time in the simulation. A PLECS model example for this tutorial is given in Fig. 2.20

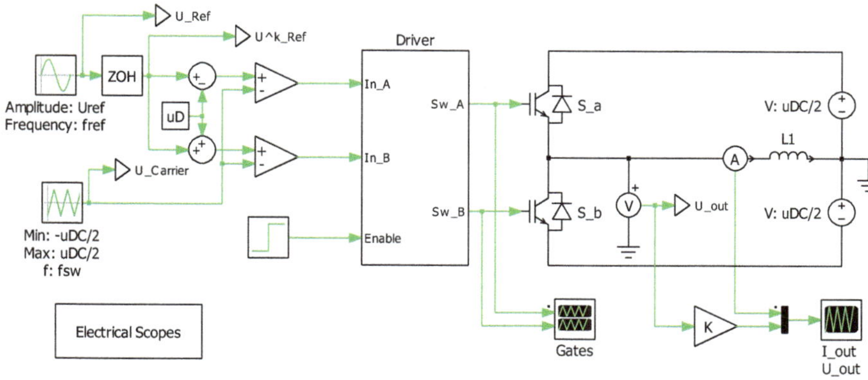

Fig. 2.20 PLECS simulation of half-bridge converter with PWM and dead-time effects

and is unchanged with respect to the previous tutorial in terms of average voltage reference, assumed load, and DC supply value.

A full-bridge *logic interface module* is used to provide the appropriate switching signals for the two switches. The error input to two comparators is an error signal which is formed by the PWM signal $U(t)$ and the sampled average voltage reference value. The latter is offset by a value of ± 15 V in order to realize the required dead-time average voltage band value of $U_\mathrm{D} = 30$ V.

The waveforms generated by the full-bridge simulation are shown in Fig. 2.21. The waveforms given in the *Gate Signals* scope show that the switching signals for the converter are no longer complimentary. Instead a dead-time interval $\Delta t_\mathrm{D} = 100\,\mu$s is present between the two signals in order to facilitate the turn-on and turn-off sequence of the devices. The effects of including this dead-time on the load voltage while changing polarity of the load current may be observed in the *Voltages* scope, by comparing the average voltage value with the sampled reference voltage value.

2.6.3 Tutorial 3: Full-Bridge Converter with Pulse Width Modulation

This tutorial is concerned with a full-bridge converter concept as discussed in Sect. 2.3. For this purpose, the half-bridge based tutorial discussed in Sect. 2.6.1 is modified to accommodate the new converter topology and the generic modulator structure shown in Fig. 2.8. The average voltage reference, load, and supply variables remain unchanged. A PLECS based simulation model example is given in Fig. 2.22, where two half-bridges with ideal IGBT semiconductors and ideal diodes are employed. A full-bridge *logic interface* module has been introduced, which houses the logic components as required to generate the semiconductor gate

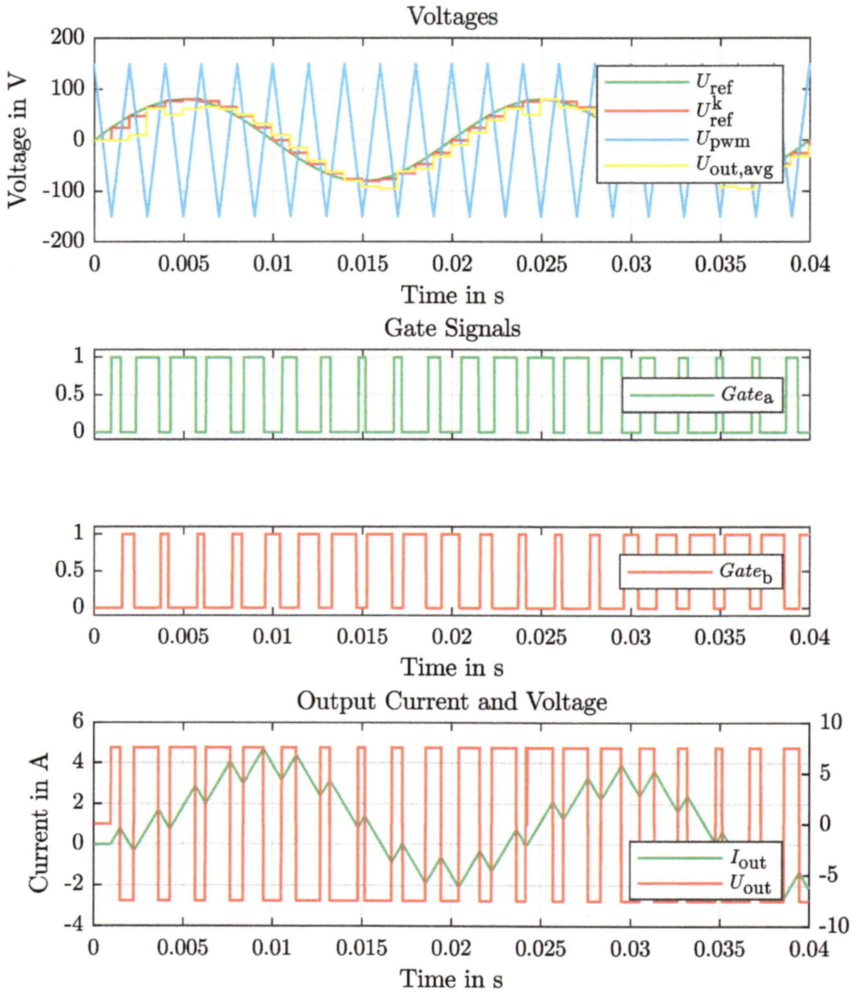

Fig. 2.21 Simulation results for half-bridge converter with PWM

signals from the comparator modules. The remaining part of the modulator circuit is in accordance with the modulator generic diagram given in Fig. 2.8.

The waveforms generated by the simulation are shown in Fig. 2.23. The load voltage toggles between ± 300 V, which is double the value that is achievable with the half-bridge converter. This also means that the maximum average voltage level U_{max} is doubled when compared to the half-bridge converter topology. It may be verified that the largest reference voltage that can be used in this tutorial without overmodulation is 300 V. It should be noted that the inductance current waveform has less ripples due to freewheeling over the diodes instead of applying a negative voltage during the off-time of the IGBTs.

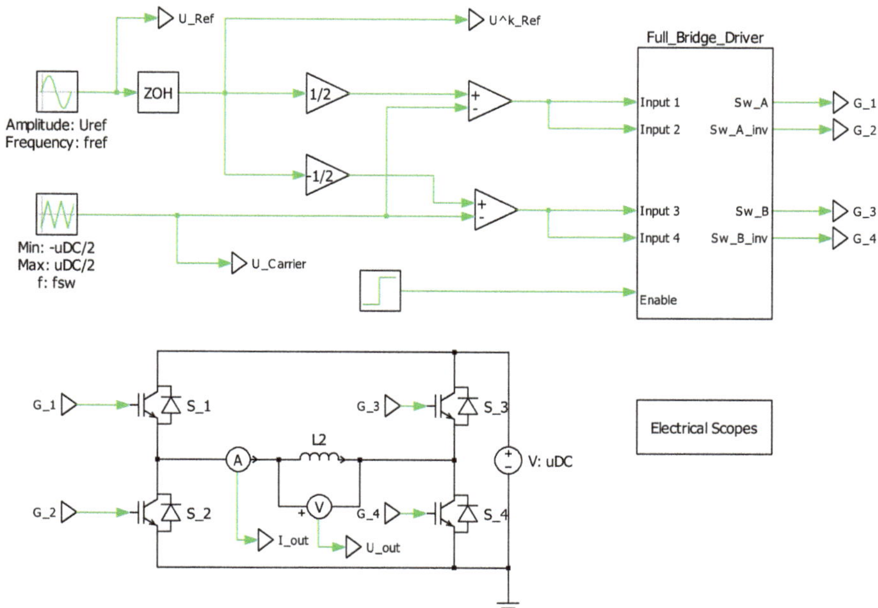

Fig. 2.22 PLECS simulation of full-bridge converter with PWM

2.6.4 Tutorial 4: Three-Phase Pulse Width Modulator with Pulse Centering

The operation of the pulse centering technique for three-phase pulse width modulators, as discussed in Sect. 2.4 is discussed in this tutorial. A simulation model of the generic structure shown in Fig. 2.12 is provided to generate the switching signals Sw^{a}, Sw^{b}, Sw^{c} on the basis of a rotating average voltage reference vector $\vec{U}^*(t)$. Furthermore, the simulation can also display the converter voltage space vectors as given in Fig. 2.15, which are used during each sample time step T_{s} for the duration of the chosen simulation time.

For this example, the average voltage reference vector at the start of the simulation is set to $\vec{U}^*(t = 0) = {}^{u_{\mathrm{DC}}}/_{2}$, corresponding to the values chosen for the example given in Figs. 2.11 and 2.13a. A sample time of $T_{\mathrm{s}} = 1$ ms and a DC-link voltage u_{DC} =300 V are assumed together with a fixed amplitude rotating (average) voltage reference vector, with an angular frequency of ${}^{\pi}/_{6T_{\mathrm{s}}}$ rad/s. The latter has been chosen to realize a 30° rotation of the vector over each sample interval, which is precisely in accordance with the example used to generate Figs. 2.11 and 2.13b. This implies that the tutorial provides an opportunity for the reader to quantitatively verify the example discussed in Sect. 2.4, with and without the use of the pulse centering technique. Note that the selected angular frequency is relatively high in comparison with those used in drive applications. The reason for this is that this

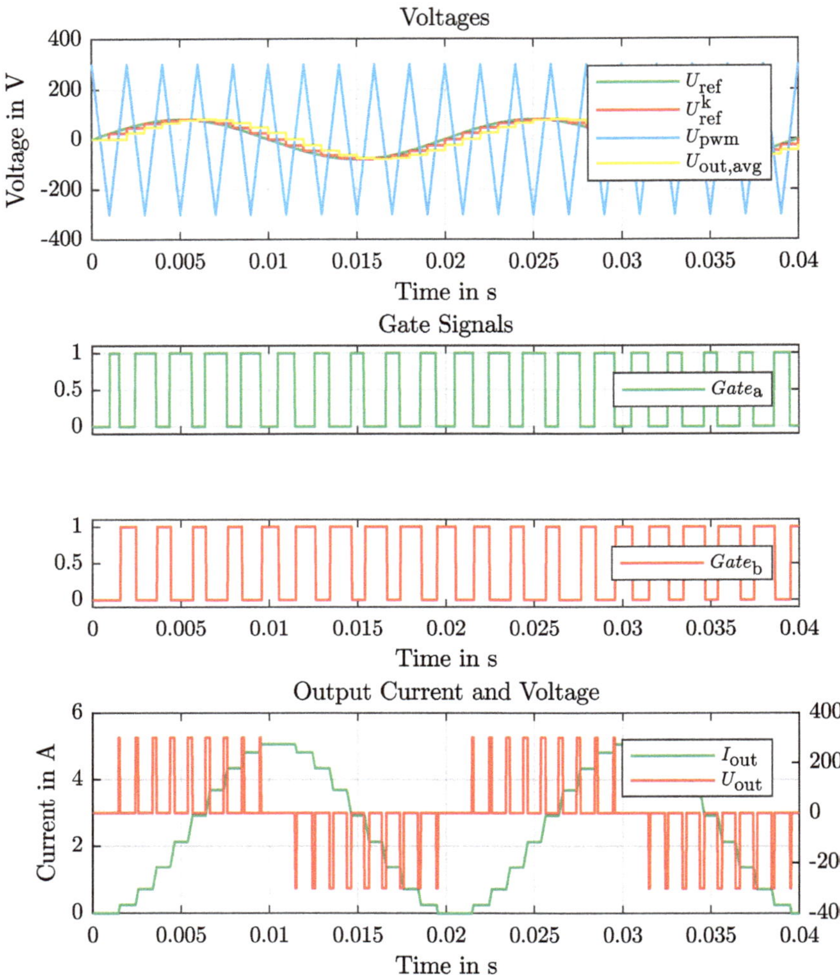

Fig. 2.23 Simulation results for full-bridge converter with PWM

tutorial is aimed at showing the user how the converter vectors are generated to deliver the required average voltage vector during each sample interval, with and without the use of pulse centering.

The PLECS based simulation model, given in Fig. 2.24, shows the three-phase modulator with an input in the form of a rotating average voltage vector. The latter is sampled and used with a two- to three-phase amplitude invariant conversion module to obtain the three-phase load average voltage reference values U_1^*, U_2^*, U_3^*. The converter reference average voltage variables U_a^*, U_b^*, U_c^* used during each interval are generated with the submodule *Pulse Center Module*, the combination of which satisfies equation set (2.23). The *Pulse Center Module* as given in Fig. 2.12 calculates the required zero-sequence reference average voltage value U_o^* which is

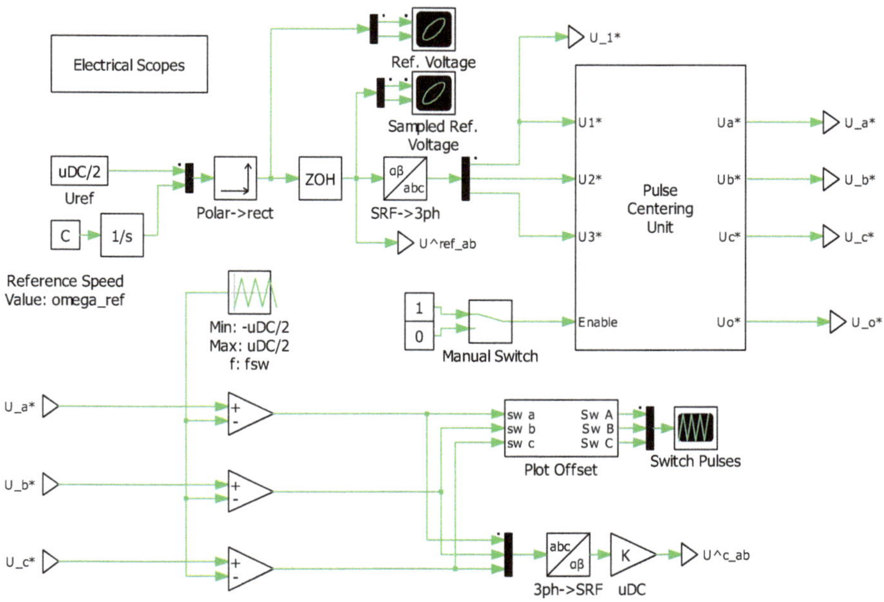

Fig. 2.24 Simulation of three-phase modulator with pulse centering

an output variable of this module. A *manual switch* allows the user to enable/disable the pulse centering function. With the value set to zero, the pulse centering module sets the zero-sequence reference average voltage to zero, in which case the output are equal to the input variables of the module.

The waveforms generated by the simulation are given in Fig. 2.25. The *Space Vectors* scope shows the reference voltage vector and the generated gate signal vector multiplied by the dc-link voltage, which represents a hypothetical converter output voltage. It can be observed which space vectors (in green) are used during each switching period to create the reference voltage vector (in red) using the *Time Range* slider of the xy scope. Note that converter *zero* vectors are shown as small dots at the origin. These change to arrows when a converter active vector is selected during the simulation.

The *Voltages* scope shows the reference voltage values of a single phase before and after the pulse centering is applied as well as the required zero-sequence reference average voltage value U_0^*. Note that the reference voltage value after pulse centering (in red) has a smaller magnitude. Therefore a higher phase voltage can be applied using the same dc-link voltage using pulse centering, as previously explained.

The *Uref* value controls the amplitude of the reference average voltage vector and can be increased up to $u_{DC}/\sqrt{3}$ using pulse centering without crossing the space vector hexagon. This hexagon represents the operational boundary where the output waveform starts to deform, i.e., the half-bridge reference value remains within the

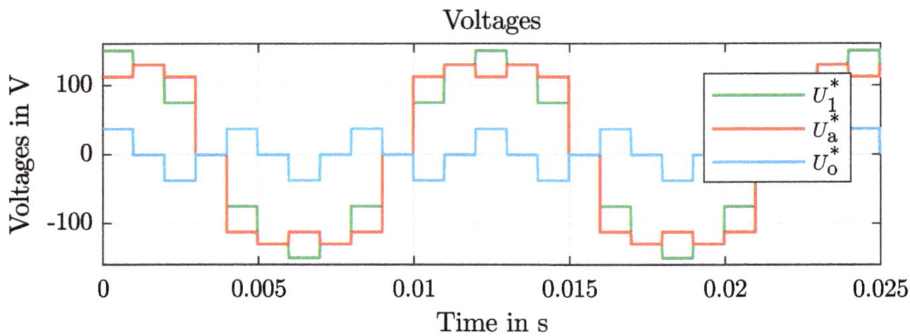

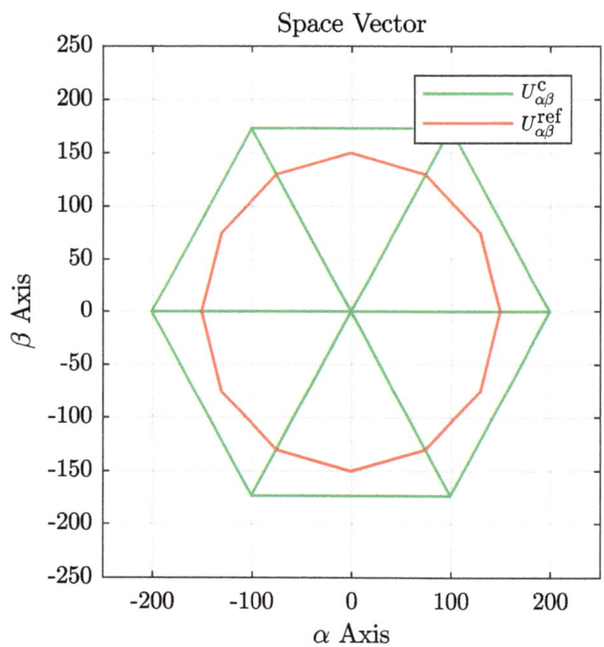

Fig. 2.25 Simulation results of three-phase modulator with pulse centering

supply limits $\pm u_{DC}/2$. Without pulse centering, reference average voltage vector magnitude is limited to $u_{DC}/2$.

2.6.5 *Tutorial 5: Three-Phase Converter with Pulse Width Modulator*

This tutorial is concerned with the implementation of a three-phase converter and modulator structure as shown in Figs. 2.9 and 2.12, respectively. The aim of this tutorial is to combine and extend on previous tutorials by including a three-phase topology and modulator structure, together with a user defined average voltage reference vector. This allows the user to examine the half-bridge voltage, load voltage, zero-sequence voltage, and load current waveforms. In this example, the frequency of the rotating reference average voltage vector is arbitrarily set to 10 Hz, with a simulation time of 500 ms to allow the simulation to settle to a steady-state over 5 periods of operation. The generic load Z is assumed to be a 2 mH inductance with a 1 Ω resistance.

A PLECS model of this tutorial is given in Fig. 2.26. A *three-phase logic interface* module has been introduced which is similar to the logic modules for the half and full-bridge converters. A toggle switch is used to allow the reader to enable/disable pulse centering.

The waveforms generated by the simulation with pulse centering are shown in Fig. 2.27. It should be noted that the current waveforms of the inductances are sinusoidal, while the average line voltages have non-sinusoidal forms. This shows that the pulse centering does not affect the phase voltages while changing the line voltage references. It is left as an exercise for the reader to reconsider the results from the simulation when the amplitude value *Uref* of the reference average voltage vector is increased to $u_{DC}/\sqrt{3}$ which represents the largest value that may be used which will ensure that the absolute average voltage half-bridge reference values are less than or equal to $u_{DC}/2$ when pulse centering is active. Disabling pulse centering at this point will lead to half-bridge average voltage reference values that exceed the supply window $\pm u_{DC}/2$ and cannot therefore be implemented by the converter, as may be observed in the simulations.

2.6.6 Tutorial 6: Three-Phase Simplified Converter Without PWM

In practice, the simulation of comprehensive models as discussed in the previous tutorials is inhibited by the need for a relatively small computation step time. In the example above, the computation step time was set to 1 μs. Consequently, such simulations are usually slow in terms of simulation run time. An often used approach which speeds up the simulation is to remove the modulator and the converter and apply the reference values received from the pulse centering unit to an algebraic model. Three *saturation* modules with outputs limited to $\pm u_{DC}/2$ are used to include the effect of limited dc-link voltage. An example simulation is given in Fig. 2.28.

In this example, the mean half-bridge voltages are used in conjunction with a three- to two-phase amplitude invariant conversion as an input to the generic model of the load. Note that the control of the current in the load is usually the aim of such simulations, as will become apparent in the next chapter. This implies that the need for a precise representation of the load voltage(s), other than their mean, may be avoided in most cases without compromising the results of the simulation.

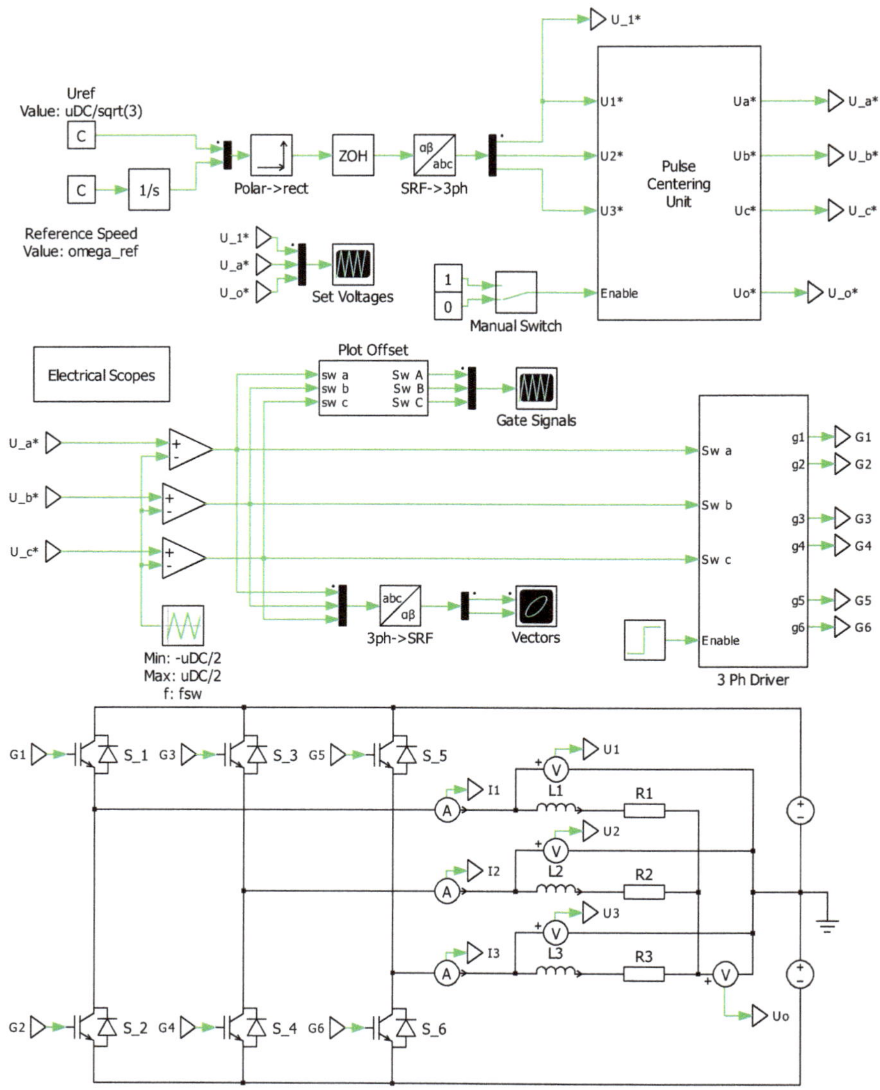

Fig. 2.26 Simulation of three-phase converter with modulator

Some of the waveforms generated by the simulation are shown in Fig. 2.29. A comparison of the load current waveform obtained with this and the previous tutorial shows that they are almost identical. The differences between the two are attributed to the PWM switching process which is no longer present in this simulation model. The operating conditions are otherwise identical as may be observed by comparing the average voltage reference waveforms of both simulations. The actual voltage waveforms are as expected vastly different, the average voltage per switching period

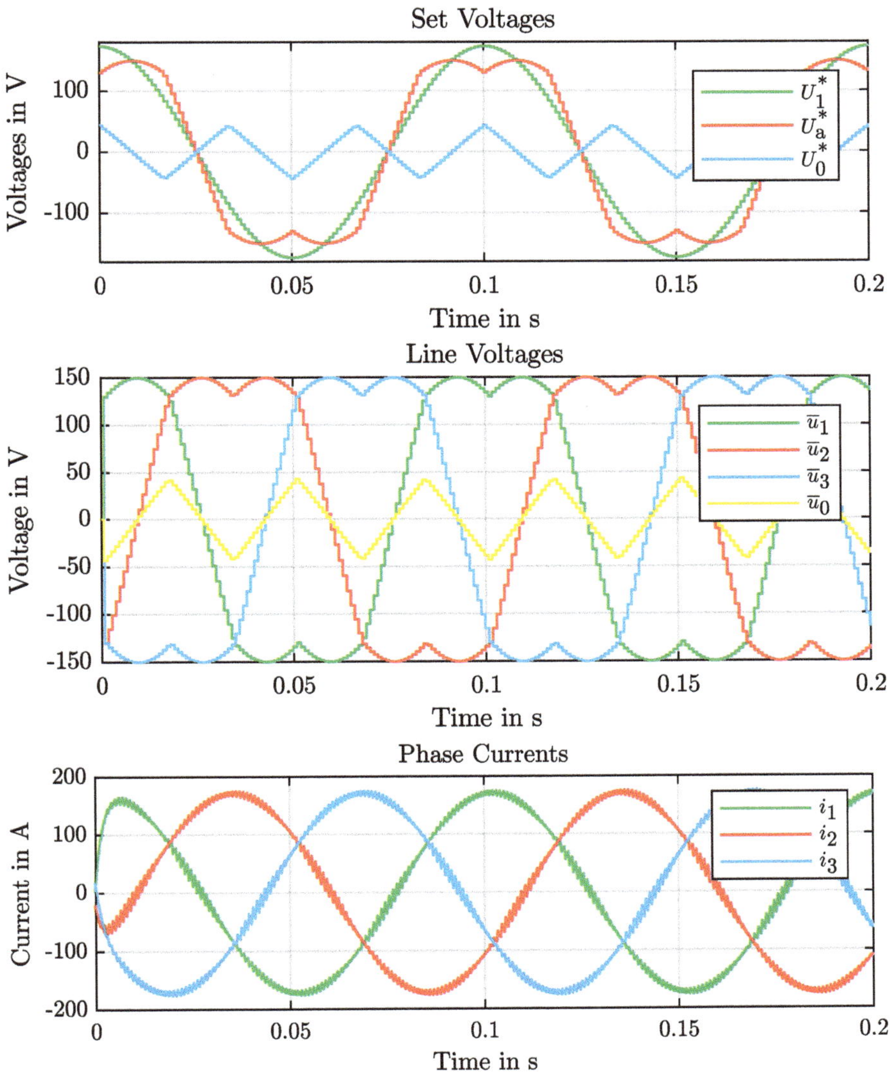

Fig. 2.27 Simulation results of three-phase converter with modulator pulse centering, switching converter

T_s is however identical for both tutorials. With the present choice of average voltage reference amplitude $u_{DC}/2$, a circular orbit of the load vector is still possible. If this reference is increased beyond the limits of the *saturation* modules, the orbit of the load current vector is no longer a circle, which implies that the phase waveforms are no longer sinusoidal.

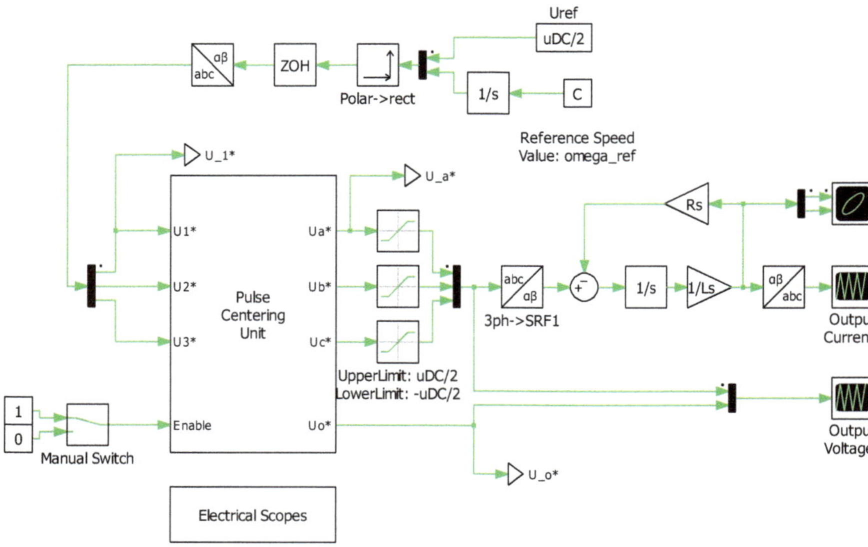

Fig. 2.28 Simulation of three-phase converter with pulse centering, simplified model

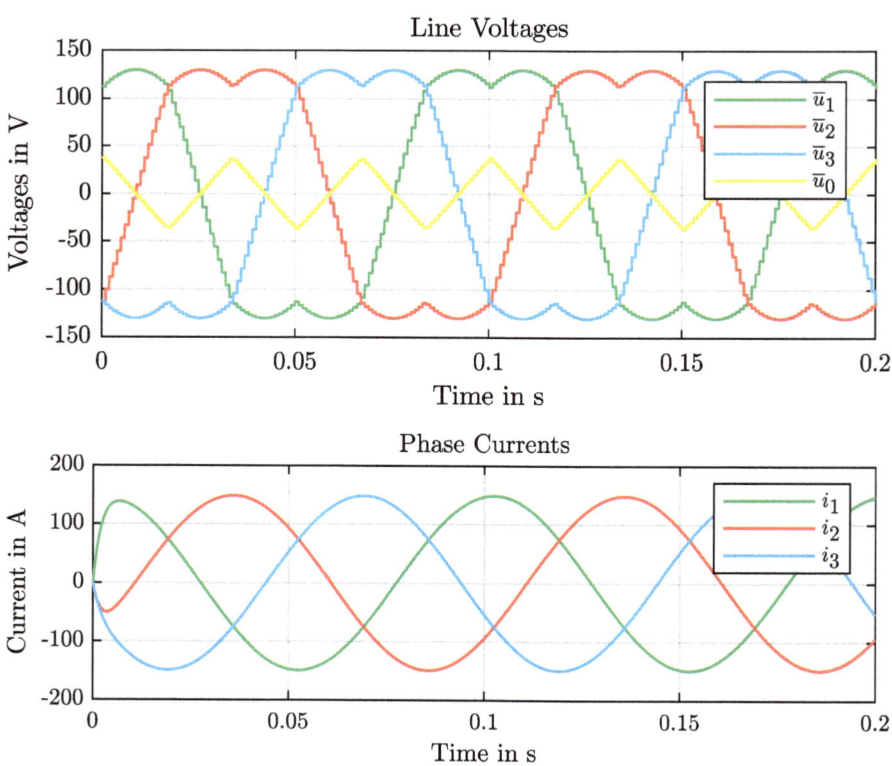

Fig. 2.29 Simulation results of three-phase converter with pulse centering, simplified model

References

1. Svensson T (1988) On modulaton and control of electronic power converters. Technical Report 186, Chalmers University of Technology, School of Electrical and Computer Engineering
2. van der Broeck H, Skudelny HC, Stanke G (1988) Analysis and realization of a pulsewidth modulator based on voltage space vectors. IEEE Trans Ind Appl 24(1):142–150. https://doi.org/10.1109/28.87265
3. Veltman A, Pulle DWJ, De Doncker R (2007) Fundamentals of electrical drives. Spinger, Berlin

Chapter 3
Current Control of Generalized Load

The ability to control the currents in an electrical machine is essential for manipulating its mechanical torque and magnetic flux, as will become apparent in Chap. 4. This chapter considers current control techniques for single- and three-phase voltage source converters which are connected to a *generalized load* as introduced in the previous chapter. The use of this type of load is instructive given that it reflects the electrical behavior of most electrical machines which are in use today. Consequently, the current control techniques discussed in this chapter may be applied for electrical machines discussed in this book. Attention has been given to the use of voltage source converters because these are widely used in low- and medium-power drive applications. However, for high power applications, current source converters are still deployed. These converters are considered outside the scope of this book.

A bewildering variety of current control concepts have been developed. In this chapter two representative techniques for single- and three-phase converters will be considered. The two techniques in question are referred to as *hysteresis* and *model based* current control. Within these categories, different implementation techniques are possible. Two practical control techniques are introduced which have been used by the authors in electrical drive applications. The tutorial section at the end of this chapter deals with a range of simulation models which provide the reader with the ability to examine the control concepts explained in this chapter.

3.1 Current Control of Single-Phase Load

3.1.1 Hysteresis Current Control

Prior to discussing this type of control it is helpful to define the so-called hysteresis concept with the aid of Fig. 3.1. Shown in Fig. 3.1 is a generic module with input

R. W. De Doncker et al., *Advanced Electrical Drives*, Power Systems,
https://doi.org/10.1007/978-3-030-48977-9_3

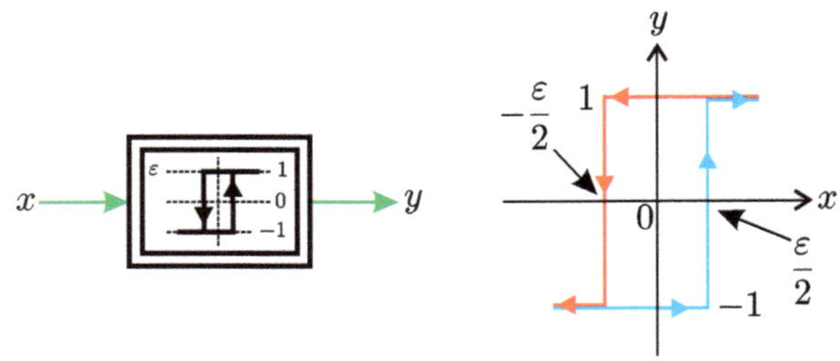

Fig. 3.1 Generic hysteresis module with hysteresis band $\Delta x*$ and transfer function

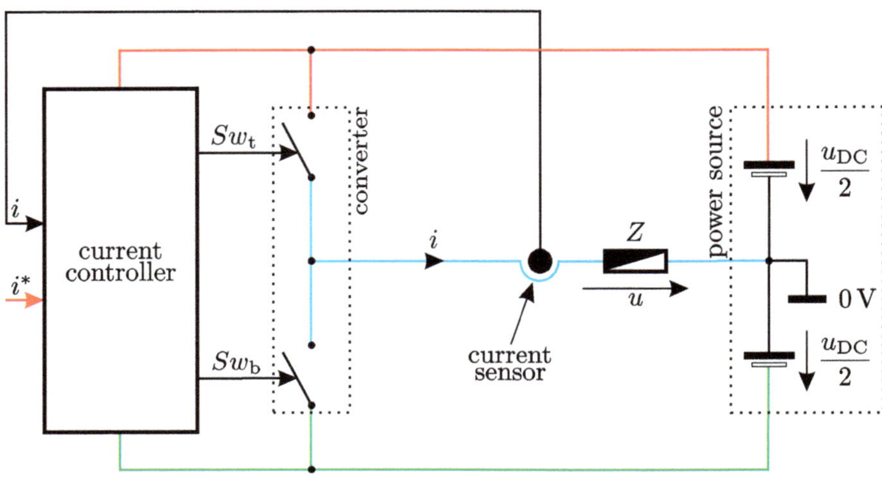

Fig. 3.2 Single-phase converter with hysteresis current controller

x and output y. The output has two states, which in this example are taken to be -1 and 1, respectively, as is apparent from the transfer function shown in Fig. 3.1. The term hysteresis is used to describe a non-singular transitional process. If, for example, the output state $y = -1$, then the output will change to $y = 1$ when the condition $x \geq \Delta x/2$ occurs. Vice versa, when the output is $y = 1$, a change to $y = -1$ will take place as soon as the condition $x \leq -\Delta x/2$ occurs. The transitions exhibit a degree of "hysteresis" defined by the variable $\Delta x*$, i.e., the hysteresis band, as shown in Fig. 3.1.

In this section, a hysteresis based current control will be discussed with the aid of the half-bridge converter topology presented in Sect. 2.2. This topology is readily adapted to hysteresis type current control by adding a current controller module and a current sensor, as indicated in Fig. 3.2. The current controller module must provide the switching signals for the converter on the basis of the instantaneous measured load current and user defined reference current.

An implementation example of such a module, in terms of its generic representation, is shown in Fig. 3.3. Figure 3.3 shows that this module contains two comparators of which the first (Comparator A) is a normal *hysteresis* type comparator with a bipolar output ± 1. The output of the comparator is fed back to the

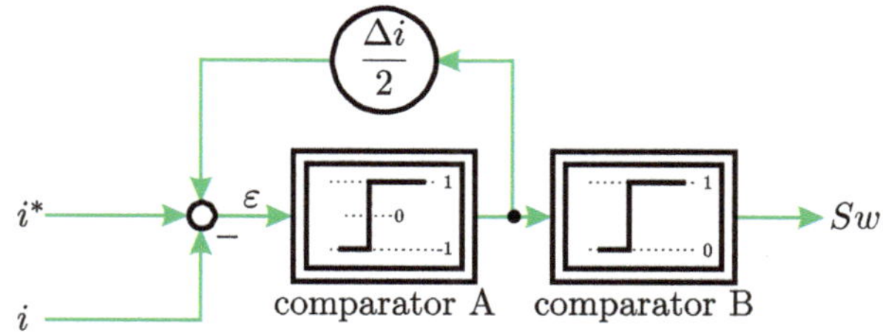

Fig. 3.3 Generic structure of hysteresis current controller

input via a gain module with gain $\Delta i/2$. Consequently, the output state of comparator A is determined by its input $i^* - i \pm \Delta i/2$. Comparator *B* generates a logic signal Sw used by the two switches, where $Sw = 1$ corresponds to a closed top switch Sw_{t} and an open bottom switch Sw_{b}. Vice versa, a state $Sw = 0$ corresponds to a closed bottom switch and an open top switch. The basic action of the controller is to maintain the load current within the limits $i^* \pm \Delta i*/2$, where $\Delta i*$ is a user defined parameter (hysteresis band) and i^* is a current command variable.

Exemplary waveforms produced by this current controller are shown in Fig. 3.4. The waveforms shown are obtained with the aid of a simulation model presented in Sect. 3.3.1. Clearly identifiable in subplot (a) are the load current (color *green*) and the reference current (color *blue*) waveforms, which confirm the ability of the controller to maintain the current within a specified hysteresis band. Also shown in subplot (b) of Fig. 3.4 are the load voltage (color *blue*) and the assumed induced voltage u_{e}. More detailed views of these two subplots are shown in subplot (c) and subplot (d). In subplot (c) the selected current band $\Delta i = 0.4\,\mathrm{A}$ used for this simulation is visible, which confirms the ability of the controller to keep the current within the set limits. Observation of the converter voltage in particular shows that the switching frequency is not constant, which is particularly uninviting in case the load is in the form of an electrical machine, given the acoustic noise signature that may appear with this type of control strategy. The frequency spectrum of the converter voltage underlines this statement. The spectrum in question is discussed (for three-phase hysteresis control) in Sect. 3.2.4 where it is compared to the spectrum generated by three-phase model based current control.

3.1.2 *Model Based Current Control*

The term *model based control* refers to a control method which assumes that the nature of the load (electrical machine in our case) is known. Consequently, a controller may be designed which, on the basis of a known current error, calculates for each sampling interval the required voltage U_{k}^* (as defined by Eq. (2.6)) needed to drive said error to zero. The basic control structure as outlined in *Fundamentals of Electrical Drives* [6] is extended here to accommodate a full-bridge converter topology as discussed in Sect. 2.3. The converter shown in Fig. 3.5 is connected to a single-phase load Z as introduced in the previous chapter.

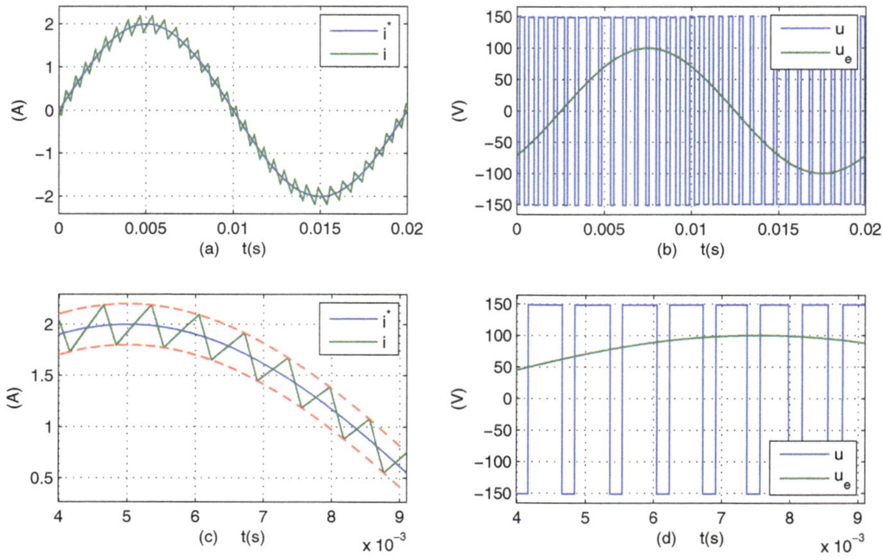

Fig. 3.4 Single-phase hysteresis control example

The modulator controls the converter switches in such a manner that the converter generates (during a given sampling interval T_s) an average voltage quantity $U(t_k)$ which corresponds to the reference average voltage value $U^*(t_k)$ that is provided by the controller module. The basic task of the controller, shown in Fig. 3.5, is therefore to calculate the required average voltage per sample at the beginning of a sampling interval t_k that is required to drive the current error to zero at the end of said interval.

The nature of this control philosophy is shown with the aid of Fig. 3.6. Shown in Fig. 3.6 are the reference current $i^*(t)$ and (typical for PWM based control) the sensed converter current $i(t)$. These currents are sampled by the controller at time marks $0,\ t_1,\ t_2, \ldots$ etc. At time $t = t_1$, a current error between the sampled reference and sampled converter current $i^*(t_1) - i(t_1)$ exists. The objective of the control approach is to determine the required average voltage reference value needed to quickly zero this error. This leads to the condition $i^*(t_1) = i(t_2)$. The control objective aimed at driving the current error to zero during each sample interval may be written as

$$i\,(t_k + T_s) = i^*\,(t_k) \tag{3.1}$$

for a regularly sampled system with sampling time T_s. The modulator will control the converter switches in a manner needed to ensure that the condition as represented by Eq. (3.2) is satisfied, as was outlined in the previous chapter (see Eq. (2.6)).

$$U^*\,(t_k) = \frac{1}{T_s}\int_{t_k}^{t_{k+1}} u\,(\tau)\,\mathrm{d}\tau \tag{3.2}$$

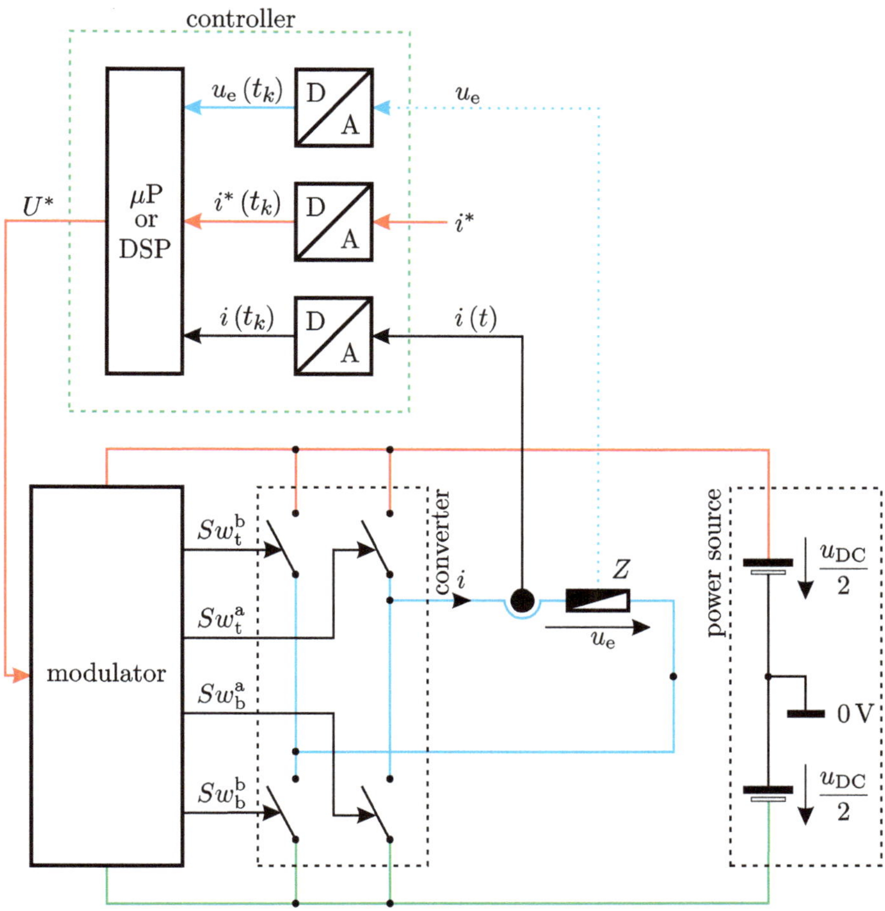

Fig. 3.5 Single-phase converter with model based current control

The variable u, shown in Eq. (3.2) represents the voltage across load Z, which may also be written as

$$u = Ri + L\frac{\mathrm{d}i}{\mathrm{d}t} + u_e \tag{3.3}$$

given that the load is formed by a series network which consists of a resistance R, inductance L, and voltage source u_e. The latter is usually referred to in machines as the back or counter electromotive force (EMF) and represents a speed, i.e., load dependent *disturbance*. In this analysis, access, by means of sensors, is assumed to the voltage u_e and the load current i, where the latter represents the control variable. When the load is in the form of an electrical machine, an estimator/observer is normally used to compute the back-EMF u_e, given that the latter cannot be measured

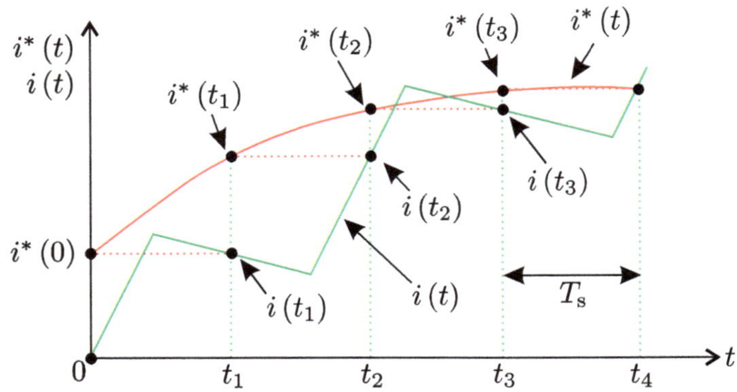

Fig. 3.6 Current waveforms of model based current controller

directly. Use of Eq. (3.3) with Eq. (3.2) allows the latter to be written as

$$U^*(t_k) = \frac{R}{T_s}\int_{t_k}^{t_k+T_s} i(\tau)\,d\tau + \frac{L}{T_s}\int_{i(t_k)}^{i(t_k+T_s)} di + \frac{1}{T_s}\int_{t_k}^{t_k+T_s} u_e(\tau)\,d\tau. \tag{3.4}$$

Equation (3.4) forms the basis for determining a generic control structure that is able to calculate the required average voltage capable of satisfying condition 3.1. It is noted that this value can only be determined on the basis of a detailed knowledge of the load Z with parameters R and L and back-EMF voltage u_e.

In practice, for a digital implementation, discretization of Eq. (3.4) is required. A first order approximation technique may be used, provided the sampling time is sufficiently small, which leads to

$$U^*(t_k) \cong Ri(t_k) + \left(\frac{L}{T_s} + \frac{R}{2}\right)\left(i^*(t_k) - i(t_k)\right) + u_e(t_k) \tag{3.5}$$

where use is made of expression (3.1). Furthermore, it is assumed that the sampling rate is sufficiently high to assure that the back-EMF can be assumed constant over the duration T_s, in which case the average voltage $U_e(t_k)$ is equal to $u_e(t_k)$ as shown in Eq. (3.5). A control structure based on Eq. (3.5) is basically a proportional type controller with a term u_e, otherwise referred to as a disturbance decoupling term. In practice, a proportional integrator type structure is preferable, because the process of zeroing the current error within each interval is in practice not achievable due to parameter discrepancies between controller and load. Furthermore, the required average voltage cannot always be met, i.e., when the converter is used outside the linear voltage operating region, as discussed in the previous chapter. The replacement of the variable $i(t_k)$ by a discrete integration term which utilizes the current error can be made plausible by considering Fig. 3.6. Observation of sampled reference and measured currents and $i(t_3)$ in particular, shows that in the ideal case, i.e., when zeroing of the current error occurs each sample, the current $i(t_3)$ may

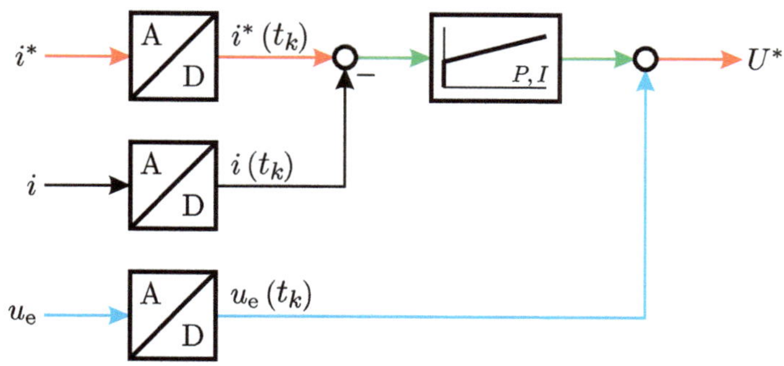

Fig. 3.7 Model based current controller structure

be represented as $i(t_3) = (i^*(t_2) - i(t_2)) + (i^*(t_1) - i(t_1)) + i^*(0)$. In general mathematical terms this error summation may be written as

$$i(t_k) \cong \sum_{j=0}^{j=k-1} \left(i^*\left(t_j\right) - i\left(t_j\right)\right) \tag{3.6}$$

with $t_0 = 0$ and $i(0) = 0$. Use of Eq. (3.6) with (3.5) leads to

$$\begin{aligned} U^*(t_k) \cong\ & R \sum_{j=0}^{j=k-1} \left(i^*\left(t_j\right) - i\left(t_j\right)\right) \\ & + \left(\frac{L}{T_s} + \frac{R}{2}\right)\left(i^*(t_k) - i(t_k)\right) + u_e(t_k)\,. \end{aligned} \tag{3.7}$$

The generic structure which corresponds to Eq. (3.7), as shown in Fig. 3.7, contains a PI controller and disturbance decoupling term $u_e(t_k)$. The proportional K_p and integral K_i gain settings for the discrete controller are in this case defined as

$$K_p = \frac{L}{T_s} + \frac{R}{2} \tag{3.8a}$$

$$K_i = \frac{R}{T_s} \tag{3.8b}$$

In practice, the condition $L/T_s > R/2$ is satisfied, which means that the proportional gain is reduced to L/T_s. Furthermore Eq. (3.7) can also be expressed in terms of a gain $K_i =\approx L/T_s$, bandwidth $\omega_i = R/L$, and current error term $\epsilon(t_k) = (i^*(t_k) - i(t_k))$, which leads to

$$U^*\left(t_k\right) \cong K_{\mathrm{p}}\left(\epsilon\left(t_k\right)+\omega_{\mathrm{i}} \sum_{j=0}^{j=k-1} \epsilon\left(t_k\right)\right)+e_{\mathrm{a}}\left(t_k\right), \tag{3.9}$$

which is the preferred method of formulating this expression for current controllers.

Note that model based current control utilizes a constant sampling and carrier frequency, which is decidedly advantageous when compared to hysteresis type control. The reason for this is that the frequency spectrum of the converter voltage is well defined (as shown in Sect. 3.2.4). Choosing the carrier frequency provides the ability to influence the acoustical spectrum, the switching losses, and provides the ability to tailor design line-side grid filters. Furthermore, the control algorithm can easily be implemented in digital controllers.

In practical implementations, the PI controller is prone to *windup*, which occurs when the voltage or current limits of the system are reached. Windup can occur in this case when the reference average voltage value generated by the PI controller exceeds the maximum value that can be delivered by the converter. Under such circumstances a current error occurs at the input of the controller, which will cause the integrator output to further ramp up or down. Practical controllers as those considered in the tutorials have an *anti-windup* feature which limits integrator action when the user defined limits are reached. For the controller in question, the limit values should be set to the $\pm$ maximum average voltage per sample.

A disadvantage of model based control is that a priori knowledge of the load is required, which is not the case for hysteresis type current control. However, in practice, its advantage of not requiring access to the machine parameters does not outweigh the acoustical noise signature that comes with the use of a variable switching frequency hysteresis current controller.

An example of model based current control is given in Fig. 3.8, which is taken from the tutorial in Sect. 3.3.2 at the end of this chapter. The current reference and the load are identical to that used for the hysteresis control example, shown in Fig. 3.4.

For more details, the reader is referred to the tutorial given in Sect. 3.3.2. Also shown in Fig. 3.8 is the converter voltage u and voltage u_{e}. The averaged (per sample interval) load voltage waveform resembles the load voltage waveform u_{e}, which is to be expected because the load is predominantly inductive. Indeed, the mean voltage across an inductance must be zero, a statement also referred to as *voltage-second balancing*. The same set of waveforms were also given for the hysteresis controller, shown in Fig. 3.4. These show a similar behavior, given that the same load model is used in both examples. Subplots (c) and (d) in Fig. 3.8 show a zoomed section of the upper two subplots, which demonstrates the ability of the controller to zero the error between sampled converter and sampled reference current at the end of each sample.

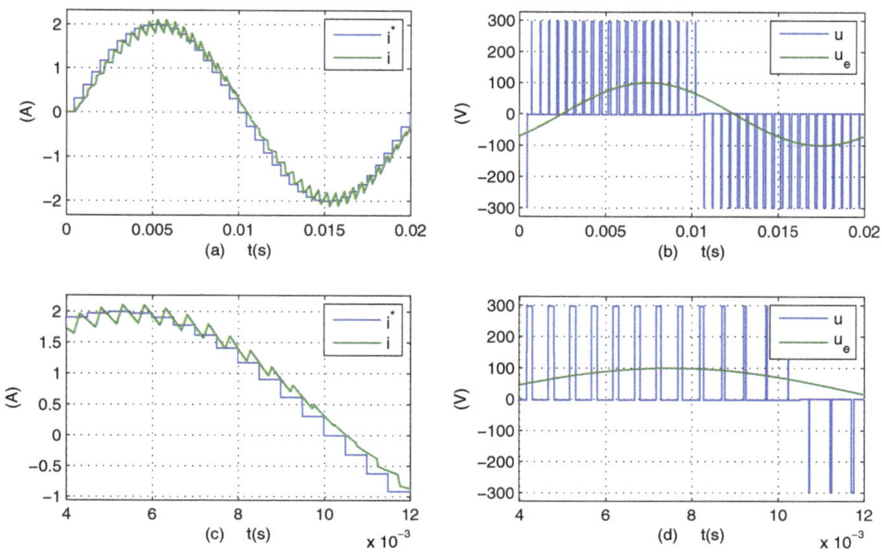

Fig. 3.8 Single-phase model based current control example

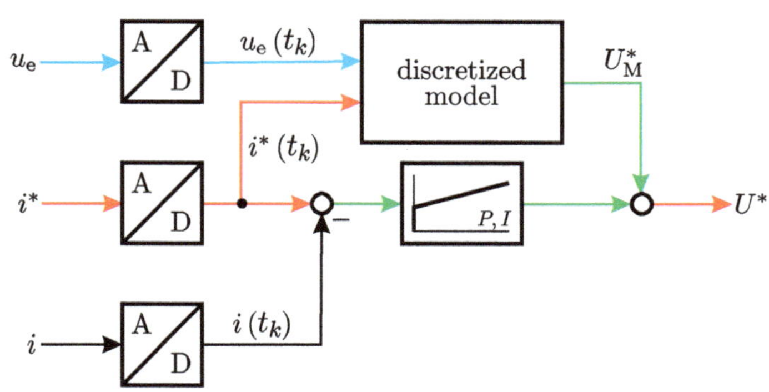

Fig. 3.9 Augmented model based controller structure

3.1.3 Augmented Model Based Current Control

In the previous section, a disturbance decoupling term u_e was added to the PI controller, of which the gains were derived using a model based control approach. A modified approach, here referred to as *augmented model based control*, is based on the use of a discretized model of the load as indicated in Fig. 3.9. Input to this model are the (sampled) reference current i^* and load voltage u_e, while the output is in the form of the average voltage reference $U^*_M(t_k)$.

The contents of the discrete model is directly based on load equation (3.4), in which the measured current is replaced by the reference value. Furthermore, the integral boundaries are taken from $t_{k-1} \rightarrow t_k$, given the need to obtain an expression in terms of the variables $i^*(t_k)$ and $i^*(t_{k-1})$. A process known as *backward time discretization*, which leads to

$$U^*_{\mathrm{M}}(t_k) \cong R\, i^*(t_k) + \left(\frac{L}{T_{\mathrm{s}}} - \frac{R}{2}\right)\left(i^*(t_k) - i^*(t_{k-1})\right) + u_{\mathrm{e}}(t_k)\,. \tag{3.10}$$

Typically, the parameters R and L and variable u_{e} are estimates of the actual load connected to the converter. Theoretically, this type of controller does not require access to the measured current, i.e., operates in an open loop control mode, assuming that there is no mismatch between the parameters used in the load and controller. In reality such errors exist in which case a PI controller with reduced gain (when compared to the previous case), as shown in Fig. 3.9, can be used to compensate for said errors. The gains K_{p} and K_{i}, as calculated using Eqs. (3.8) and (3.9), represent a suitable upper limit for the PI controller shown in Fig. 3.9. However, in practice lower gains can be used because time constants associated with parameter changes are low.

The use of an augmented model based current control approach must be considered with caution as the model typically contains differential terms, which require a rate-of-change limiter of the input commands to avoid hitting the voltage or current limits of the converter. Furthermore, differential terms can be prone to noise. However, note that in model based controllers the different terms are in the open loop command path, which often does not exhibit noise. The latter cannot be said when using differential terms in proportional, integral, and differential (PID) feedback controllers which act on sensed, i.e., noisy, feedback signals. In the model shown here, the differential of the reference current command is required, according to Eq. (3.10). Hence, in this case, the reference current signal is not a measured quantity and its differentiation does not exhibit high noise levels. In conclusion, the augmented model based controller places the differentiating part of the controller in the so-called open loop feed-forward path and not in the feedback path, thereby avoiding amplifying noise.

3.2 Current Control of a Three-Phase Load

The extension to three-phase current control laws for voltage source converters is usually undertaken with the aid of space vectors. Such an approach is prudent on the grounds that the sum of the three-phase currents is zero, in which case the number of degrees of freedom to control current reduces to two. A space vector representation of the variables takes this into account and is therefore convenient. A further step in this process is to consider the control process in a synchronized reference frame linked to a suitable load vector as indicated in Fig. 3.10.

Shown in Fig. 3.10 is the load voltage vector $\vec{u}_{\mathrm{e}}$, which in turn can be linked to a flux vector $\vec{\psi}_{\mathrm{e}}$. If the load is in the form of an electrical machine, this vector may, for example, represent the rotor flux vector. At constant rotational speed ω_{e},

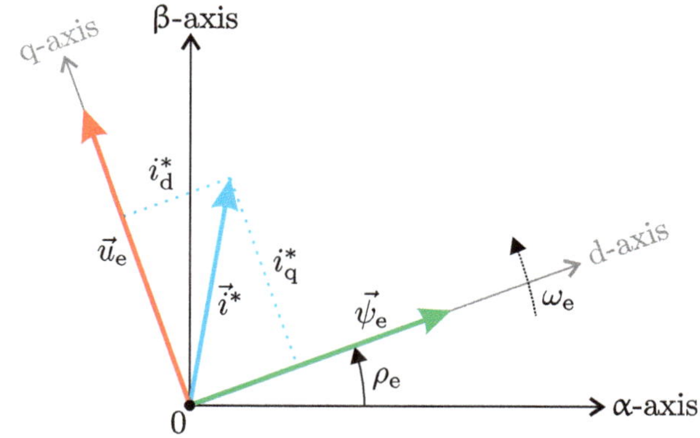

Fig. 3.10 Synchronous reference frame for three-phase current control

the amplitude of the flux vector $\vec{\psi}_e$ is equal to $\hat{u}_e/\omega_e$, where ω_e represents the rotational electrical speed of the vector. The *direct d*-axis of the *dq* coordinate system is tied to the flux vector, which is displaced by an angle ρ_e with respect to the stationary reference frame. The *quadrature q*-axis is tied to the load vector $\vec{u}_e$. A user defined current reference vector $\vec{i}^*$ is also shown in Fig. 3.10, which can be expressed in terms of the direct and quadrature reference current components i_d^* and i_q^*, respectively.

Prior to discussing two commonly used current control strategies, it is instructive to consider the basic principles which govern the current control process. A suitable starting point for this analysis is to consider Fig. 2.15 which shows the eight possible converter voltage vectors $\vec{u}_{\{Sw^a,Sw^b,Sw^c\}}$ that can be realized with a three-phase voltage source converter. The converter in question is connected to a three-phase generalized load, as shown in Fig. 2.9. Each load phase consists of an inductance L, resistance R, and phase load voltage u_e^i, with $i = 1, 2, 3$. The three load phases are represented in space vector format, for example, by the vector $\vec{u}_e = \hat{u}_e\, e^{j\omega_e t}$. The corresponding load currents may be found by considering the terminal equation, in space vector format, namely

$$\vec{u}_{\{Sw^a,Sw^b,Sw^c\}} = L\frac{d\vec{i}}{dt} + R\vec{i} + \vec{u}_e. \tag{3.11}$$

For the loads considered in this book, the term $R\vec{i}$ is small in comparison to the term $L\; d\vec{i}/dt$ and is neglected in this analysis. A first order approximation of Eq. (3.11) gives

$$\Delta\vec{i}_{\{Sw^a,Sw^b,Sw^c\}} \simeq \frac{\Delta T}{L}\left(\vec{u}_{\{Sw^a,Sw^b,Sw^c\}} - \vec{u}_e\right), \tag{3.12}$$

where ΔT represent the time interval in which the voltage vector $\vec{u}_{\{Sw^a,Sw^b,Sw^c\}}$ is active. Equation (3.12) is significant as it shows that there are only a discrete set of possibilities for changing the instantaneous current vector from $\vec{i} \rightarrow \vec{i} + \Delta\vec{i}$ over

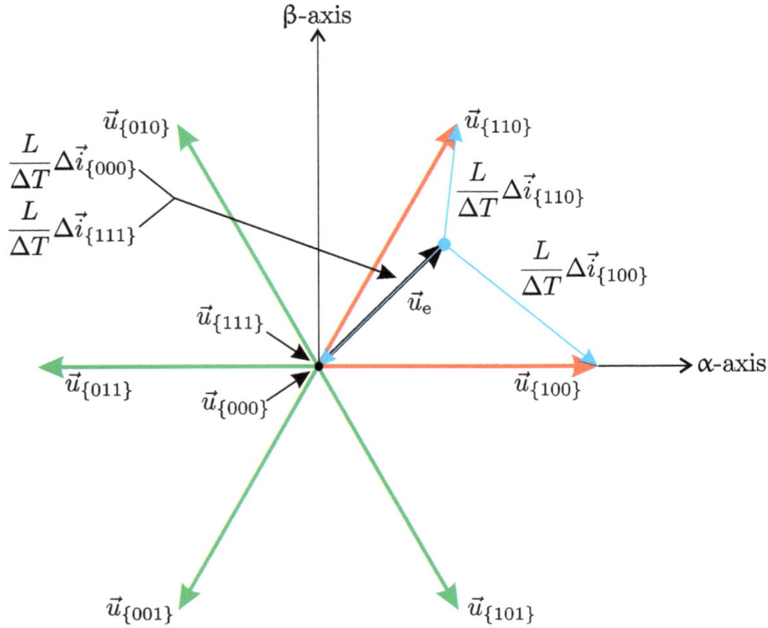

Fig. 3.11 Choice of incremental current $\Delta\vec{i}$ vectors

a time interval ΔT in terms of the direction and amplitude. This may be observed from the example given in Fig. 3.11. This figure shows a rotating voltage vector $\vec{u}_{\mathrm{e}}$ at a given point in time, which is arbitrarily positioned between two (adjacent) active converter vectors $\vec{u}_{\{100\}}$ and $\vec{u}_{\{110\}}$ (shown in *red*). Under these conditions, there are four converter vectors (which include two zero vectors) that can be utilized by the current controller to change the direction of the instantaneous current. Note that the variable ΔT determines the actual magnitude of current change in a particular direction as may be observed from Eq. (3.12) and Fig. 3.11.

In terms of achieving three-phase current control, two specific techniques are possible. The first, as used for hysteresis type control, is to track the locus of the measured current and compare its trajectory relative to the reference current vector trajectory. The error current vector $(\vec{i}^* - \vec{i})$ may be used in conjunction with a user defined boundary box to select the appropriate active or zero voltage vector needed to hold to the absolute current error within user defined limits. An alternative approach for regularly sampled systems, i.e., those which utilize a fixed sampling time T_{s}, is to identify the vector current error at the start of the interval and calculate the required average voltage reference vector $\vec{U}^*(t_k)$ needed to drive this error to zero at the end of said interval. This model based current control type of approach, as applied to a single-phase system in the previous section, can be readily extended to three-phase systems. The three-phase PWM modulator as discussed in the previous chapter identifies the required active and zero vectors as well as the duration for which these should be active during a sampling interval T_s, as will become apparent in Sect. 3.2.2.

3.2.1 Three-Phase Hysteresis Current Control

For this type of control, a current control module is required as given in Fig. 3.12, which has as inputs the measured current vector $\vec{i}$, reference current vector $\vec{i}^*$, and load voltage vector $\vec{u}_e$. The *load* module consists of three star connected load phases. Each phase is represented by a resistive/inductance network and load voltage u_e, as discussed in the previous chapter. In this case, the load voltage vector $\vec{u}_e$, which represents the three-phase shifted load phase voltages is used as a reference for the synchronous current controller (see Fig. 3.10). However, in drive applications an estimated flux vector is commonly used for this purpose, as will become apparent at a later stage. The controller outputs are the three converter switch signals Sw^a, Sw^b, Sw^c (as defined in the previous chapter), which in effect identify the voltage vector $\vec{u}_{\{Sw^a,Sw^b,Sw^c\}}$ and its required duration ΔT to minimize the error between measured and reference current vectors.

The subject of hysteresis type current control for three-phase voltage source converters has received considerable attention in literature [1–3]. Consequently, there are a variety of algorithms which are able to perform hysteresis type current control. In this book, a hysteresis control approach is considered which is referred to as the *box* method [5]. This method utilizes the current error $\Delta\vec{i}$ in a synchronous reference frame that is tied to the orthogonal vectors $\vec{u}_e$, $\vec{\psi}_e$, as shown in Fig. 3.10.

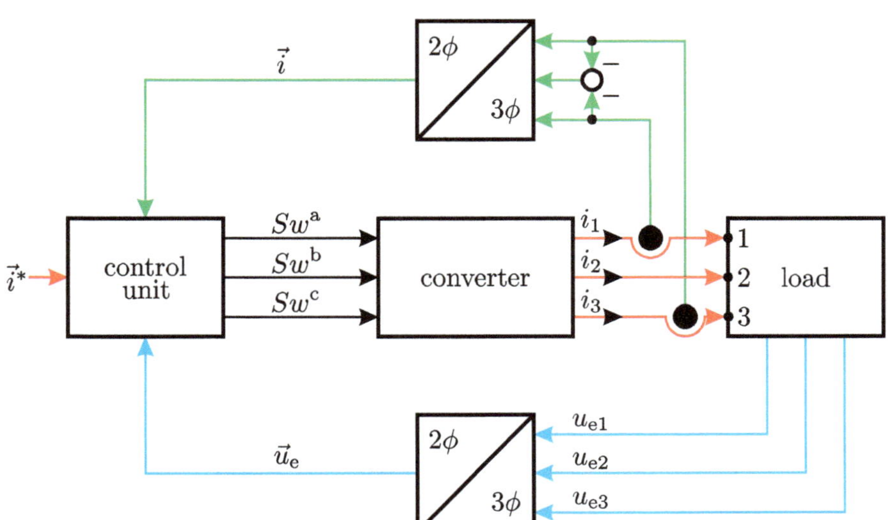

Fig. 3.12 Three-phase hysteresis current control

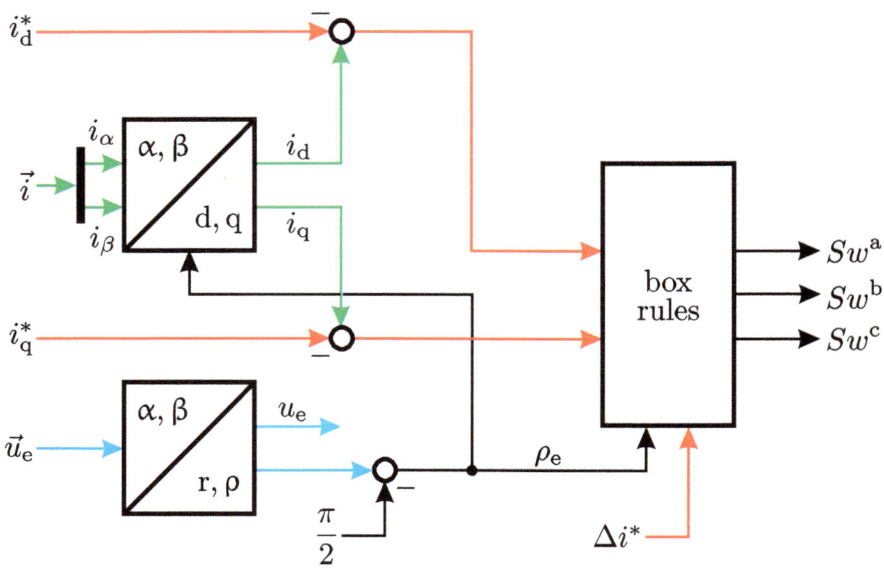

Fig. 3.13 Generic diagram of the hysteresis, *box method* type controller

The generic representation of this control concept, as given in Fig. 3.13, shows the coordinate conversion of the measured current vector to a synchronous reference frame. The orientation of this reference frame is realized with the aid of the voltage vector $\vec{u}_e$ and a Cartesian to polar conversion module which identifies the instantaneous angle of vector $\vec{u}_e$ with respect to a stationary reference frame. The imaginary axis of the synchronous frame controller is to be aligned with the voltage vector $\vec{u}_e$. Consequently, a phase angle shift of $-\pi/2$ rad is introduced in the generic module to arrive at the required reference angle ρ_e for the direct axis of the synchronous reference frame. Also shown in Fig. 3.13 is a *box rules* module which generates the required converter vector $\vec{u}_{\{Sw^a,Sw^b,Sw^c\}}$ (in terms of the required converter switch states Sw^a, Sw^b, Sw^c) on the basis of a direct and quadrature current error, defined as $(i_d - i_d^*)$ and $(i_q - i_q^*)$, respectively.

The process of selecting the appropriate voltage vector $\vec{u}_{\{Sw^a,Sw^b,Sw^c\}}$ may be explained with the aid of Fig. 3.14. Shown in Fig. 3.14 are the eight converter voltage vectors, together with an arbitrarily chosen voltage vector $\vec{u}_e$, which in turn is linked to the flux vector $\vec{\psi}_e$.

Tied to the current reference vector is a *box* shaped contour with sides numbered 1–4. These sides represent the boundary within which the instantaneous measured current vector $\vec{i}$ should be held by the controller.

The *box* is oriented with respect to the synchronous reference frame, as may be seen by the presence of the direct and quadrature axes shown in the box. These dq-axes match the orientation of the vectors $\vec{u}_e$ and $\vec{\psi}_e$ as can be observed in Fig. 3.14. Of key importance for the box method is to consider the vector endpoint of the error vector $\Delta\vec{i}$ with respect to its location within the box, the size of which is determined

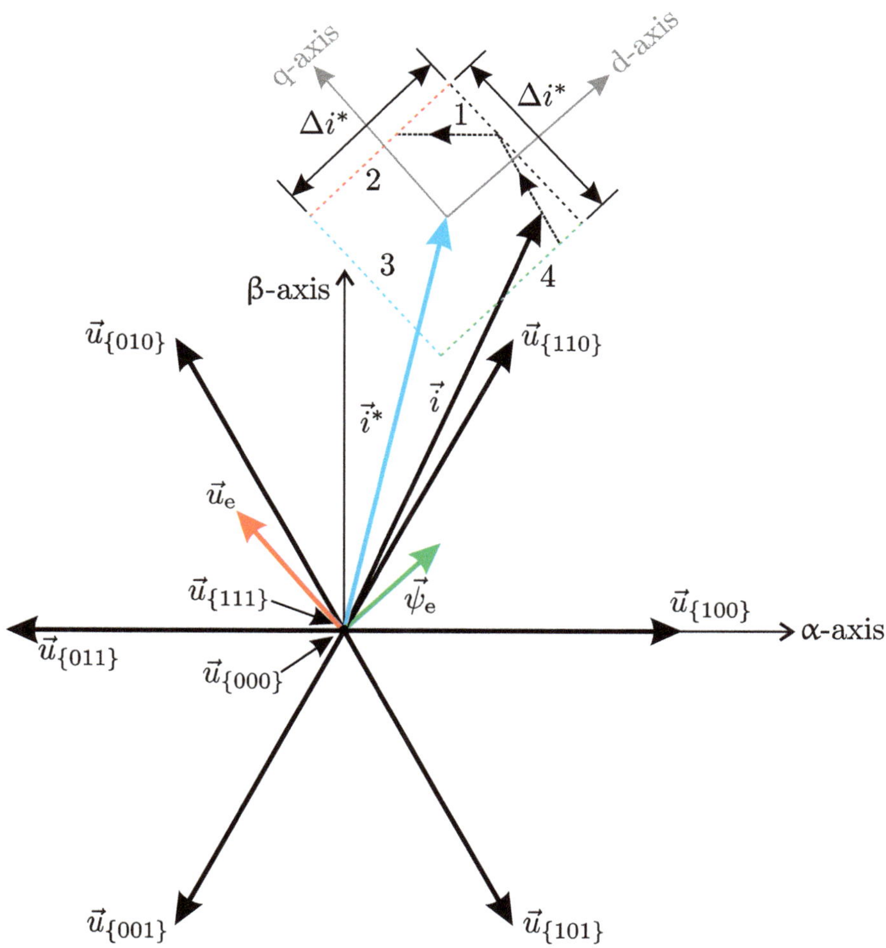

Fig. 3.14 Switching algorithm for a hysteresis, *box method* type controller

by the variable Δi^*. If the error vector endpoint is located within the box, no action of the controller is taken. However, if the error vector endpoint meets or exceeds one of the four numbered boundaries of the box, specific action is undertaken, namely at:

- Boundary 1: Check if the active vector $\vec{u}_{\{Sw^a, Sw^b, Sw^c\}}$ currently in use *lags* the vector u_e. If this is the case, select the next *counter clockwise* active vector. If, for example, vector $\vec{u}_{\{010\}}$ is the active vector in use, the controller would switch to vector $\vec{u}_{\{011\}}$ when boundary 1 was encountered by the error vector endpoint.
- Boundary 2: Check which active vector $\vec{u}_{\{Sw^a, Sw^b, Sw^c\}}$ is in use and switch to the nearest (with the minimum number of switching actions) *zero* vector. If, for

example, vector $\vec{u}_{\{010\}}$ is the active vector, then the controller would switch to zero vector $\vec{u}_{\{000\}}$ when boundary 2 was encountered by the error vector endpoint.

- Boundary 3: Check if the active vector $\vec{u}_{\{Sw^a, Sw^b, Sw^c\}}$ currently in use *leads* the vector u_e. If this is the case, select the next *clockwise* active vector. If, for example, vector $\vec{u}_{\{011\}}$ is the active vector in use, the controller would switch to vector $\vec{u}_{\{010\}}$ when boundary 3 was encountered by the error vector endpoint.
- Boundary 4: Check which active vector $\vec{u}_{\{Sw^a, Sw^b, Sw^c\}}$ was used last and *reactivate* this vector. For example, if vector $\vec{u}_{\{010\}}$ was active prior to encountering the zero vector $\vec{u}_{\{000\}}$, then the controller would switch to vector $\vec{u}_{\{010\}}$ when boundary 4 was encountered by the error vector endpoint.

The motivation behind the *box rules* can be understood by considering the current trajectory that will occur when, for example, an active or zero converter vector $\vec{u}_{\{Sw^a, Sw^b, Sw^c\}}$ is selected. From Fig. 3.11 it may be observed that the incremental current vector $\Delta\vec{i}$ is proportional to the vector $(\vec{u}_{\{Sw^a, Sw^b, Sw^c\}} - \vec{u}_e)$. This implies that the use of the active vector $\vec{u}_{\{010\}}$ (see Fig. 3.14) would cause the current vector $\vec{i}$ to move towards boundary 1. When boundary 1 is reached, the current direction must be changed, which for $\omega_e > 0$ means activating converter vector $\vec{u}_{\{011\}}$. This is precisely the action taken by box boundary rule 1. Note that, for this rule, action is only undertaken provided that the currently active converter vector is lagging. The reason for this rule is that lagging vectors are capable of yielding an incremental current vector direction which will head towards the general direction of boundary 1. The reasoning outlined above can be similarly applied to understand the motivation for the remaining box rules.

The tutorial in Sect. 3.3.3 demonstrates this type of current control. The reader is urged to examine this tutorial and in particular view the reference current vector locus and actual current vector locus which appears during the course of this simulation. An example of the results achieved with this simulation is shown in Fig. 3.15. In this example, the direct axis and quadrature reference current were set to $i_d^* = 0\,\text{A}$ and $i_q^* = 15\,\text{A}$, respectively, while the error reference value Δi^* was taken to be 4.4 A. Hence, in this example, the reference current vector $\vec{i}^*$ is aligned with the vector $\vec{u}_e$. Furthermore, a constant amplitude voltage vector $\vec{u}_e$ is assumed which rotates at a speed of 3000 rpm. Shown in Fig. 3.15 are the current locus $\vec{i}(t)$ and reference current locus $\vec{i}^*(t)$ (*red* trace) for one 20 ms cycle of operation. An observation of Fig. 3.15 shows that the current controller is able to maintain the current within the limits of the box.

The example discussed here assumes an analog control structure, but a digital implementation is equally feasible. With a digital implementation, the error vector can exceed the limits of the box, because changes to the trajectory will take place once during each sample. Furthermore, the measured current locus is built from a set of sub-trajectories which have a specified direction, as discussed earlier (see Fig. 3.11). It is instructive to consider the current error vector $\Delta\vec{i}$, for this example in a synchronous reference frame, i.e., with respect to the box shown in Fig. 3.14.

Figure 3.16 shows part of the 20 ms current error vector trace (*green* trace). Also shown in Fig. 3.16 is part of this trace (*red* trace), which is numbered to explain

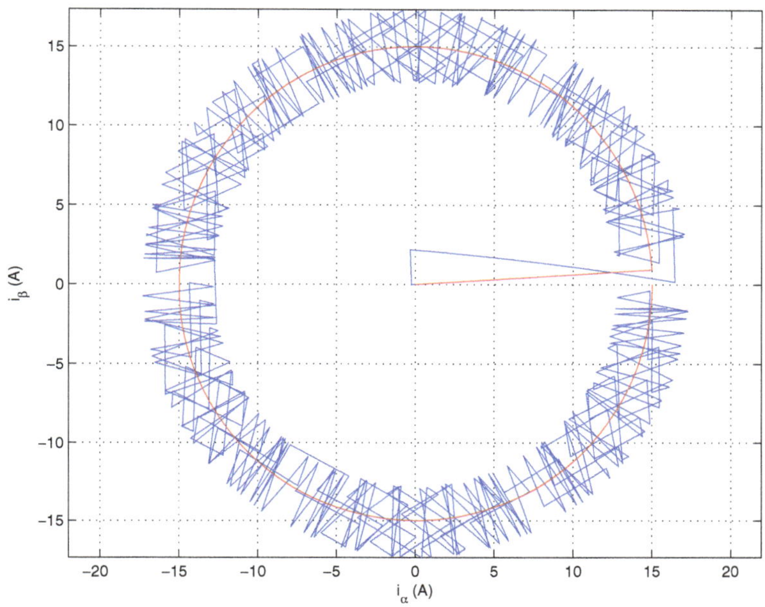

Fig. 3.15 Current locus of the hysteresis *box method* example

the sequence of actions undertaken by the controller. It is helpful in this context to note again that the error reference value Δi^* was taken to be 4.4 A and this determines the size of the box. Clearly observable from Fig. 3.16 is the fact that the measured current trace is kept within the box. The actions of the controller are in accordance with the *box rules* given earlier. Nonetheless it is instructive to examine the numbered trajectory in some detail. During current locus leg 1, an active vector has been selected which drives the error vector $\Delta\vec{i}$ to the *top* limit (boundary 2 in Fig. 3.14) of the box. When this boundary is reached, a zero vector is activated, which results in current locus leg 2. When the *bottom* limit (boundary 4 in Fig. 3.14) of the box is reached, the controller switches back to the last active vector, i.e., the vector which was active for locus leg 1. Locus leg 3 is now active and the sequence of events described above repeats. This implies that during locus leg 4 a zero vector is active and the last active vector is selected during leg locus 5. This locus leg meets the *left* box limit (boundary 3 in Fig. 3.14) where the active voltage vector is switched clockwise by $\pi/3$ rad, which results in locus leg 6. This locus leg meets the *right* box limit (boundary 1 in Fig. 3.14) in which case the active vector is switched anti-clockwise by $\pi/3$ rad leading to locus leg 7.

In the following, attention is given to the fact that, when applying zero vectors, the trajectories are slanted relative to the vertical axis of the plot shown in Fig. 3.16. A detailed explanation of the phenomenon is given below for those readers which have a desire or need to comprehend this issue. Note, however, that this effect does not affect the hysteresis control concept in a significant way. There may be

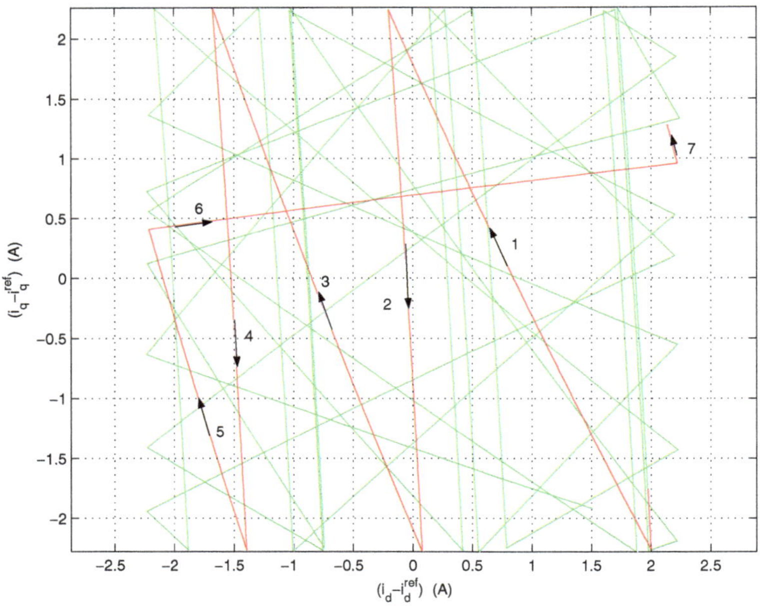

Fig. 3.16 Current locus, hysteresis, *box method* example, current error

an expectation that the trajectories in question should coincide with the *vertical* (q-axis) of the plot because the use of a zero vector leads to an error current trajectory which is in the direction of $-\vec{u}_{\mathrm{e}}$, as indicated in Fig. 3.11. From a stationary reference frame perspective this is indeed the case. However, in a synchronous reference frame, the trajectory differs as may be explained with the aid of Eq. (3.11). This expression may also be written in terms of a synchronous reference frame, namely

$$\vec{u}^{dq}_{\{Sw^{\mathrm{a}},Sw^{\mathrm{b}},Sw^{\mathrm{c}}\}} = L\frac{\mathrm{d}\vec{i}^{\mathrm{dq}}}{\mathrm{d}t} + \mathrm{j}\omega_{\mathrm{e}}L\left(\vec{i}^{*}\right)^{\mathrm{dq}} + R\left(\vec{i}^{*}\right)^{\mathrm{dq}} + \vec{u}^{\mathrm{dq}}_{\mathrm{e}} \tag{3.13}$$

where $\vec{u}^{\mathrm{dq}}_{\mathrm{e}}$ is equal to $\mathrm{j}u_{\mathrm{e}}$ and ω_{e} represents the rotational frequency of the synchronous reference frame as mentioned earlier. To examine the direction of the incremental current for the case when the zero vector is active, it is helpful to rewrite $(\vec{i}^{*})^{\mathrm{dq}}$ in the form of its direct and quadrature components i^{*}_{d} and i^{*}_{q} and ignore the resistive component (as undertaken with Eq. (3.11)), which after some manipulation gives

$$\Delta\vec{i}^{\mathrm{dq}}_{0} \simeq \frac{\Delta T}{L}\left[\omega_{\mathrm{e}}Li^{*}_{\mathrm{q}} - \mathrm{j}\left(\omega_{\mathrm{e}}Li^{*}_{\mathrm{d}} + u_{\mathrm{e}}\right)\right] \tag{3.14}$$

In Eq. (3.14), $\Delta\vec{i}_0^{\rm dq}$ represents the current trajectory (in dq coordinates) which will occur when a zero vector is selected. Equation (3.14) shows that this current trajectory will be in the direction of the negative quadrature axis when $i_{\rm q}^* = 0$. However, in the general case, the trajectory will be rotated counter clockwise with respect to the *negative* quadrature axis by an angle ρ_0 which, according to Eq. (3.14), equals

$$\rho_0 \simeq \arctan\left(\frac{Li_{\rm q}^*}{Li_{\rm d}^* + \psi_{\rm e}}\right), \tag{3.15}$$

where $\psi_{\rm e} = {}^{u_{\rm e}}/_{\omega_{\rm e}}$ represents the amplitude of the flux vector $\vec{\psi}_{\rm e}$ (see Fig. 3.14). In the example shown in Fig. 3.16, the direct current reference value was set to zero $i_{\rm d}^* = 0$ and $Li_{\rm q}^*$ was chosen small in comparison to $\psi_{\rm e}$. Consequently, the locus rotation as defined by Eq. (3.15) is relatively small. However, in some applications this rotation ρ_0 of the current trajectory may become large. In this case, additional measures (which are outside the scope of this book) need to be taken in terms of the orientation of the box (see Fig. 3.14) in order for the *box method*, described in this section, to work satisfactorily.

3.2.2 Model Based Three-Phase Current Control

The approach used to obtain the generic controller structure for a three-phase load is very similar to the method described in Sect. 3.1.2, which was concerned with single-phase loads. The drive structure, given in Fig. 3.17, shows a three-phase load, modulator/converter, and control module. The first two modules have been discussed. It is noted that a three-phase star connected load is assumed because this provides direct access to the load currents, which can be measured with only two current transducers (a third unit is not needed given that the sum of the currents must be zero for an unconnected star point). The three induced voltages $u_{\rm e1}$, $u_{\rm e2}$, $u_{\rm e3}$ are also measured as their inputs are required for the controller module. Two *three-phase to two-phase* converter modules are used to transform the measured variables to space vector form.

A discrete controller is assumed which should drive the error $|\vec{i}(t_k+T_{\rm s}) - \vec{i}^*(t_k)|$ to zero over a sample interval $T_{\rm s}$. The design follows the single-phase approach where condition 3.1 was introduced, which may be rewritten in space vector form as

$$\vec{i}\,(t_k + T_{\rm s}) = \vec{i}^*\,(t_k)\,. \tag{3.16}$$

The task of the controller module is to determine the average voltage reference space vector $\vec{U}^*(t_k)$ that is needed to satisfy condition 3.16. In Sect. 2.4, attention was given to the modulator/converter module with respect to determining the switching

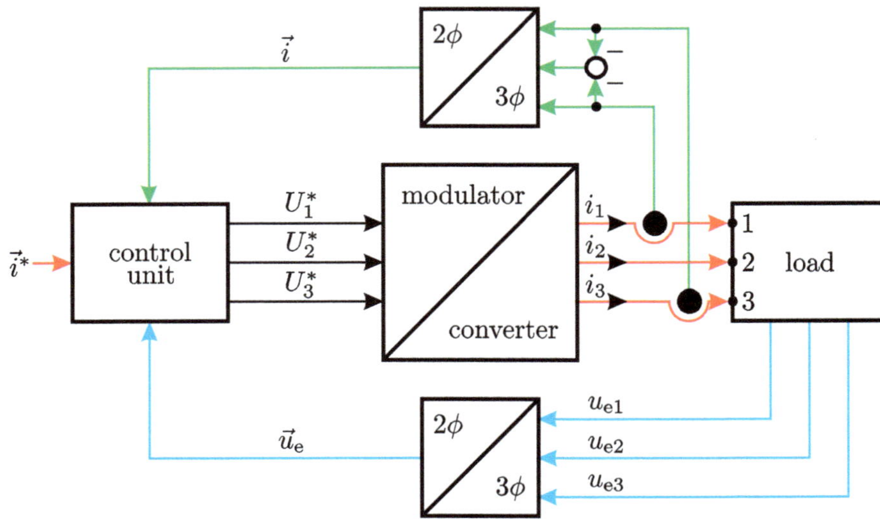

Fig. 3.17 Model based three-phase current control

algorithm needed to satisfy the following condition:

$$\vec{U}^*(t_k) = \frac{1}{T_s}\int_{t_k}^{t_k+T_s} \vec{u}(\tau)\,d\tau, \tag{3.17}$$

where $\vec{u}(t)$ represents the load voltage space vector of a star connected load. Each load phase is represented by a generic impedance Z, which represents a symmetric R, L, u_e circuit, similar as was discussed in the previous chapter. This implies that the load space vector may be expressed as

$$\vec{u} = R\,\vec{i} + L\frac{d\vec{i}}{dt} + \vec{u}_e. \tag{3.18}$$

The introduction of a synchronous reference frame, as shown in Fig. 3.10, is helpful from a control perspective, as was mentioned at the beginning of this chapter. Transformation of the vectors $\vec{u}$, $\vec{i}$, $\vec{u}_e$ to this complex plane requires use of the general vector transformation $\vec{x}^{dq} = \vec{x}\,e^{-j\rho_e}$, in which case equation (3.17) can be rewritten as

$$\left(\vec{U}^{dq}\right)^*(t_k)\,\frac{1}{1-j\omega_e T_s} = \frac{1}{T_s}\int_{t_k}^{t_k+T_s} \vec{u}_s^{\,dq}(\tau)\,d\tau. \tag{3.19}$$

The sampling frequency $1/T_s$ is normally much higher than the vector rotation speed ω_e in which case the term $1/1-j\omega_e T_s$ can be taken at unity value. Transformation of the load equation (3.18) to a synchronous reference frame leads to

$$\vec{u}^{\mathrm{dq}}(t) = R\,\vec{i}^{\mathrm{dq}} + L\frac{\mathrm{d}\vec{i}^{\mathrm{dq}}}{\mathrm{d}t} + \vec{u}_{\mathrm{e}}^{\mathrm{dq}} + \mathrm{j}\omega_{\mathrm{e}}L\,\vec{i}^{\mathrm{dq}} \tag{3.20}$$

Substituting Eq. (3.20) into Eq. (3.19) and combining the real and imaginary terms of the latter yields

$$\begin{aligned} U_{\mathrm{d}}^{*}(t_k) = {} & \frac{R}{T_{\mathrm{s}}}\int_{t_k}^{t_k+T_{\mathrm{s}}} i_{\mathrm{d}}(\tau)\,\mathrm{d}\tau + \frac{L}{T_{\mathrm{s}}}\int_{i_{\mathrm{d}}(t_k)}^{i_{\mathrm{d}}(t_k+T_{\mathrm{s}})} \mathrm{d}i_{\mathrm{d}} \\ & - \frac{1}{T_{\mathrm{s}}}\int_{t_k}^{t_k+T_{\mathrm{s}}} \omega_{\mathrm{e}}\,L\,i_{\mathrm{q}}(\tau)\,\mathrm{d}\tau, \end{aligned} \tag{3.21a}$$

$$\begin{aligned} U_{\mathrm{q}}^{*}(t_k) = {} & \frac{R}{T_{\mathrm{s}}}\int_{t_k}^{t_k+T_{\mathrm{s}}} i_{\mathrm{q}}(\tau)\,d\tau + \frac{L}{T_{\mathrm{s}}}\int_{i_{\mathrm{q}}(t_k)}^{i_{\mathrm{q}}(t_k+T_{\mathrm{s}})} \mathrm{d}i_{\mathrm{q}} \\ & + \frac{1}{T_{\mathrm{s}}}\int_{t_k}^{t_k+T_S} \left(\omega_{\mathrm{e}}\,L\,i_{\mathrm{d}}(\tau) + u_{\mathrm{e}}(\tau)\right)\mathrm{d}\tau. \end{aligned} \tag{3.21b}$$

A comparison of Eqs. (3.4) and (3.21) shows that the three-phase control problem has been reduced to two single-phase control problems. This means that the discretization technique developed in Sect. 3.1.2 for the single-phase case may be directly applied to Eq. (3.21), which gives

$$\begin{aligned} U_{\mathrm{d}}^{*}(t_k) \cong {} & R\sum_{j=0}^{j=k-1}\left(i_{\mathrm{d}}^{*}(t_j) - i_{\mathrm{d}}(t_j)\right) \\ & + \left(\frac{L}{T_{\mathrm{s}}} + \frac{R}{2}\right)\left(i_{\mathrm{d}}^{*}(t_k) - i_{\mathrm{d}}(t_k)\right) \\ & - \omega_{\mathrm{e}}\,L\,i_{\mathrm{q}}(t_k), \end{aligned} \tag{3.22a}$$

$$\begin{aligned} U_{\mathrm{q}}^{*}(t_k) \cong {} & R\sum_{j=0}^{j=k-1}\left(i_{\mathrm{q}}^{*}(t_j) - i_{\mathrm{q}}(t_j)\right) \\ & + \left(\frac{L}{T_{\mathrm{s}}} + \frac{R}{2}\right)\left(i_{\mathrm{q}}^{*}(t_k) - i_{\mathrm{q}}(t_k)\right) \\ & + \omega_{\mathrm{e}}\,L\,i_{\mathrm{d}}(t_k) + u_{\mathrm{e}}(t_k). \end{aligned} \tag{3.22b}$$

The generic diagram which corresponds to the control equation set (3.22a) and (3.22b) is given in Fig. 3.18. The diagram in question represents an embodiment of the control unit shown in Fig. 3.17.

The PI gain settings K_{p} and K_{i} for the two current controllers are, as may be observed from Eqs. (3.22a) and (3.22b), identical to those found for the single-phase case (see Eq. (3.10)). However, the *dq* disturbance decoupling terms are different, as may be observed from Fig. 3.18. Clearly identifiable are the terms with gain

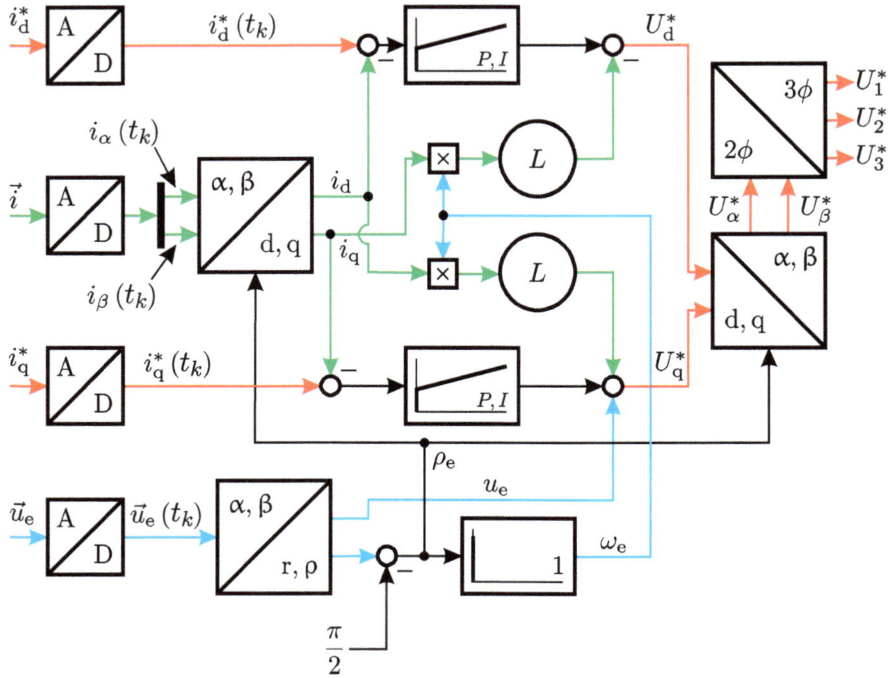

Fig. 3.18 Model based three-phase controller structure

L which serve to decouple the direct axis (active) and quadrature axis (reactive) current components. Also present is the back-EMF voltage term u_e which appears in the quadrature axis. Furthermore, a differentiator module is used to determine the frequency ω_e of the load vector $\vec{u}_e$ or flux vector $\vec{\psi}_e$. In a grid connected configuration, the fixed grid frequency ω_e is known and the controller structure can be simplified by omitting the discrete differentiator module. Also shown in Fig. 3.18 are the two conversion modules used to generate the three average voltage phase references, which form the inputs to the modulator module.

It is noted that the current reference $\vec{i}^*$ shown in Fig. 3.17 is formed by the direct and quadrature reference components i_d^* and i_q^* as shown in Fig. 3.10. Access to the direct and quadrature reference values (i_d^*, i_q^*) provides the user with the ability to control the position of the reference current vector (and thereby the actual current given condition 3.16) with respect to the flux vector $\vec{\psi}_e$. From a power definition perspective, control of the variables (i_d^*, i_q^*) gives the ability of controlling the real and reactive power components with respect to the voltage vector $\vec{u}_e$. For most controller structures to be considered for electrical machines, the use of a flux vector will be of paramount importance, as will become apparent in this book. Given its perceived importance, the generic structure of the flux $\vec{\psi}_e$ oriented model based current controller is shown in Fig. 3.19. A comparison between the voltage and flux

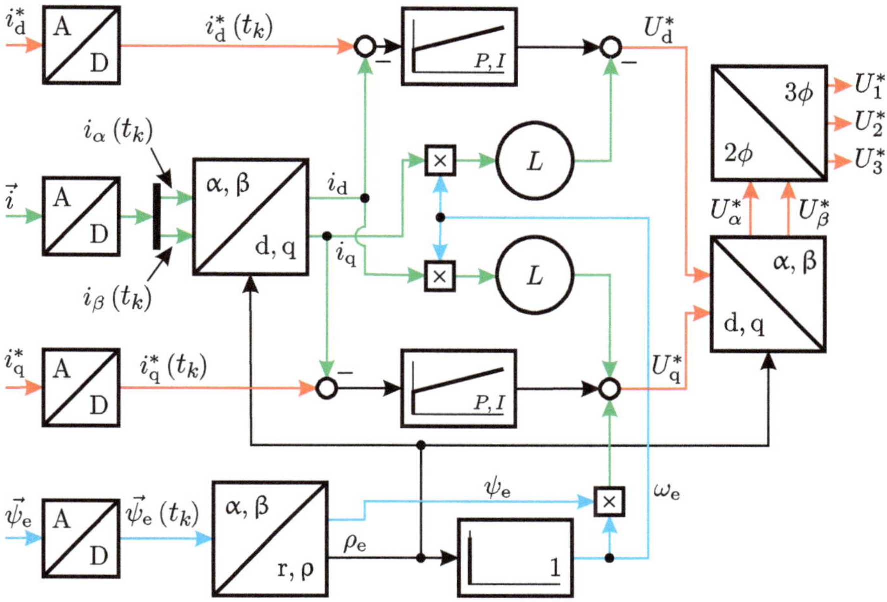

Fig. 3.19 Model based three-phase, flux oriented controller structure

oriented control structures shows that in the latter case the back-EMF voltage term u_e (see Eq. (3.22b)) is computed using $u_e = \psi_e \, \omega_e$.

A detailed tutorial of the concepts outlined in this section is given in Sect. 3.3.4. This example, which is based on the model according to Fig. 3.18, is identical to that discussed for the three-phase hysteresis case, in terms of the load and direct/quadrature reference current setting, namely $i_d^* = 0\,A$, $i_q^* = 15\,A$ (reference vector $\vec{i}^*$ aligned with the vector $\vec{u}_e$, which rotates at 3000 rpm). However, in this case, a modulator structure and a controller structure were added as represented by Figs. 2.12 and 3.18, respectively. The sample and PWM carrier frequencies were set to 5 and 2.5 kHz, respectively. The values chosen are lower than normally used in practice, to clearly visualize the current endpoint trajectories. An example of the reference and actual current vector endpoint trajectories is given in Fig. 3.20 for one period of the fundamental component (one 20 ms rotation of the voltage vector $\vec{u}_e$).

It is instructive to compare the results achieved with this type of control, as indicated by Fig. 3.20, with those obtained with the three-phase hysteresis type controller (see Fig. 3.15). With the model based controller, the current error is not specifically defined, but instead is governed by the load parameters and choice of sample and carrier frequencies. The present combination of these variables is such that an absolute current error is apparent in Fig. 3.20 which is larger than the one that was seen in Fig. 3.15. The other noticeable difference between the two current $\vec{i}$ endpoint trajectories lies with the sequence and duration of active and zero voltage vectors for both methods. This may be demonstrated by considering the current

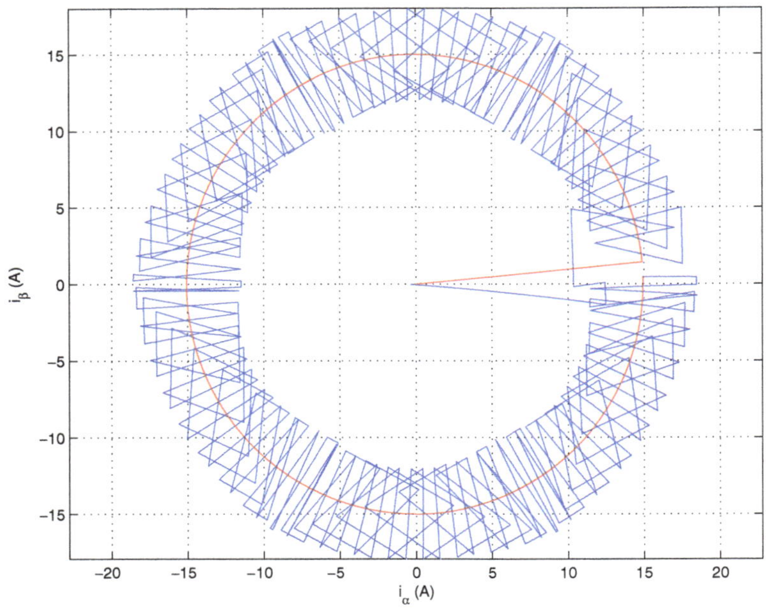

Fig. 3.20 Current locus, model based control example

error vector $\Delta\vec{i} = \vec{i} - \vec{i}^*$ in a synchronous reference frame, as undertaken for the hysteresis type control approach. The endpoint locus for the current error $\Delta\vec{i}$, as shown in Fig. 3.21, is decidedly different in comparison with the hysteresis type result given in Fig. 3.16.

Figure 3.21 shows the first part of the 20 ms current error vector trace (*green* trace). Also shown in Fig. 3.21 are the trajectories over a time interval equal to $2\,T_s$ of this simulation (*red* trace). The trajectories are numbered to explain the sequence of actions undertaken by the controller. The subinterval shown in *red* has been judicially chosen in such a manner that it approximately coincides with the location of the load vector as shown in Fig. 3.11. The reader is reminded that the q-axis of the synchronous reference is aligned with the vector $\vec{u}_e$. Furthermore, in this particular example, the current reference vector $\vec{i}^*$ is also aligned with this vector. The direction in which the error current endpoint will travel is dictated by the relative position of the load vector $\vec{u}_e$ with respect to the two active vectors in use. During locus leg 1, the voltage vector $\vec{u}_{\{110\}}$ is active for a given time interval after which vector $\vec{u}_{\{100\}}$ is activated which leads to the trajectory that corresponds to locus leg 2. With this type of control, typically two active vectors are sequentially activated after which a zero vector time interval occurs (as may be observed in Fig. 2.11b during a sample time interval $t_k \rightarrow t_{k+1}$). This sequence of events is also identifiable in Fig. 3.21 in terms of the trajectory identified as locus leg 3. During this leg a zero vector is active, which in turn is followed by the reactivation of the last active vector $\vec{u}_{\{100\}}$ during locus leg 4. The second active vector $\vec{u}_{\{110\}}$ is then

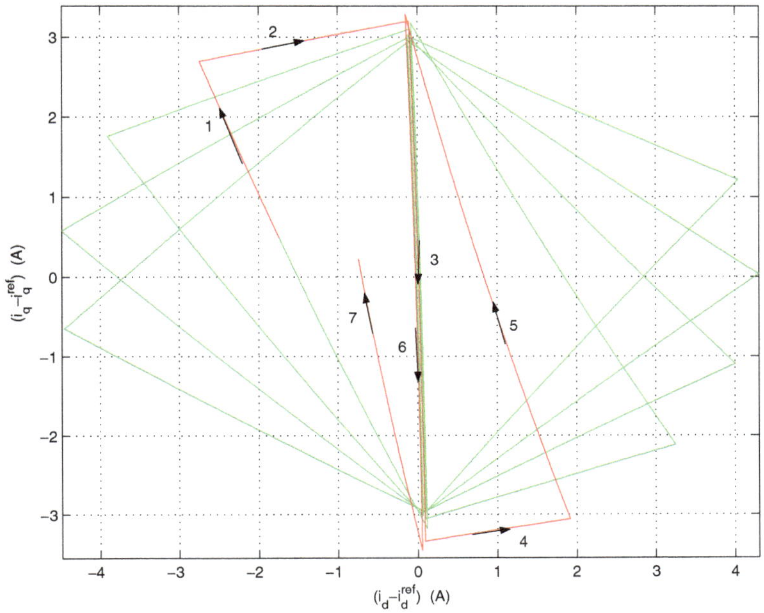

Fig. 3.21 Current locus, model based control example: current error

activated, which gives the trajectory path for locus leg 5. A zero vector is again initiated after every one or two active vectors which results in locus leg 6, which must coincide with leg three, given that the q-axis of the synchronous frame is tied to the voltage vector $\vec{u}_e$. After this zero vector has been applied, the active vector $\vec{u}_{\{110\}}$ is selected, which leads to the trajectory locus leg 7. Note that the current error trajectories of, for example, locus legs 1 and 7 are not identical, despite the fact that they both utilize the same active vector. The reason is that during the course of this time interval the load vector $\vec{u}_e$ has rotated clockwise with respect to the two active vectors shown in Fig. 3.11. Consequently, the error current direction must change as may be observed from Fig. 3.11. A final observation with respect to Fig. 3.21 is concerned with the trajectory undertaken when a zero vector is activated, i.e., locus legs 3 and 6. A detailed observation of Fig. 3.21 reveals that the trajectory is slanted relative to the vertical axis of the plot by an angle ρ_0. A similar observation was also made for the hysteresis controller and an explanation of this phenomenon was discussed with the aid of Eq. (3.15).

3.2.3 Augmented Three-Phase Model Based Current Control

The single-phase augmented model based control concept as discussed in Sect. 3.1.3 is extended to the three-phase case in this section. Accordingly, the approach

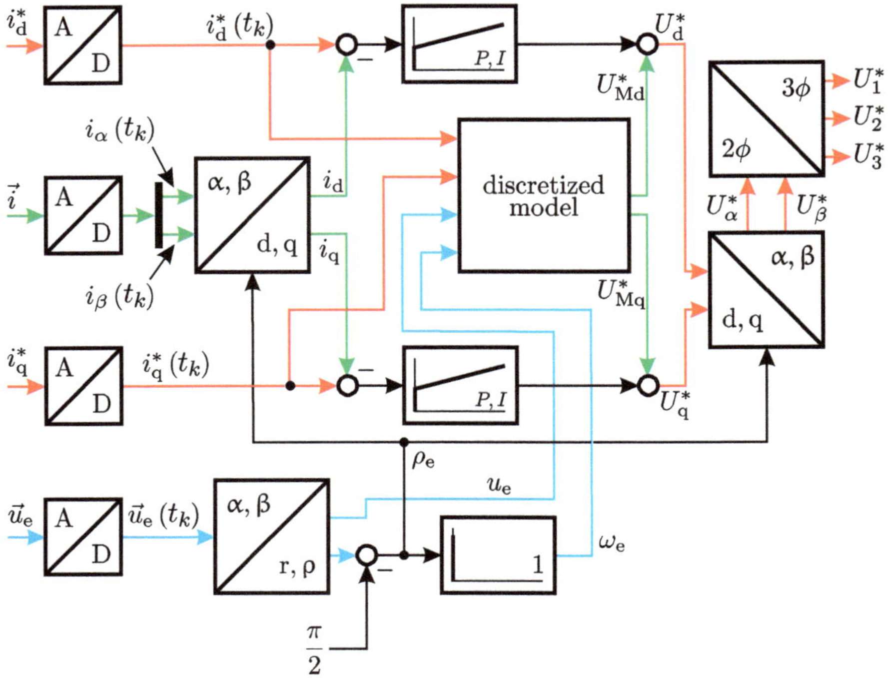

Fig. 3.22 Augmented model based three-phase controller structure

taken to find a discrete representation of the model is similar. In this case, it is advantageous to make use of the synchronous model equation set (3.21). Backward discretization of equation set (3.21) along the lines undertaken for the single-phase case leads to

$$U_{\mathrm{Md}}^*(t_k) \cong R\, i_{\mathrm{d}}^*(t_k) + \left(\frac{L}{T_{\mathrm{s}}} - \frac{R}{2}\right)\left(i_{\mathrm{d}}^*(t_k) - i_{\mathrm{d}}^*(t_{\mathrm{k}-1})\right) - \omega_{\mathrm{e}} L\, i_{\mathrm{q}}^*(t_k)\,, \tag{3.23a}$$

$$U_{\mathrm{Mq}}^*(t_k) \cong R\, i_{\mathrm{q}}^*(t_k) + \left(\frac{L}{T_{\mathrm{s}}} - \frac{R}{2}\right)\left(i_{\mathrm{q}}^*(t_k) - i_{\mathrm{q}}^*(t_{\mathrm{k}-1})\right) + \omega_{\mathrm{e}} L\, i_{\mathrm{d}}^*(t_k) + u_{\mathrm{e}}(t_k)\,. \tag{3.23b}$$

Equation set (3.23) form the basic algorithm which is located in the discretized model shown in Fig. 3.22. Output of this module are the direct and quadrature average voltage references $U_{\mathrm{Md}}^*(t_k)$ and $U_{\mathrm{Mq}}^*(t_k)$ which are used, together with the two PI controllers, to calculate the average voltage references $U_{\mathrm{d}}^*(t_k)$ and $U_{\mathrm{q}}^*(t_k)$.

Deviations between the load parameters used in the controller and those actually present are compensated by the two PI controllers. The dynamics which need to be handled by these PI controllers are rather limited because they relate to parameter changes due to, for example, temperature. Hence, the gains can be chosen relatively low. An upper limit for these gains settings are given in equation set (3.22), respectively. A comparison between equation sets (3.23) and (3.22) shows that the discrete model given in Fig. 3.22 contains a number of elements which were also used in the controller shown in Fig. 3.18. However, the discrete model uses four additional terms $R\ i_{\mathrm{d}}^{*}(t_k)$, $R\ i_{\mathrm{q}}^{*}(t_k)$ and $(L/T_{\mathrm{s}} - R/2)(i_{\mathrm{d}}^{*}(t_k) - i_{\mathrm{d}}^{*}(t_{\mathrm{k}-1}))$, $(L/T_{\mathrm{s}} - R/2)(i_{\mathrm{q}}^{*}(t_k) - i_{\mathrm{q}}^{*}(t_{\mathrm{k}-1}))$ of which the latter two are current difference type equations which may be susceptible to noise. Consequently, bandwidth restrictions on the direct and quadrature reference current values should be imposed to avoid unwarranted excursions in the variables $U_{\mathrm{Md}}^{*}(t_k)$, $U_{\mathrm{Mq}}^{*}(t_k)$ due to the use of the differential terms present in equation set (3.23). Similarly, the process of generating the estimated rotational speed ω_{e} from the rotor angle ρ_{e} by way of a differentiator module (see Fig. 3.22) may also cause unwarranted noise related disturbances in the outputs of the discrete model. Note that in three-phase model based control the distinction between an augmented and non-augmented approach is not significant as both implement key decoupling elements of the same discrete model.

3.2.4 Frequency Spectrum of Hysteresis and Model Based Current Controllers

Both hysteresis and model based current control techniques are used widely in industry. It is deemed to be helpful for the reader to gain an appreciation of a typical frequency spectrum of the converter phase voltage for both current control methods.

For this purpose, the three-phase hysteresis and model based tutorials, discussed in Sects. 3.3.3 and 3.3.4, respectively, are analyzed. The load, supply, and direct/quadrature reference current conditions for both examples have been chosen identical to compare the frequency spectra of the voltage across one phase of the load. Figure 3.23 shows the spectrum for both control techniques of the phase voltage with respect to the amplitude of the fundamental component over a frequency range $0 \rightarrow 10\,\mathrm{kHz}$.

A general observation of Fig. 3.23 shows that the spectrum, in addition to the 50 Hz fundamental component, of the hysteresis controller approach is wide spread in terms of the frequency components. The spectrum which will appear is dependent on the size of the box and the amplitude of the voltage vector $\vec{u}_{\mathrm{e}}$. The spectrum of the fixed frequency model based control approach is more discrete in nature. In practice, this is reflected by an acoustical noise signature that is more predictable and can therefore be tuned to the application, for example, machine-load combinations, as to avoid specific resonance frequencies [2, 4]. In the example given in Fig. 3.23 the PWM carrier and sample frequencies were set to 2.5 and 5.0 kHz, respectively, which is reflected in the frequency spectrum shown in Fig. 3.23b.

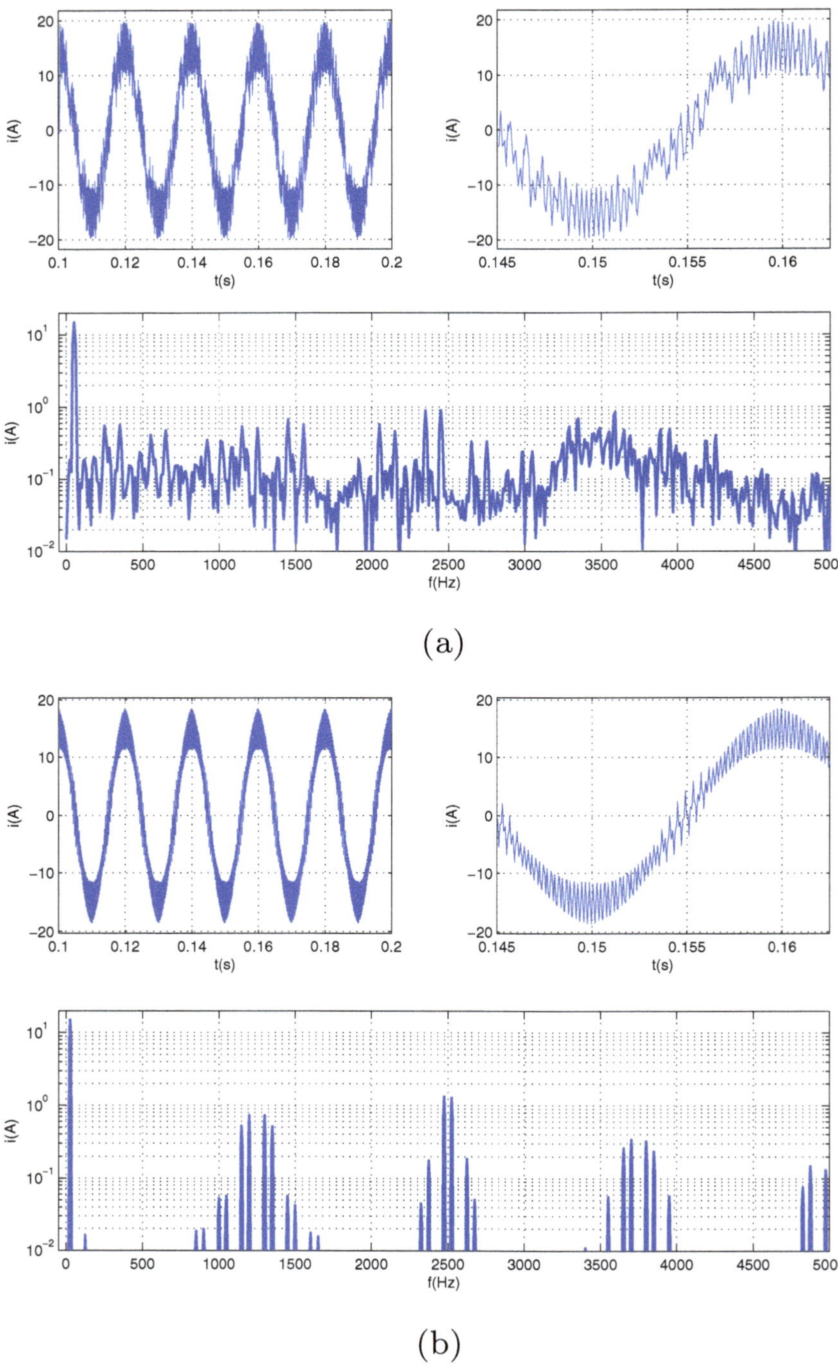

Fig. 3.23 Frequency spectrum of the converter phase voltage for a three-phase drive operating under hysteresis and model based control. (**a**) Hysteresis control. (**b**) Model based control

3.3 Tutorials

3.3.1 *Tutorial 1: Single-Phase Hysteresis Current Control*

In this tutorial, a single-phase half-bridge converter connected to a load is implemented. The load is formed by a series connected resistance $R = 20\,\mathrm{m}\Omega$, inductance $L = 100\,\mathrm{mH}$, and a sinusoidal voltage source $u_\mathrm{e} = \hat{u}_\mathrm{e}\sin(\omega_\mathrm{e}t)$, with $\hat{u}_\mathrm{e} = 100\,\mathrm{V}$ and $\omega_\mathrm{e} = 100\pi$ rad/s. The half-bridge converter bus voltage u_DC is set to 300 V. The aim is to build a hysteresis type current controller with a hysteresis band setting of $\Delta i* = 0.4\,\mathrm{A}$ in accordance with the generic model shown in Fig. 3.3, so that the user can examine the various waveforms and switching activities as discussed in Sect. 3.1.1.

A simulation model example for this tutorial is given in Fig. 3.24, while simulation results are provided in Fig. 3.25. An input current reference waveform of $i^* = \hat{i}\sin\omega_\mathrm{e}t$, with $\hat{i} = 2\,\mathrm{A}$ is chosen so that the output current waveform is in phase with the voltage source at the load. Therefore, the converter feeds the ac supply at the load with unity power factor. This mode of operation is employed in uninterruptable power supplies (UPS). The reader is encouraged to change the hysteresis bandwidth, current reference amplitude, and inductance values to observe the effects on the output current waveform.

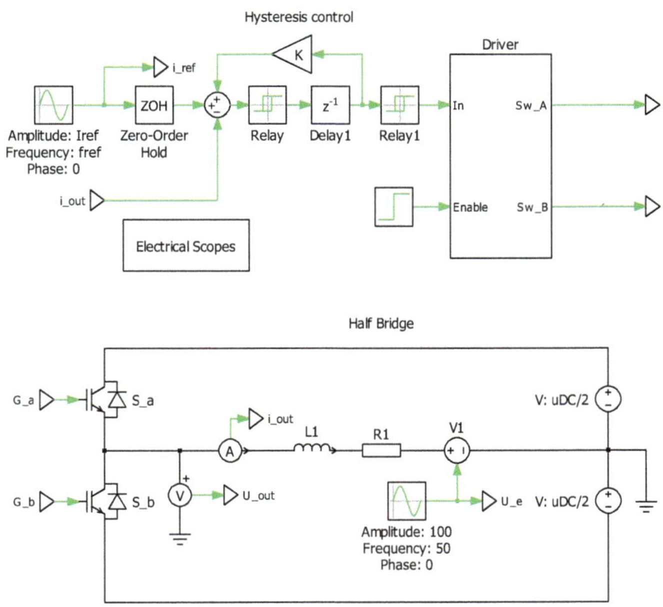

Fig. 3.24 Simulation of single-phase hysteresis current control

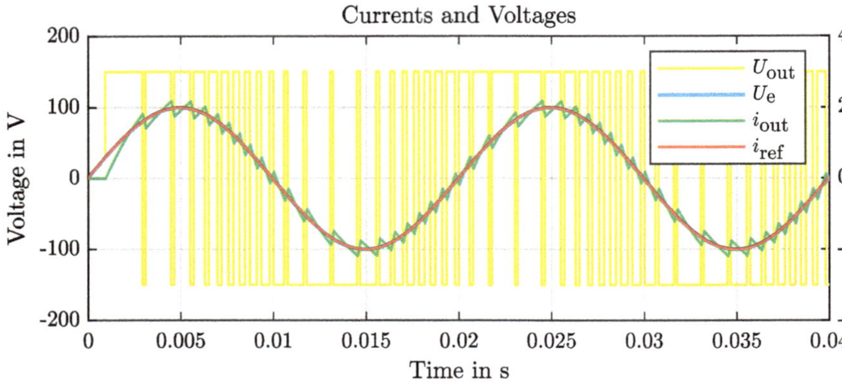

Fig. 3.25 Simulation result of single-phase hysteresis current control

3.3.2 *Tutorial 2: Single-Phase Model Based Current Control*

A single-phase model based controller is to be considered with a full-bridge converter topology. The simulation parameters defined in the previous tutorial are also used without any changes in this example. A sampling time of $T_s = 0.5$ ms is selected, which corresponds to a PWM carrier frequency of 1 kHz.

An example of a simulation model with the required controller structure employing a PI controller is given in Fig. 3.26. In the example, an anti-windup PI controller is introduced, where the outputs of the integral part of the PI controller are limited to ±300 V, which is the largest average voltage per sample that can be realized by the converter. Inputs to the anti-windup PI controller are the current error and the disturbance decoupling signal. The gain and bandwidth of the controller are now specified by Eq. (3.9). Note that these values represent theoretical upper limits and in practice, the value of K_p is typically chosen to be 25% of L/T_s while the bandwidth ω_i is limited to 2000 rad/s. In many applications, the use of a feedback term u_e is either not possible or not recommended because the signal is derived from a measurement. Measurement noise can be injected into the controller through the feedback term and disturb the control. Without the feedback term, the integrator will be forced to compensate for the additional error instead.

The results as shown by scope modules in Fig. 3.27 are not significantly different than those obtained with the previous simulation (see Sect. 3.3.1). The control techniques used, on the other hand, are significantly different.

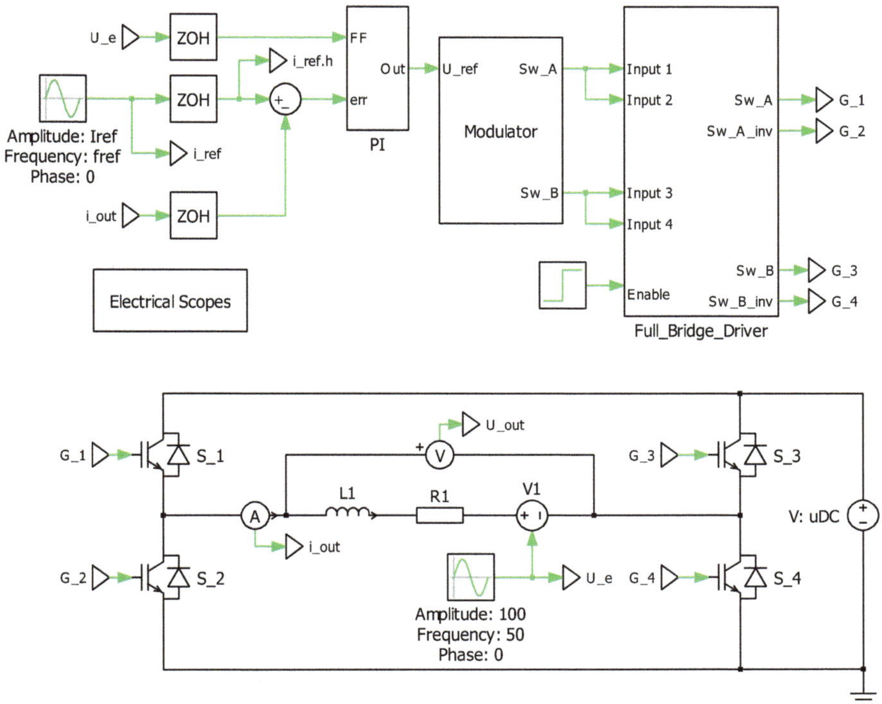

Fig. 3.26 Simulation of single-phase model based current control

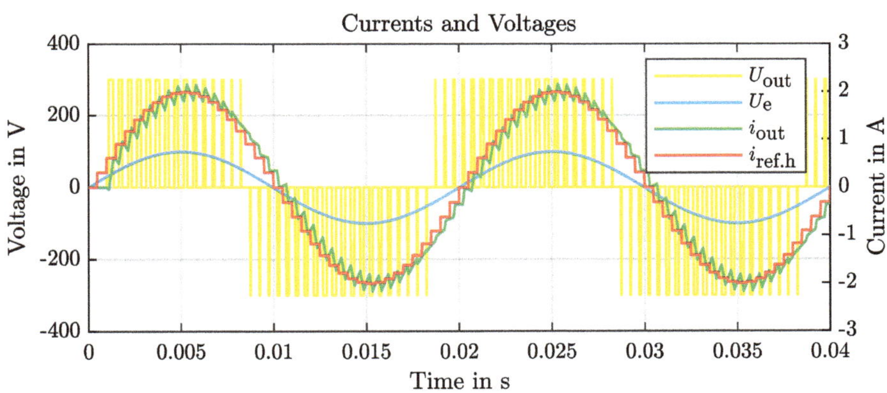

Fig. 3.27 Simulation results of single-phase model based current control

3.3.3 *Tutorial 3: Three-Phase Box Method Type Hysteresis Current Control*

The simulation of a three-phase converter with a *box method* hysteresis type current control is developed in this tutorial. The converter drives a symmetrical star connected load. Each load phase consists of a resistance $R = 20\,\text{m}\Omega$, an inductance $L = 3.4\,\text{mH}$, and a sinusoidal voltage source with a peak value of 100 V and frequency $\omega_\text{e} = 100\pi$ rad. An analog hysteresis *box method* type current controller is implemented as described in Sect. 3.2.1. Output of this controller is a set of half-bridge control signals Sw^a, Sw^b, Sw^c, which can be used to control the converter switches. The load is modeled using the *Electrical* library of PLECS. A DC bus voltage of $u_\text{DC} = 600\,\text{V}$ is used with the converter. The direct and quadrature current reference values are taken to be $i_\text{d}^* = 0\,\text{A}$ and $i_\text{q}^* = 15\,\text{A}$, respectively; whereas the box size is set to $\Delta i^* = 4.4\,\text{A}$.

A PLECS based implementation of this tutorial is illustrated in Fig. 3.28. The load voltages are transformed to vector form $\vec{u}_\text{e}$ inside the *Load* subsystem. A vector to polar conversion module is used to derive the instantaneous argument of the vector $\vec{u}_\text{e}$, from which an angle $\pi/2$ is subtracted to arrive at the reference angle ρ_e used for the synchronous reference conversion process. The line currents are

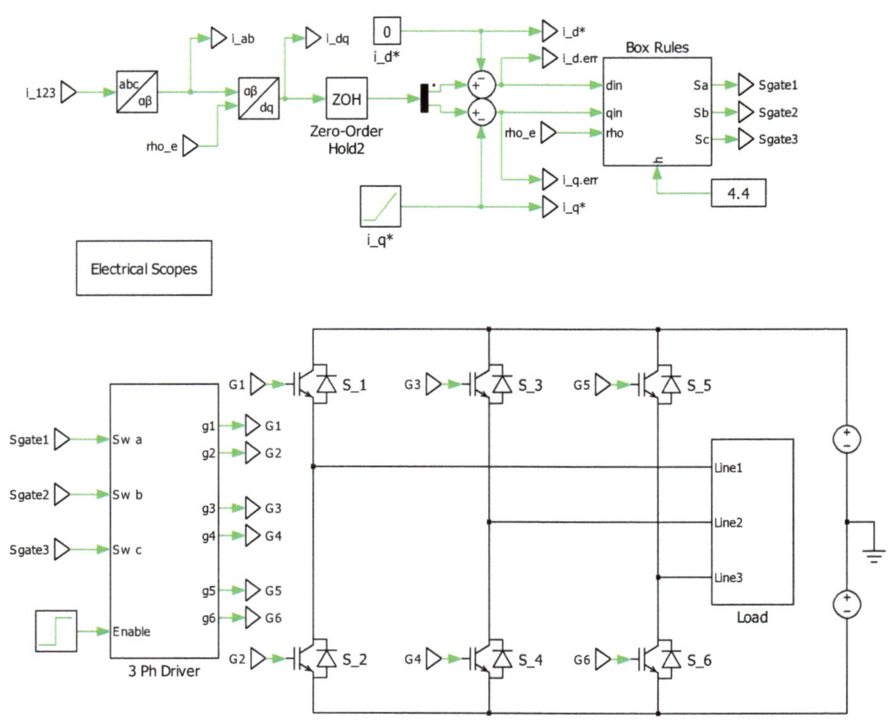

Fig. 3.28 Simulation of three-phase *box method* type hysteresis current control

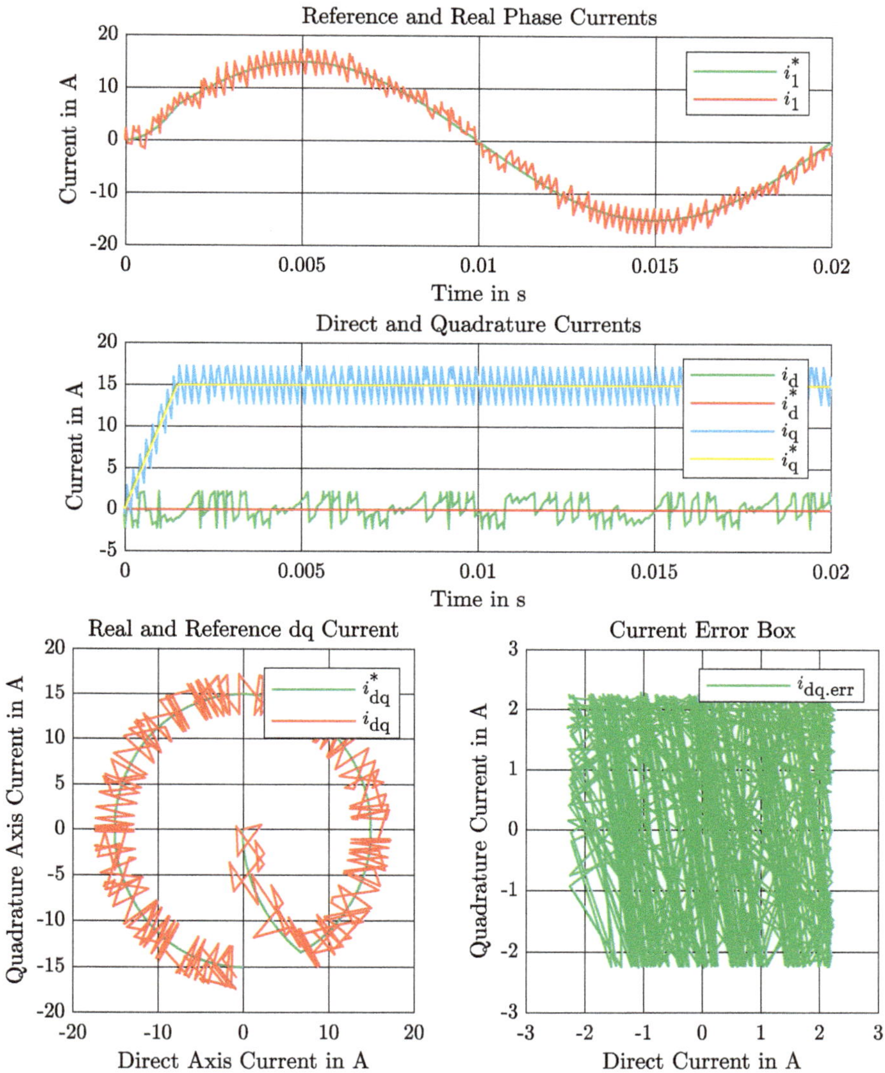

Fig. 3.29 Simulation results of three-phase *box method* type hysteresis current control

also measured in this subsystem, which are converted to a synchronous reference frame with variables i_d and i_q. The synchronous frame based reference currents are subtracted from these values to calculate the current error vector $\Delta\vec{i}$ (represented by "i_dq.err" in the simulation), which is used by the *box rules* module as indicated in the generic diagram in Fig. 3.13. This module requires two additional inputs of the hysteresis box size and the angle ρ_e. The latter is used for the space vector selection algorithm as discussed in Sect. 3.2.1. The space vector currents $\vec{i}$, $\vec{i}^*$ and error current vector plots are given in Fig. 3.29.

3.3.4 Tutorial 4: Three-Phase Model Based Current Control

The aim of this tutorial is to replace the hysteresis based current controller discussed in previous subsection with a model based controller as outlined in Sect. 3.2.2. The same inverter and load models from the previous tutorial are also employed in this simulation model. The load parameters, reference current settings, and converter supply voltage level remain unchanged. A PWM switching frequency of 2.5 kHz is employed.

The PLECS model given in Fig. 3.30 includes the converter and load components from the previous tutorial. The three-phase modulator structure, which was outlined in Sect. 2.6.6, is added to the system. Most importantly, a model based current controller is implemented according to the generic structure shown in Fig. 3.18. The reference angle ρ_e is generated from the voltage vector $\vec{u}_e$ as outlined in the previous section. However, in this tutorial a set of sample-and-hold elements are introduced to represent the behavior of a digital controller. This implies that the controller utilizes the sampled reference angle $\rho_e(t_k)$. A direct consequence of this approach is the fact that the controller based synchronous reference frame lags on average with respect to the actual reference frame by an angle $\omega_e T_s/2$ rad. This must be compensated in the controller by introducing a phase angle shift of the same value, derived from different signals as shown in Fig. 3.31 yielding a compensated reference angle ρ_{ec}.

It is *crucial* to note that the compensated reference angle ρ_{ec} is used only for the reverse dq transformation of the reference voltage vector, since this angle represents the average reference angle throughout the switching period. The dq transformation of the current vector is still carried out using the sampled reference angle $\rho_e(t_k)$, since this is the correct angle at the instant of the current measurement.

The simulation results including current locus plots and current error plots are given in Fig. 3.31.

3.3.5 Tutorial 5: Three-Phase Model Based Current Control Without PWM, Using Simplified Converter Model

In the previous tutorial, the model based control concept was explored with the aid of a three-phase converter model with ideal semiconductor devices. This simulation approach may be simplified by omitting the modulator/converter structure, using the approach outlined in the tutorial given in Sect. 2.6.6. The sampling time, reference current waveforms, and the load parameters remain unchanged in comparison with the previous tutorial.

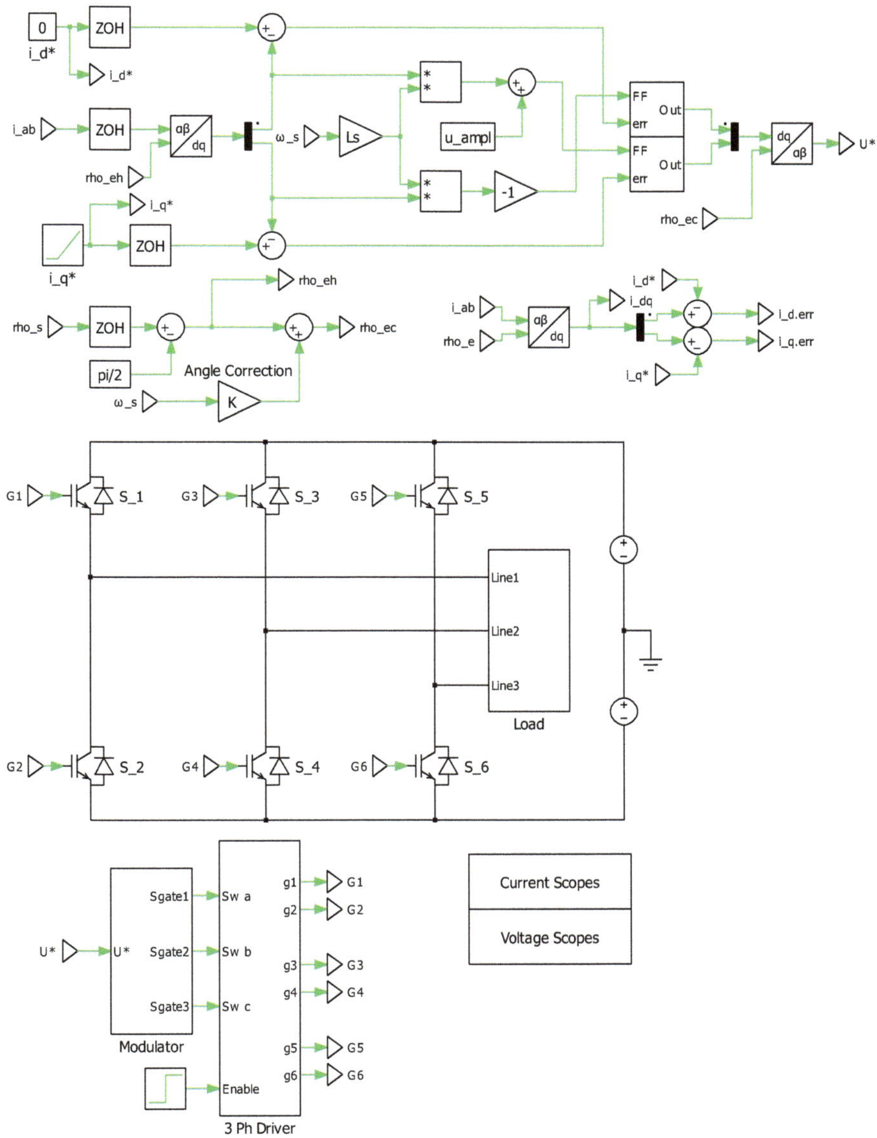

Fig. 3.30 Simulation model of synchronous current controller

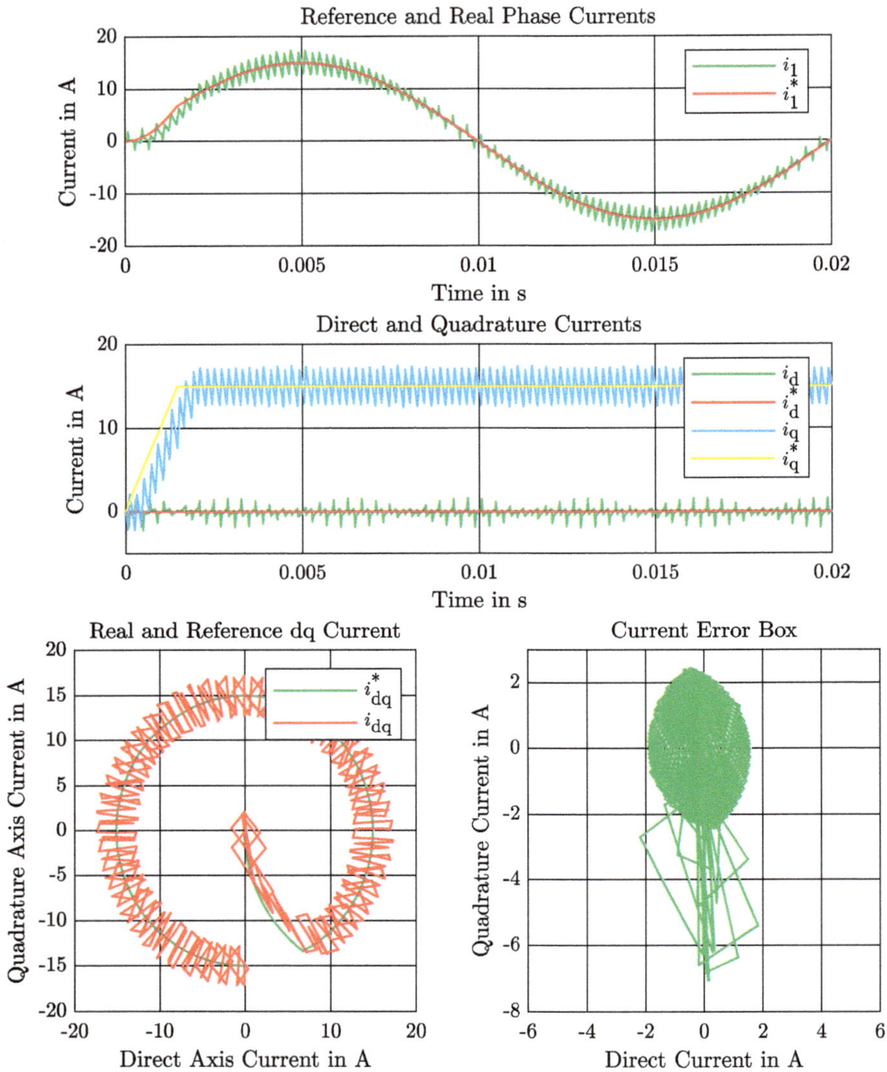

Fig. 3.31 Simulation results of model based synchronous current controller

The PLECS model for this tutorial is shown in Fig. 3.32. A *Modulator + Simplified Inverter* module, which contains the pulse centering module and a supply limiter, is created to calculate line voltages for the load from the reference voltage

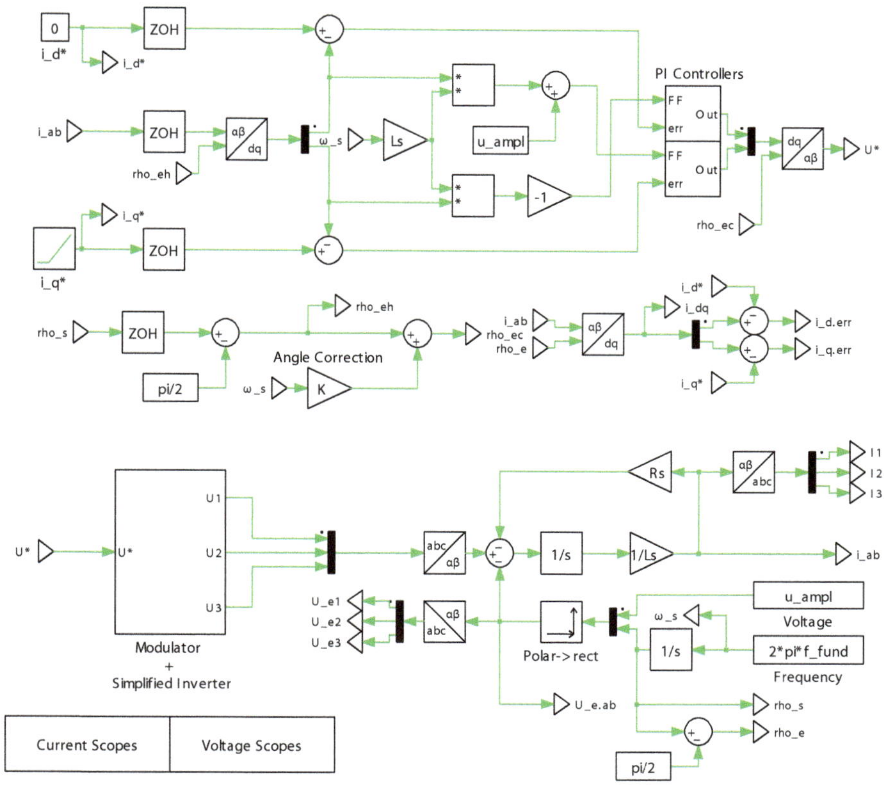

Fig. 3.32 Simulation model of synchronous current controller

vector generated by the controller. The line voltages are assumed to be equal to the switching period average of the line voltages generated by the converter. Consequently, a three-to-two phase conversion module is used to generate the load voltage vector $\vec{u}$ for the generic space vector based model of the load.

The results obtained with this model, as shown in Fig. 3.33, are comparable to those given in Fig. 3.30. The current ripples caused by converter switching are missing from the figures, as the converter is replaced by a simplified inverter model. The influence of the controller on the current waveforms may be observed more clearly in such a simulation.

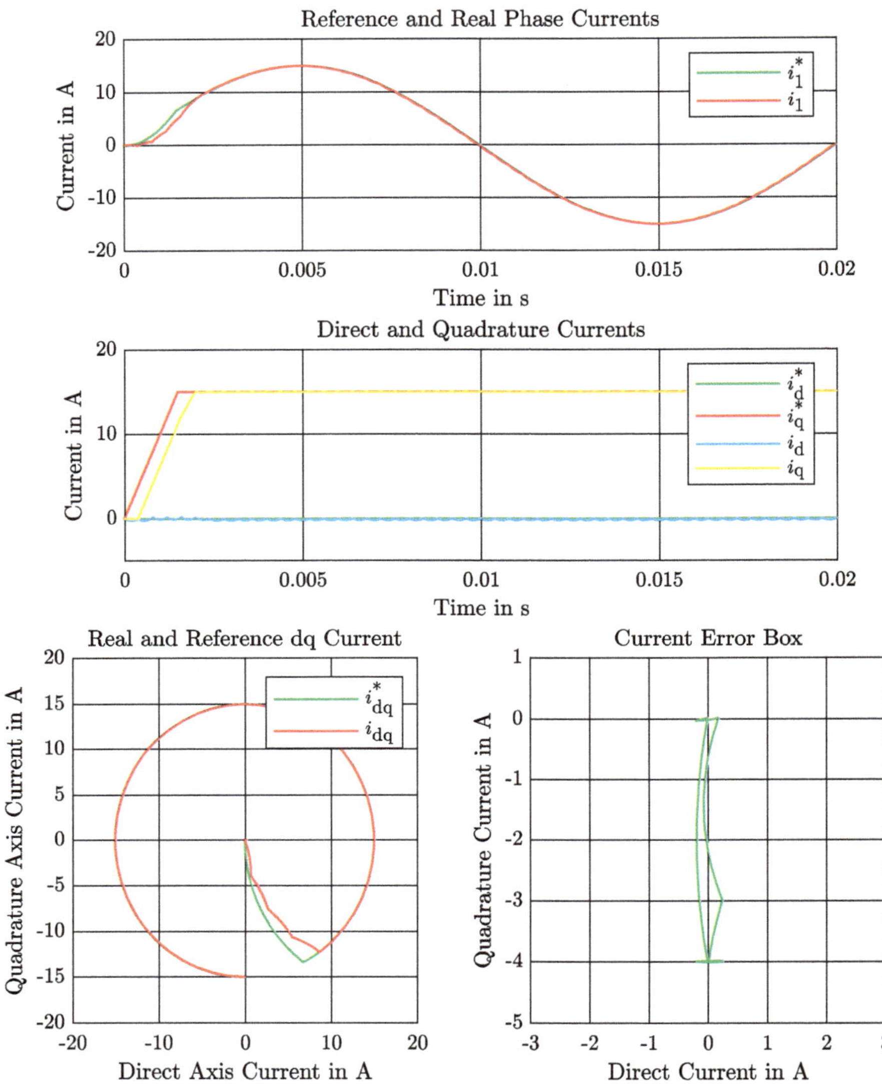

Fig. 3.33 Simulation results of model based synchronous current controller

References

1. Brod DM, Novotny DW (1985) Current control of VSI-PWM inverters. IEEE Trans Ind Appl 21(3):562–570. https://doi.org/10.1109/TIA.1985.349711
2. Habetler T, Divan D (1989) Performance characterization of a new discrete pulse-modulated current regulator. IEEE Trans Ind Appl 25(6):1139–1148. https://doi.org/10.1109/28.44257
3. Malesani L, Tenti P (1990) A novel hysteresis control method for current-controlled voltage-source pwm inverters with constant modulation frequency. IEEE Trans Ind Appl 26(1):88–92. https://doi.org/10.1109/28.52678

4. Mohan N, Undeland TM, Robbins WP (2002) Power electronics: converters, applications, and design, 3rd edn. Wiley, Hoboken
5. Veltman A (1993) The fish method: interaction between ac-machines and switching power converters. PhD Thesis, Delft University
6. Veltman A, Pulle DWJ, De Doncker R (2007) Fundamentals of electrical drives. Spinger, Berlin

Chapter 4
Drive Principles

In successive chapters extensive attention will be given to the modeling and control of rotating field machines. Rotating field machines can be conveniently modeled with the aid of a so-called ideal rotating transformer (IRTF). The initial part of this chapter explores the IRTF concept. It will be shown that torque production may be described mathematically by the cross product of a flux and current space vector. In the previous chapter, three-phase current control was introduced with the precise aim of being able to manipulate the current space vector. The reason for this approach is to develop a set of fundamental drive concepts which aim to, at an elementary level, control torque in drive systems based on rotating field machines such as synchronous or asynchronous machines. Application of the IRTF concept for brushed DC machines is also possible, though it is not widely used and will not be discussed in detail. Note that switched reluctance machines do not embrace the Lorentz force based concept, which implies that they do not follow the IRTF model and therefore they are treated separately in this book at a later stage.

4.1 ITF and IRTF Concepts

The *ideal transformer* (ITF) and *ideal rotating transformer* (IRTF) concepts have been discussed extensively in the book *Fundamentals of Electrical drives* [5]. The IRTF was first used in [4]. The introduction of these concepts has proven to be effective for electrical machine modeling purposes. In this book, the ITF/IRTF concepts will be extended further. Hence, it is in the interest of readability to provide a brief review of these concepts prior to considering the machine models in this and following chapters.

Symbolic ITF Model

The symbolic ITF concept as shown in Fig. 4.1a represents a magnetically and electrically ideal transformer, i.e., without leakage inductances, copper or core

R. W. De Doncker et al., *Advanced Electrical Drives*, Power Systems,
https://doi.org/10.1007/978-3-030-48977-9_4

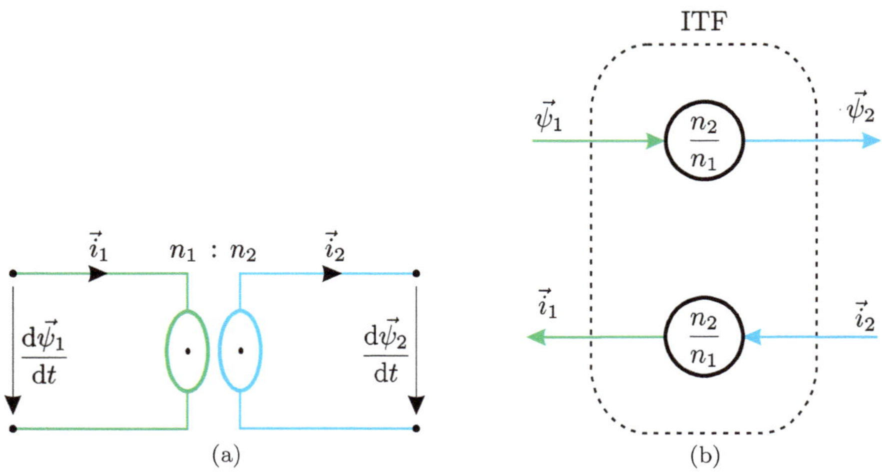

Fig. 4.1 Symbolic and generic space vector based ITF models. (**a**) Symbolic model. (**b**) Generic model

losses and with a primary (subscript 1) to secondary winding ratio of $n_1 : n_2$. The ideal transformer requires no magnetizing current and can thus be regarded to have an infinite magnetizing inductance.

The space vector equation set which corresponds with this model is of the form

$$\vec{\psi}_2 = \left(\frac{n_2}{n_1}\right)\vec{\psi}_1 \tag{4.1a}$$

$$\vec{i}_1 = \left(\frac{n_2}{n_1}\right)\vec{i}_2. \tag{4.1b}$$

Flux- and Current-Based ITF Representation
The flux/current equation set (4.1) forms the basis for the generic model given in Fig. 4.1b. Note that the generic model shown in Fig. 4.1b represents the so-called *ITF-flux* version, because the primary flux vector $\vec{\psi}_1$ is designated as an input. The alternative so-called *ITF-current* version utilizes the primary current vector $\vec{i}_1$ as an input. The selection of a version depends on the nature of the machine model in which it is applied. It is emphasized that the ITF model is based on the use of flux linkages and currents instead of voltages and currents.

Symbolic IRTF Module
The *ideal rotating transformer* (IRTF) module, as given in Fig. 4.2, is a three-port transducer, i.e., electrical machine, that describes the interactions between electrical quantities in stator and rotor (flux and current) and the mechanical quantities (torque and speed) on its shaft. The IRTF contains no means to store any energy, because there is neither mechanical inertia nor inductance. The IRTF describes how torque results from current and flux and how the moving shaft influences the relations

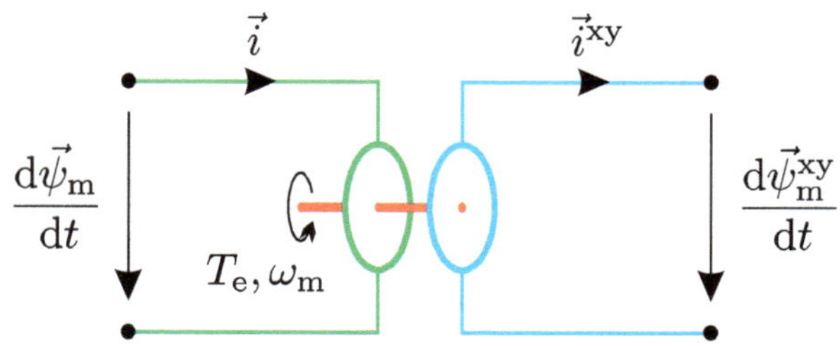

Fig. 4.2 Symbolic IRTF representation

between rotor and stator quantities. The IRTF can be regarded as a model for the air-gap quantities in rotating machines. A realistic dynamic machine model, valid for all electrical waveforms and speed transients, can be constructed by adding elements such as mechanical inertia, main inductance, leakage inductance, stator and rotor resistance to an IRTF.

In Fig. 4.2 a symbolic shaft (shown in *red*) enables the coupling to the mechanical world. The rotor port (shown in *blue*) allows electrical rotor components to be added. An observer linked to the rotor port describes the rotor port variables in a rotating coordinate system, indicated with superscripts "xy." The stator port (shown in *green*) provides the interface to connect stator related components.

The magnetic air-gap flux linked with the stator and rotor is equal to $\vec{\psi}_{\mathrm{m}}$ and can be expressed in terms of the components, i.e., coordinates, seen by each winding, namely

$$\vec{\psi}_{\mathrm{m}} = \psi_{\mathrm{m}\alpha} + \mathrm{j}\,\psi_{\mathrm{m}\beta} \tag{4.2a}$$

$$\vec{\psi}_{\mathrm{m}}^{\mathrm{xy}} = \psi_{\mathrm{mx}} + \mathrm{j}\,\psi_{\mathrm{my}}. \tag{4.2b}$$

Space Vectors in the IRTF

An illustration of the flux linkage seen by the rotor and stator winding is given in Fig. 4.3a.

The relationship between the stator and rotor oriented flux linkage and corresponding current space vectors, as shown in Fig. 4.3, may be written as

$$\vec{\psi}_{\mathrm{m}}^{\mathrm{xy}} = \vec{\psi}_{\mathrm{m}}\,\mathrm{e}^{-\mathrm{j}\theta} \tag{4.3a}$$

$$\vec{i} = \vec{i}^{\,\mathrm{xy}}\,\mathrm{e}^{\mathrm{j}\,\theta}. \tag{4.3b}$$

Flux Linkage and Current Distribution in AC Machine

Figure 4.3 emphasizes the fact that there is only one flux linkage and one current space vector present in the IRTF. This fact is underlined by Fig. 4.4. which shows the cross-section of a typical AC machine with a three-phase sinusoidally distributed winding on the rotor and stator.

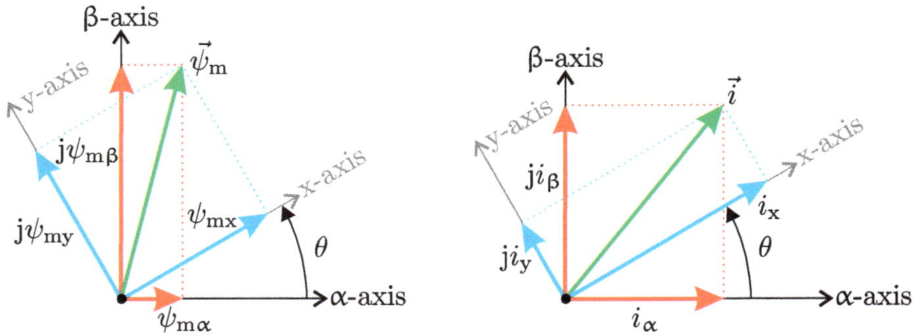

(a) Flux linkage space vector (b) Current space vector

Fig. 4.3 Flux linkage and current space vector diagrams. (**a**) Flux linkage space vector. (**b**) Current space vector

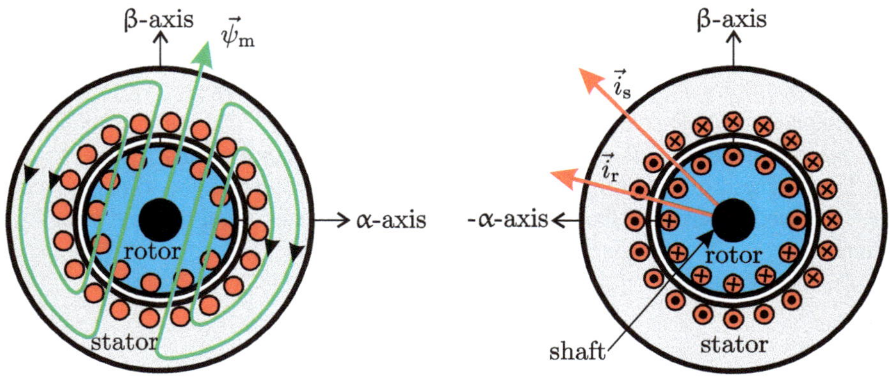

(a) Flux linkage distribution (b) Current distribution

Fig. 4.4 Flux linkage and current distribution in a typical wound rotor machine, showing flux and current space vectors. (**a**) Flux linkage distribution. (**b**) Current distribution

During operation the three-phase flux linkage and current contributions in a typical wound rotor AC machine can be represented by a single flux linkage and two current distributions as shown in Fig. 4.4. These may in turn be represented by a space vector which for the flux is aligned with the resultant two-pole magnetic flux axis. The space vector $\vec{i}_{\mathrm{s}}$ is aligned with the current distribution of the stator, while the space vector $\vec{i}_{\mathrm{r}}$ is aligned with the current distribution in the rotor bars. When considering the current vector seen by the rotor, the IRTF module uses by convention the shown vector, while in reality the current distribution in the rotor is reversed in polarity. The reason for this is that the sum of stator and rotor magneto-motive forces (MMF) approaches zero when the permeability of the magnetic material of the IRTF model is taken towards infinity and the air-gap is taken to be very small.

Torque Production in the IRTF

The components of the space vectors shown in Fig. 4.3 can be projected onto a rotating orthogonal frame ($\Re^{xy}, \Im^{xy}$) or stationary orthogonal reference frame ($\Re, \Im$). If the angle θ is set to zero, the IRTF model mirrors the ITF module with the exception that the IRTF has a unity winding ratio. According to the torque principles explained in the next section, the electrodynamic torque T_e produced by the two-pole IRTF module is of the form

$$T_e = \frac{3}{2}\Im\left\{\left(\vec{\psi}_m\right)^* \vec{i}\right\} \tag{4.4}$$

if the space vectors $\vec{\psi}_m$ and $\vec{i}$ are interpreted as complex numbers or

$$\vec{T}_e = \frac{3}{2}\left(\vec{\psi}_m \times \vec{i}\right) \tag{4.5}$$

in vector form. Note that the space vector $\vec{i}$ represents either the stator current space vector $\vec{i}_s$ or the rotor current space vector $\vec{i}_r$. The factor 3/2 is valid only for 3-phase systems with amplitude invariant space vectors. Hence, the torque acting on the rotor is at its maximum value when the two vectors $\vec{\psi}_m$ and $\vec{i}$, shown in Fig. 4.3, are perpendicular with respect to each other. The generic diagram of the IRTF module that corresponds to the symbolic representation shown in Fig. 4.2 is based on the use of Eqs. (4.3a), (4.3b), and (4.5).

Generic IRTF Module

The generic IRTF module as given in Fig. 4.5a is shown with a stator-to-rotor coordinate flux conversion module and rotor-to-stator current conversion module which complies with Eq. (4.3). The two coordinate conversion modules can also be reversed as shown in Fig. 4.5b. The IRTF version used is application dependent as will become apparent in subsequent sections. The torque computation is not affected by the choice of coordinate system because both vectors use the factor $e^{j\theta}$ and its conjugate $e^{-j\theta}$. Hence, these coordinate transformations in the form of vector rotation cancel out in expression (4.5).

The rotor angle θ required for the IRTF module can be derived from the mechanical equation set of the (two-pole) machine, which is of the form

$$T_e - T_l = J\frac{d\omega_m}{dt} \tag{4.6a}$$

(continued)

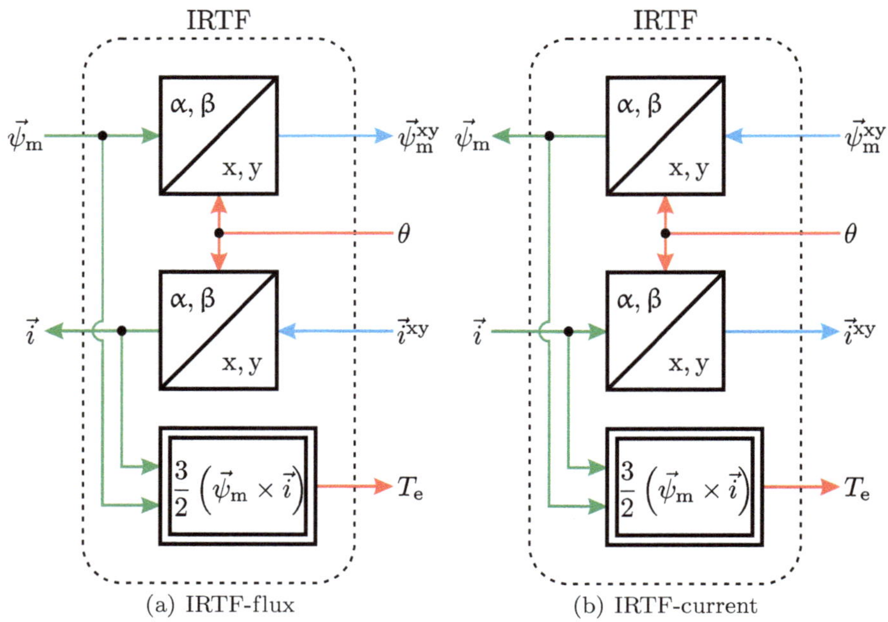

Fig. 4.5 Generic representations of IRTF module (for three-phase, two-pole machines). (**a**) IRTF-flux. (**b**) IRTF-current

$$\omega_m = \frac{d\theta}{dt} \tag{4.6b}$$

with T_l and J representing the load torque and inertia of the rotor/load combination, respectively.

4.2 Electromagnetic Torque Control Principles

In the previous chapter considerable attention was given to three-phase current control and the ability to generate a user defined current vector. In this section, it will be shown how such a current vector can be used for electromagnetic torque control in simplified AC and DC machines. The aim of this approach is to allow the reader to become more familiar with the use of the IRTF module for machine modeling and control design prior to undertaking more detailed studies in subsequent chapters.

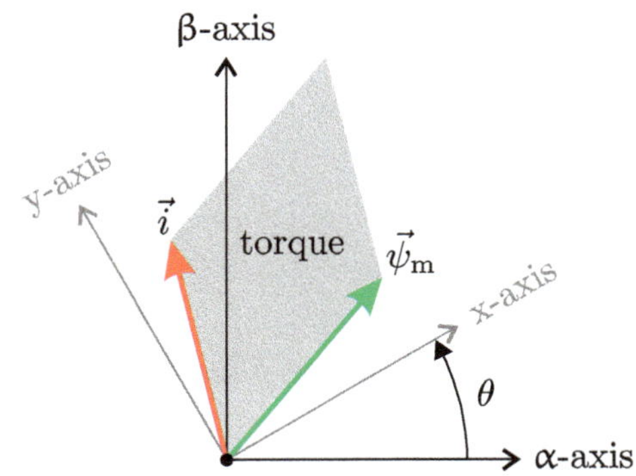

Fig. 4.6 Relationship between IRTF current/flux linkage space vectors and torque

Furthermore, the introduction of simplified models will serve to emphasize the analogy that exists between the DC, synchronous, and induction machine concepts.

Central to electromagnetic torque production is the interaction between the magnetic flux and the current sheet in the machine according to Eq. (4.5). The torque magnitude can be computed by the cross product of the IRTF vectors $\vec{\psi}_m$ and $\vec{i}$ as shown in Fig. 4.6.

This figure shows that torque control in an electrical machine may be realized by manipulating the current vector relative to the flux vector. The *gray* shaded area in Fig. 4.6 represents the torque magnitude. It may be readily deduced that maximum torque (largest *gray area*) for given flux and current magnitudes will be realized when the two vectors are orthogonal. For steady-state operation, i.e., constant torque, the two vectors must remain stationary with respect to each other. Changes in torque are usually instigated by changes of either the magnitude or the orientation of the current vector relative to the flux vector. The reason for this is that the dynamics linked to the flux vector are usually lower compared to the current vector. This explains why current control techniques, as discussed in the previous chapter, play a key role in electrical drives. Note that orientation of both vectors relative to the xy rotor reference frame (also shown in Fig. 4.6) is machine dependent, as will become apparent shortly.

In subsequent subsections, simplified IRTF based synchronous and asynchronous machine models will be introduced to which a space vector current source is connected. In practice, the latter is provided by controlling current of voltage source converters as discussed in the previous chapter. For each specific machine model additional components need to be introduced. In all cases, a magnetizing inductance L_m is required because machines are not magnetically ideal. For example, a finite air-gap between rotor and stator can be represented by a reluctance which is inversely proportional to the magnetizing inductance.

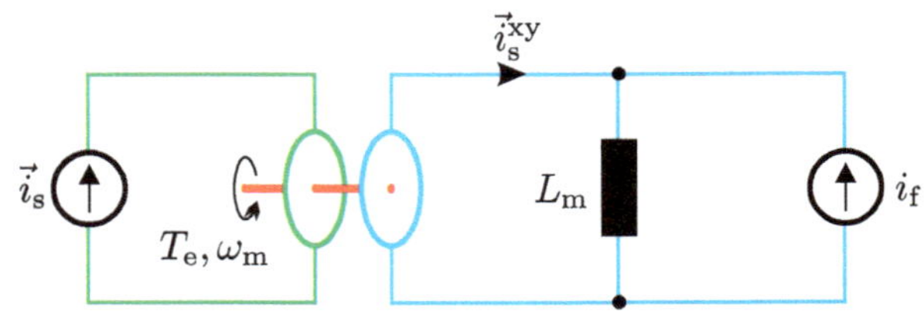

Fig. 4.7 IRTF based synchronous drive

4.2.1 *Synchronous Machine*

Synchronous machines which utilize a field winding are often connected to an excitation source via a set of slip rings. In the IRTF based symbolic model given in Fig. 4.7, this excitation process is represented by the DC current source i_f. A magnetizing inductance L_m is also shown which has been arbitrarily positioned on the rotor side (this is allowed as the transformation ratio of the IRTF equals unity). The *converter* is represented by the current source $\vec{i}_s$ and the question arises how this vector can be chosen to achieve torque control of the drive.

Observation of Fig. 4.7 and use of the IRTF equation set, as given in the previous section, shows that the following set of equations apply, namely

$$\vec{\psi}_m^{xy} = L_m \vec{i}_s^{xy} + L_m i_f \tag{4.7a}$$

$$\vec{T}_e = \frac{3}{2}\left(\vec{\psi}_m^{xy} \times \vec{i}_s^{xy}\right). \tag{4.7b}$$

The term $L_m i_f$ represents the flux linkage ψ_f due to the excitation of either a field winding or permanent magnet. This variable is linked to a space vector $\vec{\psi}_f^{xy} = \psi_f$, which is tied to the real axis of the rotor coordinate system, as shown in Fig. 4.8. Use of the flux variables $\vec{\psi}_f^{xy} = \psi_f$ with Eq. (4.7) leads to

$$\vec{\psi}_m^{xy} = L_m \vec{i}_s^{xy} + \vec{\psi}_f^{xy} \tag{4.8a}$$

$$T_e = \frac{3}{2}\left(\psi_f \Im\left\{\vec{i}_s^{xy}\right\}\right). \tag{4.8b}$$

From Eq. (4.8b), optimum torque control (highest torque for the lowest current) is achieved by choosing the current vector $\vec{i}_s^{xy} = jI$, as shown in Fig. 4.8, where I represents the current magnitude of said vector. The corresponding electromagnetic torque and current vector to be generated by the converter can be written as

$$T_e = \frac{3}{2}\psi_f I \tag{4.9a}$$

$$\vec{i}_s = j\, I\, e^{j\theta}. \tag{4.9b}$$

Fig. 4.8 Current and flux linkage space vectors: synchronous drive

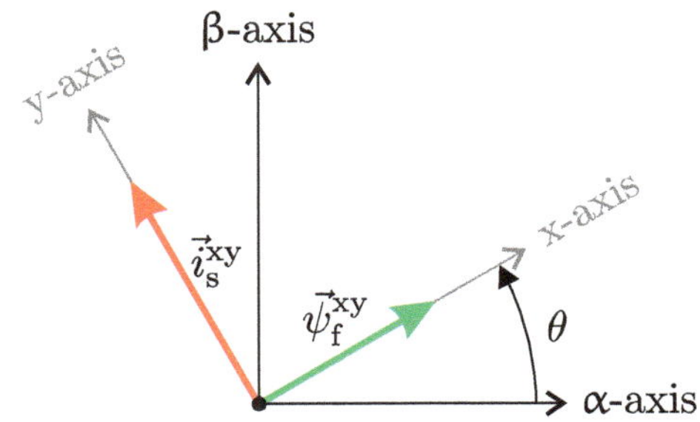

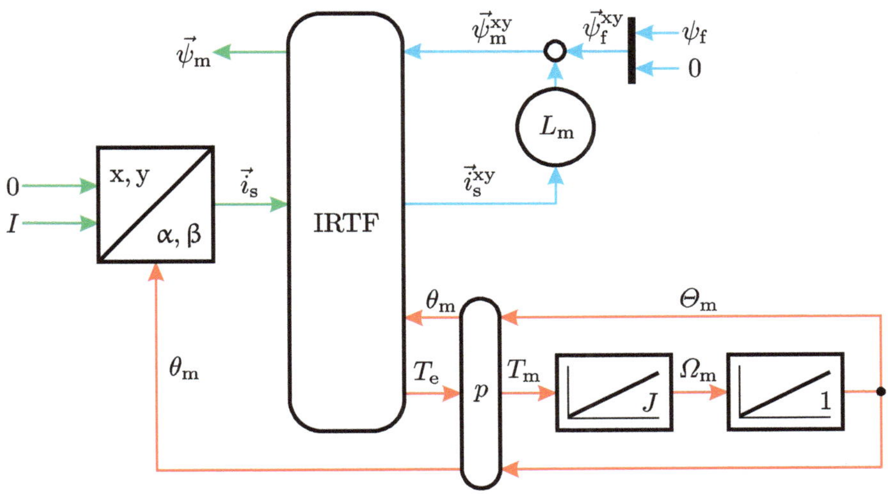

Fig. 4.9 Generic IRTF model of a synchronous machine drive at no-load

An IRTF based generic model of the proposed synchronous drive structure given in Fig. 4.8 shows the IRTF module and a coordinate conversion module (x, y $\rightarrow$ α, β) which implements expression (4.9b). Input to this module is the rotor shaft angle θ which is derived from the torque using equation set (4.6) with $T_l = 0$, which implies that the machine is not connected to a load.

For the computation of the torque, the IRTF makes use of the vectors $\vec{i}_s$ and $\vec{\psi}_m$, where the latter is found using Eq. (4.8a). Note that with the present choice of the current vector (orthogonal to the flux vector), the flux contribution $L_m \vec{i}_s^{xy}$ will not affect the torque given that it is in phase with the current. Furthermore, it is emphasized that this type of drive requires access to the measured or estimated (using electrical sensors and knowledge of the model) shaft angle Θ_m.

Note that the torque production process, as discussed in this subsection, is remarkably similar to that shown for the simplified DC machine drive. The key difference is that the rotor and stator functions in the machine module have been exchanged. In the DC machine the excitation is provided by the stator, while in the synchronous machine it is the rotor which handles this task.

A tutorial based on Fig. 4.9 is given in Sect. 4.5.1 which underlines the concepts discussed in this subsection. The reader is reminded of the fact that the synchronous

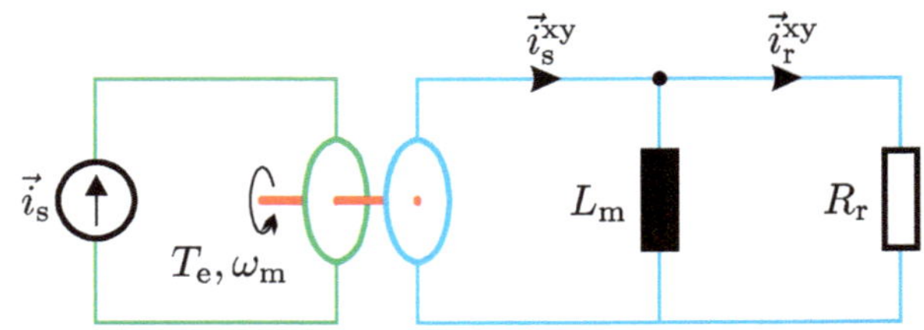

Fig. 4.10 IRTF based induction machine model

drive concept presented here is designed to provide a basic insight only. In reality, the machine design is more complex and, correspondingly, the control techniques to be deployed must be more extensive as will be shown in Chaps. 6 and 7, respectively.

4.2.2 Induction Machine

An induction machine which utilizes a squirrel-cage rotor [5] can in elementary form be represented with an IRTF based symbolic model as shown in Fig. 4.10. Readily apparent are the rotor resistance R_r of the squirrel cage and the magnetizing inductance L_m. The *converter* of the drive is represented by the current source $\vec{i}_s$ which must be manipulated to achieve torque control. Unlike the synchronous drive discussed in the previous subsection, the flux ψ_m must be provided by the converter.

The equation set related to the flux and current can, with the aid of Fig. 4.10, be expressed as

$$\frac{d\vec{\psi}_m^{xy}}{dt} = R_r\,\vec{i}_r^{xy} \tag{4.10a}$$

$$\vec{\psi}_m^{xy} = L_m\big(\vec{i}_s^{xy} - \vec{i}_r^{xy}\big) \tag{4.10b}$$

$$\vec{T}_e = \frac{3}{2}\left(\vec{\psi}_m \times \vec{i}_s\right). \tag{4.10c}$$

Also shown in equation set (4.10) is the torque expression which makes use of the flux space vector $\vec{\psi}_m = \psi_m\,e^{j\theta_\psi}$, where θ_ψ represents the instantaneous angle between said vector and real axis of the stationary reference frame. This vector is shown in Fig. 4.11 together with a new orthogonal so-called *synchronous* coordinate system. The real axis $\Re^{dq}$ of said coordinate system is tied to the flux vector in which case the latter can be expressed as $\vec{\psi}_m^{dq} = \psi_m$. The superscript dq identifies the reference frame in use. Also shown in Fig. 4.11 is the stator current vector $\vec{i}_s$, which can also be expressed in the synchronous reference coordinate frame as $\vec{i}_s^{dq} = i_d + j i_q$, where i_d and i_q are known as the direct axis and quadrature axis current components, respectively. Subsequent transformation of equation set (4.10) to synchronous coordinates allows the flux and torque equations to be written as

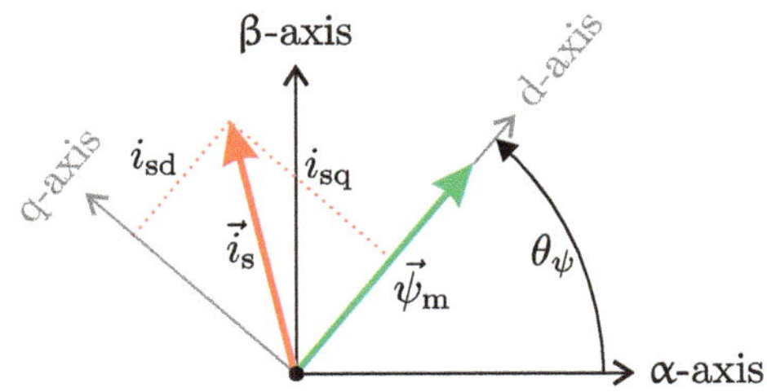

Fig. 4.11 Current and flux linkage space vectors: induction machine drive

$$\left(\frac{L_m}{R_r}\right)\frac{d\psi_m}{dt} + \psi_m = L_m i_d \tag{4.11a}$$

$$T_e = \frac{3}{2}\psi_m\, i_q. \tag{4.11b}$$

Equation set (4.11) is significant for this type of drive given that it shows that the direct axis component defines the flux level in the machine. Furthermore, changes to the current i_d will not lead to an instant change of the flux level, given that this is governed by the time constant L_m/R_r. Under quasi-steady-state conditions, the flux level will be equal to $\psi_m \cong L_m\, i_d$. On the other hand, torque control can be virtually instantaneous (subject to the dynamics of the current controller), because it can be achieved by varying the value of the quadrature current i_q of the current source, which is set to the reference value I in this example. The corresponding electromagnetic torque and current vector to be generated by the converter for the induction machine drive can be written as

$$T_e = \frac{3}{2}\psi_m\, I \tag{4.12a}$$

$$\vec{i}_s \simeq \left(\frac{\psi_m^*}{L_m} + j\, I\right) e^{j\,\theta_\psi} \tag{4.12b}$$

where ψ_m^* represents the reference flux value. A generic representation of the drive structure is illustrated in Fig. 4.12, showing the IRTF module at the center of the induction machine model. A Cartesian to polar conversion module is used to determine the required flux angle θ_ψ that is required for the coordinate conversion module (d, q $\rightarrow \alpha, \beta$), which generates the current vector $\vec{i}_s$.

The tutorial given in Sect. 4.5.2 proves the opportunity to explore the drive concept as discussed in this subsection. It should come as no surprise that, in practice, induction machine models are more complex than shown here. This implies that more extensive control techniques are needed to achieve torque and flux control. Chapters 8 and 9 cover the advanced modeling and control aspects, respectively.

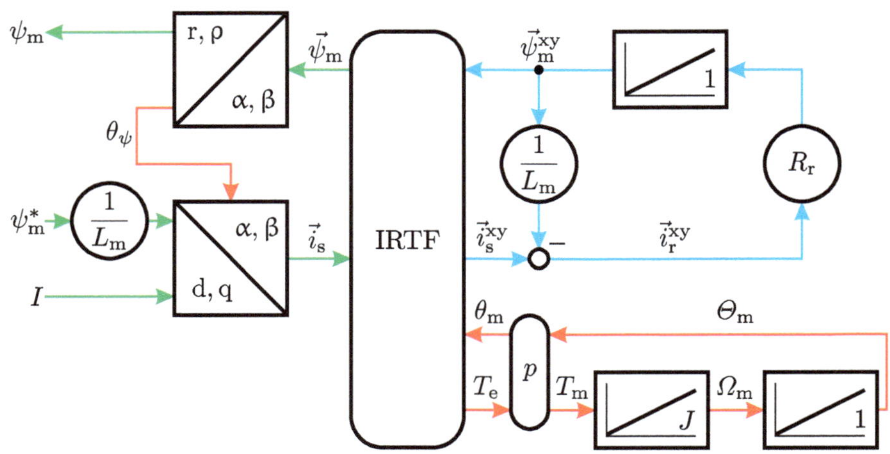

Fig. 4.12 Generic model IRTF based induction machine drive

4.3 Drive Dynamics

This section examines the mechanical interaction between the electrical machine and load. More specifically, Newton's laws of motion are introduced for linear and rotational based drive systems [3]. In addition, attention is given when an electrical machine is connected to a rotating and translatory load. Furthermore, the mechanical dynamics of a machine connected to a load are examined with the aid of a tutorial example outlined in Sect. 4.5.3.

4.3.1 Linear and Rotational Motion

Linear Motion

Prior to discussing rotational motion it is helpful to consider the linear or translatory motion of a point mass with the aid of Fig. 4.13. In this example, a point mass has been purposely chosen, so that the size of the body can be ignored for the ensuing analysis. Furthermore, the point mass is shown in a two-dimensional complex plane given that the rotational analysis will also be undertaken in such a reference frame. The reason for this is that rotating machines utilize rotors which exhibit a radial and axial symmetry, which justifies the use of a two-dimensional analysis.

The displacement vector $\vec{s} = s_\alpha + \mathrm{j}\, s_\beta$ defines the orientation of the point mass relative to the complex plane. The instantaneous velocity $\vec{v} = v_\alpha + \mathrm{j}\, v_\beta$ is defined according to

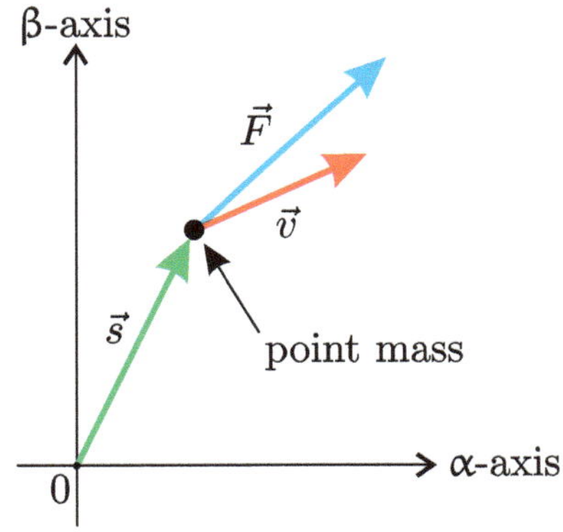

Fig. 4.13 Linear motion of a point mass

$$\vec{v} = \frac{d\vec{s}}{dt}. \tag{4.13}$$

Newton's first law known as the *law of inertia* states that a body (the point mass in this case) will be either at rest ($\vec{v} = 0$) or moving at a constant velocity, when no net external forces are applied. If an external net force $\vec{F}$ (as shown in Fig. 4.13) is applied to the point mass, an acceleration $\vec{a}$ will occur according to

$$\vec{F} = m\vec{a} \tag{4.14a}$$

$$\vec{a} = \frac{d\vec{v}}{dt} \tag{4.14b}$$

where m represents the mass of the body, shown in Fig. 4.13. Expression (4.14a) is known as *Newton's second law*. Note that the force $\vec{F}$ represents the vector sum of all applied forces. Consequently, a zero net force yields zero acceleration, hence $d\vec{v}/dt = 0$, which implies motion at either constant velocity or rest (Newton's first law).

Rotational Motion

The analysis given above can be readily extended to a rotating body such as the rotor of an electrical machine. Figure 4.14 shows a particle dm that is part of a rotating body (typically radially symmetric) which rotates around a fixed axis that is orthogonal to the origin 0 of the non-rotating complex plane.

The orientation of the mass particle is defined by the displacement vector $\vec{s}$, which can be expressed in its polar form as

$$\vec{s} = r\,e^{j\theta_m} \tag{4.15a}$$

$$\omega_m = \frac{d\theta_m}{dt} \tag{4.15b}$$

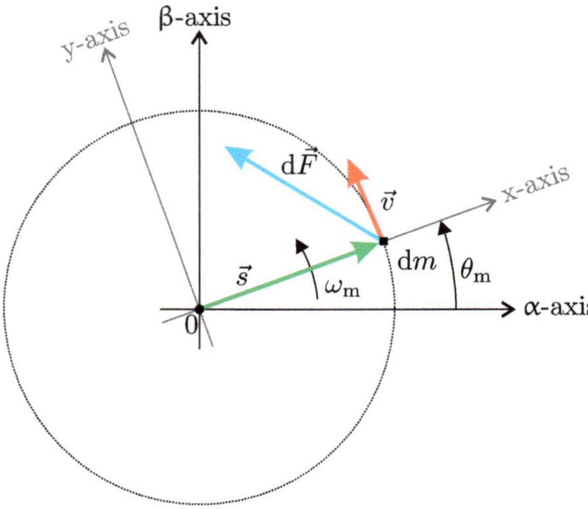

Fig. 4.14 Non-uniform circular motion of a mass particle

where $|\vec{s}| = r$ represents the distance of the mass particle relative to the point of rotation, which is constant in this case, because it is part of a larger (solid) rotating body. Also shown in Eq. (4.15) is the angular frequency ω_m of the mass particle which can, with the aid of Eqs. (4.13) and (4.15a), be expressed as $\vec{v} = j\vec{s}\omega_m$. Note that the velocity is tangential to the circular orbit of the mass particle. It is also helpful to consider the speed in terms of an xy reference plane tied to the vector $\vec{s}$ by making use of the conversion $v^{xy} = \vec{v}\, e^{-j\theta_m}$. Subsequent analysis shows that the rotating body referred velocity vector v^{xy} is equal to $v^{xy} = \omega_m r$, which is aligned with the $\Im^{xy}$ axis of the rotating coordinate frame.

The incremental force $d\vec{F}$ may be found with the aid of Eq. (4.14) (where m must be replaced by dm) and Eq. (4.15), which gives

$$d\vec{F} = -\vec{s}\,\omega_m^2\, dm + j\vec{s}\, dm\, \frac{d\omega_m}{dt}. \tag{4.16}$$

Equation (4.16) demonstrates that the incremental force consists of a so-called centripetal component which is oriented towards the point of rotation. Furthermore, a tangential force component exists, when non-uniform rotation $d\omega_m/dt \neq 0$ occurs. The force equation can also be written in a rotor reference frame format as

$$d\vec{F}^{xy} = \underbrace{-r\,\omega_m^2\, dm}_{dF_c} + \underbrace{j\, r\, dm\, \frac{d\omega_m}{dt}}_{dF_t} \tag{4.17}$$

where dF_c and dF_t represent the incremental centripetal and tangential force components, respectively. Note that solid bodies that rotate around a fixed axis, the sum of all centripetal forces acts on this axis, i.e., the shaft of the rotor, which is kept in place by bearings. In electrical machines the bearings need to counteract these centripetal forces. To avoid vibrations and bearing wear out rotors are carefully balanced during production of the machines. Hence, in practice the sum of all centripetal forces equals zero and only the tangential forces need to be considered further to determine the shaft torque T.

The incremental torque $\mathrm{d}T$ can be found using $\mathrm{d}T = r\,\mathrm{d}F_{\mathrm{t}}$, which, with the aid of Eq. (4.17), may be written as $\mathrm{d}T = r^2\,\mathrm{d}m\,{}^{\mathrm{d}\omega_{\mathrm{m}}}/_{\mathrm{d}t}$. Computation of the torque associated with the entire rotating body volume V requires three-dimensional integration of the incremental torque expression, namely

$$T = \underbrace{\iiint_V r^2\,\mathrm{d}m}_{J}\,\frac{\mathrm{d}\omega_{\mathrm{m}}}{\mathrm{d}t} \tag{4.18}$$

where J is referred to as the *moment of inertia*. This allows expression (4.18) to be written as

$$T = J\,\frac{\mathrm{d}\omega_{\mathrm{m}}}{\mathrm{d}t} \tag{4.19}$$

which is the rotational equivalent of the linear expression $\vec{F} = m\,{}^{\mathrm{d}\vec{v}}/_{\mathrm{d}t}$. The torque T represents the net torque. In an electrical drive it is formed by the difference of the electromagnetic torque T_{e} produced by the electrical machine and the load torque T_{l}. The latter may be due to friction and/or mechanical loads, attached to the electrical machine. Expression (4.18) can be used to calculate the moment of inertia of rotating bodies. For example, the inertia of a solid disk with radius R and mass M is found to be equal to $J = {}^1/_2\,M\,R^2$. Following this section, a number of transmission examples will be discussed which makes use of the theory presented in this subsection.

4.3.2 Rotational to Translational Transmission

For applications which require linear motion, often a *rack and pinion* set of gears, as shown in Fig. 4.15, is used. The pinion, with radius r, is connected to the shaft of the rotating electrical machine which generates a torque T_{e} with shaft speed ω_{m}. The combined inertia of the electrical machine rotating parts connected to the shaft and pinion is defined as J. The rack is connected to the translatory load, which is represented in Fig. 4.15 by a force F_{l}. The total mass of rack and load is defined as m.

The force exerted by the pinion gear on the rack is defined as F_{a}. Vice versa, the rack exerts an equal but opposite force F_{r} on the teeth of the pinion which gives

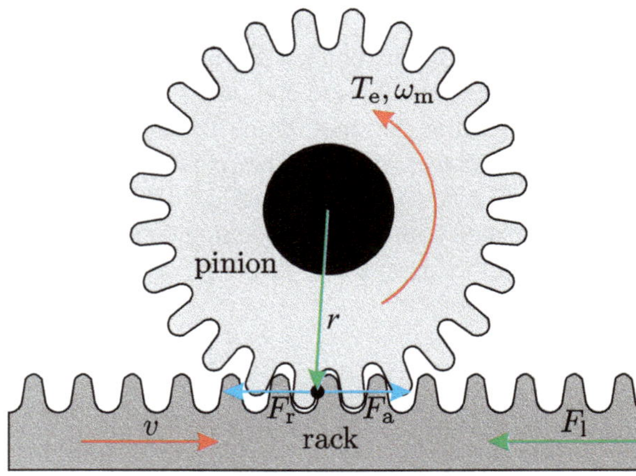

Fig. 4.15 Rotational to translational transmission example: rack and pinion

$$F_a = F_r. \tag{4.20}$$

Relationship (4.20) is known as Newton's third law which states that *to every action there is an equal but opposite reaction*. Application of Newton's second law to the set of gears shown in Fig. 4.15 leads to

$$T_e - r F_r = J \frac{d\omega_m}{dt} \tag{4.21a}$$

$$F_a - F_l = m \frac{dv}{dt}. \tag{4.21b}$$

Subsequent use of Eq. (4.20) with Eq. (4.21) and taking into account that the speed of the rack can be written as $\vec{v} = \omega_m r$ gives

$$T_e - r F_l = \Big(\underbrace{mr^2}_{J_e} + J\Big) \frac{d\omega_m}{dt}. \tag{4.22}$$

Expression (4.22) shows that the introduction of a translatory load leads to an increase of the inertia seen by the rotating machine. The additional inertia J_e is defined as

$$J_e = mr^2 = m \left(\frac{v}{\omega_m}\right)^2. \tag{4.23}$$

This implies that a translational load which requires a large linear velocity v will lead to a substantial increase in the inertia experienced by the rotating machine/pinion combination. Note that Eq. (4.22) can be derived easily by considering the change of total kinetic energy stored in the system.

4.3.3 Gear Transmission

A wide range of industrial rotational applications use a transmission device between load and electrical machine. Reasons for this may be due to physical load enclosure constraints, i.e., where the machine cannot be directly attached. In other instances, the mismatch between optimum load and machine speeds must be resolved by making use of a rotational to rotational transmission device. For this purpose gears or pulley's are used. Gears are wheels with teeth which mesh with each other, as shown in the example given in Fig. 4.16. The use of gears allows the transfer of forces without slippage, which is in contrast with pulley/belt systems, where this phenomenon can occur.

The relationship that exists between the torque and force of the two gears shown in Fig. 4.16 may be considered by application of Newton's second and third laws. In this example, the electrical machine is arbitrarily connected to gear 1 with radius r_1 and provides a shaft torque T_e with rotational speed $\omega_{m\,1}$. The load is connected to gear 2 with radius r_2 to which the load torque T_l is applied.

The relationship between the rotational speeds $\omega_{m\,1}$ and $\omega_{m\,2}$ follows from the observation that the tangential speed of both gears must be equal (no slippage), which gives

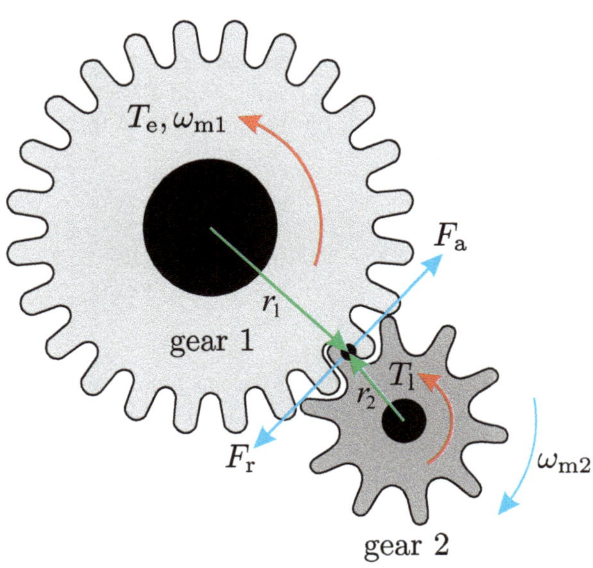

Fig. 4.16 Gear transmission example: spur gearing

$$\omega_{m\,1} r_1 = \omega_{m\,2} r_2. \tag{4.24}$$

The relationship between machine driving torque and load torque may be found by taking into account that the driving gear (gear 1 in this example) exerts a force F_a (as shown in Fig. 4.16) on the teeth of gear 2. According to Newton's third law an equal but opposite force F_r will be exerted on gear 1. If the gears are replaced by a set of pulleys, this force will be transferred via the belt. Observation of Fig. 4.16 and application of Newton's second law gives

$$T_e - r_1 F_r = J_1 \frac{d\omega_{m\,1}}{dt} \tag{4.25a}$$

$$r_2 F_a - T_l = J_2 \frac{d\omega_{m\,2}}{dt} \tag{4.25b}$$

where J_1 and J_2 represent the inertia of the electrical machine (with gear 1) and load (with gear 2), respectively

Subsequent elimination of the force variables F_a and F_r from Eq. (4.25) and using $F_a = F_r$ (Newton's third law), as well as Eq. (4.24) gives

$$T_e - \left(\frac{r_1}{r_2}\right) T_l = \left(J_1 + \underbrace{\left(\frac{r_1}{r_2}\right)^2 J_2}_{J_{eq}} \right) \frac{d\omega_{m\,1}}{dt}. \tag{4.26}$$

Expression (4.26) represents Newton's second law expressed in terms of the machine variables T_e, $\omega_{m\,1}$. The result shows that the load inertia J_2 appears as an equivalent inertia J_e on the drive side of the transmission, which is computed according to

$$J_{eq} = \left(\frac{r_1}{r_2}\right)^2 J_2. \tag{4.27}$$

Hence, the inertia J_{eq} *seen* at the machine side of the transmission will be greater than the actual load inertia J_2 in case $r_1 > r_2$, which is the case shown in Fig. 4.16. Note that Eq. (4.26) can be derived easily by considering the change of kinetic energy stored in the system.

4.3.4 Dynamic Model of a Drive Train

The process of transmitting power from the electrical machine to the load is considered in this subsection with the aid of Fig. 4.17. Shown in Fig. 4.17 are two rotating masses with inertia J_l and J_2, which are assigned to the rotor of the electrical machine and load, respectively. A coupling of some type, which may simply be a shaft, is used to link the two masses. If the coupling is sufficiently stiff, the two inertias may be simply represented by the sum of the two inertias J_{total}. In this case, Newton's second law for the drive train is given as

$$T_e - T_l = J_{\text{total}} \frac{d\omega_m}{dt} \tag{4.28}$$

with $J_{\text{total}} = J_1 + J_2$ and $\omega_m = \omega_{m\,1} = \omega_{m\,2}$. Furthermore, the relationship between angular frequency and shaft angle is reduced to $\omega_m = {}^{d\theta}/_{dt}$, with $\theta_{m\,1} = \theta_m = \theta_{m\,2}$. The power p_e delivered by the machine and the power supplied to the load p_l are given as $p_e = T_e\, \omega_{m\,1}$ and $p_l = T_l\, \omega_{m\,2}$, respectively.

If the coupling between load and machine cannot be considered as stiff, a certain amount of twisting (torsion) of said coupling about its axis can occur, which depends on the torque ΔT applied and the properties of the coupling. The coupling properties typically are the torsion coefficient κ, also known as the *spring constant*, and damping coefficient c. Both coefficients are related to the applied torsional torque ΔT, the instantaneous frequencies, and rotor angles of the drive train according to Eq. (4.29).

$$\Delta T = \kappa\ (\theta_{m1} - \theta_{m2}) + c\,(\omega_{m1} - \omega_{m2})\,. \tag{4.29}$$

The formulation of Newton' second law for the drive train in question must take into account the twisting torque of the coupling, which leads to

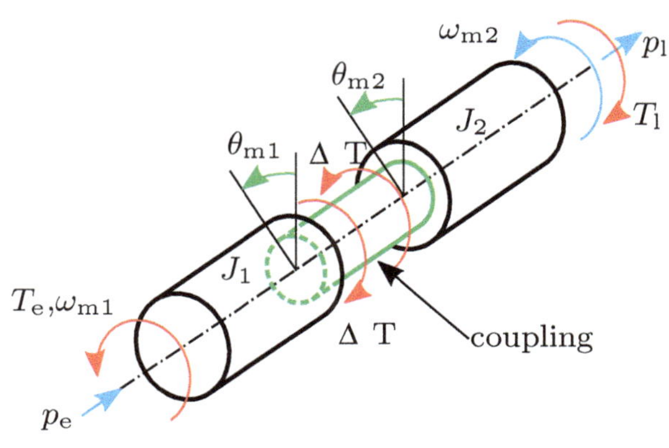

Fig. 4.17 Dynamic model of a two-mass drive train

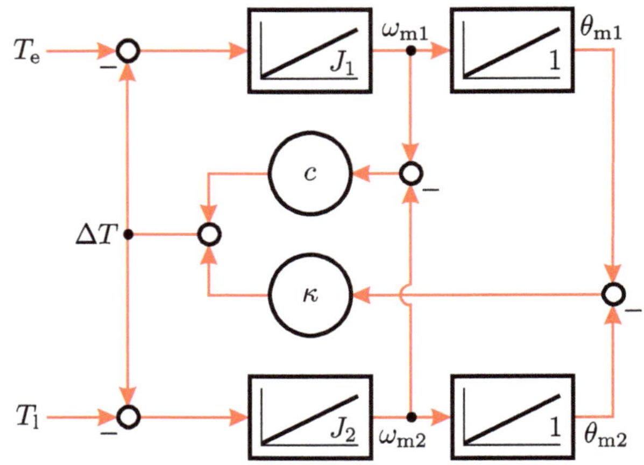

Fig. 4.18 Generic model of a two-mass drive train

$$T_e - \Delta T = J_1 \frac{d\omega_{m1}}{dt} \tag{4.30a}$$

$$\Delta T - T_l = J_2 \frac{d\omega_{m2}}{dt}. \tag{4.30b}$$

To study the interaction between load and machine under dynamic conditions it is helpful to develop a generic model of this system with the aid of Eqs. (4.30) and (4.29) and the following expressions:

$$\omega_{m1} = \frac{d\theta_{m1}}{dt} \tag{4.31a}$$

$$\omega_{m2} = \frac{d\theta_{m2}}{dt}. \tag{4.31b}$$

The latter defines the relationship between instantaneous shaft frequencies and shaft angles of the drive train under consideration. A generic diagram which satisfies the equation set given above is shown in Fig. 4.18.

The tutorial given in Sect. 4.5.3 provides a numerical example which is directly based on the generic model given in Fig. 4.18. An example of the results obtained with this simulation model is given in Fig. 4.19.

Shown in Fig. 4.18 are the torque variable T_e, ΔT as function of time in the event that a shaft torque step of $T_e = 20\,\text{Nm}$ is applied at $t = 0\,\text{s}$. Also shown in Fig. 4.18 is the angular frequency difference $(\omega_{m1} - \omega_{m2})$ versus time, which must inevitably yield a zero steady-state difference given that both rotating masses ultimately accelerate towards the same speed. The load torque in this example has been purposely set to zero. Note that the approach discussed in this section can be readily extended to more complex drive train structures. Note that in the tutorial the difference angle $(\theta_{m1} - \theta_{m2})$ is calculated directly by integration of the speed

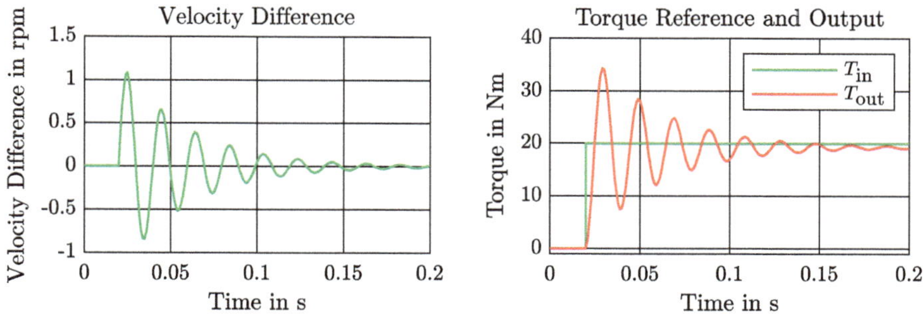

Fig. 4.19 Dynamic response of a two-mass drive system with $\kappa = 5000\,\mathrm{Nm/rad}$, $c = 2.5\,\mathrm{Nm\,s/rad}$, $J_1 = 0.051\,\mathrm{kg\,m^2}$ and $J_2 = 1.35\,\mathrm{kg\,m^2}$

difference. In this way, no special measures need to be taken to prevent increased rounding errors by otherwise subtracting two angles that would never stop growing.

4.4 Shaft Speed Control Loop Design Principles

In previous sections it was shown that torque control can be achieved by manipulating a current variable I which can be determined from the required reference torque T_e^* using

$$I = \frac{T_e^*}{\psi_m} \tag{4.32}$$

where ψ_m represents the magnetizing flux in the machine. However, in many applications control of the shaft speed is required which means that an additional outer control loop must be deployed. This so-called cascaded control approach [1, 2] is typified by an *inner* current control loop and *outer* speed control loop. The time constant associated with the inner loop is usually small in comparison with the outer loop, which is dictated by mechanical time constants, as was mentioned earlier. Consequently, for the purpose of dimensioning the speed controller an ideal current controller may be assumed, which implies that the reference torque will be equal to the output torque of the drive. Under these conditions the drive is reduced to the generic model given in Fig. 4.20.

The drive, shown in none-discrete form, consists of a proportional-integral speed controller, of which the Laplace transform may be written as

$$\frac{T_e^*}{\Delta\omega_m} = K_p\left(1 + \frac{1}{\tau_i s}\right) \tag{4.33}$$

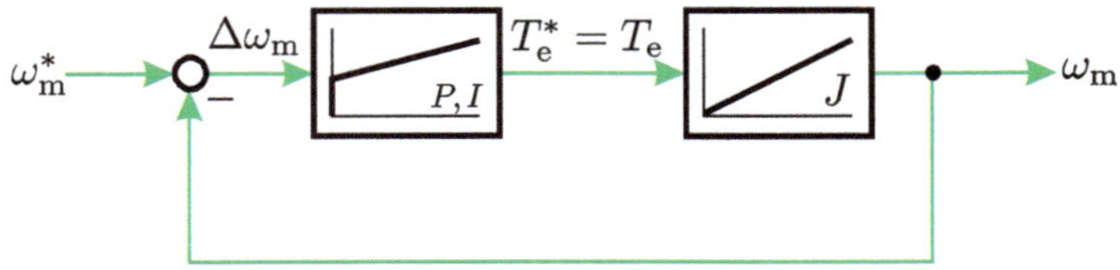

Fig. 4.20 Continuous time domain block diagram of speed controller with simplified generic drive model

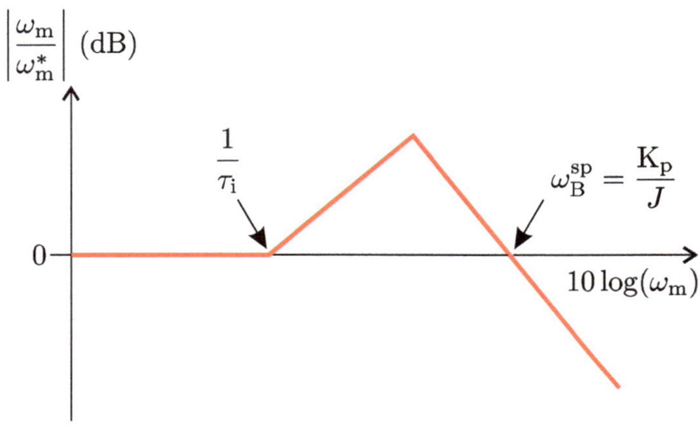

Fig. 4.21 Bode amplitude plot example, with real poles and zeros

where s is the Laplace operator. In steady-state sinusoidal analysis, s can be considered as a complex variable, i.e., the *complex frequency* $s = j\omega$. Furthermore, $\Delta\omega_m$ is defined as $\Delta\omega_m = \omega_m^* - \omega_m$. In this case, the controller is represented in terms of a proportional gain K_p and integral time constant τ_i instead of an integral gain $K_i = K_p/\tau_i$, as used for the current controller. From a control perspective the use of variables K_p, τ_i as opposed to K_p, K_i is often preferable, because they can be chosen independently.

The process of determining the controller parameters may be undertaken by considering the Laplace transform of the drive representation according to Fig. 4.20 which is of the form

$$\frac{\omega_m}{\omega_m^*} = \frac{K_p\,(\tau_i s + 1)}{J\tau_i s^2 + K_p\tau_i s + K_p}. \tag{4.34}$$

A Bode amplitude plot of expression (4.34) is given in Fig. 4.21, using straight line asymptotic approximations. In the example shown the poles and zeros are assumed to be real, while the gradient of the linear functions is equal to 20 dB/decade.

Observation of Eq. (4.34) and Fig. 4.21 demonstrates that for high frequencies the transfer function (in terms of its amplitude) may be written as

$$\left|\frac{\omega_m}{\omega_m^*}\right| \simeq \frac{K_p}{J\omega}. \tag{4.35}$$

The frequency $\omega = \omega_B^{sp}$ (as shown in Fig. 4.21) at which the transfer function, according to Eq. (4.35), reaches unity gain effectively defines the bandwidth of the speed control loop. Accordingly, the proportional gain of the speed controller is given as

$$K_p = \omega_B^{sp} J \tag{4.36}$$

where J represents the total inertia of the drive.

Consequently, it is necessary to have knowledge of the mechanical load that is attached to the machine, given the need to estimate the combined inertia J. Furthermore, a value for the effective bandwidth of the speed controller must be provided by the user. For typical high-performance drives the effective speed control bandwidth ω_B^{sp} is in the order of $\omega_B^{sp} = 100\,\text{rad/s}$, which, together with the inertia J, fully defines the proportional gain of the controller.

Computation of the integrator time constant τ_i may be undertaken by reconsidering Eq. (4.34), which upon substitution of Eq. (4.36) may be rewritten as

$$\frac{\omega_m}{\omega_m^*} = \left(\omega_B^{sp}\right) \frac{s + \frac{1}{\tau_i}}{s^2 + \omega_B^{sp} s + \omega_B^{sp}/\tau_i}. \tag{4.37}$$

The denominator of Eq. (4.37) can also be written as $s^2 + 2\zeta\omega_o + \omega_o^2$, where ζ, ω_o represent the damping and natural frequency, respectively, and which may be expressed as

$$\zeta = \frac{1}{2}\sqrt{\omega_B^{sp}\tau_i} \tag{4.38a}$$

$$\omega_o = \sqrt{\frac{\omega_B^{sp}}{\tau_i}}. \tag{4.38b}$$

The poles of the transfer function as defined by expression (4.37) determine the behavior of the model and in this context a value for the damping factor needs to be selected. According to general linear control theory principles [1, 2], two options are normally entertained, namely:

- $\zeta = 1$, which according to Eq. (4.38) corresponds to $\tau_i = 4/\omega_B^{sp}$, $\omega_o = \omega_B^{sp}/2$. This option gives two poles s_1, s_s located in the complex s plane at $s_{1,2} = -\omega_B^{sp}/2$.
- $\zeta = 1/\sqrt{2}$, which according to Eq. (4.38) corresponds to $\tau_i = 2/\omega_B^{sp}$, $\omega_o = \omega_B^{sp}/\sqrt{2}$. This option yields two complex conjugate poles located in the complex plane at $s_{1,2} = -\omega_B^{sp}/2 \pm j\omega_B^{sp}/2$.

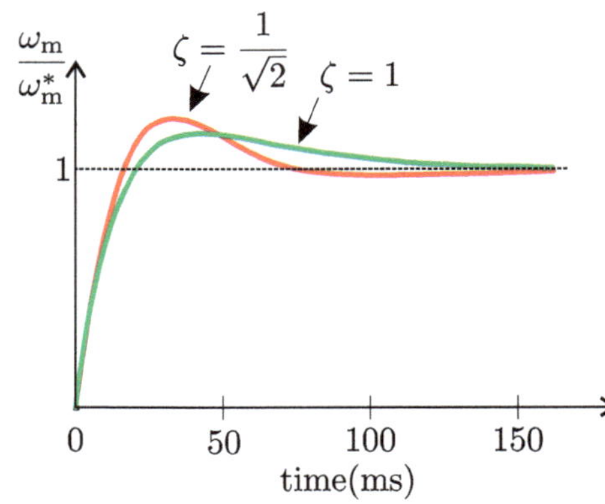

Fig. 4.22 Step response (ω_m/ω_m^*), with $\omega_B^{sp} = 100\,\text{rad/s}$

The choice of damping value must be considered in conjunction with the drive configuration in use and the nature of the transient response which will appear. For example, the step response of Eq. (4.37) for the two values of ζ considered above is shown in Fig. 4.22 for the speed bandwidth frequency of $\omega_B^{sp} = 100\,\text{rad/s}$.

A relatively high damping factor of $\zeta = 1$ is particularly important for drives which are not capable of accepting energy from the mechanical side. Given that such drives must rely on the mechanical time constant of the motor/load to recover from an overspeed (speed in excess of the reference value) condition. For four quadrant drives a critical $\zeta = 1/\sqrt{2}$ or underdamped $\zeta < 1/\sqrt{2}$ response is warranted, particularly when short settling times are required, for example, in servo applications. Given these considerations, a damping factor $\zeta = 1$ is a prudent choice, in which case the resulting set of controller parameters may be written as

$$K_p = \omega_B^{sp} J, \tag{4.39a}$$

$$\tau_i = \frac{4}{\omega_B^{sp}}. \tag{4.39b}$$

Note that it is often useful to use Eq. (4.33) in the form

$$T_e^* = \left(K_p + \frac{K_i}{s}\right) \Delta\omega_m \tag{4.40}$$

in which case the proportional gain will be equal to

$$K_p = J\omega_B^{sp} \tag{4.41a}$$

$$K_i = \frac{J\omega_B^{sp2}}{4} \tag{4.41b}$$

where the ratio between K_i and K_p is equal to $\omega_B^{sp}/4$.

In a practical drive environment, as discussed in the accompanying tutorial (see Sect. 4.5.4), a discrete PI controller with *anti-windup* is required. This so-called *windup* effect occurs in all control systems which utilize an integrator to nullify steady-state errors. An explanation of this phenomenon may be undertaken with the aid of Fig. 4.20 and imposing a practical constraint $\pm T_{\mathrm{e}}^{\max}$ on the torque level that can be delivered by the drive. If the torque level of the drive reaches the drive limit value, a situation can arise where a speed error occurs which causes the controller output to increase beyond the limit values $\pm T_{\mathrm{e}}^{\max}$ of the drive. An anti-windup controller has the capability to counteract the integrator action of the controller when the drive reaches a torque boundary.

A tutorial is given in Sect. 4.5.4 which outlines the concepts shown in this subsection. Notably this tutorial also demonstrates integrator windup effect and examines how this phenomenon can be countered.

4.5 Tutorials

4.5.1 Tutorial 1: Elementary Synchronous Drive

The purpose of this tutorial is to build a simulation model of the basic synchronous drive based on the generic model shown in Fig. 4.9. A simplified current source, IRTF based two-pole machine model is to be developed with a set of parameters as given in Table 4.1.

In this example, a machine without an external load is deployed to keep the tutorial as simple as possible. A control input i_{q}^{*} is created to generate a partially linear *quadrature current* set value signal for the drive with a range of $\pm$ 10 A. In addition, provide a diagram which shows the torque T_{e} for the duration of the simulation, which must be set to $T = 100$ ms.

The simulation model given in Fig. 4.23 satisfies the tutorial requirements outlined above. A simplified controller structure is shown which makes use of the *measured* shaft angle θ and the reference current I to generate the reference current $\vec{i}_{\mathrm{s}}$. Two scope modules are used to observe the relation between the quadrature current set value i_{q}^{*}, torque and speed, as well as the phase currents. The simulation results are given in Fig. 4.24. Note that the speed is not a sinusoidal waveform. It is linear when torque is constant and quadratic when the torque is linear.

Table 4.1 Synchronous machine parameters

Parameters	Value
Field flux ψ_{f}	1.0 Wb
Magnetizing inductance L_{m}	100 mH
Inertia J	10 m kg m^2

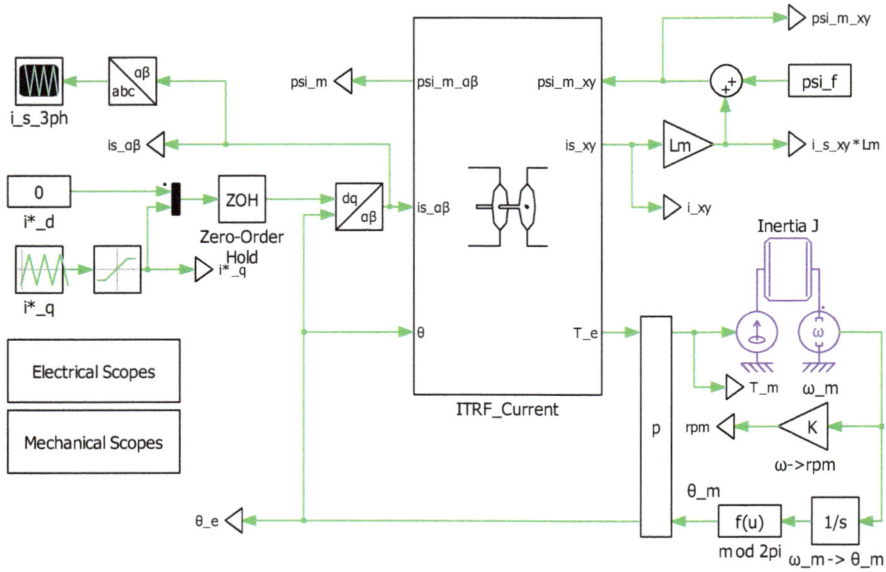

Fig. 4.23 Simplified synchronous drive simulation model

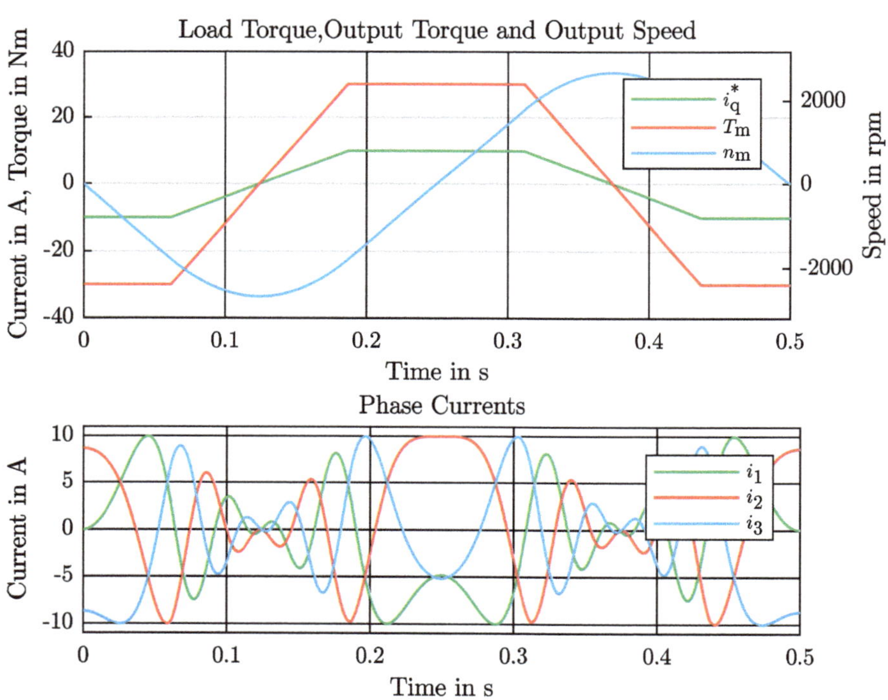

Fig. 4.24 Simulation results of simplified synchronous drive simulation model

4.5.2 Tutorial 2: Elementary Asynchronous (Induction) Drive

A simulation model of the basic asynchronous drive based on the generic model shown in Fig. 4.12 is generated in this tutorial. The drive uses a current source connected to a two-pole induction machine with parameters as given in Table 4.2. The PLECS model for the tutorial is given in Fig. 4.25. The drive control is similar to the previous tutorial. A second control input is introduced to set the reference flux ψ_{m}^{*} in the range of $\psi_{\mathrm{m}}^{*} = 0 \rightarrow 1\,\mathrm{Wb}$ in addition to quadrature current set value i_{q}^{*}. A delay is placed after the set value i_{q}^{*} to allow time for the mutual flux linkage to build in the machine. The effect of having only direct axis current in an induction machine can be observed in this period.

The dq transformation of the reference current vector is carried out using the instantaneous mutual flux linkage angle θ_{ψ}, while the transformations in the IRTF are carried out using the rotor position in (Figs. 4.26, 4.27, and 4.28). Therefore, the direct and quadrature axis are no longer aligned with the rotor, as was the case in the synchronous machine example.

Table 4.2 Induction machine parameters

Parameters	Value
Rotor resistance R_{r}	$10.0\,\Omega$
Magnetizing inductance L_{m}	$100\,\mathrm{mH}$
Inertia J	$10\,\mathrm{m\,kg\,m^2}$

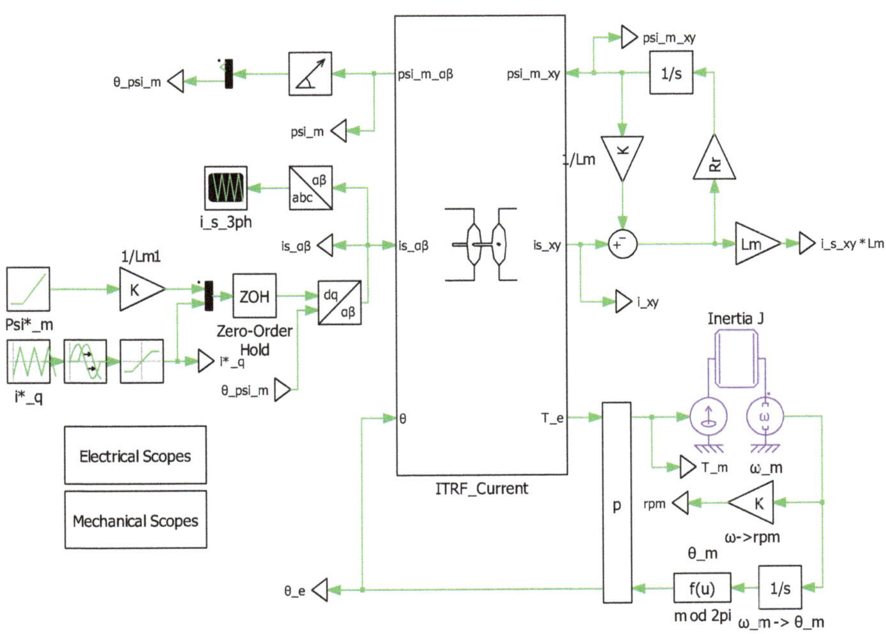

Fig. 4.25 Simplified asynchronous drive simulation model

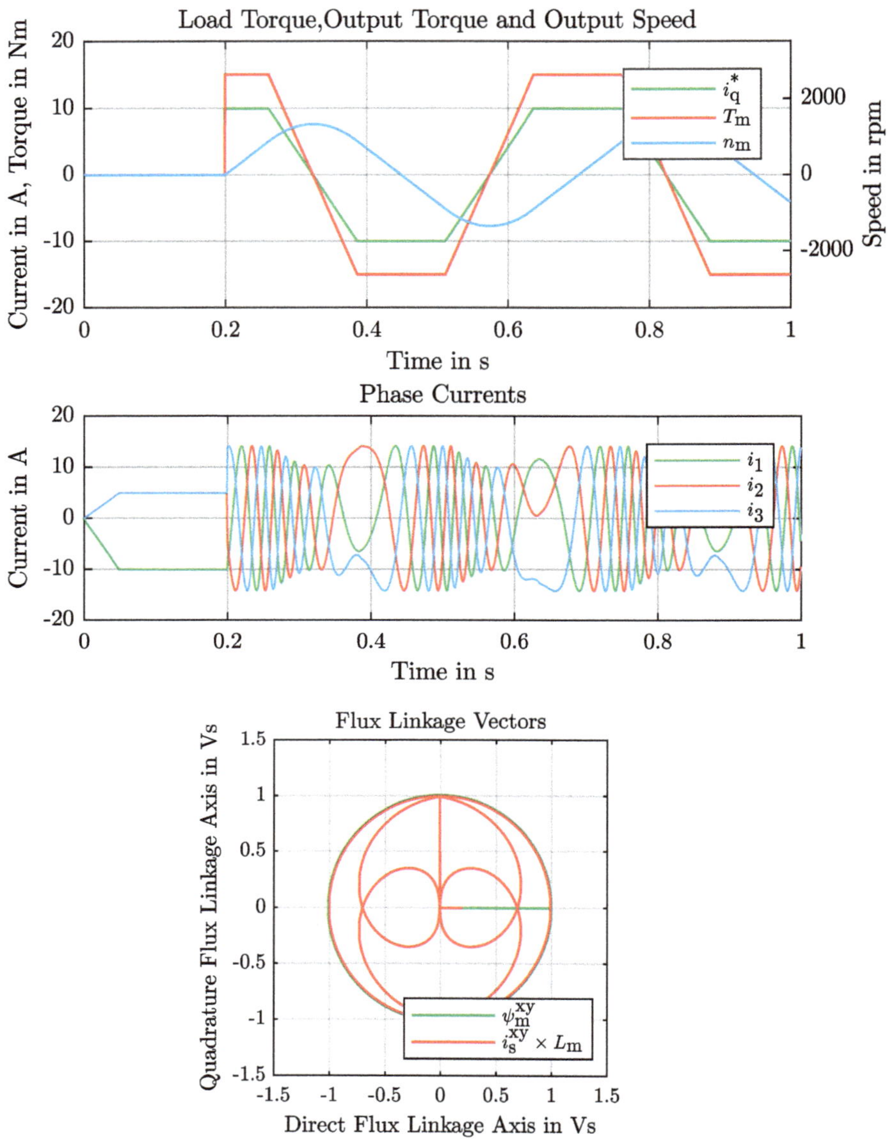

Fig. 4.26 Simulation results of simplified asynchronous drive simulation model

It is instructive to observe that in the beginning of the simulation, when quadrature current set value i_q^* is zero, the stator current vector $\vec{i}_s$ is aligned with the flux vector $\vec{\psi}_m$. This implies that the torque will be zero and the steady-state flux will be equal to $\psi_m = L_m \, i_d$. Furthermore, it may be observed that step changes in ψ_m^* elicit a first order response in the machine flux ψ_m which is governed by the time constant L_m/R_r.

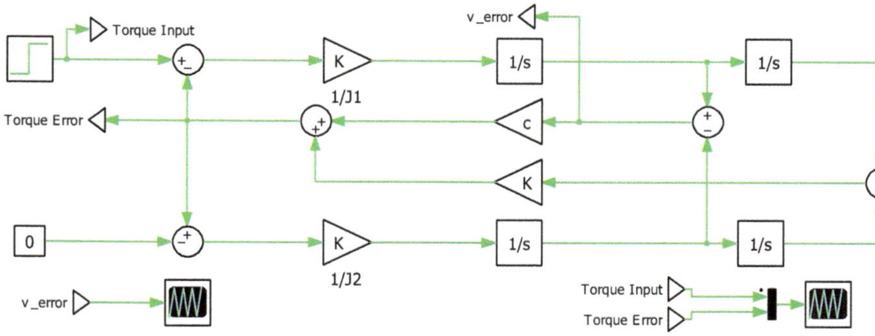

Fig. 4.27 Dynamic model of two-mass drive train

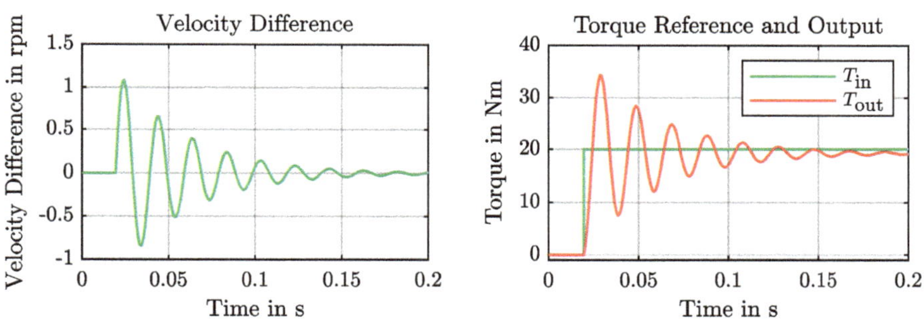

Fig. 4.28 Scope signals from two-mass drive train model

It is also worth noting that under constant flux linkage, step changes in quadrature current set value i_q^* are directly reflected in the torque. Furthermore, the spatial relationship between the flux linkage and current vectors is maintained with changes to the quadrature current which is consistent with the theory presented in Sect. 4.2.2.

4.5.3 Tutorial 4: Drive Dynamics Example

In this tutorial, the dynamic model of a two-mass drive train is examined. The electrical machine and load inertias are chosen to be $J_1 = 0.051\,\mathrm{kg\,m^2}$ and $J_2 = 1.35\,\mathrm{kg\,m^2}$, respectively. The coupling between the two inertias has a torsion coefficient and damping coefficient of $\kappa = 5000\,\mathrm{Nm/rad}$ and $c = 2.5\,\mathrm{Nms/rad}$, respectively. A mechanical shaft torque step of $T_e = 20\,\mathrm{Nm}$ is applied at $t = 0\,\mathrm{s}$. The load torque T_l is assumed to be zero.

A convenient approach to solving this problem is the use of the generic model discussed in Sect. 4.3.4. Simulation results based on the parameters and excitation conditions described above are shown in Sect. 4.3.4.

4.5.4 Tutorial 5: Speed Control Loop Design Example

This tutorial considers the design and implementation of a speed control loop according to the approach discussed in Sect. 4.4. A simulation model according to the drive model shown in Fig. 4.20 is developed in PLECS, as given in Fig. 4.29. The "electrical drive" is modeled by a single saturation block, assuming that an ideal electrical drive generates the reference torque at its output, as long as the reference torque is within its operating range. The mechanical model consists of an integrator with gain $1/J$ and a constant load. A speed feedback loop is created and fed to the PI controller to control the machine speed. A speed reference sets the speed to $n_{\mathrm{m}}^{*} = 1500$ rpm at $t = 200$ ms, then to -1500 rpm at $t = 0.5$ s to observe step responses of the controller. The drive torque limit as well as the inertia and speed loop bandwidth are given in Table 4.3.

The gain K_{p} and K_{i} of the speed controller are calculated using Eq. (4.41), with $\omega_{\mathrm{B}}^{\mathrm{sp}} = 100$ rad/s (see Sect. 4.4), which gives $K_{\mathrm{p}} = 0.5$ Nm/(rad/s) and $K_{\mathrm{i}} = J\omega_{\mathrm{B}}^{\mathrm{sp}2}/4 = 12.5$ Nm/rad, respectively. A first order discretization of Eq. (4.41) is used in this tutorial in accordance with the approach used for the PI current controller for this example, the sampling time for the speed controller is set to $\omega_{T}^{\mathrm{s}} = 1$ ms

The issue of *integrator windup* is explored in detail in this tutorial. The integral part of a PI controller acts against steady-state errors that cannot be compensated by a proportional controller. The output of the integral keeps increasing as long as an error is present at the input of a PI controller. During the ramp up of the machine, a large speed error occurs for a long period compared to the bandwidth of the PI controller. This causes the integrator output to increase far beyond reasonable values even though the drive is already producing maximum torque. This increase of integrator sum is called *integrator windup*.

An *anti-windup PI controller* limits both the state of the internal integrator and the controller output to the drive limits. When proper controller limits are set, the

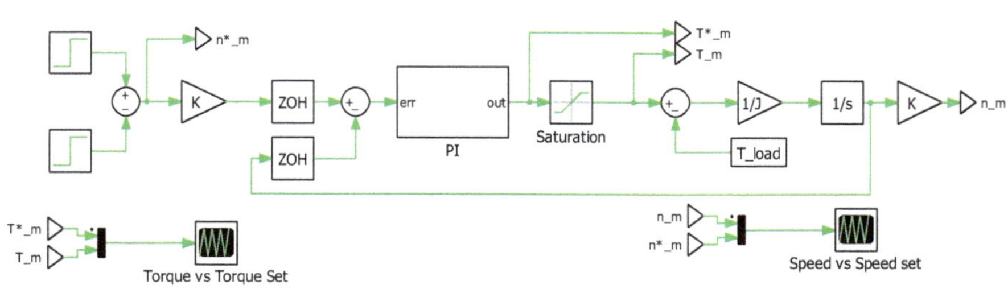

Fig. 4.29 Simulation of speed control example

Table 4.3 Drive parameters

Parameters	Value
Machine torque limits $T_{\mathrm{e}}^{\max}$	± 8 Nm
Inertia J	0.005 kg m^2
Speed loop bandwidth $\omega_{\mathrm{B}}^{\mathrm{sp}}$	100 rad/s

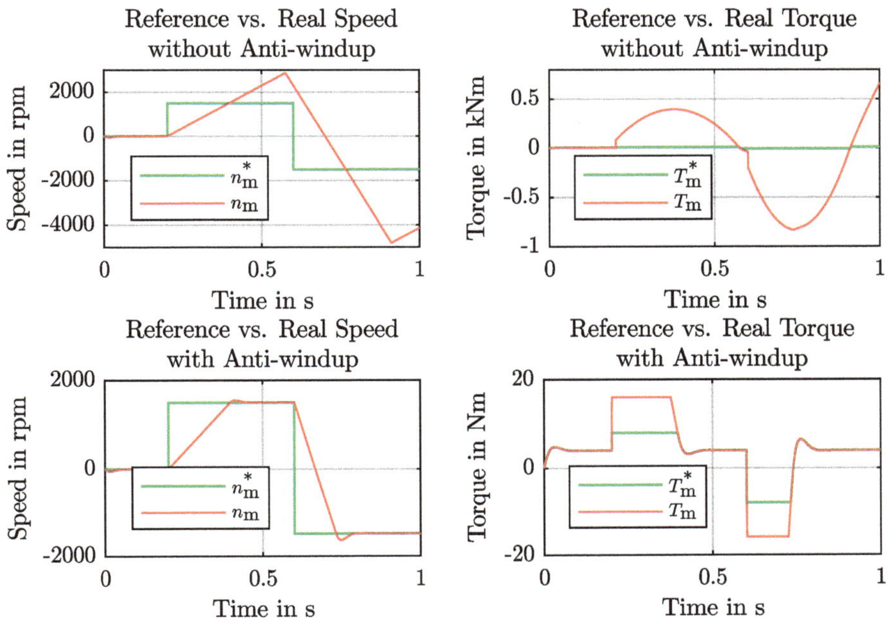

Fig. 4.30 Simulation results for speed and reference torque versus time without and with anti-windup

integrator will stop integrating when it reaches the drive limits. The speed and torque waveforms of two simulations with and without anti-windup are given in Fig. 4.30.

References

1. Dorf RC, Bishop RH (2007) Modern control systems, 11th edn. Prentice Hall, Upper Saddle River
2. Friedland B (2005) Control system design: an introduction to state-space methods. Dover Publications, New York
3. Isermann R (2005) Mechatronic systems: fundamentals, 1st edn. Springer, Berlin
4. Veltman A (1993) The fish method: interaction between AC-machines and switching power converters. PhD Thesis, Delft University
5. Veltman A, Pulle DWJ, De Doncker R (2007) Fundamentals of electrical drives. Spinger, Berlin

Chapter 5
Modeling and Control of DC Machines

The brushed DC machine, which derives its excitation flux from a field winding or permanent magnets, remains commercially relevant in the field of drives. This despite the influx of brushless drive technologies which offer a maintenance-less alternative to the commutator/brush assembly, which is an inevitable component of the brushed DC machine. Well-established motor manufacturing techniques, together with low-complexity power electronic converters, have been instrumental in retaining its popularity in a diverse range of applications. For household goods and automotive products, the use of low cost, brushed permanent magnet motors remains virtually unchallenged. In particular, the brushed DC series machine known as the *universal machine* is widely deployed in domestic appliances and starter motors. In the field of manufacturing automation, small high-dynamic brushed servo drives continue to play an essential role. Furthermore, due to the apparent simplicity, the brushed DC machine can still be found in some medium-power applications where dynamic performance is considered not to be a key issue.

Given these considerations, it is prudent to consider the modeling and control of these machines in some detail. From a didactic perspective, there is decided merit in examining the brushed drive concept first. This approach provides the opportunity to introduce and demonstrate some basic modeling and model based control aspects, which are highly relevant for AC drives that utilize rotating field machines. The inversion of the current source model, an approach which is one of the cornerstones of this book, is initially demonstrated for a brushed DC drive. In addition, attention will be given to the development of control strategies which ensure that drive operation is kept within the drive envelope dictated by voltage or current constraints. In this context, field weakening strategies for separately excited machines are introduced, together with a set of PLECS based tutorials.

R. W. De Doncker et al., *Advanced Electrical Drives*, Power Systems,
https://doi.org/10.1007/978-3-030-48977-9_5

5.1 Modeling of Brushed DC Machines

To give a brief review of DC machine concepts, the cross-sectional views as well as the symbolic and generic models are presented together with the relevant equations. A more extensive examination can be found in [3]. A comprehensive treatment of basic machine concepts for the uninitiated reader is given, for example, in [1] and [2]. Cross-sectional views of two typical one pole pair brushed DC machine examples, namely the separately excited DC machine and the permanent magnet DC machine, are given in Fig. 5.1. Common to both machines is the armature, which is the rotational component of the motor that is linked to the brush/commutator assembly which ensures that the current distribution in said armature is stationary with respect to the αβ coordinate system that in turn is tied to the stator of the machine. The magnetic excitation for both types is noticeable different: the separately excited machine carries a field winding, which implies that the excitation air-gap flux ϕ_f that is aligned with the α axis can be altered using the field current i_f. The high field strength of today's permanent magnet materials allows the permanent magnet machine to be constructed with a smaller outside diameter, which results in more compact machine design as compared to DC machines with excitation windings. However, this advantage of the permanent magnet design comes at the expense of losing one degree of freedom, namely the ability to alter the excitation flux level,which may be required at high speed operation. Prior to presenting a symbolic and generic model of these machines it is helpful to analyze the machine with the aid of the vector diagram shown in Fig. 5.2.

For convenience the armature is assumed to be connected via the brush and commutator assembly to a current source with amplitude i_a. The resultant armature current distribution is such that a positive (i.e., anti-clockwise) torque

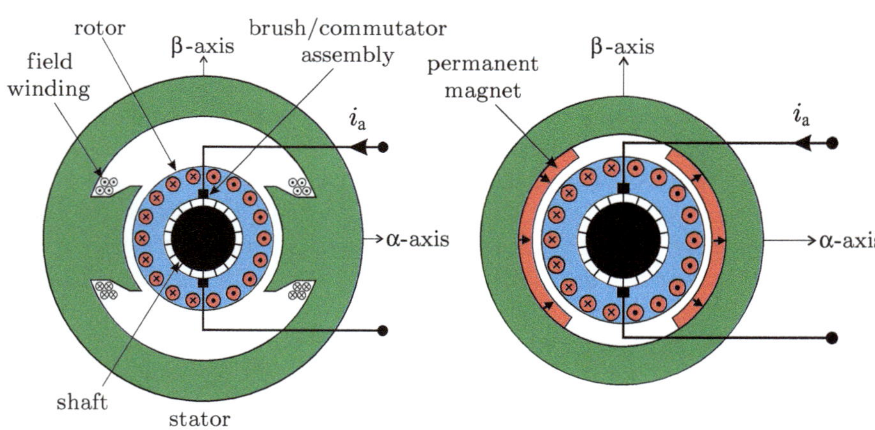

Fig. 5.1 Cross-sectional view of a brushed separately excited and permanent magnet DC machine [4]

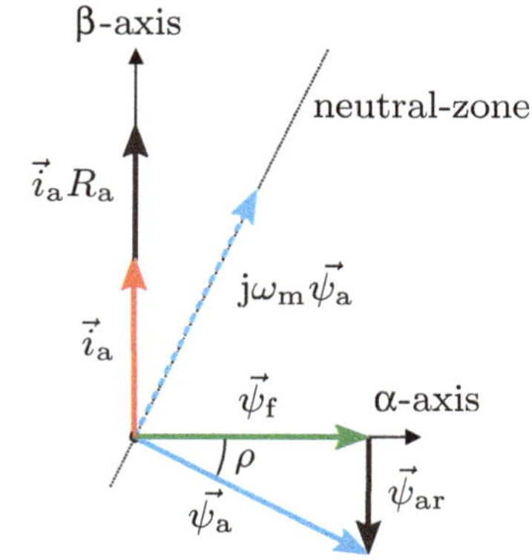

Fig. 5.2 Vector diagram for brushed DC machine with constant armature current i_a

$T_e = \Im\left\{\left(\vec{\psi}_f\right)^* \vec{i}_a\right\}$ is generated due to the interaction between winding current i_a and air-gap (magnetic) flux ϕ_f, which in the armature causes a flux linkage $k\phi_f = \psi_f$. The factor k is a machine design factor, which depends on many parameters, among others, radius of the rotor, air-gap length, number of turns connected in series of the armature winding, number of pole pairs, etc.

In the vector diagram shown in Fig. 5.2, the armature current distribution is represented by the vector $\vec{i}_a$. Note that the current vector is oriented along the *positive* β axis, while the magnetic field $\vec{\psi}_{ar}$ associated with this current distribution is in the *negative* direction of the β axis.

Hence, the total armature flux linkage is represented by the vector $\vec{\psi}_a$ and is displaced by an angle ρ relative to the field flux vector $\vec{\psi}_f$ due to the effect of this so-called *armature reaction*. The magnetic flux associated with the armature current distribution can be expressed as $\vec{\psi}_{ar} = -L_a\vec{i}_a$, where L_a is the armature inductance. The resultant magnetic field seen by the armature is therefore equal to

$$\vec{\psi}_a = \vec{\psi}_f \underbrace{-L_a\vec{i}_a}_{\vec{\psi}_{ar}}. \tag{5.1}$$

For motor operation, as is assumed here, the resultant flux is rotated by an angle $\rho < 0$ which is *in opposition* (i.e., clockwise) to the direction of positive motor rotation. The EMF $\vec{e}_a$ associated with the armature rotating at a speed ω_m is defined as $\vec{e}_a = j\omega_m\vec{\psi}_a$ and is therefore also displaced by the angle ρ from the β axis. This displacement of flux and EMF results in an equal displacement of the so-called neutral zone from the β axis and therefore also from the brush/commutator assembly, where the current reversal in the armature winding takes place.

The effects of this angular displacement of the resultant armature flux due to armature reaction are twofold, namely:

- The neutral zone is no longer aligned with the β axis of the machine and therefore current commutation takes place outside the neutral zone. This can result in severe arcing across the commutator segments due to the so-called *under*

commutation. This refers to the current not being fully reversed after the armature passing through the brush assembly. The resultant current discontinuity will lead to voltage spikes that cause arcing under the brushes [1].
- For machines with poles, the pole regions located in the second and fourth quadrant of the separately excited motor shown in Fig. 5.1 are exposed to a higher flux density, which causes saturation and may cause local higher induced voltages in the conductors that traverse these regions. This in turn can cause arcing across the affected commutator segments.

Due to the effects described above, the effect of armature reaction should be reduced. This can be achieved in three ways, namely:

- By shifting the commutator/brush assembly such that it is aligned with the neutral zone. This approach is used in Fig. 5.4a, which is why the current vector $\vec{i}_a$ is not aligned with the β axis. This method is not ideal as the orientation of the neutral zone is a function of the armature current. Hence, in practice the angle ρ must be set to a value that corresponds to the current associated with the load connected to the machine. Note that for generator operation the angle ρ will be positive.
- For machines with a field winding, a so-called *compensation winding* can be added. These extra windings are located between the poles of the stator and carry the armature current i_a. The flux generated by the compensation windings will be such that it compensates the armature reaction flux $\vec{\psi}_{ar}$ in which case the neutral zone remains aligned with the geometrical β axis. The displacement angle now is $\rho = 0$ and the armature flux is $\vec{\psi}_a = \vec{\psi}_f$. Note that the compensation winding is magnetically coupled with the armature. Due to this mutual coupling the resulting inductance L_a reduces to a leakage inductance. However, adding such compensation windings in the poles is expensive and therefore only used for larger machines and servo machines, i.e., machines that are exposed to fast changing load conditions. Such a machine is shown in Fig. 5.3 (left) and the resulting vector diagram is shown in Fig. 5.4b.
- A set of so -called *interpoles* [1], in effect similar to compensation windings, may be introduced to improve the commutation process. Interpoles are additional poles with windings that also carry the armature current i_a. These poles are positioned on the β axis of the machine and also serve to compensate the armature reaction flux $\vec{\psi}_{ar}$. The cross-section of such a machine is shown in Fig. 5.3 (right).

5.1.1 Symbolic and Generic Model of the Brushed DC Machine

The task of developing a set of representative models of the DC machine with either a field winding or permanent magnets is undertaken with the aid of Fig. 5.4. The underlying assumption in this case is that the machine has sufficient commutator

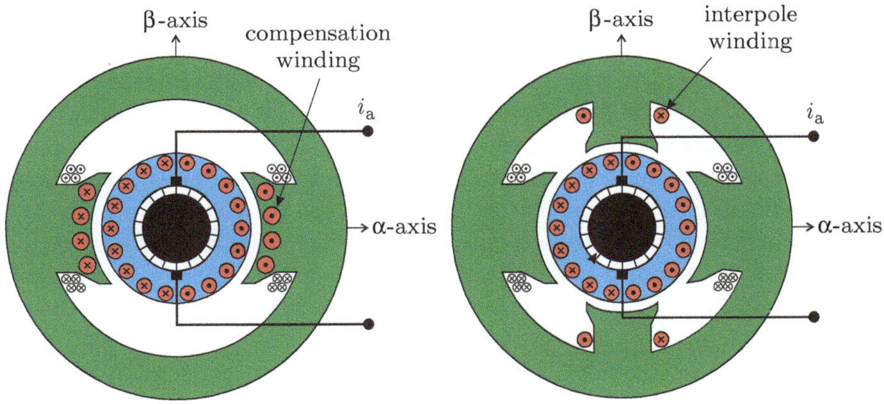

Fig. 5.3 Cross-sectional view of a brushed separately excited DC machine with compensation windings (left) and interpoles (right)

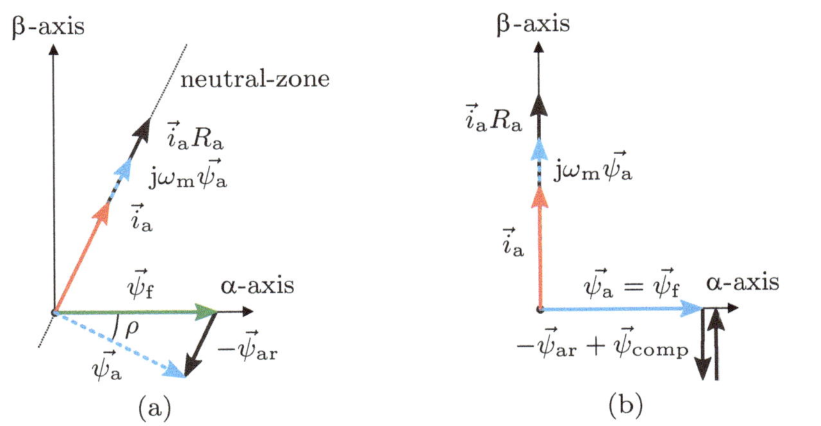

Fig. 5.4 Vector diagram for brushed DC machine with current commutation in the neutral zone. (**a**) Rotated commutator and brush assembly. (**b**) Compensation windings or interpoles

segments to ensure that the current vector $\vec{i}_a$ remains stationary and stays aligned with the neutral zone of the machine. Under these circumstances the scalar voltage/current variables are simply equal to $i_a = \Im\left\{\vec{i}_a\right\}$, $u_a = \Im\{\vec{u}_a\}$ and $e_a = \Im\left\{\vec{e}_a = j\omega_m \vec{\psi}_a\right\}$. The resultant armature flux linkage equals $\psi_a = \Re\left\{\vec{\psi}_a\right\}$, which may also be written as

$$\psi_a = \psi_f \sqrt{1 - \left(\frac{i_a}{i_{sc}}\right)^2} \tag{5.2}$$

where the variable $i_{sc} = \frac{\psi_f}{L_a}$ can be interpreted as the neutral zone armature short-circuit current of the DC machine (neglecting the armature resistance). For machines with permanent magnets this value is constant, however, with separately excited machines the flux ψ_f can be altered by varying the field current, in which case the short circuit current value also changes. In most cases the term $\frac{i_a}{i_{sc}}$ is typically small, in which case the resultant armature flux $\psi_a \approx \psi_f$. Note that for DC machines with compensation or interpole windings $\frac{i_a}{i_{sc}}$ equals zero, i.e., $\psi_a = \psi_f$.

The complete armature voltage equation set, which also includes the voltage across the series connected armature and compensation winding resistances R_a and the inductance L_a in the event of a variable armature current i_a, may be written as

$$u_a = i_a R_a + L_a \frac{di_a}{dt} + e_a \tag{5.3a}$$

$$e_a = \omega_m \psi_f \tag{5.3b}$$

$$T_e = i_a \psi_f \tag{5.3c}$$

$$T_e - T_l = J \frac{d\omega_m}{dt} \tag{5.3d}$$

which is valid under the assumption $0 \leq \frac{i_a}{i_{sc}} < 1$, with e_a the back-EMF (induced voltage), ω_m the rotational speed, T_e the electromagnetic produced torque , T_l the load torque, and J the inertia of the machine. When using SI-units throughout all equations the electromagnetic torque expression (5.3c) can be easily derived from the back-EMF by power balancing, as the electromagnetic air-gap power P_{em} equals the mechanical output power P_m.

$$P_{em} = i_a e_a = T_e \omega_m = P_m. \tag{5.4}$$

Observation of equation set (5.3) shows that the complete generic model of the current source connected DC machine (without the mechanical load equations) can be directly derived without the need for additional modules or use of an IRTF module. The underlying assumption is the presence of a neutral zone brush/commutator assembly or a DC machine with compensation windings. A symbolic representation of the quadrature axis based model is given in Fig. 5.5.

For machines which carry a field winding an additional terminal voltage expression is required, which is of the form

Fig. 5.5 Symbolic field-oriented model: quadrature axis configuration

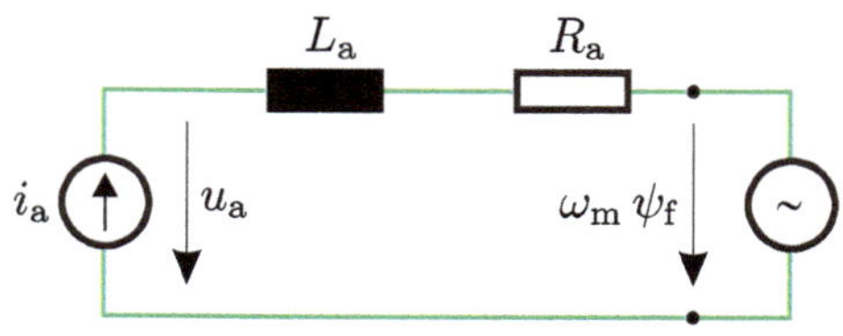

Fig. 5.6 Voltage source model of the brushed DC machine

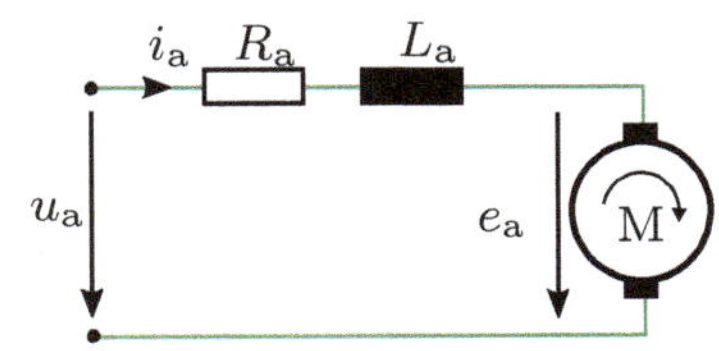

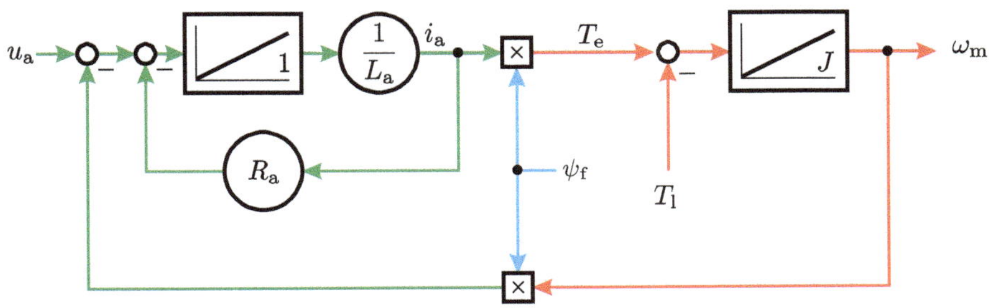

Fig. 5.7 Generic model of a brushed DC machine, connected to a voltage source u_a

Fig. 5.8 Generic model of current source connected separately excited DC machine

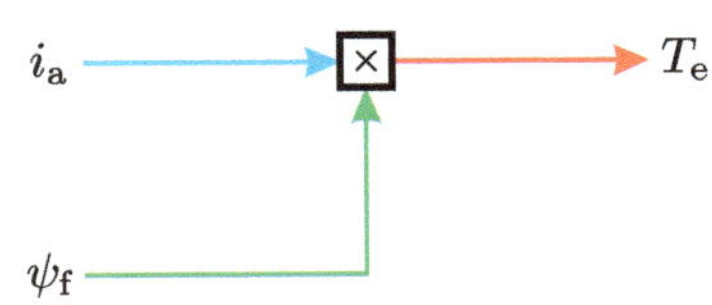

$$u_f = i_f R_f + L_f \frac{di_f}{dt} \tag{5.5}$$

where the variables R_f and L_f represent the field winding resistance and inductance, respectively. Note that typically the relationship between the flux linkage ψ_f and current i_f, referred to as the magnetization characteristic is nonlinear due to stator saturation. Correspondingly, the inductance $L_f = \frac{\psi_f}{i_f}$ is only defined for the initial linear part of the magnetization characteristic. The symbolic model of the brushed DC machine shown in Fig. 5.6 is the embodiment of expressions (5.3a)–(5.3d).

The corresponding generic model of the brushed DC machine as shown in Fig. 5.7 can be readily constructed with the aid of Eq. (5.3), where the excitation flux ψ_f can be provided via permanent magnets or a field winding. In the latter case an additional generic model of the field circuit must be added based on Eq. (5.5) and (if needed) a nonlinear module that represents the magnetization curve $\psi_f(i_f)$.

In some cases a current source based generic model, as shown in Fig. 5.8, may be required which can be readily derived from of Eq. (5.3c).

The relevance of this simple model is not to be underestimated as it shows the importance of using current control with a DC machine. For example, the ability of the machine to respond quickly to a step change in the torque is dictated by the armature inductance and the current controller in use. This will become apparent in the control section of this chapter. Furthermore, there is complete decoupling between the current and field flux. The latter can also be used to control the torque.

However, the time constant associated with controlling this variable is significantly larger than the time constant of the armature circuit (see Fig. 5.7). The time constant linked to the field winding is defined as L_f/R_f as can be deduced from Eq. (5.5). Typically, L_f is much larger than L_a (L_a can be an order of magnitude smaller than L_f in DC servo machines with compensation windings). Therefore, the armature time constant is considerably shorter than the time constant of the field winding. It is noted that current control leads to more efficient drive operation because the dissipative losses are reduced (when compared to a voltage-controlled machine), as may be deduced from the tutorial discussed in Sect. 5.3.1.

5.2 Control of Brushed DC Machines

The interpretation of the term *control* is primarily considered from the perspective of achieving a specified dynamic torque and excitation response. The basic approach envisaged is to *invert* the dynamic current source field-oriented model with the purpose of generating the required reference current(s) for a given user defined torque reference value. A second *control* objective is to integrate the control concept with a voltage source converter, using the current control and modulation strategies discussed in the previous two chapters. In addition to these two *control* objectives, there is a need to consider the optimum use of the drive within the constraints imposed by the maximum DC supply voltage and maximum current of machine or power electronic converter. Henceforth, attention is given to the issue of establishing the optimum performance trajectories and control techniques which may be deployed.

5.2.1 *Controller Concept*

The process of developing a DC drive controller may be initialized by using the generic model given in Fig. 5.8 and reconfiguring this topology with the torque as an input variable. Application of this *inversion* principle, using the generic model, according to Fig. 5.8 shows that it may be reduced to the configuration given in Fig. 5.9.

Input to this model is the required torque reference value T_e^* for the drive and the field flux value ψ_f^* of the DC machine in use. Output of the controller is the current

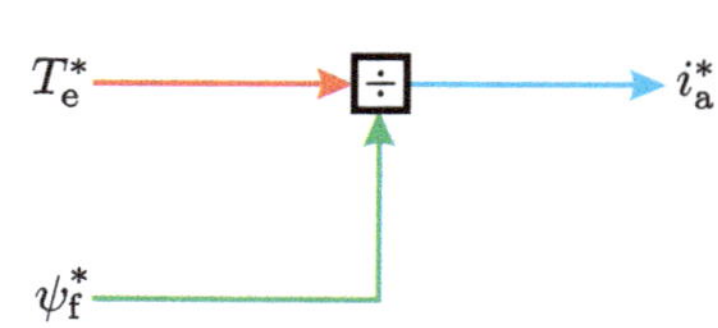

Fig. 5.9 Inverted torque model of current source fed DC machine

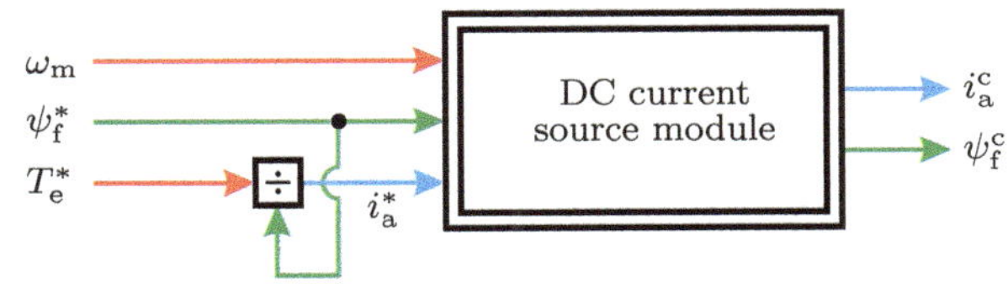

Fig. 5.10 Current source based DC machine control structure

reference value i_a^* which will act as an input for a single-phase current controller (as discussed in Sects. 3.1.1 and 3.1.2) to maintain the condition $i_a = i_a^*$. In permanent magnet DC machines, where the field excitation is provided by magnets, the flux ψ_f will inevitably be constant (at given temperature and assuming no saturation effects). In this case, the armature reference current is the only control variable available for controlling the torque. However, this value must be constrained to ensure that the maximum current limits of the machine or the converter are not exceeded. Consequently, it is prudent to introduce a current control variable i_a^c which will be equal to i_a^*, provided the machine and the converter are operating within their intended design envelopes. When specific drive limits (maximum current or voltage limits of machine or converter) are reached, the controller should act to reduce the value of i_a^c in which case its value will not be equal to i_a^*.

In electrically excited DC machines, the variable ψ_f can be set by the controller. Thereby, an additional degree of control flexibility is provided, which will prove to be beneficial at high speed. When operating within the normal base speed design envelope, the flux value will normally be set to its highest level, designated as ψ_f^{max}, which will ensure that torque production is realized with the lowest armature current level. This approach minimizes the armature copper loss, which will in turn improve the efficiency of the machine. However, in some cases, for example, at high speed or partial load operation, the controller may need to lower the value of the field by introducing a control variable ψ_f^c. The *extended* DC control structure as given in Fig. 5.10 contains an additional (compare to Fig. 5.9) control module which generates the required control variables i_a^c and ψ_f^c, which are successively used by the armature current controller and field excitation converter (if present).

The introduction of the DC control unit simply ensures that the machine can be optimally utilized within its steady-state operational limitations.

5.2.2 *Operational Drive Boundaries*

The discussion on operational drive boundaries is undertaken with the aid of a steady-state equation set that may be directly derived using Eq. (5.3) and setting the time derivative terms to zero, which gives

$$\bar{u}_a = \bar{i}_a R_a + \bar{e}_a \tag{5.6a}$$

$$\bar{e}_a = \bar{\omega}_m \bar{\psi}_f \tag{5.6b}$$

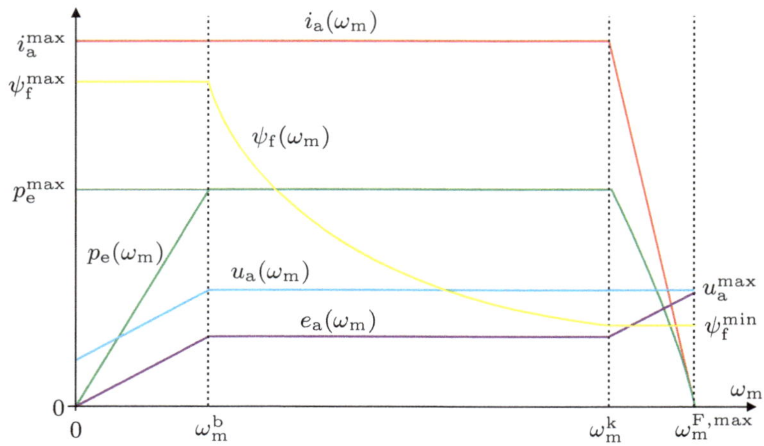

Fig. 5.11 Operational drive limits of the DC drive

$$\bar{T}_e = \bar{i}_a \,\bar{\psi}_f \tag{5.6c}$$

$$\bar{T}_e = \bar{T}_l \tag{5.6d}$$

where the variable notation $\bar{x}$ represents the steady-state (average) value of a time dependent variable x. Furthermore, a current-controlled DC drive is assumed, which is constrained by the maximum armature voltage u_a^{max}, armature current i_a^{max}, and field or permanent magnet flux ψ_f^{max}. The operational drive boundaries for the DC drive as indicated in Fig. 5.11 show how the steady variables armature current, armature voltage, EMF, and flux (for machines with a field winding) must be chosen as function of the steady-state shaft speed $\bar{\omega}_m$. Observation of Fig. 5.11 shows that there are three operating regions for the current-controlled DC drives, namely:

- *Base speed* region, speed range $0 \to \omega_m^b$: its upper limit is set by the conditions

$$\bar{i}_a = i_a^{max} \tag{5.7a}$$

$$\bar{\psi}_f = \psi_f^{max}. \tag{5.7b}$$

The corresponding steady-state equation set for this operating region may be written as

$$\bar{u}_a = i_a^{max} \,R_a + \bar{e}_a \tag{5.8a}$$

$$\bar{e}_a = \bar{\omega}_m \,\psi_f^{max} \tag{5.8b}$$

$$\bar{T}_e = \bar{i}_a \,\psi_f^{max} \tag{5.8c}$$

$$\bar{p}_e = \bar{T}_e \,\bar{\omega}_m \tag{5.8d}$$

where $\bar{p}_{\mathrm{e}}$ represents the steady-state output power of the DC machine. Analysis of Eq. (5.8) yields that (at constant current) both output power and EMF increase linearly with the shaft speed, as may also be observed from Fig. 5.11. Both armature current and flux are kept constant, hence the shaft torque for this region remains constant. Therefore, this operating range is also known as the *constant torque* or *base speed* range. The armature voltage increases at the same rate as the EMF, and the limit speed of the base speed region, known as the base shaft speed $\omega_{\mathrm{m}}^{\mathrm{b}}$, is reached when the armature voltage u_{a} reaches the maximum armature voltage u_a^{max}. At the base speed the output power of the drive reaches its rated value $p_{\mathrm{e}}^{\max} = i_{\mathrm{a}}^{\max}\psi_{\mathrm{f}}^{\max}\omega_{\mathrm{m}}^{\mathrm{b}}$, where the base speed value may be calculated using Eqs. (5.8a), (5.8b) which gives

$$\omega_{\mathrm{m}}^{\mathrm{b}} = \frac{u_{\mathrm{a}}^{\max} - i_{\mathrm{a}}^{\max} R_{\mathrm{a}}}{\psi_{\mathrm{f}}^{\max}}. \tag{5.9}$$

For PM brushed DC machines Eq. (5.9) represents the highest operating speed under full load condition, i.e., with the armature current at its maximum value. The theoretical upper speed limit $\omega_{\mathrm{m}}^{\max}$ for the PM machine is found by using Eq. (5.9) with the condition $i_{\mathrm{a}}^{\max} = 0$, which gives

$$\omega_{\mathrm{m}}^{\max} = \frac{u_{\mathrm{a}}^{\max}}{\psi_{\mathrm{f}}^{\max}}. \tag{5.10}$$

Equation (5.10) simply states that the highest possible speed of machines without field weakening capabilities (such as PM machines) is achieved when the EMF is equal to the maximum armature voltage $u_{\mathrm{a}}^{\max}$.

- *Field weakening* region, speed range $\omega_{\mathrm{m}}^{\mathrm{b}} \rightarrow \omega_{\mathrm{m}}^{\mathrm{k}}$: for operation in this region the following conditions apply

$$\bar{i}_{\mathrm{a}} = i_{\mathrm{a}}^{\max} \tag{5.11a}$$

$$\bar{u}_{\mathrm{a}} = u_{\mathrm{a}}^{\max}. \tag{5.11b}$$

Given these constraints, the steady-state equation set may be written as

$$u_{\mathrm{a}}^{\max} = i_{\mathrm{a}}^{\max} R_{\mathrm{a}} + \bar{e}_{\mathrm{a}} \tag{5.12a}$$

$$\bar{e}_{\mathrm{a}} = \bar{\omega}_{\mathrm{m}} \bar{\psi}_{\mathrm{f}} \tag{5.12b}$$

$$\bar{\psi}_{\mathrm{f}} = \psi_{\mathrm{f}}^{\max} \left(\frac{\omega_{\mathrm{m}}^{\mathrm{b}}}{\bar{\omega}_{\mathrm{m}}}\right) \tag{5.12c}$$

$$\bar{T}_{\mathrm{e}} = i_{\mathrm{a}}^{\max} \bar{\psi}_{\mathrm{f}} \tag{5.12d}$$

$$\bar{p}_{\mathrm{e}} = p_{\mathrm{e}}^{\max} \tag{5.12e}$$

where $\bar{p}_{\mathrm{e}}$ represents the steady-state electromagnetic output power of the DC machine. Analysis of Eq. (5.12) learns that the output power level remains at its rated level p_e^{max}, which is why this boundary and the region below it is also referred to as the *constant power* speed range. The armature voltage remains at its rated value, which implies that field weakening of the flux is required as function of speed, as indicated by Eq. (5.12c). Hence, flux, torque, and EMF are inversely proportional to the shaft speed in this region, as may also be observed from Fig. 5.11.

As shaft speed increases, the field flux $\bar{\psi}_{\mathrm{f}}$ decreases. In DC machines without compensation windings this affects the orientation of the brush/commutator neutral zone as may be shown with the aid of Fig. 5.4a. The orientation of the brush assembly is typically chosen such that it is aligned with the neutral zone of the machine under rated conditions, i.e., with $\bar{i}_{\mathrm{a}} = i_{\mathrm{a}}^{\mathrm{max}}$ and $\bar{\psi}_{\mathrm{f}} = \psi_{\mathrm{f}}^{\mathrm{max}}$, which corresponds to a commutator angle of

$$\rho_{\mathrm{nom}} \approx -\arctan\left(\frac{i_{\mathrm{a}}^{\mathrm{max}} L_{\mathrm{a}}}{\psi_{\mathrm{f}}^{\mathrm{max}}}\right). \tag{5.13}$$

The effect of field weakening is to *reduce* the flux, hence the neutral zone will rotate, in the opposite direction of the shaft speed (motoring operation assumed), which implies that the commutator angle will increase in the *negative* direction. This means that commutation effects such as commutator arcing can occur, and that these effects will progressively increase as speed increases and flux weakens. For this reason motor manufacturers will usually set a minimum flux level $\psi_{\mathrm{f}}^{\mathrm{min}}$ or highest allowable shaft speed $\omega_{\mathrm{m}}^{\mathrm{k}}$. The relation between these two variables can, with the aid of Eq. (5.12c), be expressed as

$$\omega_{\mathrm{m}}^{\mathrm{k}} = \frac{\psi_{\mathrm{f}}^{\mathrm{max}}}{\psi_{\mathrm{f}}^{\mathrm{min}}} \omega_{\mathrm{m}}^{\mathrm{b}}. \tag{5.14}$$

- *Maximum speed* region, speed range $\omega_{\mathrm{m}}^{\mathrm{k}} \rightarrow \omega_{\mathrm{m}}^{\mathrm{F,max}}$: for operation in this region the following conditions apply:

$$\bar{i}_{\mathrm{a}} = i_{\mathrm{a}}^{\mathrm{max}} \left(\frac{\bar{\omega}_{\mathrm{m}} - \omega_{\mathrm{m}}^{\mathrm{F,max}}}{\omega_{\mathrm{m}}^{\mathrm{k}} - \omega_{\mathrm{m}}^{\mathrm{F,max}}}\right) \tag{5.15a}$$

$$\bar{\psi}_{\mathrm{f}} = \psi_{\mathrm{f}}^{\mathrm{min}}. \tag{5.15b}$$

Given these constraints, the steady-state equation set may be written as

$$\bar{u}_{\mathrm{a}} = \bar{i}_{\mathrm{a}} R_{\mathrm{a}} + \bar{e}_{\mathrm{a}} \tag{5.16a}$$

$$\bar{e}_{\mathrm{a}} = \bar{\omega}_{\mathrm{m}} \psi_{\mathrm{f}}^{\mathrm{min}} \tag{5.16b}$$

$$\bar{T}_{\mathrm{e}} = i_{\mathrm{a}}^{\max} \psi_{\mathrm{f}}^{\min} \left(\frac{\bar{\omega}_{\mathrm{m}} - \omega_{\mathrm{m}}^{\mathrm{F,max}}}{\omega_{\mathrm{m}}^{\mathrm{k}} - \omega_{\mathrm{m}}^{\mathrm{F,max}}} \right) \tag{5.16c}$$

$$\bar{p}_{\mathrm{e}} = \bar{T}_{\mathrm{e}} \, \bar{\omega}_{\mathrm{m}}. \tag{5.16d}$$

The maximum speed $\omega_{\mathrm{m}}^{\mathrm{F,max}}$ obtainable under minimum flux conditions is found using Eqs. (5.16b), (5.10) and the condition $\bar{e}_{\mathrm{a}} = u_{\mathrm{a}}^{\max}$ (which corresponds to $\bar{i}_{\mathrm{a}} = 0$). This leads to

$$\omega_{\mathrm{m}}^{\mathrm{F,max}} = \omega_{\mathrm{m}}^{\max} \left(\frac{\omega_{\mathrm{m}}^{\mathrm{k}}}{\omega_{\mathrm{m}}^{\mathrm{b}}} \right) \tag{5.17}$$

. Note that the speed $\omega_{\mathrm{m}}^{\max}$ represents the highest achievable shaft speed of the machine operating at no-load with a fixed field flux set to $\psi_{\mathrm{f}}^{\max}$. Use of field weakening allows a substantial increase in the operation speed range of the drive as defined by the ratio $\left(\frac{\omega_{\mathrm{m}}^{\mathrm{k}}}{\omega_{\mathrm{m}}^{\mathrm{b}}} \right)$.

Particular care should be taken to ensure that the mechanical and electrical limitations specified by the motor manufacturer are not exceeded when operating beyond the base operating speed $\omega_{\mathrm{m}}^{\mathrm{b}}$ of the drive. For these and safety reasons, also separately excited DC machines with compensation windings require a minimal $\psi_{\mathrm{f}}^{\min}$ to avoid runaway beyond a prescribed $\omega_{\mathrm{m}}^{\mathrm{F,max}}$.

5.2.3 *Use of a Current Source DC Model with Model Based Current Control*

The process of integrating the control concepts with the electrical drive may be initially undertaken with the aid of a current source model of the DC machine. This approach, which is also used for synchronous and induction machine drives, is effective as long as the dynamics of the current controller do not influence the drive. Ignoring (initially) the dynamics and implementation details of the current controller is a valid simplification because the electrical time constant associated with current control is usually much shorter than those linked with the mechanical side of the drive (at least for machines which utilize a compensation winding). A generic representation of the current-controlled DC machine (see Figs. 5.7 and 5.8), with a control structure as given by Fig. 5.10, is shown in Fig. 5.12. This controller has torque T_{e}^{*} as input control variable. Outputs of the DC control module are the current reference $i_{\mathrm{a}}^{\mathrm{c}}$ and field flux reference value $\psi_{\mathrm{f}}^{\mathrm{c}}$. For the current based approach considered in this section, the current reference value $i_{\mathrm{a}}^{\mathrm{c}}$ will be equal to the current in the machine. Likewise, the field flux ψ_{f} is deemed to be equal to the flux reference $\psi_{\mathrm{f}}^{\mathrm{c}}$. In this example the number of magnetic pole pairs has been set to p, given

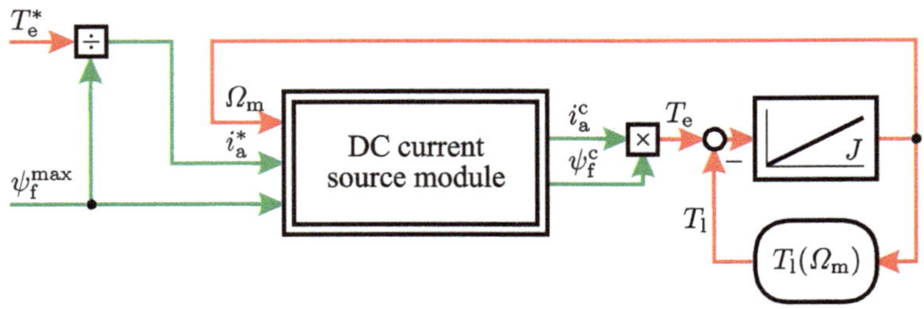

Fig. 5.12 Model based controller of DC drive with current source machine model

that most practical DC machines have multiple poles. For didactical reasons, the theoretical analysis in this chapter has been presented using two poles, i.e., $p = 1$.

In this drive controller example, the shaft speed, which is considered from a control perspective a disturbance, is assumed to be measurable, i.e., a speed or position encoder is attached to the shaft, as shown in Fig. 5.12. Consequently, the load characteristics are decoupled from the controller. This approach may not be viable in some applications, either because only one shaft end is available (for the load $T_l\,(\Omega_m)$), or because robustness and reliability considerations may preclude the use of a shaft sensor, in which case position sensorless techniques must be used. In the accompanying tutorial given in Sect. 5.3.2, a drive model according to Fig. 5.7 which exemplifies the controller concepts outlined in this and the previous section is discussed.

5.2.4 Use of a Voltage Source DC Model with Model Based Current Control

In modern drives voltage source power converters are used as discussed in Sect. 2.3. In terms of current control, a hysteresis type controller as discussed in Sect. 3.1.1, or a model based current regulator approach (see Sect. 3.1.2) may be used. Today, regularly sampled controllers are most commonly used, which is why the model based current control technique discussed in the previous section is adopted in this section. A voltage source connected model is required which can be derived from the stator voltage equation (5.3a). The complete electrical drive connected to a mechanical load $T_l\,(\Omega_m)$, with the DC controller as shown in Fig. 5.13, is shown in Fig. 5.14. It brings together a range of concepts introduced in this and the previous two chapters.

The required proportional and integral gain values are calculated using Eq. (3.8), where the variables L and R represent the armature inductance and armature resistance, respectively. In addition to the above, it is favorable to generate the term $u_e = \psi_f^c\,\omega_m$, i.e., the induced voltage or back-EMF of the machine, for the current controller (see Eq. (3.7)). This is realized by making use of the DC controller

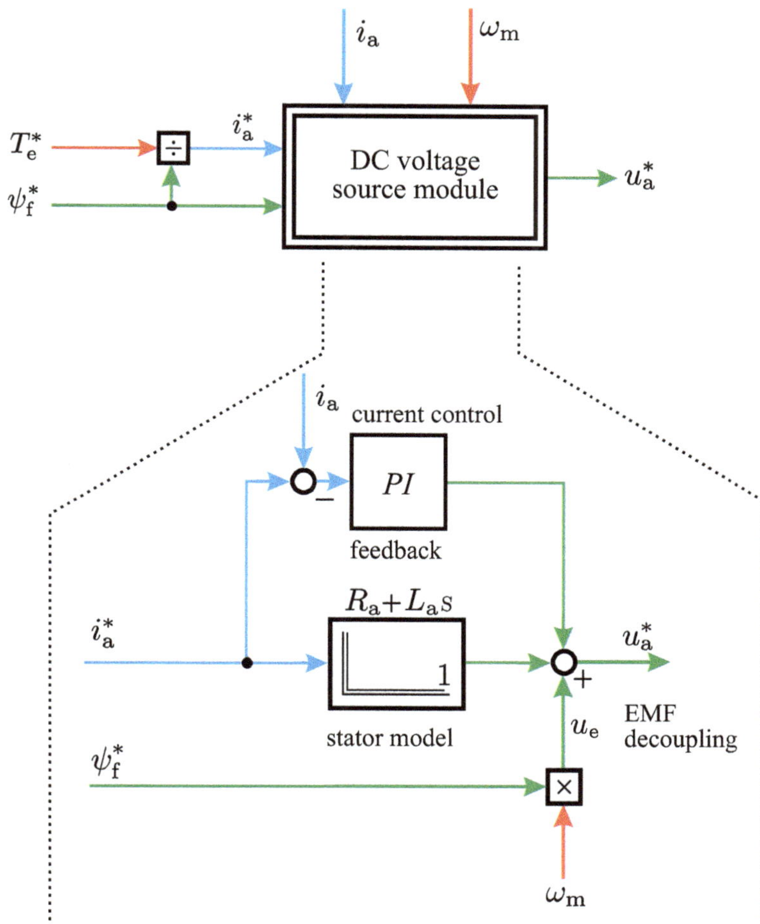

Fig. 5.13 Voltage source based DC machine control structure

field flux value ψ_f^c and measured shaft speed ω_m. In most cases the controller flux reference ψ_f^c is used and converted to a field current reference value i_f^c, which requires knowledge of the inverse magnetization characteristic $i_f^c(\psi_f^c)$. A second current controller is then used to control the field current in the machine. A tutorial is given in Sect. 5.3.3 which gives the reader the opportunity of examining in detail the waveforms and variables introduced in the drive model discussed in this section. Note that the effects of converter *dead time*, as mentioned in Sect. 2.5, are not included in these models.

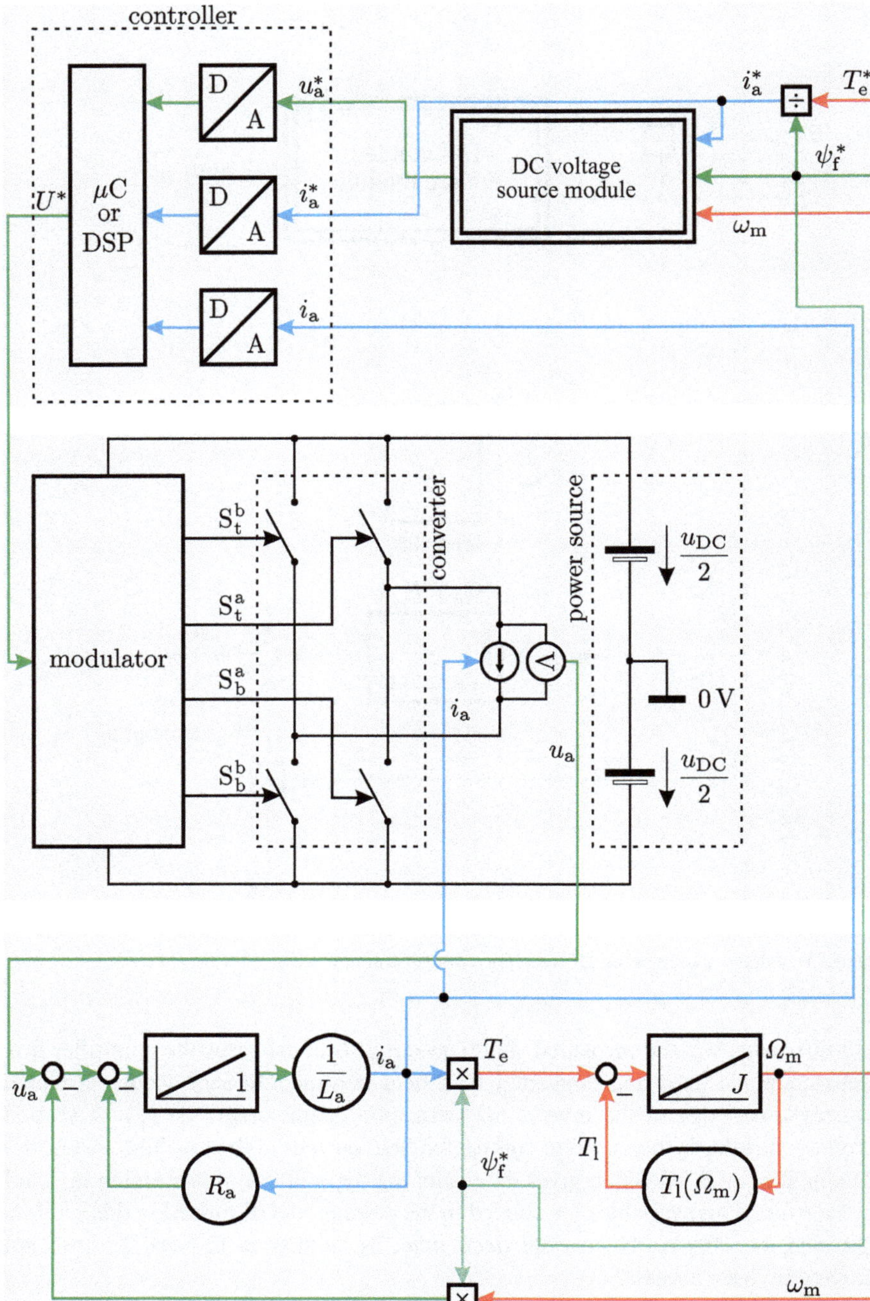

Fig. 5.14 DC drive with model based control and voltage source based machine model

5.3 Tutorials

5.3.1 *Tutorial 1: Modeling of a Current and Voltage Source Connected Brushed DC Motor*

This tutorial is concerned with comparing a voltage-controlled and identical current-controlled brushed DC machine, with a view to determine which is preferable in terms of operating efficiency. For this purpose, a machine with parameters according to Table 5.1 is to be deployed. It is to be used under no-load conditions, i.e., no external mechanical load is connected.

A voltage step of $u_a^* = 80\,\mathrm{V}$ is to be applied at $t = 10\,\mathrm{ms}$ to the voltage source connected machine which is assumed to be initially at standstill. The no-load shaft speed ω_m^v can be obtained at $t = 200\,\mathrm{ms}$, which represents the time mark where steady-state operation is achieved for the machine in question. The no-load machine shaft speed ω_m^v of the voltage source connected machine is to be used to calculate the required armature current i_a^* which must be applied at $t = 10\,\mathrm{ms}$ to the current source connected machine to achieve the same shaft speed as the voltage source connected machine at $t = 200\,\mathrm{ms}$. Determine for both machines the dissipated energy over the designated run period and on the basis of these results, identify which mode of operation is preferred in terms of operating efficiency. To obtain an answer to the posed question, a voltage source based generic model according to Fig. 5.7 may be used as shown in the simulation given in Fig. 5.15.

In terms of the current-controlled model topology, we use Fig. 5.8, together with the appropriate mechanical modules needed to obtain the shaft speed ω_m. Computation of the required current i_a^* needed for the current source connected machine to reach the shaft speed ω_m^v may be undertaken with the aid of Eq. (5.3d), with $T_l = 0$ and Eq. (5.3c), which gives

$$i_a^* = \frac{J\,\omega_m^v}{\psi_f\,\Delta T} \tag{5.18}$$

where ΔT represents the run time of machine, which in this case, equals $\Delta T = 190\,\mathrm{ms}$. The energy dissipation during the run-up interval may be found by evaluation of the integral $E_d = \int R_a\, i_a^2\, \mathrm{d}t$. This task is carried out in the simulation via the sub-module named *Energy module*.

Table 5.1 DC machine parameters

Parameters	Value
Armature inductance L_a	0.05 H
Armature resistance R_a	10 Ω
Field flux ψ_f	1.0 Wb
Inertia J	0.005 kg m^2
Initial rotor speed ω_m^o	0 rad/s

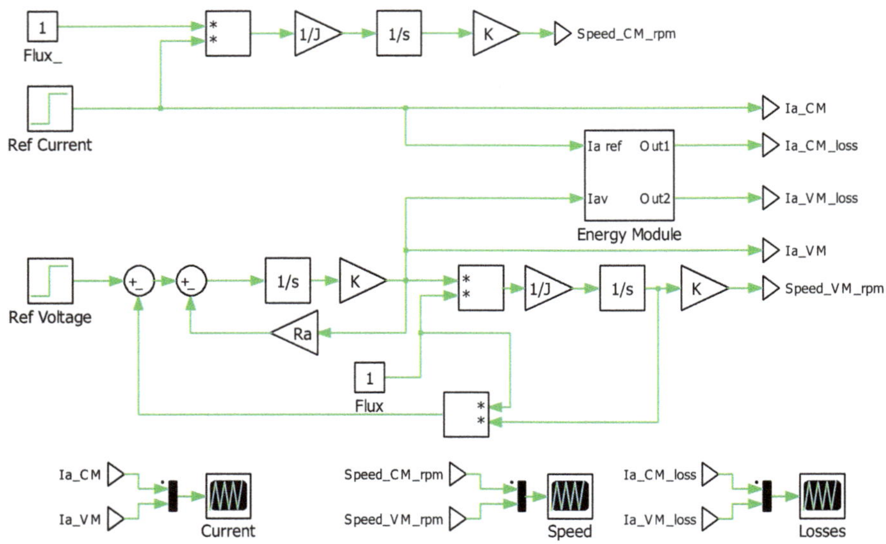

Fig. 5.15 PLECS simulation model of current and voltage source connected brushed DC motor

The results obtained with the aid of the simulation, as given in Fig. 5.15, show the shaft speed, current, and energy dissipated for both machines. Observation of Fig. 5.16 shows that both machines reach the same shaft speed of $\omega_{\mathrm{m}}^{\mathrm{v}} = 751\,\mathrm{rpm}$, which is approximately equal to the theoretical no-load speed of $\omega_{\mathrm{m}}^{\mathrm{o}} = u_{\mathrm{a}}^{*}/\psi_{\mathrm{f}}$, namely $\omega_{\mathrm{m}}^{\mathrm{o}} = 763\,\mathrm{rpm}$. A comparison between the dissipated energy levels of the two machines at the end of the operating sequence demonstrates that the current source connected machine is able to reach the required operating speed with a reduction in dissipated energy of 46% relative to using a constant voltage-source connected machine. Apart from improved dynamic performance, this is another important reason, why current control of electrical machines is the preferred control technique in drive applications.

5.3.2 Tutorial 2: Current Source Connected Brushed DC Motor with Field Weakening Controller

This tutorial aims to examine the functioning of a current controller which operates in accordance with the control laws discussed in Sect. 5.2.2. For this purpose the brushed DC machine, as introduced in Sect. 5.3.1, is to be connected to a DC current controller module which requires as inputs the reference armature current i_{a}^{*} and shaft speed ω_{m}. Control input for the drive should be a torque reference value which should be adjustable between the limits $\pm T_{\mathrm{e}}^{\max}$. The operating drive parameters for the drive are given in Table 5.2.

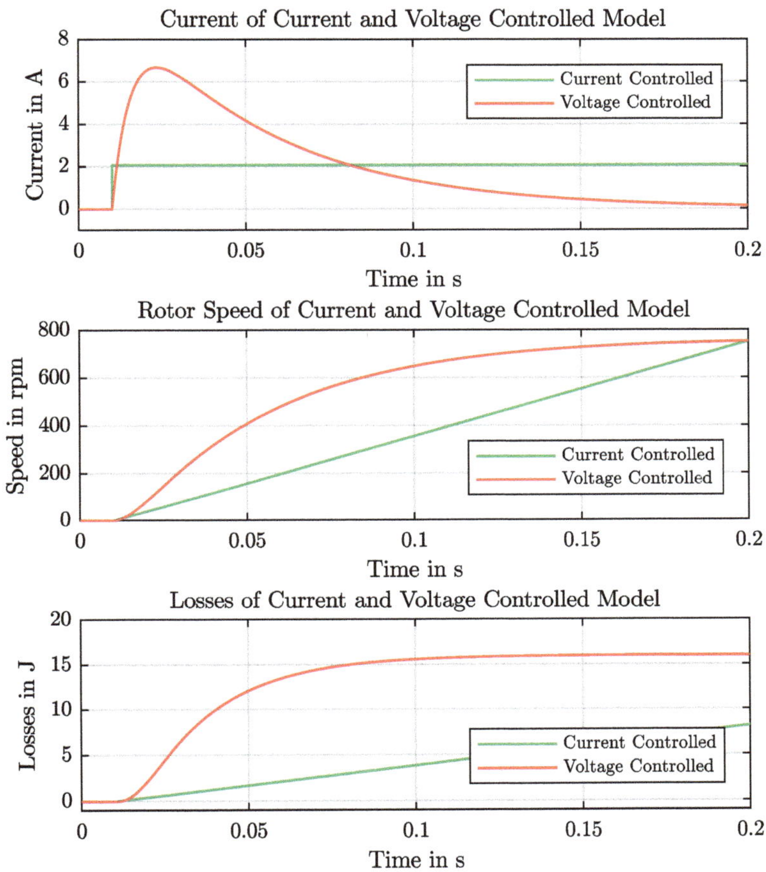

Fig. 5.16 Voltage (red) and current (green) controlled DC machines: armature current i_{a}, mechanical speed ω_{m}, and dissipated energy E_{d} during run-up from standstill

Table 5.2 DC drive parameters

Parameters	Value
Maximum armature current $i_{\mathrm{a}}^{\mathrm{max}}$	8.0 A
Maximum supply voltage $u_{\mathrm{a}}^{\mathrm{max}}$	200.0 V
Maximum field flux $\psi_{\mathrm{f}}^{\mathrm{max}}$	1.0 Wb

Calculate the base speed $\omega_{\mathrm{m}}^{\mathrm{b}}$ of the drive, using the data provided in Tables 5.1 and 5.2. Observe the shaft torque and shaft speed of the drive. A quadratic load torque versus speed characteristic is assumed. This load produces a torque of 30 Nm at 3000 rpm.

Calculation of the base speed may be undertaken with the aid of Eq. (5.9), which shows that its value is equal to $n_{\mathrm{m}}^{\mathrm{b}} = 1146\,\mathrm{rpm}$. The simulation model as given in Fig. 5.17 shows the current model of the machine together with the current controller. The torque limits of this drive lie within the range $\pm 8\,\mathrm{Nm}$. The DC

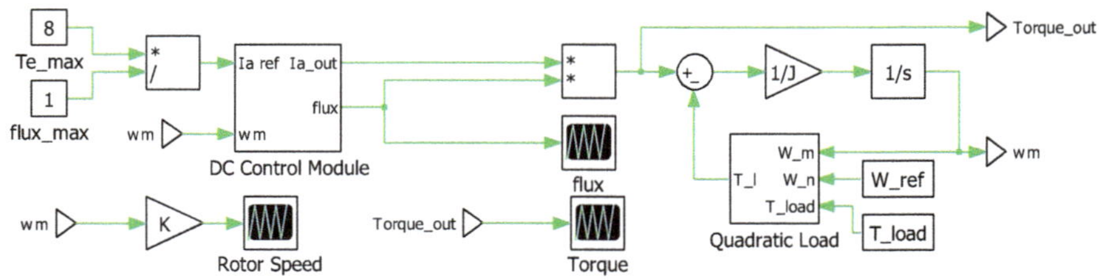

Fig. 5.17 Current source connected brushed DC motor with field weakening controller

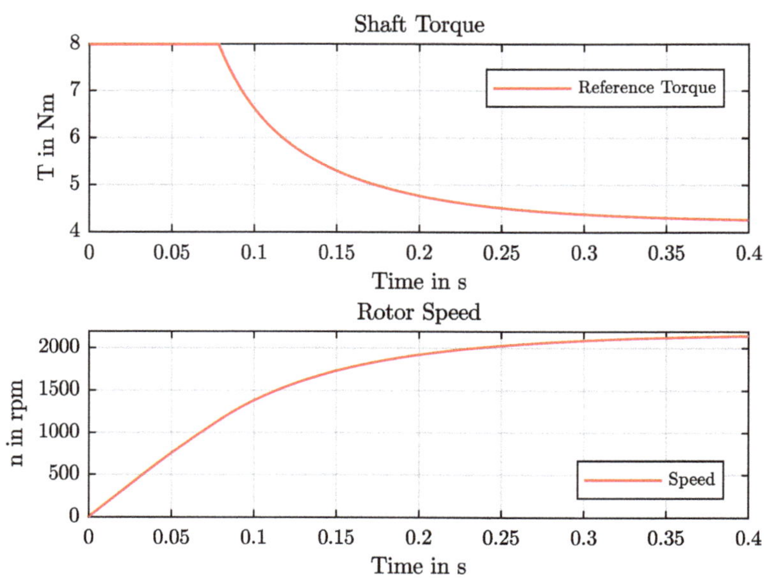

Fig. 5.18 Simulation results for current source connected DC drive with field weakening

drive module main outputs are the field flux $\psi_{\mathrm{f}}^{\mathrm{c}}$ and armature current $i_{\mathrm{a}}^{\mathrm{c}}$, which in this example serve as inputs to the current based model (ideal current control is assumed). The controller module controls the flux in accordance with the control law defined by Eq. (5.12c). Note that the *over-speed* mode of operation is not implemented in the DC controller module used in this tutorial.

The results given in Fig. 5.18 show the shaft torque and shaft speed of the drive for an operating speed above and below the base speed of $n_{\mathrm{m}}^{\mathrm{b}} = 1146\,\mathrm{rpm}$.

These results clearly show that field weakening is active above the base speed of the drive. For completeness the maximum operating speed of the drive operating without field weakening may also be computed using Eq. (5.10), this value is equal to $n_m^{max} = 1910\,\mathrm{rpm}$.

5.3.3 Tutorial 3: DC Drive Operating Under Model Based Current Control and a Field Weakening Controller

A single-phase model based current controller is to be considered with a full-bridge converter/modulator topology as discussed in tutorial Sect. 3.3.2. The controller is connected to a brushed DC machine. The machine parameters and operating drive voltage/current limits to be used are in accordance with those discussed in the previous tutorials. Asymmetric pulse width modulation with a sampling time of $T_s = 0.5$ ms is to be used, which corresponds with a PWM carrier frequency of 1 kHz. Modify the model given in tutorial Sect. 3.3.2 in accordance with the DC drive concept presented in Sect. 5.2.4. In the context of this exercise, determine the new gains settings for the current controller and plot the reference torque and machine torque in a single diagram as well as the shaft speed over the simulation period to be specified. An anti-windup PI controller is to be used, for which the output limit values must be found. The machine is deemed to operate without any external mechanical load. A torque reference step of $T_e^* = 4$ Nm (50% rated torque) is applied at $t = 10$ ms and the control value is reversed at $t = 100$ ms. Examine the behavior of the drive over a simulation period of $T = 200$ ms.

An example of a simulation model, as given in Fig. 5.19, shows the required controller structure with the PI controller and back-EMF term u_e (marked as U_e in Fig. 5.19) as calculated using the measured shaft speed and DC controller field flux value ψ_f^c (marked as flux_out in Fig. 5.19). Note, that in this example, the proportional gain $L/T_s + R/2$ of the controller is dominated by the inductance term given that $L/R > T_s/2$. Hence, its value is equal to $K_p = 105.0$, while the bandwidth value was found to be $\omega_i = 200$ rad/s. A discrete anti-windup PI controller is introduced which has as inputs the disturbance decoupling signal and reference/measured current signals. The output voltage limits for this controller should be set to the maximum average voltage per sample values that may be realized. In this case the limits should be set to ±200 V given that the supply voltage is set to 200 V.

The IGBT based H-bridge four quadrant DC-DC converter shows a circuit-generic model interface unit that provides the voltage input to the generic model of the separately excited brushed (with ideal commutator) DC machine. Output of the model is the armature current i_a, that is fed back to the circuit model and the current controller, as may be observed from Fig. 5.19.

The resulting waveforms obtained from the simulation are collected with the aid of scope modules as shown in Fig. 5.20. Observation of Fig. 5.20 shows that the drive is able to deliver the required torque response. A torque ripple component due to the use of a regularly sampled switched converter can be observed. The maximum operating shaft speed, shown in Fig. 5.20, is in this example below the base speed. This implies that the DC controller maintains the field flux ψ_f^c at its maximum value of 1 Wb for the entire simulation. The reader is encouraged to reconsider this tutorial with a hysteresis type current controller, as discussed in Sect. 3.3.1.

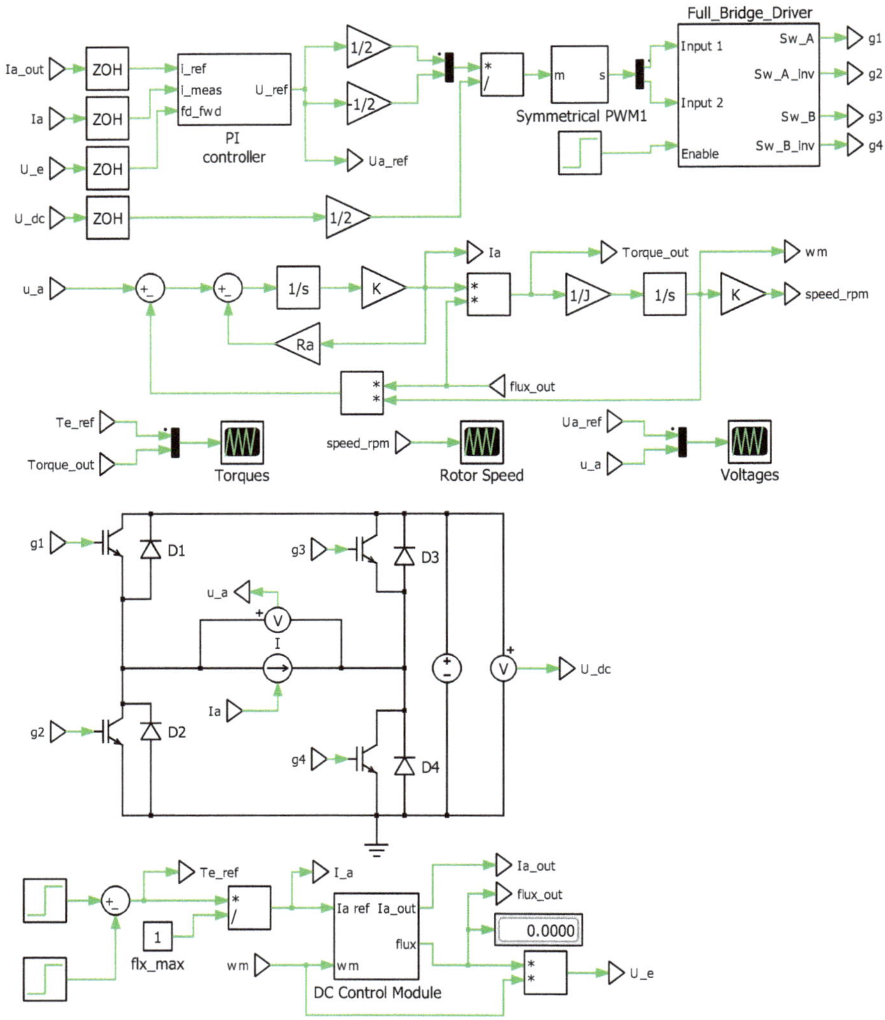

Fig. 5.19 Simulation of current-controlled brushed DC motor with field weakening controller

5.3.4 *Tutorial 4: DC Drive with Model Based Current Control and Shaft Speed Control Loop*

This tutorial aims to extend the previous model by introducing an outer shaft speed control loop as discussed in Sect. 4.4. For this purpose, add a discrete *anti-windup* PI speed controller to the drive and calculate the required proportional gain and time constant values. Set the output limits of the speed controller to $\pm T_{\text{e}}^{\max}$. Choose a simulation run time of 0.5 s and apply a step $n_{\text{m}}^{\text{ref}} = 0 \rightarrow 1500\,\text{rpm}$ at $t = 10\,\text{ms}$. Reverse the speed reference at $t = 0.25$ s and plot the measured and reference shaft

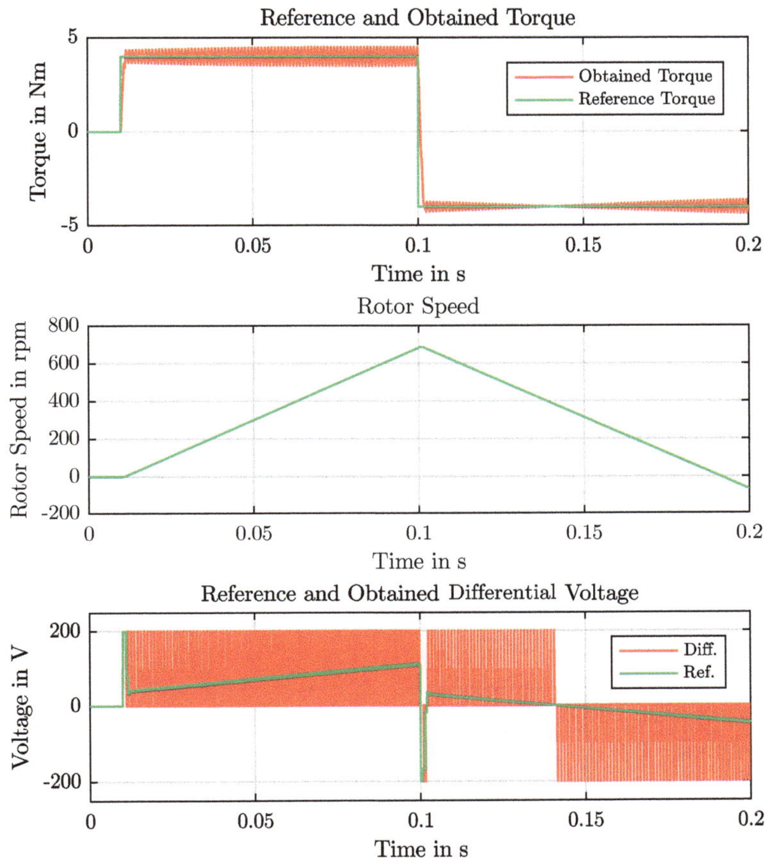

Fig. 5.20 Simulation results for DC drive with model based current control

speed, reference armature voltage and flux over the specified time interval. Likewise, plot the measured and reference torque versus time waveforms. The remaining parameters and simulation setting remain unchanged.

The simulation model, shown in Fig. 5.21, contains a speed control loop that features a discrete anti-windup PI control module, which has as inputs the sampled reference speed ω_{m}^{*} and sampled measured shaft speed ω_{m}. The gain K_{p} and time constant τ_{i} of the speed controller are calculated using Eq. (4.39), with $\omega_{\mathrm{B}}^{\mathrm{sp}} = 100\,\mathrm{rad/s}$, as mentioned in Sect. 4.4. The resulting speed gain and time constant values, for the given drive parameters J and T_{s}, are $K_{\mathrm{p}} = 0.5\mathrm{Nm\,s/rad}$ and $\tau_{\mathrm{i}} = 0.04\,\mathrm{Nm/rad}$, respectively. The values found are identical to those found in the tutorial shown in sect. 4.5.4, with the important change that the speed controller variables are now defined in terms of gain and bandwidth $\omega_i = \frac{1}{\tau_i}$ (rad/s).

The results from the simulation as given in Fig. 5.22 show the reference armature voltage, reference field flux, reference/actual shaft speed, and torque over a 0.5 s interval. The results, shown in Fig. 5.22, provide a good overview of drive control. Recall that the base operating speed of this drive is equal to $n_{\mathrm{m}}^{\mathrm{b}} = 1146rpm$,

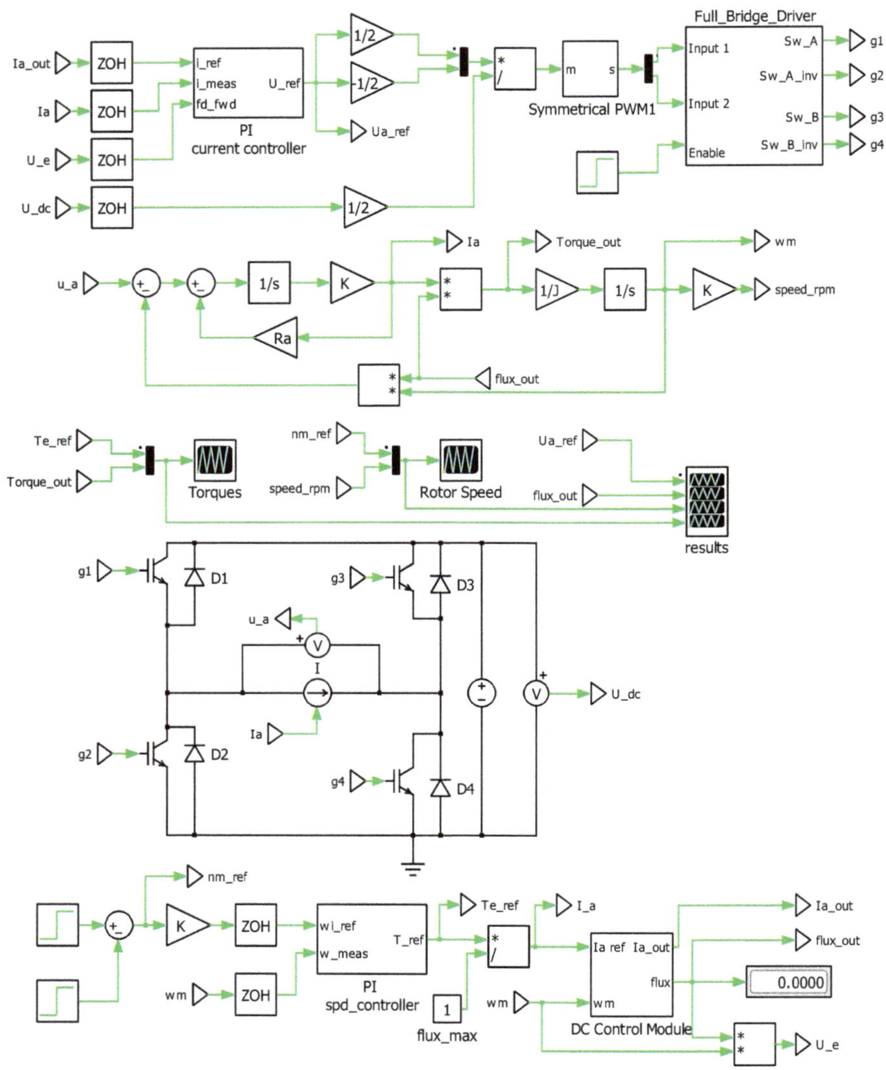

Fig. 5.21 Simulation of DC drive with model based current control and shaft speed control loop

which coincides with time mark I at $t = 0.0846$ s in Fig. 5.22. For the time interval $< 0t < 0.0846$ the drive operates in the base speed mode, with the field at its maximum value. When the drive reaches its base speed (which coincides with maximum armature voltage and maximum current), field weakening starts and this mode is present in the time range between time mark I and II, which is at $t = 0.277$ s. Note that during the torque reversal at $t = 0.25$ s the drive is still in field weakening mode hence the actual torque will be less than the reference value, gives that the flux is lower than rated. During the time period $0.277 < t < 0.427$ s the

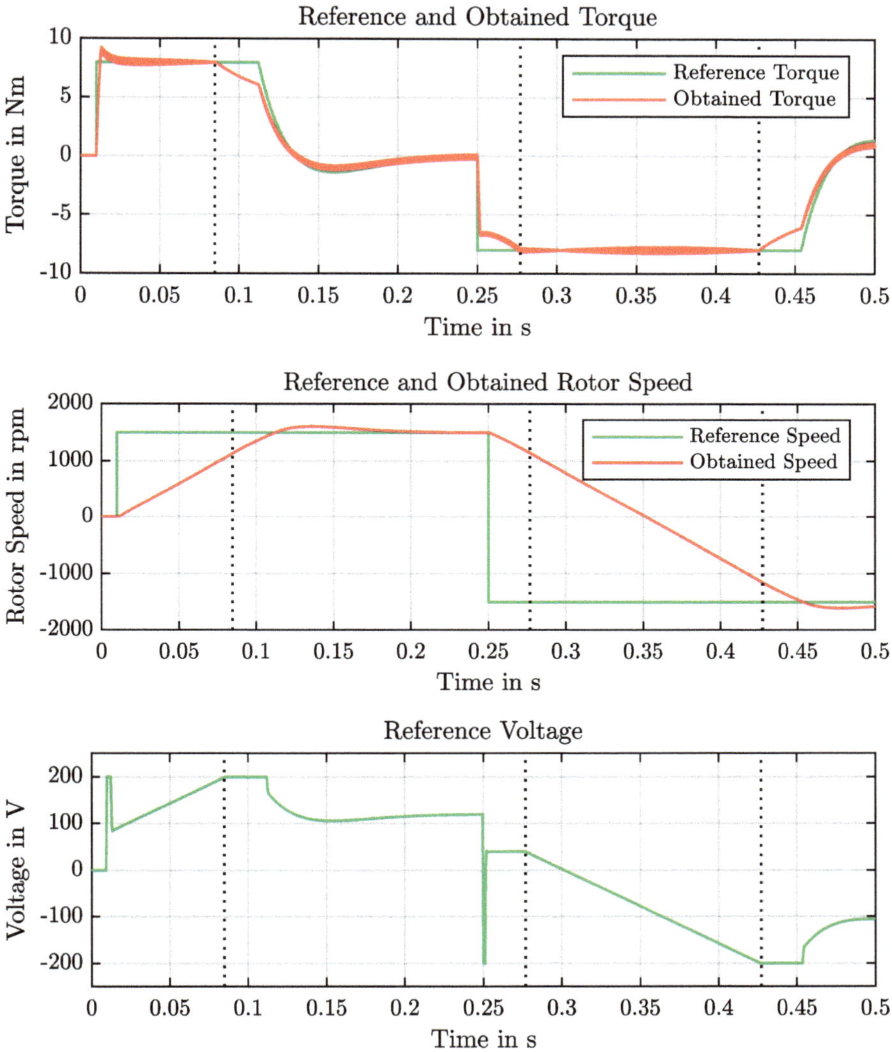

Fig. 5.22 Simulation results for DC drive with model based current control and shaft speed control loop

drive is once again operating in base speed model and for $t > 0.427$ field weakening is again active, which is why the actual torque becomes less than the reference value for part of this region.

References

1. Hughes A (2006) Electric motors and drives: fundamentals, types and applications, 3rd edn. Newnes, Oxford
2. Toliyat HA, Kliman GB (2004) Handbook of electric motors, 2nd edn. CRC Press, Boca Raton
3. Veltman A, Pulle DWJ, De Doncker R (2007) Fundamentals of electrical drives. Spinger, Berlin
4. Veltman A, Pulle DW, Doncker RWD (2016) Fundamentals of electrical drives. Springer International Publishing, Berlin. https://doi.org/10.1007/978-3-319-29409-4

Chapter 6
Synchronous Machine Modeling Concepts

In this chapter, attention is given to modeling three-phase AC synchronous machines with quasi-sinusoidally distributed stator windings. The excitation may be provided by a rotor winding which is connected via a brush/slip-ring set to an electrical excitation source [1–3]. Alternatively, permanent magnets which produce a quasi-sinusoidal magnetic flux density distribution in the air-gap can be used. Both non-salient and salient IRTF based machine models are considered in this chapter from a dynamic and quasi-steady-state perspective.

The field-oriented modeling approach, briefly introduced in Chap. 4, is extended to encompass a rotor flux oriented model of the non-salient and salient machine. Such an approach is deemed to be important in the context of the *model inversion* control principle embraced in this book. In contrast to the previous chapter, the modeling and control sections have been separated to facilitate readability. Subsequent to this chapter, a comprehensive set of tutorials is introduced to provide the reader with the opportunity to interactively examine the key concepts discussed in this chapter.

6.1 Non-salient Machine

In this section, a brief review of the symbolic and generic models and the relevant equations of a *non-salient* synchronous machine is given. Furthermore, attention is given to the steady-state characteristics of a voltage source connected machine. For a detailed discussion the reader may refer to our book Fundamentals of Electrical Drives [4].

In a synchronous machine, the excitation may be provided by permanent magnets or by a rotor based excitation winding which carries a field current i_f. In both cases, the flux density distribution due to the excitation is assumed to be sinusoidal. Furthermore, it is assumed that the magnetizing inductance is equal along both axes

R. W. De Doncker et al., *Advanced Electrical Drives*, Power Systems,
https://doi.org/10.1007/978-3-030-48977-9_6

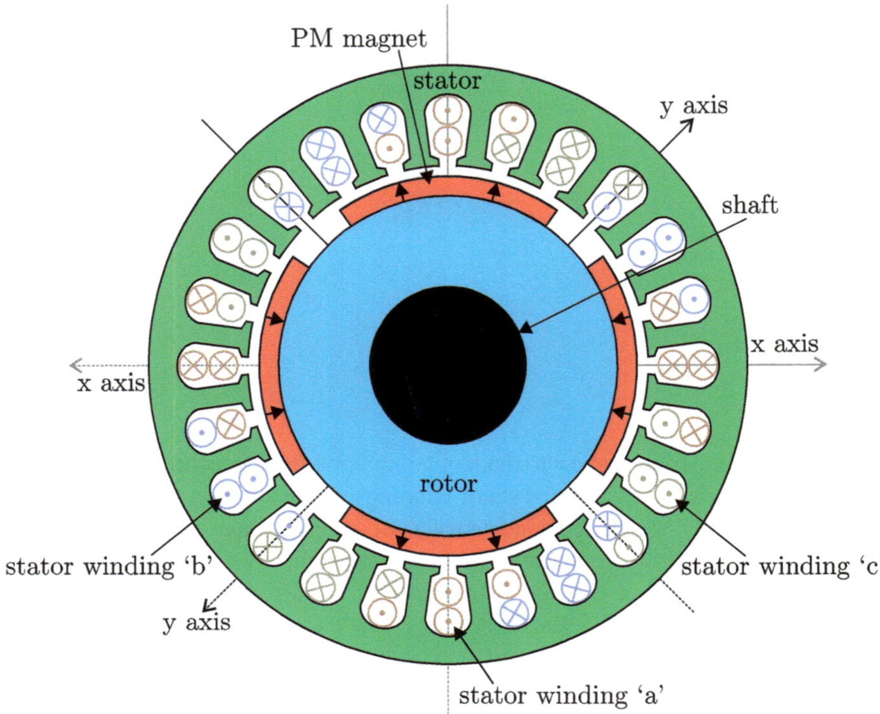

Fig. 6.1 Non-salient four-pole PM synchronous machine, showing a two-layer, three-phase stator winding and surface magnets on the rotor

of the xy plane that is linked to the rotor. A cross-sectional view of a four-pole PM non-salient machine is given in Fig. 6.1. It shows the stator and the rotor with a set of surface mount magnets. Note the presence of a dual set of xy axes, because a four-pole machine is shown in Fig. 6.1.

The machine does not carry any damper windings given that these are normally not found in inverter fed servo drive applications (to avoid losses) which are predominantly considered in this book.

6.1.1 Symbolic Model of a Non-salient Machine

The machine can be described by an IRTF based model as shown in Fig. 6.2. The model was derived from the elementary model introduced in Sect. 4.2.1 by accommodating the stator resistance R_s and the stator leakage inductance $L_{\sigma s}$. The magnetizing inductance L_m is shown on the rotor side of the machine. In general, inductances of non-salient machines may be placed on either side of the IRTF module. For salient machines the magnetizing inductance depends on the orientation

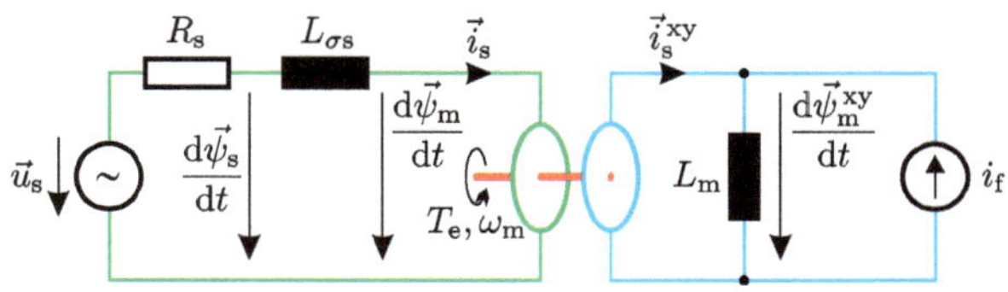

Fig. 6.2 Non-salient synchronous machine

relative to the rotor and thus it is prudent to place it on the rotor side. For this reason, it is beneficial to locate the magnetizing inductance of the non-salient machine on the rotor side.

The equation set which corresponds to this machine is given by

$$\vec{u}_s = R_s \vec{i}_s + \frac{d\vec{\psi}_s}{dt} \tag{6.1a}$$

$$\vec{\psi}_s = L_{\sigma s} \vec{i}_s + \vec{\psi}_m \tag{6.1b}$$

$$\vec{\psi}_m^{xy} = L_m \left(\vec{i}_s^{xy} + i_f \right). \tag{6.1c}$$

In the symbolic model in Fig. 6.2, a linear relationship $\psi_f = L_m i_f$ between the field flux linkage ψ_f and excitation current i_f is assumed, as was done earlier in Sect. 4.2.1. Note that in case of saturation, ψ_f is a nonlinear function of current i_f which can be experimentally derived from the open circuit (back-EMF) stator voltage, i.e. when i_s equals zero.

Using the transformation $\vec{A} = \vec{A}^{xy} e^{j\theta}$, Eq. (6.1b) can be written in rotor coordinates $\vec{\psi}_s^{xy} = L_{\sigma s} \vec{i}_s^{xy} + \vec{\psi}_m^{xy}$. Combining this equation with Eq. (6.1c), grouping the two inductances $L_{\sigma s}$ and L_m and replacing $L_m i_f$ by ψ_f, the stator flux linkage can be written as

$$\vec{\psi}_s^{xy} = \underbrace{(L_{\sigma s} + L_m)}_{L_s} \vec{i}_s^{xy} + \psi_f. \tag{6.2}$$

The sum of the two inductances in Eq. (6.2) is the stator inductance L_s. Equation (6.2) is the general flux linkage relationship including the field current i_f. In case of permanent magnet excitation, the flux linkage ψ_f is converted to an equivalent current $i_f = \psi_f / L_m$, because a flux linkage source cannot be modeled in the symbolic model directly. This equation will be used in the development of the generic model in the next section.

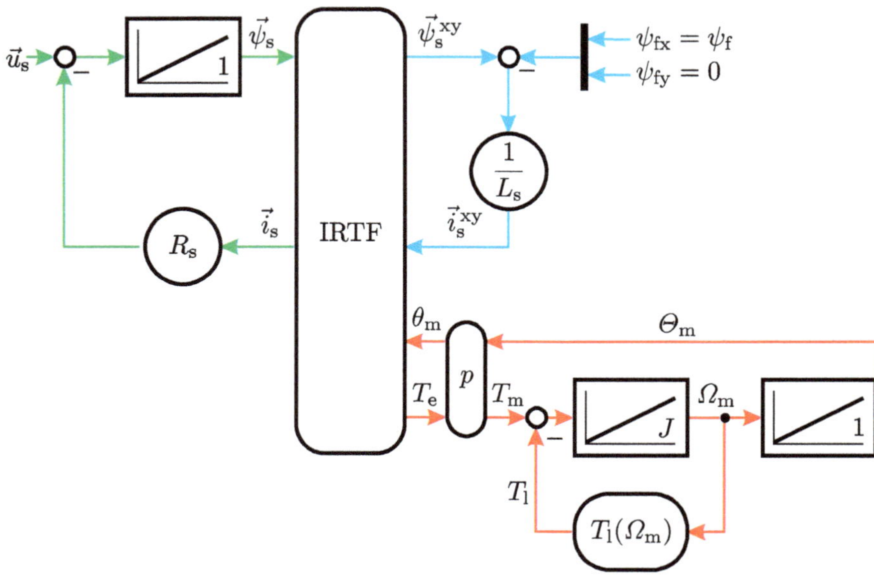

Fig. 6.3 Generic model of a synchronous machine which corresponds with equation set (6.1) and Eq. (6.2)

6.1.2 Generic Model of Non-salient Synchronous Machines

A generic model of the non-salient synchronous machine can be built by either using a current based or a flux based IRTF model. In this section, a flux based IRTF model is used. The generic machine model is directly found by using Eq. (6.1a) at the stator side and Eq. (6.2) at the rotor side of the IRTF. The resulting model is shown in Fig. 6.3.

At the stator side, an integrator is used to calculate the stator flux linkage $\vec{\psi}_s$. This integrator must be provided with an initial value $\vec{\psi}_s(t = 0) = \psi_f e^{j\theta}$ in the (normal) event that excitation of the machine is provided at the start of machine operation.

The machine torque can be derived from IRTF equation (4.5), using the (amplitude invariant) space vectors of the stator flux linkage $\vec{\psi}_s$ and the stator current $\vec{i}_s$ resulting in $\vec{T}_e = \frac{3}{2}\vec{\psi}_s \times \vec{i}_s$. The use of Eq. (6.1b) with said torque equation gives

$$\vec{T}_e = \frac{3}{2}\left(\vec{\psi}_m \times \vec{i}_s + L_{\sigma s}\underbrace{\vec{i}_s \times \vec{i}_s}_{=0}\right) \tag{6.3}$$

of which the second term equals zero. This implies that the torque is not affected by the leakage inductance $L_{\sigma s}$. Hence, either the stator flux linkage vector $\vec{\psi}_s$ or the magnetizing flux linkage vector $\vec{\psi}_m$ can be used for torque calculation. Also shown in Fig. 6.3 is a mechanical load module which represents the relationship between load torque and shaft speed. A tutorial based on the generic model shown in Fig. 6.3 is discussed in Sect. 6.3.1.

6.1.3 Rotor-Oriented Model: Non-salient Synchronous Machine

In the machine models discussed previously, the current, voltage, and flux linkage space vectors were defined with respect to a stationary or a shaft oriented reference frame. Note that the *direct* and *quadrature* axis terminology is normally used to represent a complex plane which is oriented with respect to a designated flux linkage vector.

In this section, a so-called *rotor-oriented* dq-transformation is introduced where the stator and rotor based space vector equations are tied to the rotor flux linkage vector. For synchronous machines, this designated flux linkage vector is normally chosen to be the field flux linkage vector $\vec{\psi}_{\mathrm{f}}^{\mathrm{xy}} = \vec{\psi}_{\mathrm{f}}^{\mathrm{dq}} = \psi_{\mathrm{f}}$, which is aligned with the x-axis as shown earlier in Sect. 4.2. Hence, for the synchronous machine, the complex plane formed by the direct and quadrature axes is aligned with the rotor based xy complex plane.

The development of a rotor-oriented model is important for creating a field-oriented control for the non-salient synchronous machine as will be done in the next chapter. The equation set for the non-salient rotor flux based model may be written as

$$\vec{u}_{\mathrm{s}}^{\mathrm{dq}} = R_{\mathrm{s}}\vec{i}_{\mathrm{s}}^{\mathrm{dq}} + \frac{\mathrm{d}\vec{\psi}_{\mathrm{s}}^{\mathrm{dq}}}{\mathrm{d}t} + \mathrm{j}\omega_{\mathrm{s}}\vec{\psi}_{\mathrm{s}}^{\mathrm{dq}} \tag{6.4a}$$

$$\vec{\psi}_{\mathrm{s}}^{\mathrm{dq}} = L_{\mathrm{s}}\vec{i}_{\mathrm{s}}^{\mathrm{dq}} + \psi_{\mathrm{f}} \tag{6.4b}$$

$$T_{\mathrm{e}} = \frac{3}{2}\psi_{\mathrm{f}}\, i_{\mathrm{sq}}. \tag{6.4c}$$

The coordinate transformation which leads to a symbolic and generic representation of the rotor-oriented model is initiated by transforming Eq. (6.1a) in a rotor-oriented dq-reference frame using $\vec{A} = \vec{A}^{\mathrm{dq}}\mathrm{e}^{\mathrm{j}\theta}$. Due to the coordinate transformation, a term $\mathrm{j}\omega_{\mathrm{s}}\vec{\psi}_{\mathrm{s}}^{\mathrm{dq}}$ appears in the voltage Eq. (6.4a) as a result of using the product rule on $\mathrm{d}(\vec{\psi}_{\mathrm{s}}\mathrm{e}^{\mathrm{j}\theta})/\mathrm{d}t$, with $\mathrm{d}\theta/\mathrm{d}t = \omega_{\mathrm{s}}$. The flux linkage expression (6.4b) is unchanged from Eq. (6.2), apart from replacing the superscript, given that the xy and dq coordinate reference frames are fully aligned in this case, as may be observed from Fig. 6.4. The general torque expression $\vec{T}_{\mathrm{e}} = \frac{3}{2}\vec{\psi} \times \vec{i}$ along with $\vec{\psi} = \vec{\psi}_{\mathrm{s}}^{\mathrm{dq}}$ and $\vec{i} = \vec{i}_{\mathrm{s}}^{\mathrm{dq}}$ leads directly to Eq. (6.4c).

Figure 6.4 graphically depicts the process of constructing the stator flux linkage vector based on a given stator current $\vec{i}_{\mathrm{s}}^{\mathrm{dq}}$, as described by Eq. (6.4b). Note that, in case of permanent magnet machines, the stator current vector $\vec{i}_{\mathrm{s}}^{\mathrm{dq}}$ normally may

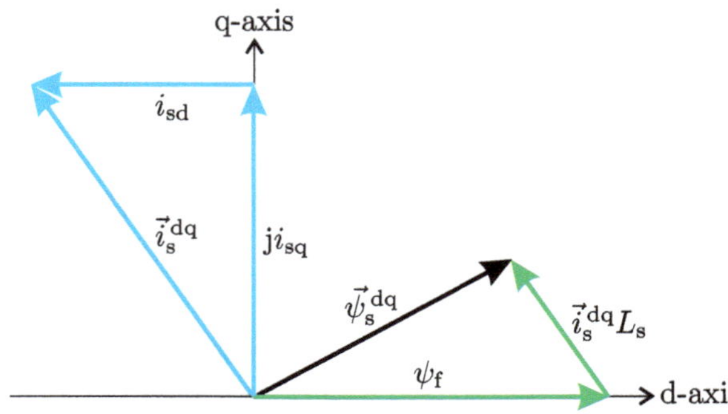

Fig. 6.4 Vector diagram with direct and quadrature axis for non-salient machine

have a negative direct current component for reasons that will become apparent in the next chapter.

The process of finding a symbolic rotor flux model representation may be further pursued by combining the voltage and flux linkage expressions shown in equation set (6.4). Separating the real and imaginary components leads to

$$u_{sd} = R_s i_{sd} - \omega_s L_s i_{sq} + L_s \frac{di_{sd}}{dt} \tag{6.5a}$$

$$u_{sq} = R_s i_{sq} + \omega_s L_s i_{sd} + L_s \frac{di_{sq}}{dt} + \omega_s \psi_f. \tag{6.5b}$$

Equation set (6.5) shows the existence of cross-coupling between the phases. For example, the q-axis current i_{sq} causes a d-axis voltage component u_{sd}. One important feature of a machine control is to compensate or to decouple the cross-coupling, as will be described in the next chapter. From equation set (6.5), a symbolic representation for the rotor flux based model can be derived, as shown in Fig. 6.5.

A current based generic model for a rotor flux based machine, which corresponds to Fig. 6.5, is given in Fig. 6.6 and directly derived from the torque equation (6.4c). From Eq. (6.4c), it can be seen that in a flux based reference system, torque is produced by the i_{sq} component of the current, allowing a simple torque control similar to the DC-machine without compensation winding. Indeed the stator voltage equation (Eq. (6.5b)) shows that any change of stator current i_{sq} requires a stator voltage that depends on the synchronous inductance L_s of the machine.

6.1.4 *Steady-State Analysis*

A brief overview of the steady-state operation of non-salient machines is presented here. The reader is referred to [4] for a more extensive analysis. The aim of this section is to derive an operational diagram referred to as *Blondel diagram* which can be used to show the impact of changing the mechanical load on the current

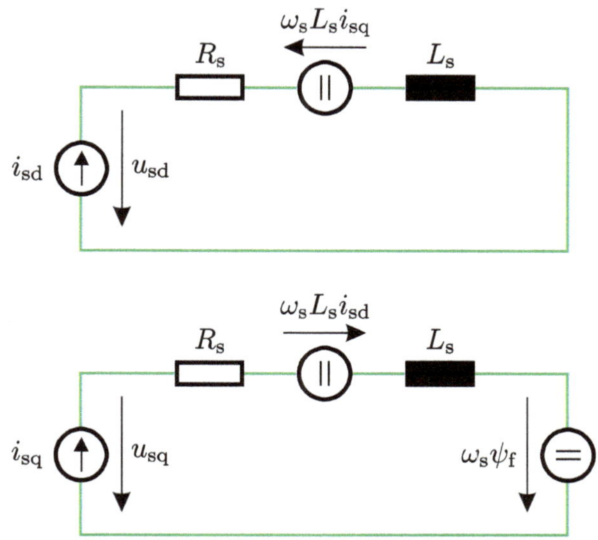

Fig. 6.5 Symbolic rotor-oriented model of a non-salient machine

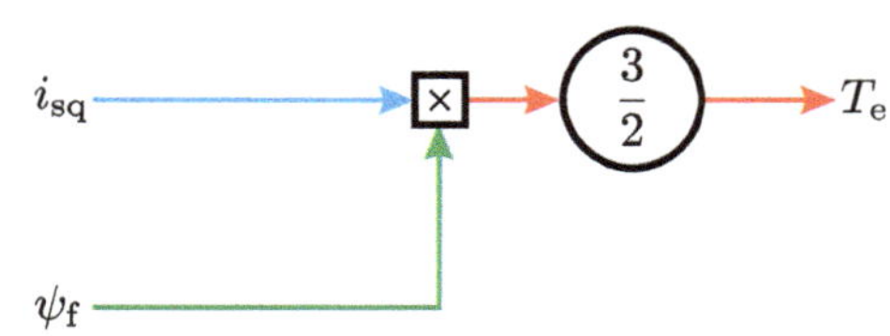

Fig. 6.6 Generic current based rotor-oriented non-salient synchronous machine model

phasor locus $\underline{i}_s$. Operational diagrams of this type utilize a complex plane of which the real axis is aligned with the terminal voltage phasor $\underline{u}_s$.

The behavior of the machine in steady-state operation can be derived from equation set (6.4) taking into account that time-dependent derivatives are zero in steady state.

$$\vec{u}_s^{\,dq} = R_s \vec{i}_s^{\,dq} + j\omega_s \vec{\psi}_s^{\,dq} \tag{6.6a}$$

$$\vec{\psi}_s^{\,dq} = L_s \vec{i}_s^{\,dq} + \psi_f. \tag{6.6b}$$

In the following, it is assumed that the machine is connected to a (three phase) voltage source with constant amplitude $\hat{u}_s$ and adjustable angular frequency ω_s. In (amplitude invariant) space vector terms that voltage source may be represented by $\vec{u}_s = \underline{u}_s e^{j\omega_s t} = \hat{u}_s e^{j\omega_s t}$. In steady state all space vectors in equation set (6.6) are rotating at the same speed, which is zero in dq coordinates and equals ω_s in stator coordinates. Therefore, the system is fully described by the relative position of the space vectors to each other and phasors (here using peak values) can be used to describe the machine variables, resulting in the following equation set:

$$\underline{u}_s = R_s \underline{i}_s + j\omega_s \underline{\psi}_s \tag{6.7a}$$

$$\underline{\psi}_{\mathrm{s}} = L_{\mathrm{s}}\underline{i}_{\mathrm{s}} + \underline{\psi}_{\mathrm{f}}. \tag{6.7b}$$

The stator voltage phasor is aligned with the real axis and therefore $\underline{u}_{\mathrm{s}} = \hat{u}$. The phasor relations are shown in Fig. 6.7. Shown in this diagram are the voltage $\underline{u}_{\mathrm{s}}$, the flux linkage $\underline{\psi}_{\mathrm{f}}$, and the current $\underline{i}_{\mathrm{s}}$ which are linked by equation set (6.7). The complex phasor coordinate plane is shown in gray. The real axis is tied to the voltage phasor $\underline{u}_{\mathrm{s}}$.

The load angle ρ_{m} is defined as the angle between the voltage phasor $\underline{u}_{\mathrm{s}}$ and the back-EMF phasor $j\omega_{\mathrm{s}}\underline{\psi}_{\mathrm{f}}$. Subsequent observation of Fig. 6.7 shows that the field flux linkage phasor $\underline{\psi}_{\mathrm{f}}$ can be written as $-\mathrm{j}\psi_{\mathrm{f}}\mathrm{e}^{\mathrm{j}\rho_{\mathrm{m}}}$ resulting in the equations

$$\underline{u}_{\mathrm{s}} = R_{\mathrm{s}}\underline{i}_{\mathrm{s}} + \mathrm{j}\omega_{\mathrm{s}}\underline{\psi}_{\mathrm{s}} \tag{6.8a}$$

$$\underline{\psi}_{\mathrm{s}} = L_{\mathrm{s}}\underline{i}_{\mathrm{s}} - \mathrm{j}\psi_{\mathrm{f}}\mathrm{e}^{\mathrm{j}\rho_{\mathrm{m}}}. \tag{6.8b}$$

An expression for the current phasor $\underline{i}_{\mathrm{s}}$ can now be found by eliminating the flux linkage phasor $\underline{\psi}_{\mathrm{s}}$

$$\underline{i}_{\mathrm{s}} = \frac{\hat{u}_{\mathrm{s}} - \omega_{\mathrm{s}}\psi_{\mathrm{f}}\mathrm{e}^{\mathrm{j}\rho_{\mathrm{m}}}}{R_{\mathrm{s}} + \mathrm{j}\omega_{\mathrm{s}}L_{\mathrm{s}}}. \tag{6.9}$$

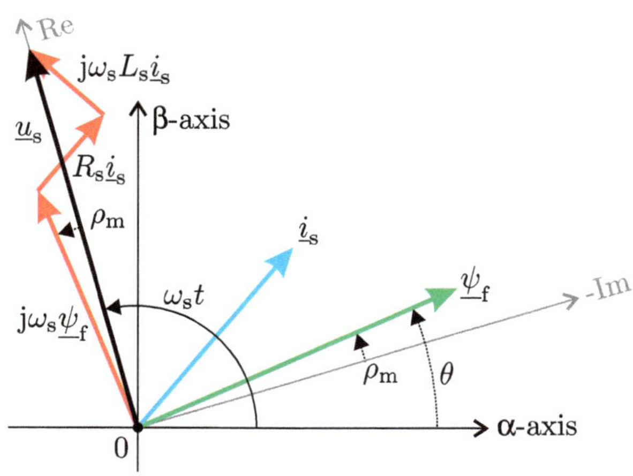

Fig. 6.7 Load angle definition (ρ_{m}) in phasor diagram

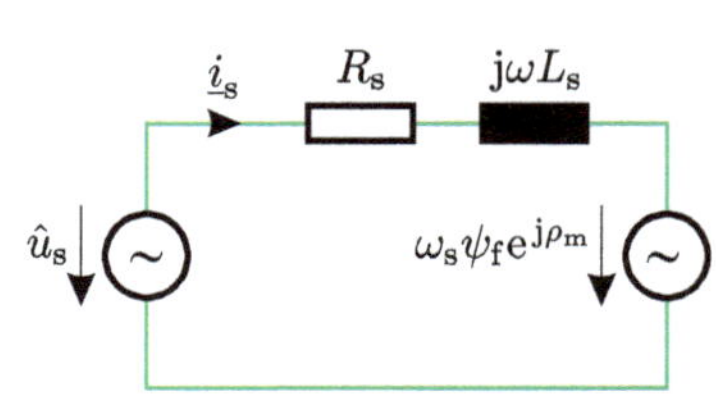

Fig. 6.8 Non-salient synchronous, phasor based machine model

Expression (6.9) may also be presented by an equivalent circuit as given in Fig. 6.8. It includes two voltage sources, the supply voltage $\underline{u}_s = \hat{u}_s$, and the back-EMF $\omega_s \underline{\psi}_f$.

Observation of the complex network shows that the current $\underline{i}_s$ can also be found by superposition of the current components for each voltage source separately, leading to

$$\underline{i}_s = \underbrace{\frac{\hat{u}_s}{R_s + j\omega_s L_s}}_{\underline{i}_{s1}} + \underbrace{\frac{-\omega_s \psi_f e^{j\rho_m}}{R_s + j\omega_s L_s}}_{\underline{i}_{s2}}. \tag{6.10}$$

A graphical representation of Eq. (6.10) is given in Fig. 6.9. When the equation is factorized it can be written as

$$\underline{i}_s = \frac{\hat{u}_s}{R_s}\left(\frac{1}{1 + j\frac{\omega_s L_s}{R_s}} + \frac{-\frac{\omega_s \psi_f}{\hat{u}_s} e^{j\rho_m}}{1 + j\frac{\omega_s L_s}{R_s}}\right). \tag{6.11}$$

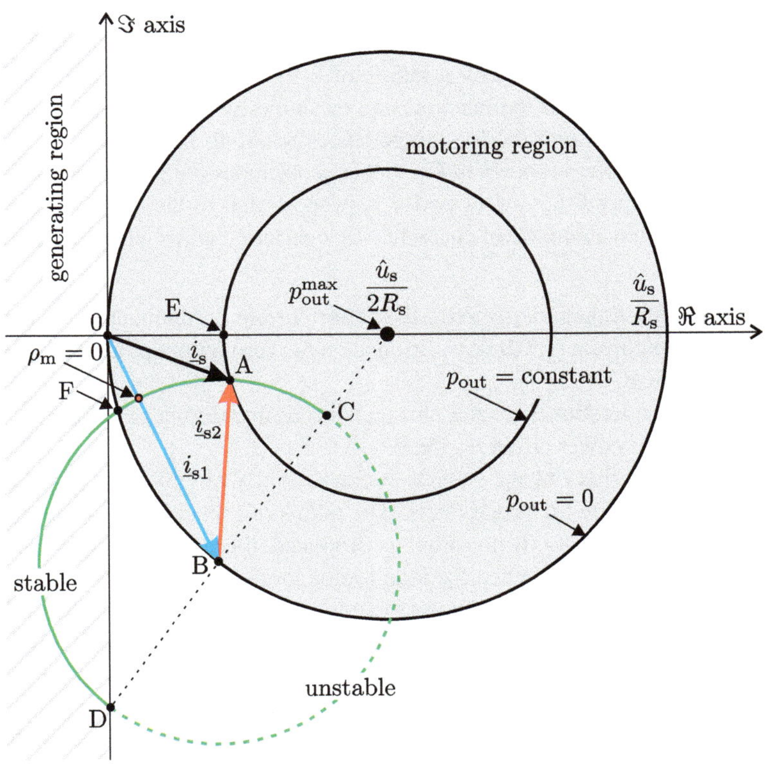

Fig. 6.9 Blondel diagram of a non-salient synchronous machine

In Fig. 6.9, $\omega_s\psi_f/\hat{u}_s$ was set at 0.7 and $\omega_s L_s/R_s$ was set at 1.96. The figure shows the two current components $\underline{i}_{s1}$ and $\underline{i}_{s2}$ for a given stator voltage, together with circles of constant output power p_{out}. Note that in practical machines the stator current $\underline{i}_s$ is limited. This limit can be represented by a circle around the origin of the Blondel plot as will be explained in detail in the next chapter. This constraint is not taken yet into consideration for the analysis presented here. The lines of constant output power p_{out} can be derived by using the energy balance equation and assuming that the machine losses are purely ohmic losses.

$$\underbrace{\frac{3}{2}\Re\{\underline{u}_s\underline{i}_s{}^*\}}_{\text{electrical power}} - \underbrace{\frac{3}{2}R_s\Re\{\underline{i}_s\underline{i}_s{}^*\}}_{\text{(ohmic) losses}} = \underbrace{p_{out}}_{\text{mechanical output power}} = \text{const.} \tag{6.12}$$

After some mathematical handling, Eq. (6.12) gives

$$\left(\Re\{\underline{i}_s\} - \frac{\hat{u}_s}{2R_s}\right)^2 + \Im\{\underline{i}_s\}^2 = \left(\left(\frac{\hat{u}_s}{2R_s}\right)^2 - \frac{2}{3}\frac{p_{out}}{R_s}\right). \tag{6.13}$$

Equation (6.13) states that the lines of constant voltage and power are circles in the complex plane with their centers on the real axis at $(\hat{u}_s/2R_s, 0)$ and with radiuses equal to $\sqrt{(\hat{u}_s/2R_s)^2 - 2p_{out}/3R_s}$. Circles of constant power are shown in Fig. 6.9. The zero output power circle has a radius of $\hat{u}_s/2R_s$. Furthermore, the theoretical maximum output power of the machine can be found by matching the source, i.e., the electrical machine, and the load impedances resulting in $p_{out}^{max} = 3\hat{u}_s^2/8R_s$. The maximum output power is shown in Fig. 6.9 at coordinates $(\hat{u}_s/2R_s, 0)$. It is noted that for a given shaft speed the output power is proportional to the output torque. The Blondel diagram has a number of characteristic operating points which are discussed below:

- If the field flux linkage ψ_f is zero, the stator current component $\underline{i}_{s2}$ is zero and thus the stator current $\underline{i}_s$ will be positioned on the zero output power circle $p_{out} = 0$, e.g. at point B.
- During motor operation the stator current is inside and during generator operation outside the zero power circle $p_{out} = 0$.
- For a given field flux linkage ($\omega_s\psi_f/\hat{u}_s$ = const.) and a variable load angle ρ_m, the stator current lies on the green circle. The radius of this circle is proportional to the field flux linkage ψ_f. If no torque is produced, the operating point is located at point F with $\rho_m = 0$. When the load torque increases ($\rho_m < 0$), the motoring region is entered and the output power increases. The distance from the operating point to the maximum output power point p_{out}^{max} determines the output power. The shorter this distance, the higher the output power. This means that operation with a current phasor at point C gives the highest output power achievable for a given field flux linkage. This operating point also represents the limit for stable (steady state) operation.
- For a given output power at, for example, point A, it is possible to minimize the stator current by changing the excitation flux linkage. Minimum stator current is achieved at point E, which corresponds to unity power factor.

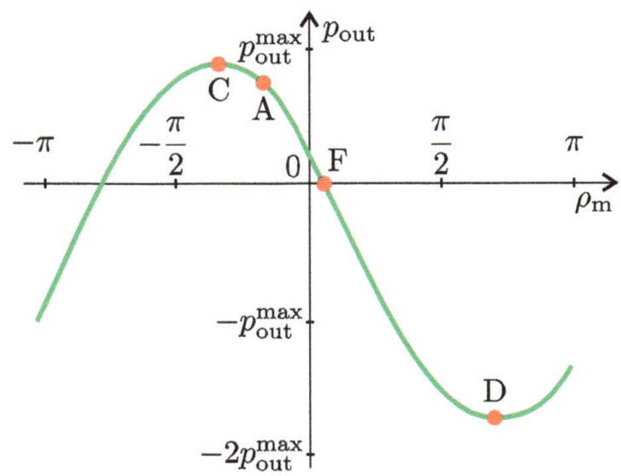

Fig. 6.10 Output power versus load angle curve

- The Blondel diagram according to Fig. 6.9 is shown with a ratio $\omega_s \psi_f/\hat{u}_s = 0.7$ and $\omega_s L_s/R_s = 1.96$. The current endpoint trajectory can be made to intersect with the maximum power point in case the ratio $\omega_s \psi_f/\hat{u}_s$ is increased to a value of 1.1.

The output power as a function of the load angle can be found using Eqs. (6.7b) and (6.9) in conjunction with the torque expression $\vec{T}_e = \frac{3}{2}\left(\underline{\psi}_s \times \underline{i}_s\right)$. Alternatively, the power balance equation can be used, as defined by expression (6.12). The resulting expression is normalized using $p_{out}^{n} = p_{out}/p_{out}^{max}$. A graphical representation of the normalized output power as a function of the load angle ρ_m is shown in Fig. 6.10, with $\omega_s \psi_f/\hat{u}_s = 0.7$ and $\omega_s L_s/R_s = 1.96$. The parameters deployed here are identical to those used for Fig. 6.9. Also, the same operating points are shown in Fig. 6.10 for comparative purposes. A tutorial example which demonstrates the use of the Blondel diagram is given in Sect. 6.3.2.

6.2 Salient Synchronous Machine

Salient synchronous machines have a reluctance that is rotor position dependent. A common type of salient synchronous machines is the *interior permanent magnet machine* which carries magnets within the rotor lamination. A cross section of an interior magnet machine with four poles is shown in Fig. 6.11. In the direction of the x-axis, the magnets are in the flux path, increasing the effective air-gap due to their low permeability. In the direction of the y-axis, the magnets have no effect and thus the effective air-gap is unchanged [3]. Consequently, the magnetizing inductances L_{mx} and L_{my} related to the x- and y-axes are different. In the example given, L_{my} is larger than L_{mx}. This is typical for interior magnet machines and contrary to *salient pole machines* with field windings where L_{my} is usually smaller than L_{mx}.

The IRTF based model of a salient machine is given in Fig. 6.12, showing the magnetizing inductance on the rotor side. The inductance is composed of the two components L_{mx} and L_{my} which act on the real and imaginary part of the current. It is emphasized that the freedom of locating these inductances to the stator side of

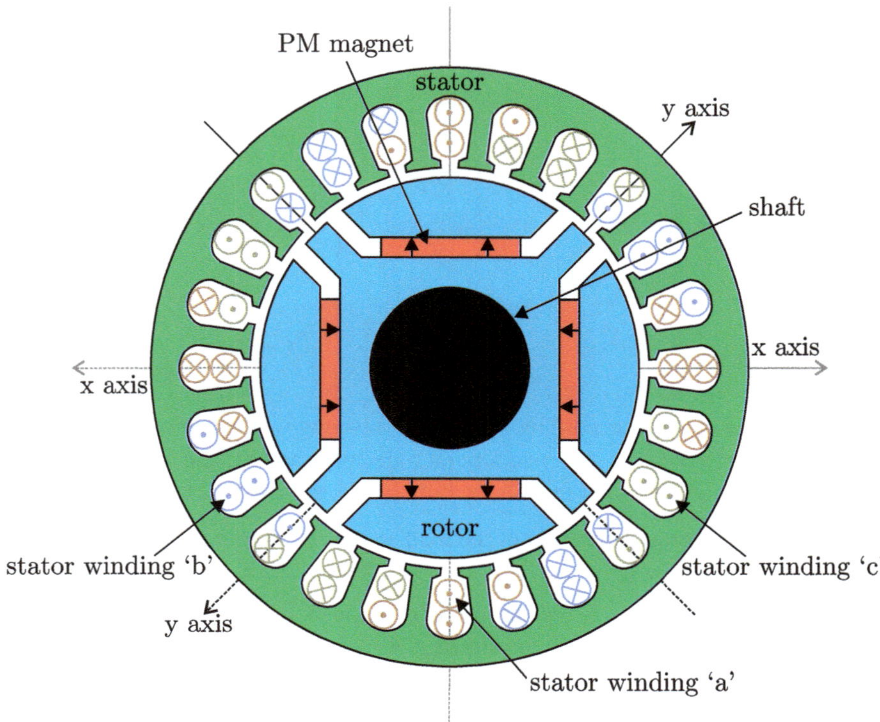

Fig. 6.11 Four pole, interior permanent magnet synchronous machine, showing two-layer stator winding and rotor saliency

Fig. 6.12 Synchronous machine, with a salient rotor

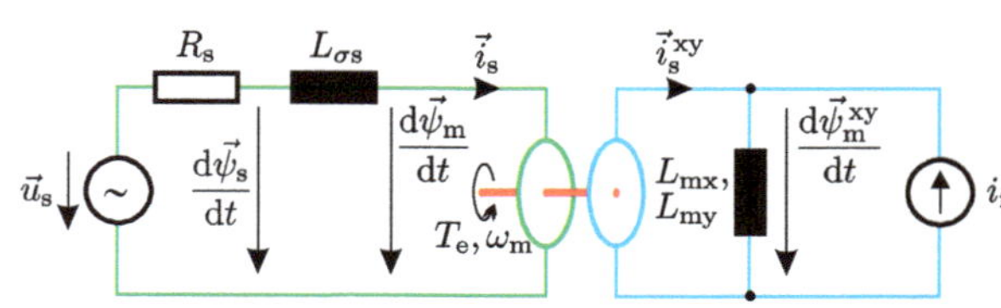

the IRTF is lost due to the rotor saliency. Transferring the inductance to the stator side would result in an angle-dependent inductance, which is rotating with the rotor, making the model unnecessarily complex. No other changes compared to the non-salient IRTF model, shown in Fig. 6.2, are needed and the stator resistance R_s and stator leakage inductance $L_{\sigma s}$ remain in place.

The equations which correspond to this machine are given by equation set (6.14). Equation (6.14b) was derived from Eq. (6.1b) by transforming it to rotor coordinates using $\vec{A} = \vec{A}^{xy} e^{j\theta}$ for any vector $\vec{A}$.

$$\vec{u}_s = R_s \vec{i}_s + \frac{d\vec{\psi}_s}{dt} \tag{6.14a}$$

$$\vec{\psi}_s^{xy} = L_{\sigma s}\vec{i}_s^{xy} + \vec{\psi}_m^{xy} \tag{6.14b}$$

$$\psi_{mx} = L_{mx}(i_{sx} + i_f) \tag{6.14c}$$

$$\psi_{my} = L_{my} i_{sy}. \tag{6.14d}$$

As with the non-salient case, some simplification of the symbolic model according to Fig. 6.12 may be achieved by rearranging the inductances $L_{\sigma s}$, L_{mx}, and L_{my} and by introducing the variable $\psi_f = L_{mx} i_f$, which allow Eqs. (6.14c), (6.14d), and (6.14b) to be written as

$$\psi_{sx} = \underbrace{(L_{mx} + L_{\sigma s})}_{L_{sd}} i_{sx} + \psi_f \tag{6.15a}$$

$$\psi_{sy} = \underbrace{\left(L_{my} + L_{\sigma s}\right)}_{L_{sq}} i_{sy}. \tag{6.15b}$$

The two parameters L_{sd} and L_{sq} represent the direct and quadrature synchronous inductances, respectively.

As explained in Sect. 6.1.3, the dq coordinate system, which is linked to the rotor flux linkage vector, is identical to the chosen xy coordinate system that is linked to the rotor of the synchronous machine. Hence, the direct and quadrature axis terminology is used to represent a complex plane which is oriented with respect to a designated flux linkage vector.

6.2.1 *Generic Model of Salient Synchronous Machine*

The generic model of the salient synchronous machine follows directly from the non-salient model, according to Fig. 6.3. For the salient model, the gain module $1/L_s$ shown in Fig. 6.3 must be replaced by a gain $1/L_{sd}$, $1/L_{sq}$ as illustrated in Fig. 6.13. The input and output vectors for this gain module are unchanged, while the gain variables are defined by equation set (6.15). As with the non-salient case, the field flux linkage ψ_f may be supplied by an excitation winding which carries a current i_f or by permanent magnets, as shown in Fig. 6.11.

6.2.2 *Rotor-Oriented Model of the Salient Synchronous Machine*

The coordinate transformation process for deriving a symbolic and generic representation of the rotor-oriented model of the salient machine is similar to that undertaken for the non-salient case in Sect. 6.1.3. The process is initiated with

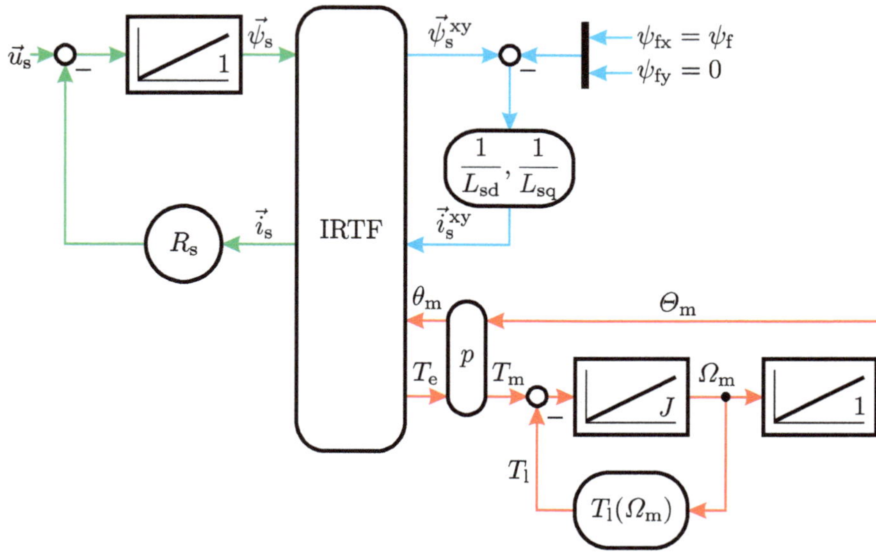

Fig. 6.13 Generic synchronous machine model, with a salient rotor

the aid of Eqs. (6.14a) and (6.15). For a rotor flux based model the synchronous reference frame with the direct and quadrature axis is linked to the field flux linkage vector ψ_f, as shown in Fig. 6.14.

The equation set for the salient rotor flux based model can be written as

$$\vec{u}_s^{dq} = R_s \vec{i}_s^{dq} + \frac{d\vec{\psi}_s^{dq}}{dt} + j\omega_s \vec{\psi}_s^{dq} \tag{6.16a}$$

$$\vec{\psi}_s^{dq} = L_{sq} \vec{i}_s^{dq} + \psi_f + \underbrace{i_{sd}\left(L_{sd} - L_{sq}\right)}_{\psi_{eq}} \tag{6.16b}$$

$$T_e = \underbrace{\frac{3}{2}\psi_f i_{sq}}_{\text{electromagnetic torque}} + \underbrace{\frac{3}{2}\psi_{eq} i_{sq}}_{\text{reluctance torque}} . \tag{6.16c}$$

The stator voltage equation (6.16a) for the salient rotor flux based model in synchronous coordinates is identical to that found for the non-salient case (see Eq. (6.4a)). The flux linkage relationship given by Eq. (6.16b) differs from equation set 6.16 due to saliency. Most importantly, salient machines are capable of producing an additional reluctance component, which is linked to a so-called equivalent flux $\psi_{eq} = i_{sd}\left(L_{sd} - L_{sq}\right)$ which is added to the flux ψ_f. For salient machines with $L_{sq} > L_{sd}$ a *negative* i_{sd} is required in order to generate an equivalent flux component that will enhance the PM flux. The vector diagram given in Fig. 6.14 shows precisely this point.

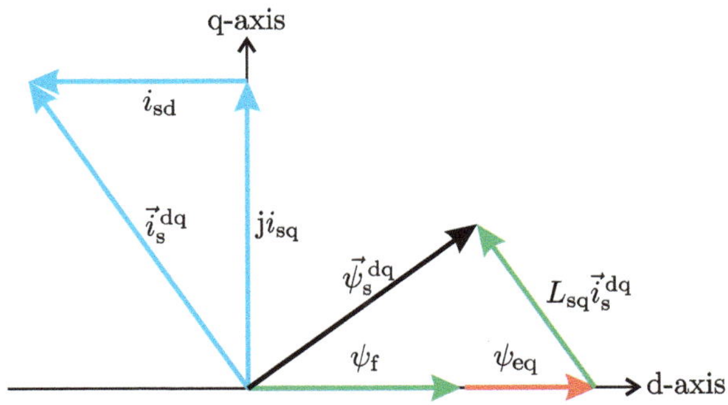

Fig. 6.14 Vector diagram with direct and quadrature axis for salient machine

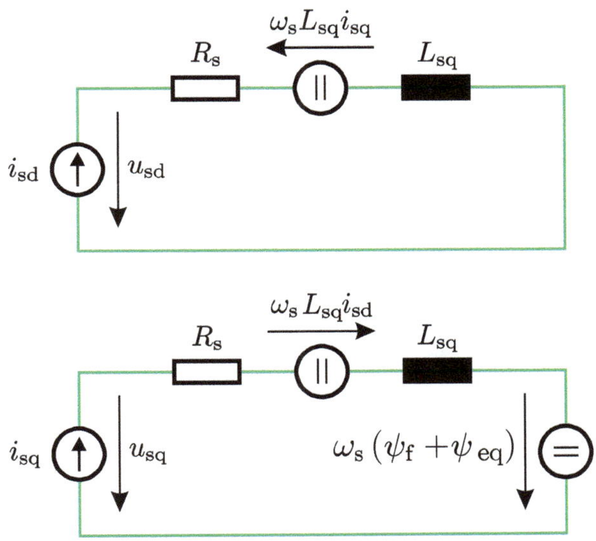

Fig. 6.15 Symbolic rotor-oriented model: salient machine

Equation (6.16c) is derived using the general torque equation $\vec{T}_e = \frac{3}{2}\vec{\psi} \times \vec{i}$ and clearly shows that the introduction of a salient rotor leads to an additional torque component $\psi_{eq} i_{sq}$.

Using Eqs. (6.16a) and (6.16b) and grouping the real and imaginary components, the following d- and q-axis voltage expressions can be derived:

$$u_{sd} = R_s i_{sd} - \omega_s L_{sq} i_{sq} + L_{sd} \frac{di_{sd}}{dt} \tag{6.17a}$$

$$u_{sq} = R_s i_{sq} + \omega_s L_{sq} i_{sd} + L_{sq} \frac{di_{sq}}{dt} + \omega_s \left(\psi_f + \psi_{eq}\right). \tag{6.17b}$$

A symbolic representation of the rotor-oriented model which corresponds with equation set (6.17) is given in Fig. 6.15.

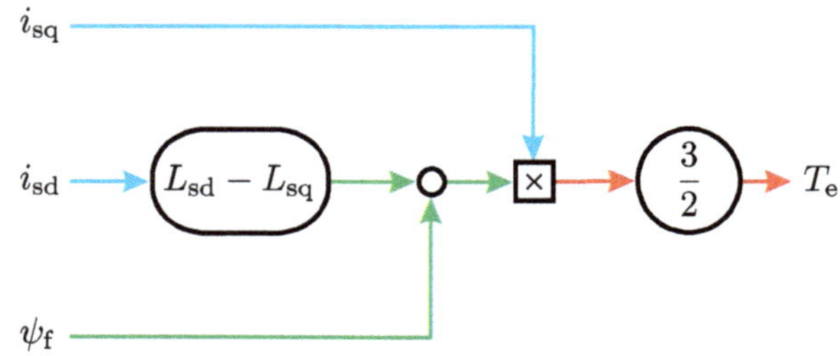

Fig. 6.16 Generic current based rotor-oriented salient synchronous machine model

Note that the model for the salient machine reduces to the model of the non-salient machine shown in Fig. 6.5, with $L_{sd} = L_{sq} = L_s$. The corresponding generic model is given in Fig. 6.16.

6.2.3 Steady-State Analysis

Similar to the non-salient machine the steady-state behavior of the salient machine is analyzed in this chapter. Consequently, only a broad outline of the steps needed is shown. It is helpful to define the saliency factor χ as

$$\chi = \frac{L_{sq} - L_{sd}}{2L_{sd}}. \tag{6.18}$$

Use of Eq. (6.18) with expression (6.16b) allows the stator flux linkage vector $\vec{\psi}_s^{dq}$ to be written as

$$\vec{\psi}_s^{dq} = L_{sd}\vec{i}_s^{dq} + j2\chi L_{sd} i_{sq} + \psi_f. \tag{6.19}$$

This expression reduces to the non-salient case as in Eq. (6.2), with $L_{sd} = L_{sq} = L_s$, i.e., the saliency factor χ equals zero. The torque expression can likewise be rewritten as

$$T_e = \frac{3}{2}\left(\psi_f i_{sq} - 2\chi L_{sd} i_{sd} i_{sq}\right). \tag{6.20}$$

If $\chi > 0$, then a field weakening current component ($i_{sd} < 0$) leads to an additional reluctance torque component as in interior magnet machines. In salient pole machines with field excitation the saliency factor χ is negative. The current components in Eq. (6.19) can be replaced by $\vec{i}_s^{dq}$ and its complex conjugate $(\vec{i}_s^{dq})^*$ which leads to

$$\vec{\psi}_{\rm s}^{\rm dq} = (1+\chi)\,L_{\rm sd}\vec{i}_{\rm s}^{\rm dq} - \chi L_{\rm sd}\left(\vec{i}_{\rm s}^{\rm dq}\right)^* + \psi_{\rm f}. \tag{6.21}$$

The Blondel diagram for the salient machine is found by making use of the voltage equation (6.16a) and the flux linkage expression (6.21). As for the non-salient machine both expressions are converted to their phasor form, based on the general transformation from stator to rotor coordinates $\vec{A}^{\rm dq} = \vec{A}{\rm e}^{-{\rm j}\theta}$ and from phasor to space vector quantities $\vec{A} = \underline{A}{\rm e}^{{\rm j}\omega_{\rm s}t}$. Note that the phasors are representing sinusoidal functions in time, using peak values. Using the relationship $\theta - \omega_{\rm s}t = \rho_{\rm m} - \pi/2$ (cp. Fig. 6.7), this leads to the expression $\vec{A}^{\rm dq} = \underline{A}{\rm e}^{-{\rm j}(\rho_{\rm m}-\frac{\pi}{2})}$. Inserting this in Eqs. (6.16a) and (6.21), we obtain the expression

$$\underline{u}_{\rm s} = R_{\rm s}\underline{i}_{\rm s} + {\rm j}\omega_{\rm s}\underline{\psi}_{\rm s} \tag{6.22a}$$

$$\underline{\psi}_{\rm s} = (1+\chi)\,L_{\rm sd}\underline{i}_{\rm s} + \chi L_{\rm sd}\underline{i}_{\rm s}^{*}{\rm e}^{{\rm j}2\rho_{\rm m}} - {\rm j}\psi_{\rm f}{\rm e}^{{\rm j}\rho_{\rm m}}. \tag{6.22b}$$

Elimination of the flux linkage phasor $\underline{\psi}_{\rm s}$ from equation set (6.22) leads to the following current phasor expression:

$$\underline{i}_{\rm s} + \underbrace{\left[\frac{{\rm j}\omega_{\rm s}L_{\rm sd}\chi{\rm e}^{{\rm j}2\rho_{\rm m}}}{R_{\rm s}+{\rm j}\omega_{\rm s}(1+\chi)\,L_{\rm sd}}\right]}_{\underline{K}}\underline{i}_{\rm s}^{*} = \underbrace{\frac{\hat{u}_{\rm s}-\omega_{\rm s}\psi_{\rm f}{\rm e}^{{\rm j}\rho_{\rm m}}}{R_{\rm s}+{\rm j}\omega_{\rm s}(1+\chi)\,L_{\rm sd}}}_{\underline{i}_{\rm s}^{\circ}} \tag{6.23}$$

For the non-salient case with $\chi = 0$, Eq. (6.23) reduces to expression (6.9). It is helpful to introduce the phasors $\underline{K}$ and $\underline{i}_{\rm s}^{\circ}$ as defined in Eq. (6.23). This allows the former expression to be written as

$$\underline{i}_{\rm s} + \underline{K}\underline{i}_{\rm s}^{*} = \underline{i}_{\rm s}^{\circ}. \tag{6.24}$$

The current phasor $\underline{i}_{\rm s}$ can be expressed as a function of the load angle $\rho_{\rm m}$ by rewriting the complex coefficient $\underline{K}$ as $\underline{K} = K_{\rm x} + {\rm j}K_{\rm y}$ and the phasor $\underline{i}_{\rm s}^{\circ}$ as $\underline{i}_{\rm s}^{\circ} = i_{\rm sx}^{\circ} + {\rm j}i_{\rm sy}^{\circ}$. Subsequent use of the aforementioned approach with Eq. (6.24) gives

$$\Re\{\underline{i}_{\rm s}\} = \frac{(1-K_{\rm x})\,i_{\rm sx}^{\circ} - K_{\rm y}i_{\rm sy}^{\circ}}{1-K_{\rm x}^2-K_{\rm y}^2} \tag{6.25a}$$

$$\Im\{\underline{i}_{\rm s}\} = \frac{(1+K_{\rm x})\,i_{\rm sy}^{\circ} - K_{\rm y}i_{\rm sx}^{\circ}}{1-K_{\rm x}^2-K_{\rm y}^2}. \tag{6.25b}$$

A graphical representation of equation set (6.25) as given in Fig. 6.17 shows the Blondel diagram of the machine with (green line) and without saliency (red line). For the salient example, the value of χ was arbitrarily chosen as 0.75, corresponding to a ratio $L_{\rm sq}/L_{\rm sd} = 2.5$. The parameters $\omega_{\rm s}\psi_{\rm f}/\hat{u}_{\rm s} = 0.7$ and $\omega_{\rm s}L_{\rm sd}/R_{\rm s} = 1.96$ are identical to those used for the non-salient machine to demonstrate the impact of

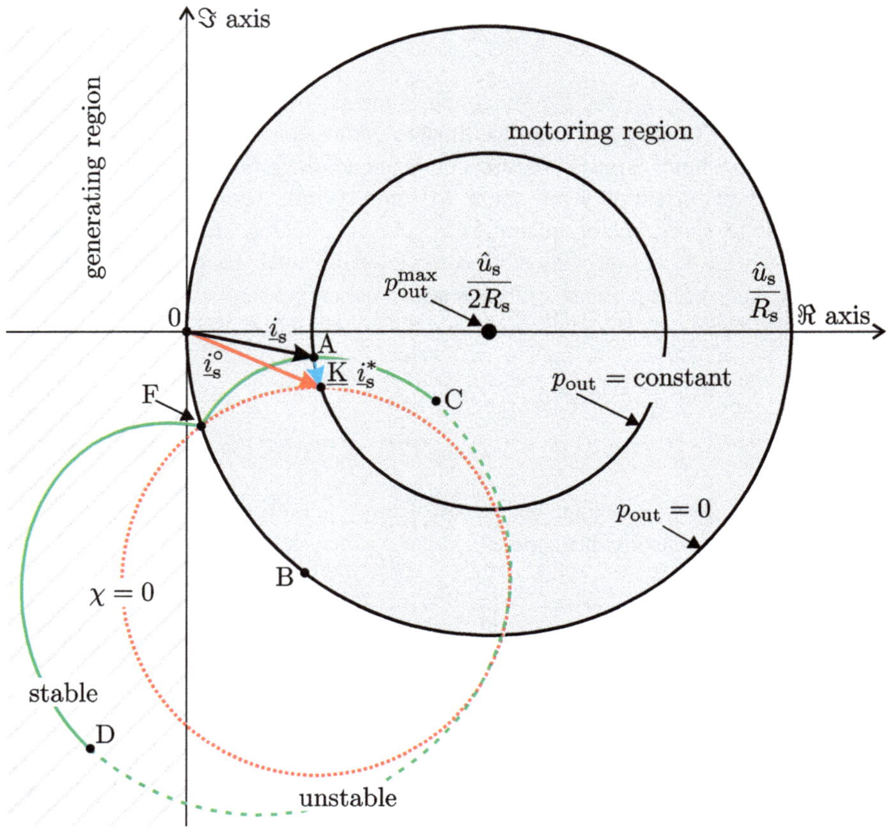

Fig. 6.17 Blondel diagram of synchronous machine with $\omega_s\psi_f/\hat{u}_s = 0.7$, $\omega_s L_{sd}/R_s = 1.96$, and $\chi = 0.75$

saliency in the Blondel diagram. As with the non-salient Blondel diagram the current locus is shown for a given shaft speed while the load angle ρ_m varies from $-\pi$ to π.

Figure 6.17 also shows the vectors $\underline{i}_s^\circ$ and $\underline{K}\underline{i}_s{}^*$ as defined by Eq. (6.24). The endpoint trajectory of the phasor $\underline{i}_s^\circ$ is circular for the non-salient case when the load angle is varied from $-\pi$ to π. For the salient case the radius of the circle will be slightly larger as may be deduced from Eq. (6.23). Also, for the salient case the current $\underline{i}_s$ is additionally determined by the phasor $\underline{K}\underline{i}_s{}^*$. Given that its

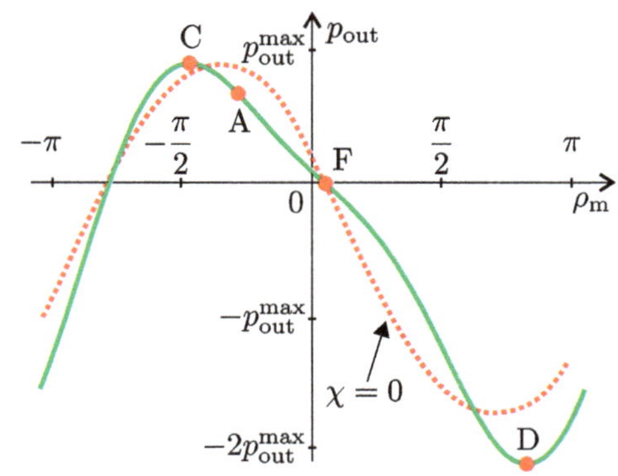

Fig. 6.18 Normalized output power versus load angle curves of salient and non-salient machine, with $\omega_s\psi_f/\hat{u}_s = 0.7$, $\omega_s L_{sd}/R_s = 1.96$, and $\chi = 0.75$

rotation is governed by the term $2\rho_m$, a deviation from the circle occurs resulting in a *cardioidal* shape, which is a special case of the *Limaçon of Pascal*. A negative saliency factor results in a stator current phasor with a differently shaped endpoint trajectory, also a member of the Limaçon class. Additionally, if stator frequency, stator voltage and machine speed are constant, the introduction of saliency increases the maximum available output power of the machine, because the current locus is closer to the theoretical power limit at p_{out}^{max}.

The output power versus load angle diagram can be found using the power equation $p_{out} = \omega_s \frac{3}{2}\Im\{\underline{\psi}_s{}^* \underline{i}_s\} = \frac{3}{2}\Re\{\underline{u}_s \underline{i}_s{}^*\} - \frac{3}{2}R_s \underline{i}_s \underline{i}_s{}^*$, together with equation set (6.25) and the stator voltage phasor $\underline{u}_s = \hat{u}_s$. As with the non-salient case a normalization with $p_{out}^{n} = p_{out}/p_{out}^{max}$ is introduced leading to Fig. 6.18. For comparison, this figure also shows the non-salient case according to Fig. 6.10 with $\omega_s\psi_f/\hat{u}_s = 0.7$ and $\omega_s L_s/R_s = 1.96$. Note that the theoretical maximum output power p_{out}^{max} is not affected by the introduction of rotor saliency, given that the voltage supply amplitude $\hat{u}_s$ and stator resistance R_s remain the same.

As can be observed from Fig. 6.18, the slope of the output power is less steep in the stable motoring area. This effect results in a reduced stiffness of the salient grid connected machine with $\chi > 0$ compared to the non-salient grid connected machine. To increase the stiffness of grid connected, electrically excited machines, i.e. machines operated at constant voltage and frequency, the machines are typically designed with a negative saliency factor. In drive applications with a power electronic converter, which enables decoupled torque and flux control, the stiffness of the drive is determined by the controller. Usually, in inverter fed permanent magnet synchronous machines a positive saliency factor is preferable, as will become apparent in the next chapter. The effect of the saliency factor on the stiffness is discussed in more detail in a tutorial given in Sect. 6.3.4.

Table 6.1 Parameters for a non-salient PM synchronous machine

Parameters	Value
Stator inductance L_s	1.365 mH
Stator resistance R_s	0.416 Ω
PM flux linkage amplitude ψ_f	0.166 Wb
Total inertia J	$3.4 \cdot 10^{-4}$ kg m^2

6.3 Tutorials

6.3.1 Tutorial 1: Dynamic Model of a Non-salient Synchronous Machine

This tutorial considers a dynamic model of a four-pole PM synchronous machine, as introduced in [4]. The set of parameters used is defined in Table 6.1. A three-phase variable frequency voltage source with $\vec{u}_s = \hat{u}_s e^{j2\pi f_s t}$ is connected to the machine model. At the start of the simulation, the frequency f_s and the amplitude $\hat{u}_s$ are ramped up to 60 Hz and 60 V, respectively, over a period of $t = 0.5$ s. A ramp load is applied to the machine shaft, which ramps up from 0 Nm at $t = 1$ s to 10 Nm at $t = 3$ s.

The simulation model according to Fig. 6.19 shows the dynamic model of the machine together with a V/f module which generates the required excitation vector $\vec{u}_s$. A vector to polar conversion module determines the instantaneous angle ρ^s of the voltage vector $\vec{u}_s$. This reference angle is used for the transformation of the current vector $\vec{i}_s$ and voltage vector $\vec{u}_s$ to the synchronous reference frame. Furthermore, the load angle is calculated in degrees by subtracting ρ^s from the electrical shaft angle θ and adding 90 °.

Results obtained with this simulation are shown in Fig. 6.20. Also shown in the simulation model are the real and reactive power levels, which show that the reactive power is capacitive with the present choice of excitation, as may also be observed in the xy plot "i_s and u_s," where current vector is leading the voltage vector after the voltage ramp is complete.

6.3.2 Tutorial 2: Steady-State Analysis of a Non-salient Synchronous Machine

A steady-state analysis of the model given in Fig. 6.19 is to be carried out using the approach outlined in Sect. 6.2.3.

The aim is to plot the Blondel diagram for the machine defined in the previous tutorial.

The plot in question should show the zero output power circle, maximum power point (with the corresponding power value), and phasors $\underline{i}_{s1}$, $-\underline{i}_{s2}$ as shown in Fig. 6.9, as well as the stator current phasor locus for operation of the machine

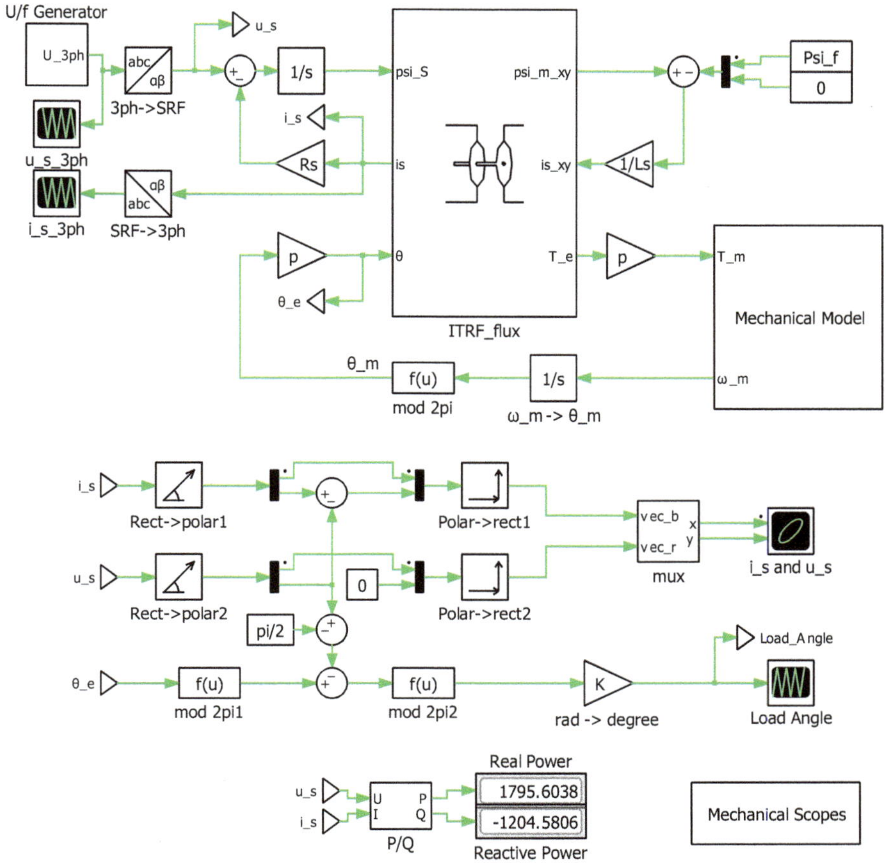

Fig. 6.19 Simulation of a non-salient synchronous machine model

for a load angle of $\rho_m = 0 \to -\pi/2$ rad. The locus is calculated using the model given in Fig. 6.8.

In addition, show in the Blondel diagram the output power circle which corresponds with the maximum achievable power level $p_{a,out}$ of the machine for the given stator voltage, frequency and excitation flux linkage value ψ_f. Finally, add to your plot the steady-state current vector $\vec{i}_s\, e^{-j\rho_m}$ endpoint coordinates for a set of load torque values (with an incremental step of 1 Nm) in the range $0 \to 10$ Nm.

A solution to this problem may be initialized by considering the zero power circle, which, according to Sect. 6.2.3, is a circle with its origin at $(\hat{u}_s/2R_s, 0)$ and radius $\hat{u}_s/2R_s$, as shown in Fig. 6.21. The maximum power p_{out}^{max} point of the machine, which is located on the origin of the output power circles, is found using Eq. (6.13) and (with the given parameters and excitation) is equal to $p_{out}^{max} = 2.16$ kW. A synchronous shaft speed of $n_s = 1800$ rpm is present, which means that the maximum achievable shaft torque that can be delivered by this machine equals

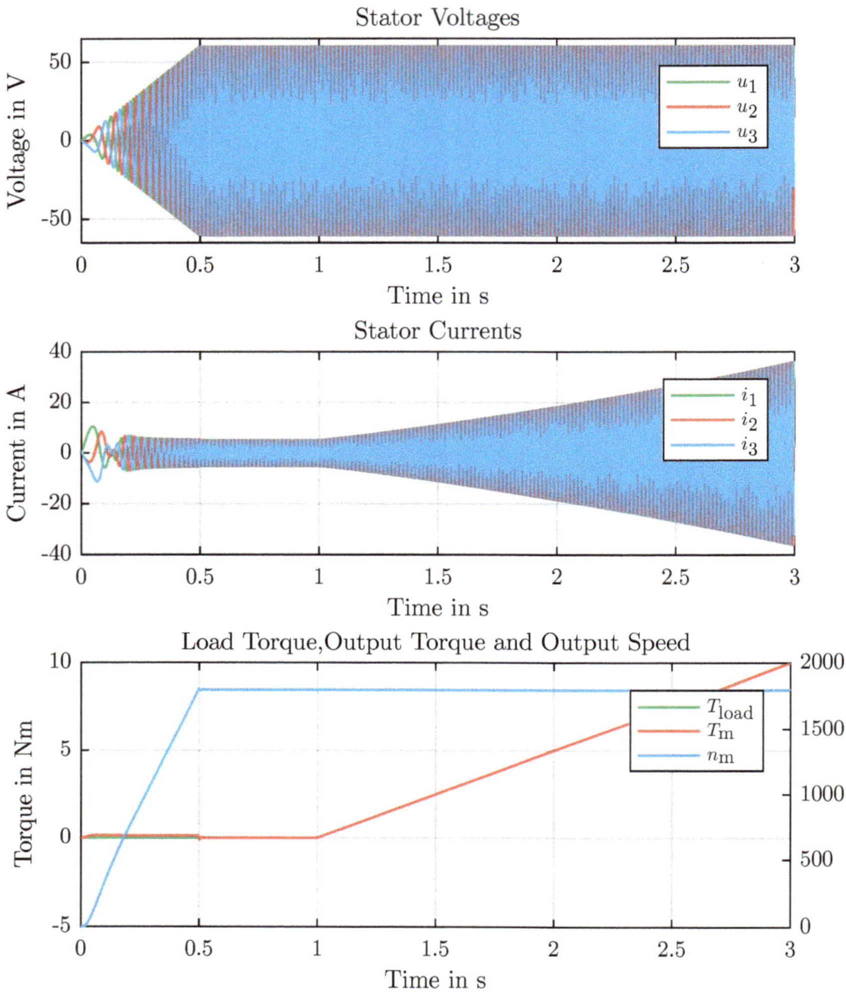

Fig. 6.20 Simulation results of the non-salient synchronous machine model

11.48 Nm. Note that these values are theoretical limit values which are normally not realizable given the thermal constraints, which limit the current to a (theoretical) maximum value $i_s^{max} = 20\,A$. The current phasors $\underline{i}_{s1}$, $-\underline{i}_{s2}$ and current phasor $\underline{i}_s$ may be found using Eq. (6.10). Changing the load angle ρ_m over the required angle range gives the current phasor locus illustrated in Fig. 6.21.

Careful observation of Fig. 6.21 shows that the current phasor endpoint for $\rho_m = 0$ is outside the zero output power circle, hence this point is not feasible for motoring operation. When a load torque is applied, the current phasor will move to a point of the locus which corresponds to the required output level. However, observation of Fig. 6.21 demonstrates that an output power level $p_{a,\,out}$ exists, which corresponds

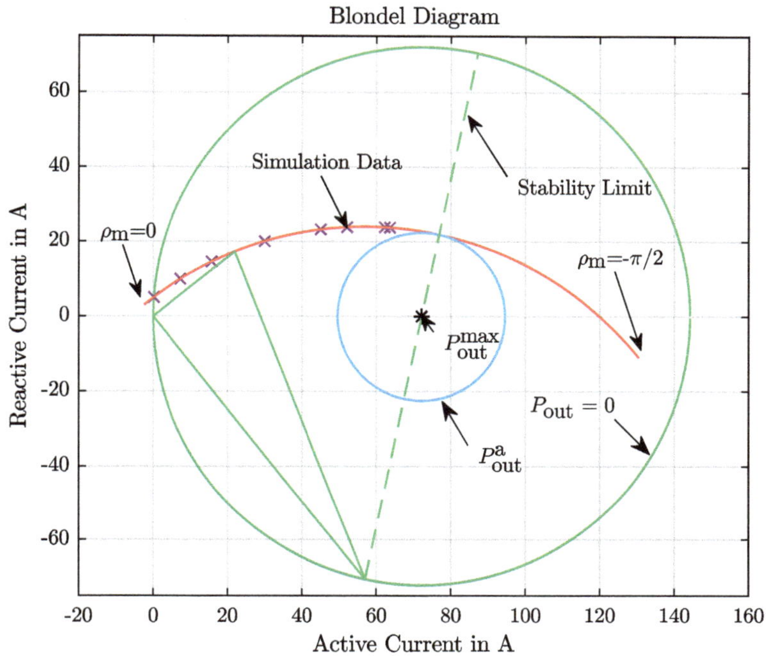

Fig. 6.21 Blondel diagram, non-salient voltage source connected machine

to a power circle that coincides with the locus at one single operating point. This limit also defines the stability range of the machine, because a load angle setting less than the value $\rho_m = -\arctan\left({}^{\omega_s L_s}/_{R_s}\right)$ will result in a power level that is less than $p_{a,\,out}$. The relationship between the radius r_p of power circle and output power is of the form $p_{out} = p_{out}^{max} - R_s\, r_p^2$, with $r_p = [0 \ldots {}^{\hat{u}}/_{2\,R_s}]$. At the maximum achievable power level, the power radius $r_p = |{}^{\omega_s\,\psi_f}/_{\sqrt{R_s^2+(\omega_s L_s)^2}} - {}^{\hat{u}}/_{2\,R_s}|$, which corresponds to $p_{a,\,out} = 1.95$ kW and a corresponding shaft torque level of 10.3 Nm. The simulation model given in the previous tutorial was used to obtain the steady-state current vector $\vec{i}_s\, e^{-j\rho_m}$ for the predetermined range of load torque settings. As expected, the resultant set of current endpoints (identified by an "asterisk" in Fig. 6.21), are located on the current locus calculated using the steady-state phasor model.

6.3.3 *Tutorial 3: Dynamic Model of a Synchronous Machine with Adjustable Saliency*

In this section the non-salient model discussed in Tutorial 1 is to be extended to accommodate rotor saliency according to the generic model shown in Fig. 6.13.

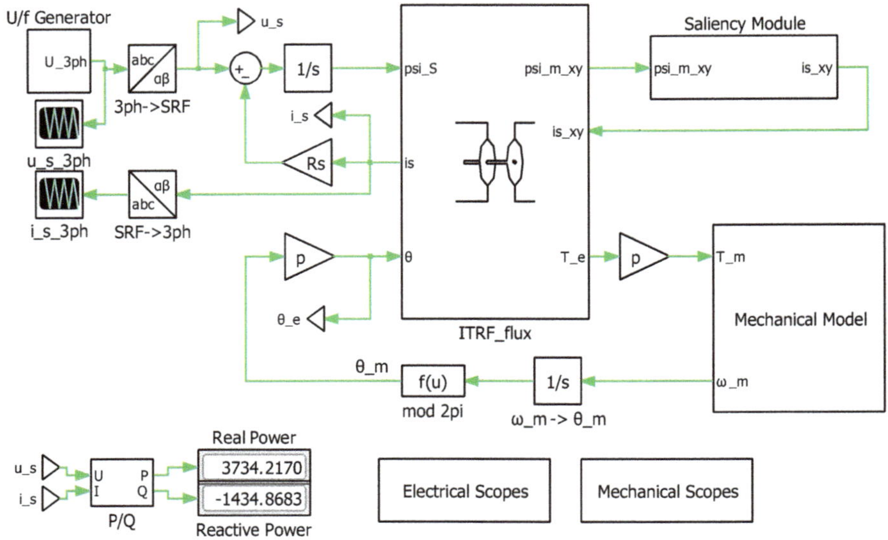

Fig. 6.22 Simulation of PM synchronous machine, full model with adjustable saliency factor

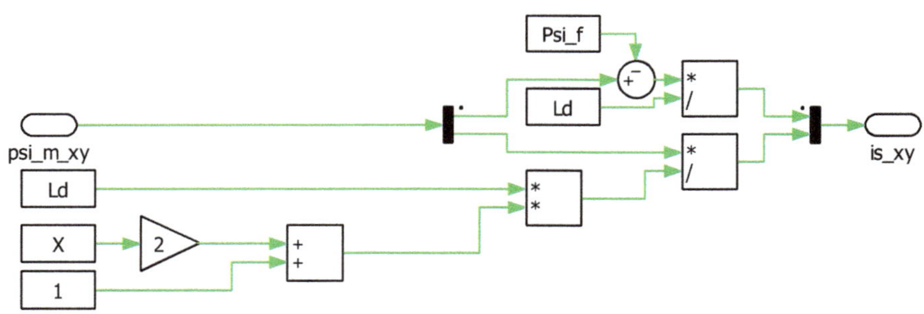

Fig. 6.23 Saliency module

A saliency module $1/L_{dq}$, which realizes the gains $1/L_{sd}$, $1/(2\chi+1)L_{sd}$ in the real and imaginary axis, respectively. The saliency factor χ can be varied over the range $-0.4 \rightarrow 0.75$ by double clicking the *Saliency Module*. The excitation and parameters as introduced in tutorial 1 remain unchanged, whereby it is noted that the inductance L_{sd} is set to L_s as defined in Table 6.1. The purpose of the exercise is to familiarize the reader with rotor saliency and to demonstrate how the voltage/current and flux linkage space vectors introduced in this tutorial will change when the load torque is adjusted. The simulation model given in Figs. 6.22 and 6.23 show the changes needed to arrive at a generalized model, where the user can choose the saliency factor of the machine.

6.3.4 *Tutorial 4: Steady-State Analysis of a Salient Synchronous Machine*

The final tutorial in this section examines the steady-state behavior of the machine that was introduced in the previous tutorial. A similar approach was taken in tutorial 2 for the non-salient machine, where the stator current phasor trajectory was calculated and compared with the simulation model operating under quasi-steady-state conditions. In this tutorial the steady-state output power versus load angle characteristic of the machine is to be calculated using the theory outlined in Sect. 6.2.3. For the purpose of this exercise restrain the load angle ρ_m range to $-\pi/2 \rightarrow 0$ rad and undertake this task for saliency factors of $\chi = -0.4, 0$ and $\chi = 0.75$, respectively. The parameters and excitation are those given in tutorials 1 and 4. In addition to the above, use the model given in Fig. 6.22 and vary the load torque in the range of $0 \rightarrow 10$ Nm in increments of 1 Nm. At each step, allow the model to settle to its steady state and then record the load angle. Add

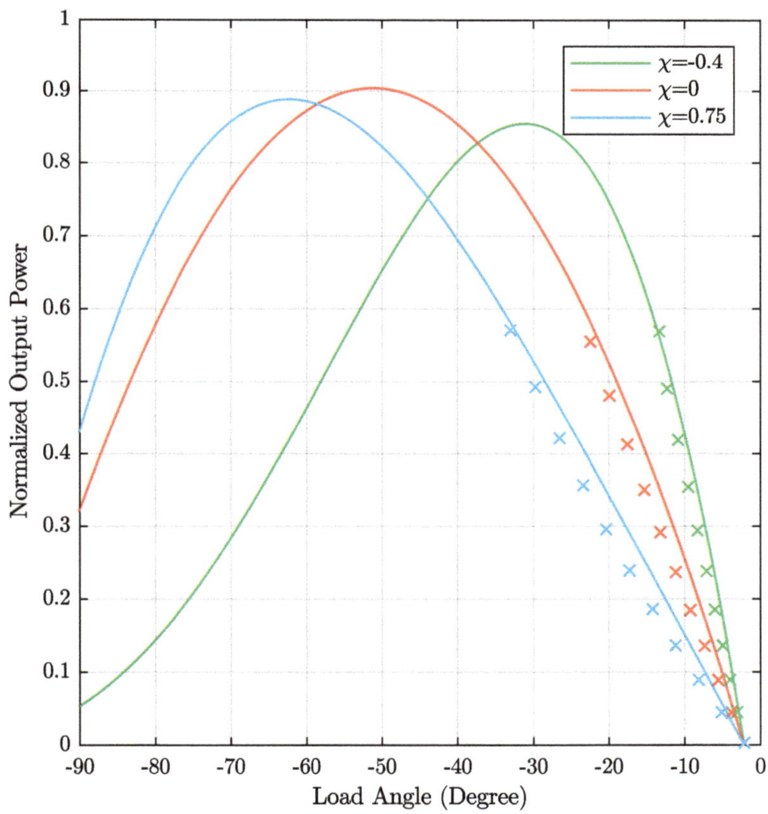

Fig. 6.24 Normalized output power versus load angle of a voltage source connected machine, with saliency factors of $\chi = -0.4, 0$ and $\chi = 0.75$, respectively

the results obtained to the calculated torque versus load angle characteristics for the three saliency factors considered.

The results shown in Fig. 6.24 confirm an earlier observation that the stiffness of the machine, i.e., its sensitivity to load angle variations when the load torque changes, is reduced when the saliency factor increases. The simulation results as obtained with the dynamic model operating under quasi-steady-state conditions are shown with asterisks (*).

References

1. Hughes A (2006) Electric motors and drives: fundamentals, types and applications, 3rd edn. Newnes, Oxford
2. Leonhard W (2001). Control of electrical drives, 3rd edn. Springer, Berlin
3. Toliyat HA, Kliman GB (2004) Handbook of electric motors, 2nd edn. CRC Press, Boca Raton
4. Veltman A, Pulle DWJ, De Doncker R (2007) Fundamentals of electrical drives. Spinger, Berlin

Chapter 7
Control of Synchronous Machine Drives

This chapter deals with field-oriented, model based control of synchronous machines. Field-oriented control has become common choice for many servo drive applications due to the availability of accurate models and affordable digital signal processors (DSPs) that execute the model based control algorithms in real time. This chapter builds on the machine models introduced in the previous chapter and extends the electromagnetic torque control principle introduced in Sect. 4.2.1. Controls are derived for both non-salient and salient synchronous machines. The operation in field weakening with constant stator flux linkage and with unity power factor is analyzed. At the end of the chapter, the controls are interfaced with a current-controlled and a voltage-controlled converter. A set of tutorials is provided to interactively illustrate the concepts and the proposed control strategies.

7.1 Controller Principles

The drive structure of a synchronous machine under field-oriented control is shown in Fig. 7.1. It consists of a field-oriented control module, a converter with a current control unit, and a synchronous machine which is connected to a mechanical load [1]. In case the machine is not permanently excited, an additional single-phase converter is connected to the excitation winding of the machine. For current control, any of the methods outlined in Chap. 3 can be used. Input to the field-oriented control module is the reference torque T_{e}^{*} which is either set by the user or by a superimposed controller (e.g., speed regulator). From the reference torque T_{e}^{*} and the mechanical rotor angle Θ_{m}, the field-oriented control module generates the command currents $i_{\mathrm{sd}}^{\mathrm{c}}$, $i_{\mathrm{sq}}^{\mathrm{c}}$, and $i_{\mathrm{f}}^{\mathrm{c}}$ and calculates the control angle θ^{c}. The rotor or shaft angle Θ_{m} is provided by a shaft encoder which is attached to the machine shaft. In an ideal case the control angle matches the shaft angle $\theta^{\mathrm{c}} = \theta_{\mathrm{m}} = p\,\Theta_{\mathrm{m}}$.

R. W. De Doncker et al., *Advanced Electrical Drives*, Power Systems,
https://doi.org/10.1007/978-3-030-48977-9_7

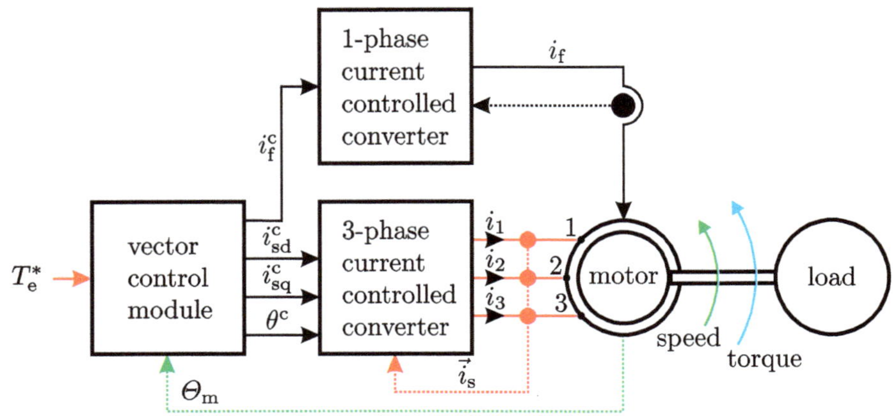

Fig. 7.1 Drive structure of a field-oriented controlled, separately excited synchronous machine

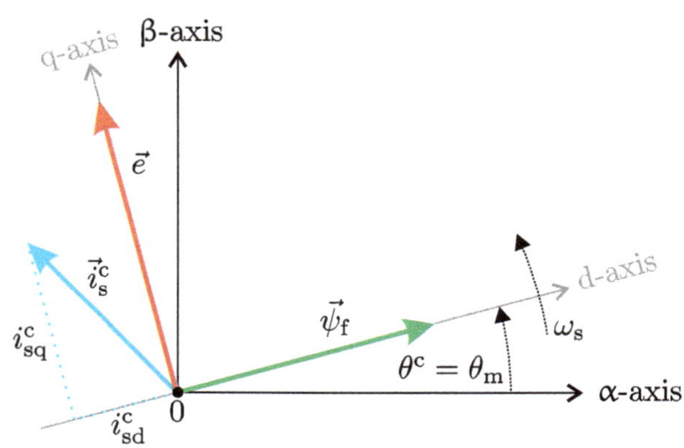

Fig. 7.2 Space vector diagram for rotor-oriented control

The field-oriented control module uses an orthogonal control grid formed by the direct axis d and the quadrature axis q, as shown in Fig. 7.2. The d-axis is aligned with the flux linkage vector $\vec{\psi}_f$, giving the field-oriented control its name. In a synchronous machine, the flux linkage vector $\vec{\psi}_f$ has a fixed orientation relative to the rotor, such that the rotor reference system (xy) matches the rotor-oriented reference system (dq). This is why this type of control is also referred to as rotor flux- or rotor-oriented control. Rotor-oriented control is most suited for synchronous drives because it has the benefit of easily decoupling the direct and quadrature currents in non-salient machines as discussed in Sect. 6.1.3. Given the above, rotor-oriented control is exclusively considered in this chapter.

Careful attention must be given to ensure that the control angle θ^c matches the rotor position θ_m. Two methods exist to achieve this, namely:

- Indirect field-oriented control (IFO), which makes use of a mechanically connected sensor to measure the shaft angle Θ_m.

- Direct field-oriented control (DFO), which estimates the angle θ^{c} by making use of the flux linkage vector $\vec{\psi}_{f}$ or the back-EMF vector $\vec{e}$. The latter can be achieved by making use of sensors in the machine or an observer which uses measured electrical quantities.

Note that position sensorless approaches have been developed to detect the rotor position by injecting high frequency signals. Depending on the method used these methods detect either the rotor position or the flux linkage. Furthermore, the reader should be aware of the fact that synchronous machines have no compensation windings as is the case in servo type DC machines. Hence, the rotor may experience the so-called armature reaction (here caused by stator currents) which leads to different saturation levels at the rotor pole tips. As such, the rotor position does not exactly match the field flux linkage vector $\vec{\psi}_{f}$. However, as synchronous machines typically have large air-gaps the effect of the armature reaction can be ignored as long as the stator currents remain below their nominal values.

In this chapter attention is given exclusively to IFO, because of its widespread use in industry.

The basic question to be addressed in the ensuing sections is how the d-axis and q-axis components i_{sd}^{c} and i_{sq}^{c} of the reference current should be chosen. In case a machine with electrical excitation is used, this question further includes the reference excitation current i_{f}^{c}. Besides the desired torque, also issues like field weakening at high speed or operation with constant stator flux linkage or unity power factor (for machines with electrical excitation) determine the variables i_{sd}^{c}, i_{sq}^{c}, and i_{f}^{c}.

7.2 Control of Non-salient Synchronous Machines

In this section, the control of permanently and electrically excited non-salient synchronous machines is considered. More specifically, the generic implementation of the field-oriented control module as given in Fig. 7.1 is discussed.

Following the key concept of this book, the control for the non-salient synchronous machine can be derived by model inversion. Starting from the current source connected machine model shown in Fig. 6.6 and Eq. (6.4c), the rotor-oriented control model can easily be derived, giving

$$i_{sq}^{*} = \frac{2}{3}\frac{T_{e}^{*}}{\psi_{f}^{*}}. \tag{7.1}$$

The resulting control structure is shown in Fig. 7.3. The control is identical to that found for the DC machine (compare Fig. 5.10). Its simplicity stresses the advantage of rotor flux orientation. The output of this model is the quadrature reference current i_{q}^{*}. It is used as an input to the non-salient control module which generates the command current values i_{sd}^{c} and i_{sq}^{c} for the current controller of the converter.

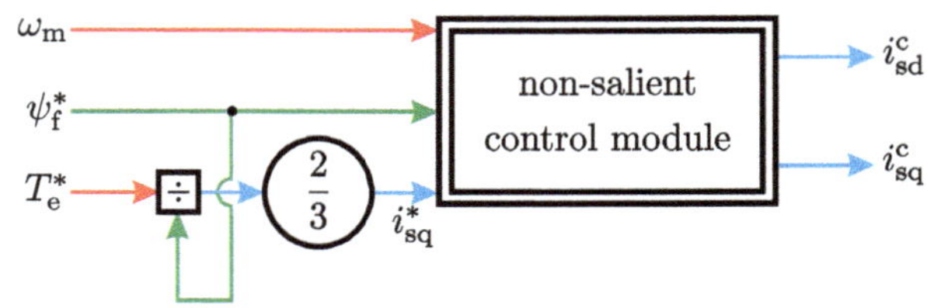

Fig. 7.3 Structure of the field-oriented control module for a non-salient machine

The purpose of the non-salient control module is to moderate the reference current values i_d^c and i_q^c. The control laws used within the control module are dependent on the type of control used and will be discussed in the next subsections. In this context, operation in field weakening under constant stator flux linkage and with unity power factor (for electrically excited machines) will be examined.

7.2.1 *Operation Under Drive Limitations*

The control has to ensure that the reference values for the controller currents are chosen in such a manner that maximum stator current i_s^{max} and maximum stator voltage u_s^{max} are not exceeded. At the same time, the reference values should be chosen to deliver the required torque with highest efficiency. In line with the approach taken for the DC drive, use is made of the current locus diagram in which the reference current vector $\vec{i}_s^{dq}$ is shown together with the operating limits of the drive (see Fig. 7.4).

Lines of Constant Torque
Equation (7.1) shows that the torque is independent of the d-axis current i_{sd} and therefore the choice of the current vector $\vec{i}_s^{dq}$ is not unique for one reference torque value.

$$i_{sq} = \frac{2}{3}\frac{T_e}{\psi_f} \tag{7.2}$$

In the current locus diagram, the lines of constant torque are horizontal lines. Four lines of constant torque are shown in Fig. 7.4 at $i_{sq} = \pm i_s^{max}$ and $i_{sq} = \pm 1/2\, i_s^{max}$ (black dotted lines) as a function of the maximum torque $T_e^{max} = \frac{3}{2} i_s^{max} \psi_f$.

By choosing the operating point dependent currents, the controller has to obey the following limits:

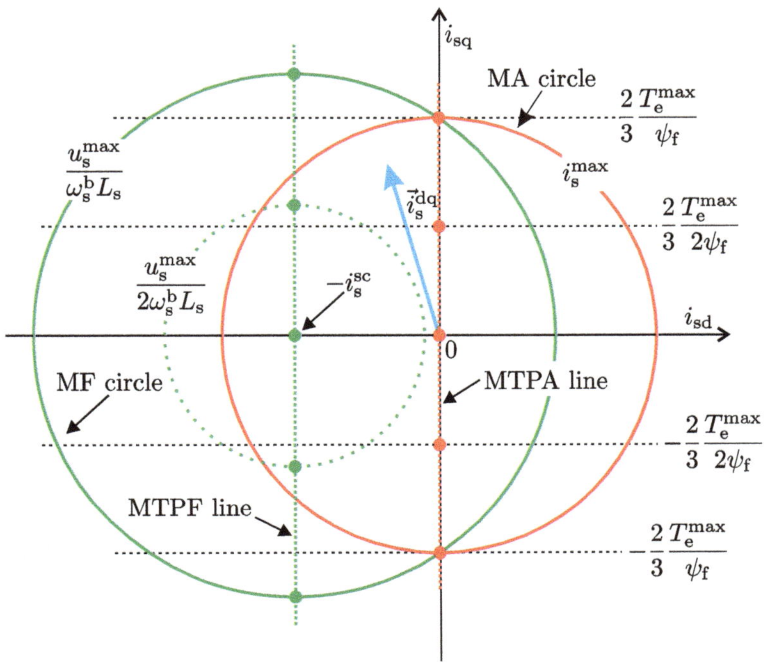

Fig. 7.4 Operational drive limits for non-salient synchronous drives

Current Limit (Maximum Ampere, MA)
A first limit is given by the maximum current of the drive. It limits the operating region to

$$|\vec{i}_s^{dq}| \leq i_s^{max} \tag{7.3}$$

as shown in the current locus diagram by the maximum ampere (MA) circle (red).

Voltage Limit (Maximum Flux Linkage, MF)
A second limit is the maximum available stator flux linkage which arises from the voltage constraint u_s^{max}. For quasi-steady-state conditions, i.e., neglecting time derivatives, equation set (6.4) may be rewritten as

$$\vec{u}_s^{dq} \simeq \vec{i}_s^{dq} R_s + j\omega_s \vec{\psi}_s^{dq} \tag{7.4a}$$

$$\vec{\psi}_s^{dq} = \psi_f + L_s \vec{i}_s^{dq}. \tag{7.4b}$$

Assuming that the dominant term in Eq. (7.4a) is formed by the induced voltage, the resistive component is ignored. Thus,

$$\vec{u}_\mathrm{s}^\mathrm{dq} \cong \mathrm{j}\omega_\mathrm{s}\vec{\psi}_\mathrm{s}^\mathrm{dq}. \tag{7.5}$$

With $|\vec{u}_\mathrm{s}^\mathrm{dq}| = u_\mathrm{s}^\mathrm{max}$, the maximum stator flux linkage can be written as

$$\psi_\mathrm{s}^\mathrm{max} \cong \frac{u_\mathrm{s}^\mathrm{max}}{\omega_\mathrm{s}}. \tag{7.6}$$

The variable $\psi_\mathrm{s}^\mathrm{max}$ represents the maximum stator flux linkage which can be realized by the converter for a given speed ω_s and a given maximum voltage $u_\mathrm{s}^\mathrm{max}$. Voltage $u_\mathrm{s}^\mathrm{max}$ is typically limited by the converter's DC bus voltage. By using $|\vec{\psi}_\mathrm{s}^\mathrm{dq}| = \psi_\mathrm{s}^\mathrm{max}$, Eqs. (7.4b) and (7.6), and by equating the real and imaginary components of $\vec{i}_\mathrm{s}^\mathrm{dq}$ the following expression is derived:

$$\left(i_\mathrm{sd} + i_\mathrm{s}^\mathrm{sc}\right)^2 + i_\mathrm{sq}^2 = \left(\frac{\psi_\mathrm{s}^\mathrm{max}}{L_\mathrm{s}}\right)^2, \tag{7.7}$$

where $i_\mathrm{s}^\mathrm{sc} = \psi_\mathrm{f}/L_\mathrm{s}$ represents the short circuit current of the machine (if R_s is neglected). Equation (7.7) describes the stator current limit due to the maximum flux linkage. It can be represented by a circle with its origin at coordinates $(-i_\mathrm{s}^\mathrm{sc}, 0)$ and a radius $(\psi_\mathrm{s}^\mathrm{max}/L_\mathrm{s} = u_\mathrm{s}^\mathrm{max}/\omega_\mathrm{s} L_\mathrm{s})$ as shown in Fig. 7.4 (green).

The circles of maximum (available) stator flux linkage will increase in radius when the shaft speed is reduced. Consequently, an operating speed will occur where the MF circle (see Fig. 7.4) will completely encompass the MA circle. This implies that the direct axis current value for the controller may be chosen freely within the area bound by the $i_\mathrm{s}^\mathrm{max}$ circle.

To simplify the equations in the following, it is helpful to introduce the ratio between the short circuit current i_s^sc and maximum current $i_\mathrm{s}^\mathrm{max}$, as well as the normalized stator current components i_sd^n and i_sq^n

$$\kappa = \frac{i_\mathrm{s}^\mathrm{sc}}{i_\mathrm{s}^\mathrm{max}} \tag{7.8a}$$

$$i_\mathrm{sd}^\mathrm{n} = \frac{i_\mathrm{sd}}{i_\mathrm{s}^\mathrm{max}} \tag{7.8b}$$

$$i_\mathrm{sq}^\mathrm{n} = \frac{i_\mathrm{sq}}{i_\mathrm{s}^\mathrm{max}}. \tag{7.8c}$$

The value of κ is determined by the current ratings of machine and converter. While the short circuit current i_s^sc is a machine parameter, usually the inverter limits the maximum current $i_\mathrm{s}^\mathrm{max}$. A value of $\kappa < 1$ signifies that short circuit operation occurs within machine and converter limits. In the example shown, a value of $\kappa = 0.66$ was assumed. Normalizing Eq. (7.7) with the maximum stator current $i_\mathrm{s}^\mathrm{max}$, the following equation is derived:

$$\left(i_{\mathrm{sd}}^{\mathrm{n}}+\kappa\right)^{2}+\left(i_{\mathrm{sq}}^{\mathrm{n}}\right)^{2}=\left(\frac{\psi_{\mathrm{s}}^{\max}}{L_{\mathrm{s}} i_{\mathrm{s}}^{\max}}\right)^{2}. \tag{7.9}$$

Equation (7.1) showed that for the non-salient synchronous machine, the control rule for maximum torque per ampere is remarkably simple: always choose the highest available q-axis current. Taking into account the current and voltage limit derived above, the following operation point dependent trajectories are derived:

Maximum Torque per Ampere (MTPA) Line
Operation with lowest ohmic losses in the machine is achieved by choosing an operating condition where the required torque is delivered with the lowest amplitude of the current vector $\vec{i}_{\mathrm{s}}^{\mathrm{dq}}$. For a non-salient machine, this may be achieved by choosing an operating trajectory along the imaginary axis of the current locus diagram, known as the maximum torque per ampere (MTPA) operating line shown in Fig. 7.4. Operation along this line corresponds to the following control law:

$$i_{\mathrm{sd}}^{\mathrm{n}}=0 \tag{7.10a}$$

$$-1<i_{\mathrm{sq}}^{\mathrm{n}}<1. \tag{7.10b}$$

The maximum speed at which nominal torque can be produced, i.e., before the voltage limit is reached, is called *base speed*. It is derived from Eqs. (7.9) and (7.10a)

$$\omega_{\mathrm{s}}^{\mathrm{b}}=\frac{u_{\mathrm{s}}^{\max}}{i_{\mathrm{s}}^{\max} L_{\mathrm{s}} \sqrt{\kappa^{2}+1}}. \tag{7.11}$$

Maximum Torque per Flux Linkage (MTPF) Line
Above base speed, the maximum torque at first occurs where the MA and the MF circles intersect. For machines-converter combinations with $\kappa<1$, maximum torque is eventually only limited by the maximum stator flux linkage when speed is increased further. Consequently, the operating points are found on the MF circle with radius $u_{\mathrm{s}}^{\max}/2\omega_{\mathrm{s}}L_{\mathrm{s}}$. They form the maximum torque per flux linkage (MTPF) line (cp. Fig. 7.4). The operating points on this line are those with the largest possible quadrature current value and the following control law applies

$$i_{\mathrm{sd}}=-i_{\mathrm{s}}^{\mathrm{sc}} \tag{7.12a}$$

$$i_{\mathrm{sq}}=\frac{u_{\mathrm{s}}^{\max}}{\omega_{\mathrm{s}} L_{\mathrm{s}}}. \tag{7.12b}$$

Operation along the MTPF line signifies drive operation far above the base speed with the highest possible torque level. Operation on the MTPF line is restricted by the maximum current constraint. This implies that operation along the MTPF line is possible for that part of the trajectory which is within the MA circle. Obviously, for drives where the short circuit current exceeds the maximum current value ($\kappa > 1$) no operation on the MTPF line is feasible. This is generally the case when the converter is not excessively overrated.

In the next three subsections, specific operating strategies are discussed that are commonly used for PM or electrically excited machines. To develop control strategies a drive sequence is used where the torque command is ramped up to $0.5\, T_{\mathrm{e}}^{\mathrm{max}}$ while accelerating from standstill to one-half rated speed. Then, the torque command stays constant and the speed is increased further well beyond the base speed. Quasi-steady operation of the drive is considered.

7.2.2 *Field Weakening Operation for PM Non-salient Drives*

For electrically excited machines, field weakening is readily achieved by reducing the field current as discussed for the DC machine. Permanent magnet machines are able to achieve field weakening by using the direct axis current to produce a flux linkage component that opposes the magnet flux linkage and thus reduces the back-EMF. The current locus and the corresponding direct and quadrature current versus speed characteristics will be examined. Two cases, namely $\kappa < 1$ and $\kappa > 1$ are considered, corresponding to machines with a short circuit current less and greater than the maximum stator current value $i_{\mathrm{s}}^{\mathrm{max}}$.

7.2.2.1 Machines with $\kappa < 1$ ($i_{\mathrm{s}}^{\mathrm{sc}} < i_{\mathrm{s}}^{\mathrm{max}}$)

The current space vector locus, which corresponds to the case $\kappa < 1$, is given in Fig. 7.5. It shows three specific operating trajectories.

First Trajectory, 0 → A, MTPA

The first trajectory, from 0 → A corresponds to the 50% maximum torque ramp applied while the machine accelerates from standstill to one-half rated speed. This operation trajectory is on the MTPA line. As speed increases, the maximum flux linkage circle radius decreases. Operation at point A is possible until speed $\omega_{\mathrm{s}}^{\mathrm{A}}$ is reached, where the MF circle intersects with said operating point. The normalized speed $\omega_{\mathrm{s}}^{\mathrm{A}}$ can be expressed with the aid of equations (7.6), (7.7), (7.10a), and (7.11):

$$\frac{\omega_{\mathrm{s}}^{\mathrm{A}}}{\omega_{\mathrm{s}}^{\mathrm{b}}} = \sqrt{\frac{\kappa^2 + 1}{\kappa^2 + \left(i_{\mathrm{sq}}^{\mathrm{n}}\right)^2}}. \tag{7.13}$$

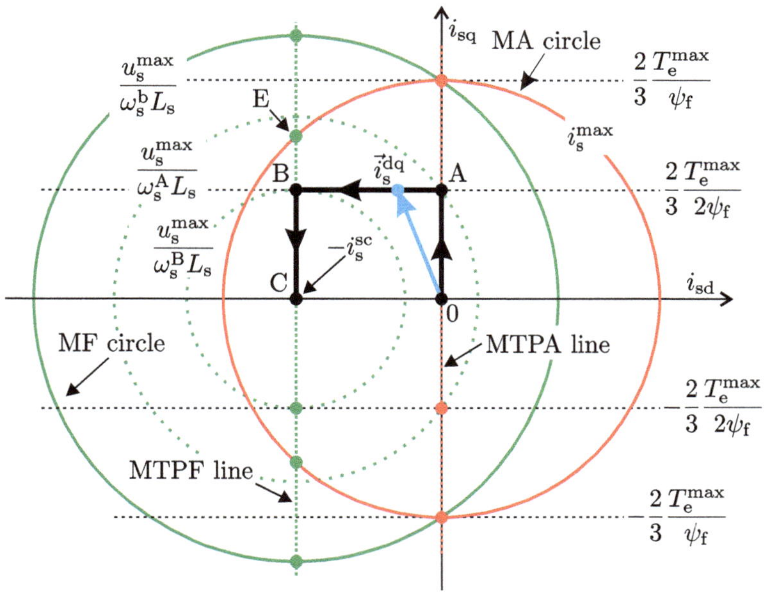

Fig. 7.5 Current locus diagram: operational drive trajectory $0 \rightarrow \omega_s^C$ for non-salient synchronous drives with $i_s^{sc} < i_s^{max}$

Second Trajectory, A $\rightarrow$ B, **Basic Field Weakening**
For shaft speeds greater than ω_s^A, field weakening must take place given the need to ensure that the current space vector remains on or within the MF circle. Note that the shaft speed at which field weakening must start is dependent on i_{sq}^n which may vary over the range $-1 \rightarrow 1$. During the operating trajectory A $\rightarrow$ B, the current space vector is kept on the maximum flux linkage circle and at the same time on the q-reference current value $i_{sq}^n = 0.5$ (for the chosen example scenario). Subsequent application of goniometric laws linked to the MF flux linkage circle and constant torque line shows that the trajectory can be expressed as

$$i_{sd}^n = -\kappa + \sqrt{\kappa^2\left(\frac{\omega_s^A}{\omega_s}\right)^2 + \left[\left(\frac{\omega_s^A}{\omega_s}\right)^2 - 1\right]\left(i_{sq}^n\right)^2} \tag{7.14a}$$

$$i_{sq}^n = \text{const.} \tag{7.14b}$$

Third Trajectory, B $\rightarrow$ C, **MTPF**
The final trajectory B $\rightarrow$ C occurs when the shaft speed ω_s^B is reached. Operation on the basic field weakening trajectory cannot be continued. For operation beyond the shaft speed ω_s^B, the operating point is on the MTPF line. In this operating mode, the currents are given by

$$i_{sd}^{n} = -\kappa \tag{7.15a}$$

$$i_{sq}^{n} = \frac{\omega_{s}^{b}}{\omega_{s}}\sqrt{\kappa^{2}+1}. \tag{7.15b}$$

Recalling that the torque is directly proportional to the q-axis current, it is obvious that the maximum available torque drops below the set value 0.5 T_{e}^{max}. The torque is readily calculated by substituting Eq. (7.15b) in the torque equation (7.1). Theoretically, the speed can be increased infinitely while the torque decreases $\propto 1/\omega_s$, assuming $R_s = 0$. The output power remains constant. Note that operation under field weakening conditions allows the drive to operate far beyond the base speed within the voltage limit. This is achieved by the use of a substantial direct axis current in the machine. If for some reason, i.e., malfunction of the drive, this current is removed, an over voltage occurs that has the potential to damage substantially the converter.

Current-Speed Diagram

An alternative approach to the use of the current locus diagram for representing the current locus is to consider the normalized direct and quadrature current components as function of the normalized (with respect to the base speed) shaft speed. The so-called *current-speed diagram* is given in Fig. 7.6. In this diagram, the green and

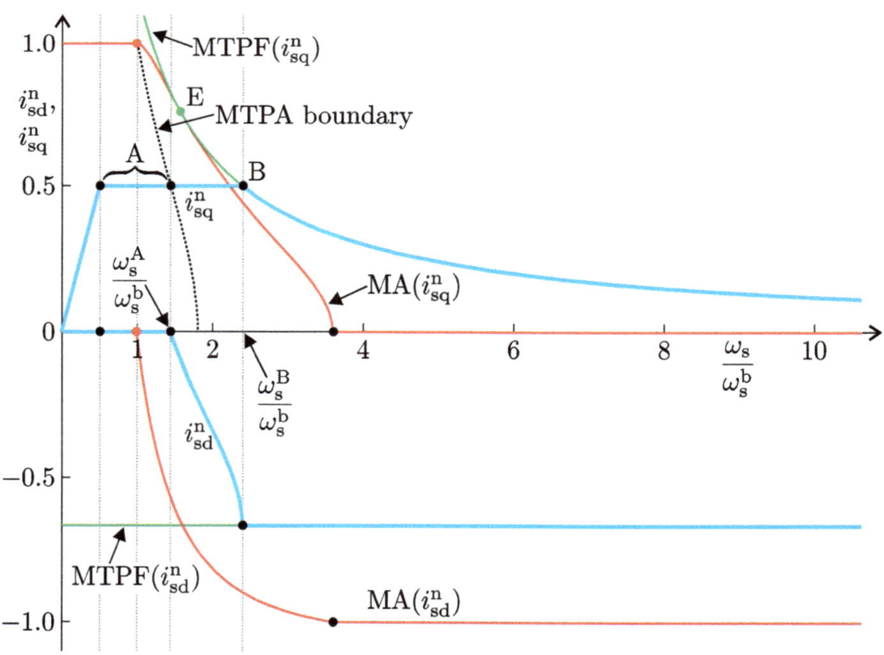

Fig. 7.6 Current-speed diagram: operational drive trajectory $0 \rightarrow \omega_{s}^{C}$ for non-salient synchronous drives with $i_{s}^{sc} < i_{s}^{max}$

red curves correspond to the maximum direct and quadrature axis currents at the given speed. They encompass the MTPA line (motoring only) up to the maximum torque point and then along the MA circle. The red dotted curve represents the base speed equation (7.13). The intersection of this line with the selected quadrature current value i_{sq}^{n} occurs at speed ω_s^A above which field weakening must be used. The green dashed curve represents the quadrature component of the MTPF line. The blue colored lines show the normalized direct current i_{sd}^{n} and the normalized quadrature current i_{sq}^{n} for currents that correspond to operation with half maximum torque. In the current locus diagram these lines correspond to the trajectory $0 \rightarrow A \rightarrow B$ and part of the trajectory $B \rightarrow C$ (C is reached when $\omega_s \rightarrow \infty$). Included in Fig. 7.6 is operating point E which identifies the speed from where operation along the MTPF trajectory is feasible. The diagram shows that the d-axis current is equal to the short circuit current for $\omega_s \geq \omega_s^B$. It further shows that the quadrature current curve i_{sq}^{n} coincides with the q-component of the MTPF trajectory as defined in current locus diagram for $\omega_s \geq \omega_s^B$.

Short Circuit Operation

In the final part of this section attention is given to operation under short circuit conditions when the operating point is point C. This happens not only at very high speeds under field weakening but also in case the supply voltage becomes zero, given that the radius of the maximum flux linkage circle is determined by the ratio u_s^{max}/ω_s. Observation of the current locus diagram shows that the machine would then operate under zero torque conditions. In practice, this is not the case because the winding resistance is non-zero. As a consequence, a relatively low, but non-zero, torque will be present in the machine at high speeds. A brief analysis of the quasi-steady-state operation under short-circuited condition may be undertaken with the aid of the synchronous, phasor based machine model from Fig. 6.8 by setting $|\underline{u}_s| = 0$. If the rotor flux linkage is represented by the phasor $\underline{\psi}_f = \psi_f$, the current amplitude $|\underline{i}_s|$ and the machine torque T_e may be rewritten as

$$|\underline{i}_s| = \frac{\omega_s \frac{\psi_f}{R_s}}{\sqrt{1 + \left(\omega_s \frac{L_s}{R_s}\right)^2}} \tag{7.16a}$$

$$T_e = -\frac{3}{2} \frac{\omega_s \frac{\psi_f^2}{R_s}}{1 + \left(\omega_s \frac{L_s}{R_s}\right)^2} \tag{7.16b}$$

An illustrative example based on the use of equation set (7.16) is shown in Fig. 7.7. Observation of Fig. 7.7 and the above equation shows that the torque versus speed characteristic is very similar to that of an induction machine operating under DC injection braking conditions (compare Eq. (9.4)). The speed which corresponds to the maximum absolute torque level $3\psi_f^2/4L_s$ is defined by the ratio R_s/L_s as may

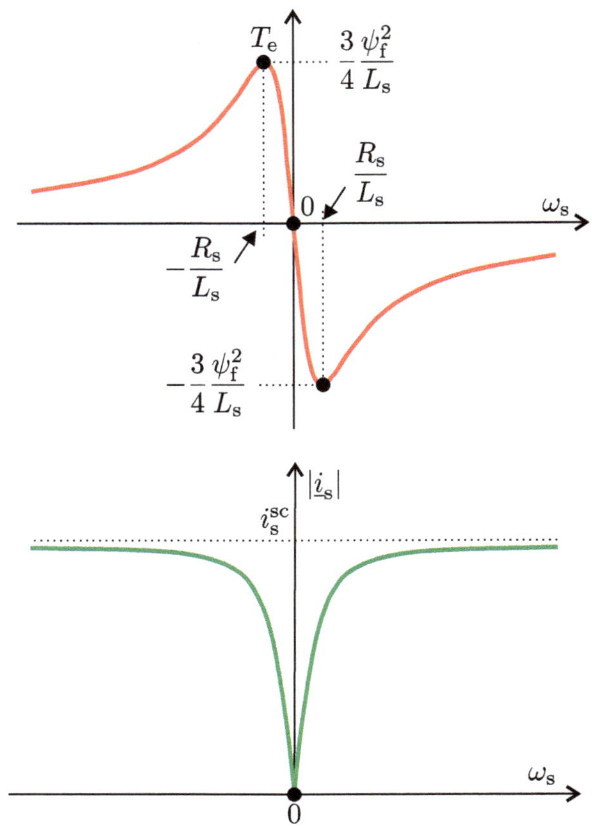

Fig. 7.7 Short circuit operation of non-salient synchronous machines

be observed from Fig. 7.7. Braking energy is in this case fully dissipated in the machine. The example also shows that the short circuit current is equal to i_s^{sc} as anticipated.

7.2.2.2 Machines with $\kappa > 1$ ($i_s^{sc} > i_s^{max}$)

The second case to be discussed in this section is a machine converter configuration with $\kappa > 1$. This is the case, when the short circuit current of the machine is larger than the maximum current rating of the converter. Shown in the diagram is the current locus for the chosen control sequence $0 \to A \to B \to C$.

First and Second Trajectory, $0 \to A \to B$
The first and second trajectory are identical to the case with $\kappa < 1$.

Third Trajectory, $B \to C$**, Maximum Current (MA)**
The final trajectory $B \to C$ occurs when the shaft speed ω_s^B is reached. In this case the operation is limited to a locus that is part of the maximum current circle. An

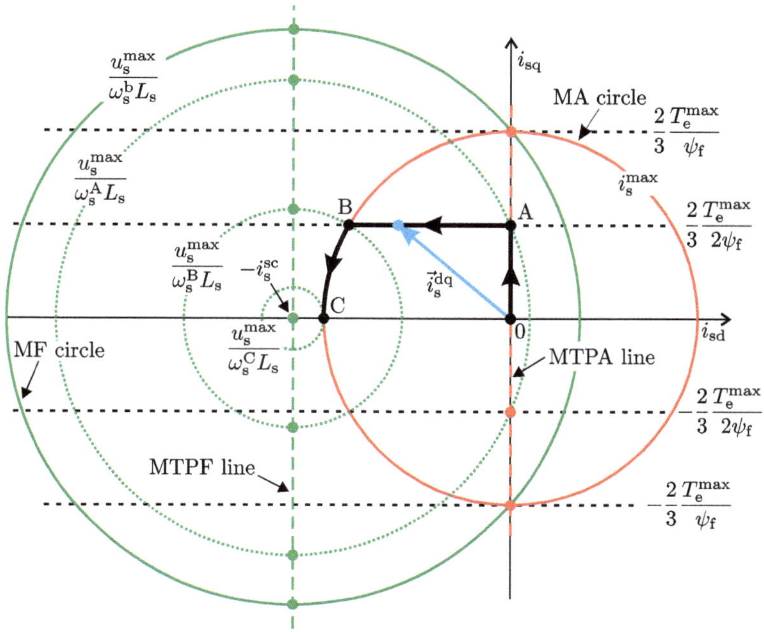

Fig. 7.8 Current locus diagram: operational drive trajectory $0 \rightarrow \omega_s^C$ for non-salient synchronous drives with $i_s^{sc} > i_s^{max}$

operation along the MTPF line is not possible, given that this trajectory is outside the maximum current circle, i.e., $\kappa > 1$. The trajectory is defined by the expressions

$$i_{sd}^n = \frac{\kappa^2 + 1}{2\kappa}\left[\left(\frac{\omega_s^b}{\omega_s}\right)^2 - 1\right] \tag{7.17a}$$

$$i_{sq}^n = \sqrt{1 - \left(i_{sd}^n\right)^2}. \tag{7.17b}$$

The highest drive operating speed ω_s^C is reached when the maximum flux linkage circle and maximum current circle coincide at a single operating point C, as may be observed from Fig. 7.8. Note that in this case torque production continues only up to the maximum speed ω_s^C, which is markedly different when compared to the case $\kappa < 1$.

Current-Speed Diagram

The corresponding current-speed diagram is given in Fig. 7.9. The colors used are identical to those in the previous case. It can be seen that operation occurs along the MA circle for $\omega_s \geq \omega_s^B$ once point B is reached. Operation along the MTPF line (green dashed curve) is not possible because it is lies outside the maximum current circle. Eventually, the maximum speed ω_s^C is reached.

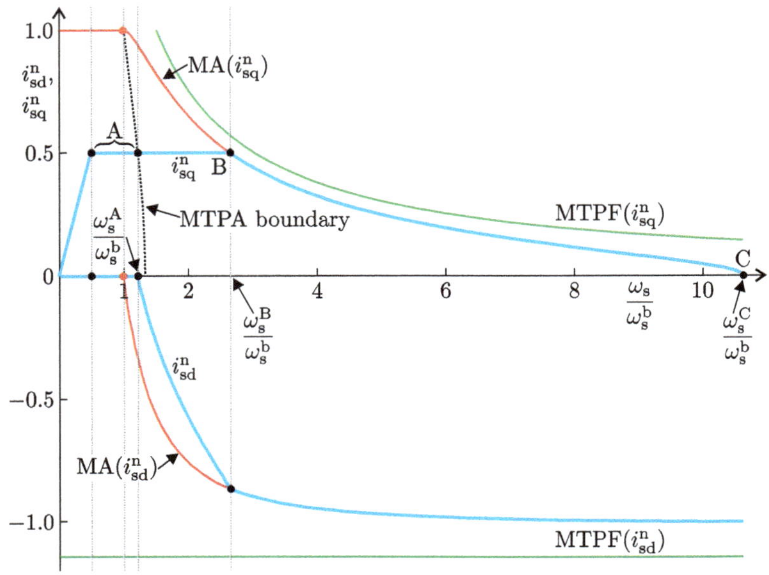

Fig. 7.9 Current-speed diagram: operational drive trajectory $0 \rightarrow \omega_s^C$ for non-salient synchronous drives with $i_s^{sc} > i_s^{max}$

A simulation-based tutorial is given in Sect. 7.6.1 in which the reader can examine the operation of the drive for both cases discussed above.

7.2.3 Field Weakening for PM Non-salient Drives, with Constant Stator Flux Linkage Control

When operating under MTPA conditions, as discussed in the previous section, the stator flux linkage will increase as the torque reference value is increased. For some machines this may lead to saturation of the stator yoke, which is not desirable. To avoid this problem, a control strategy is discussed in this section which maintains the stator flux linkage constant for a given (constant) field flux linkage ψ_f. An illustration of the proposed control strategy is given in Fig. 7.10. The direct axis current i_{sd}^c is manipulated to maintain the stator flux linkage amplitude constant for a given value of quadrature axis current i_{sq}^c. Figure 7.10 shows that the relationship between stator flux linkage, field flux linkage, and stator current components i_{sd}^c and i_{sq}^c can be written as

$$\psi_s = \sqrt{\left(\psi_f + L_s\, i_{sd}^c\right)^2 + \left(L_s\, i_{sq}^c\right)^2}. \tag{7.18}$$

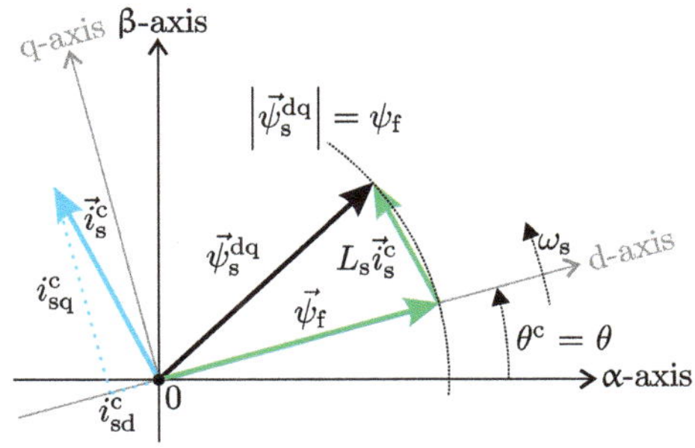

Fig. 7.10 Rotor-oriented control with constant stator flux linkage

The stator flux linkage magnitude is set to $|\vec{\psi}_s| = \psi_f$, which is the value that must be accommodated in case $i_{sd} = 0$ and $i_{sq} = 0$. Equation (7.18) can then be expressed in normalized form as

$$i_{sd}^n = -\kappa \pm \sqrt{\kappa^2 - \left(i_{sq}^n\right)^2} \tag{7.19}$$

with $\kappa = {}^{i_s^{sc}}/_{i_s^{max}}$ and $i_s^c = {}^{\psi_f}/_{L_s}$. The current locus diagram given in Fig. 7.11 shows the operating locus of the current space vector. It has three specific trajectories. The first trajectory, $0 \rightarrow$ A, represents operation as defined by the control law according to equation set (7.19). The quadrature current i_{sq}^n is increased from 0 to 0.5 and then remains constant.

Figure 7.11 shows that for shaft speeds below the speed ω_s^{bS}, constant stator flux linkage is realizable for a quadrature current range which is limited to point D. The corresponding (normalized) currents are given by equation set (7.20). They confirm that the maximum torque capability of the drive has been reduced as a result of constant stator flux linkage control strategy. The extend of this limitation is dependent on the value of κ, as may be deduced from Eq. (7.20b) (under the assumption that $\kappa > 0.5$).

$$i_{sd}^n = -\frac{1}{2\kappa} \tag{7.20a}$$

$$i_{sq}^n = \sqrt{1 - \frac{1}{4\kappa^2}} \tag{7.20b}$$

Operation under constant stator flux linkage conditions can be maintained, provided that the maximum flux linkage circle is outside the circle which coincides with trajectory $0 \rightarrow$ A. As speed increases, the maximum flux linkage circle radius reduces until the speed ω_s^{bS} is reached.

$$\frac{\omega_s^{bS}}{\omega_s^b} = \sqrt{\frac{\kappa^2 + 1}{\kappa^2}} \tag{7.21}$$

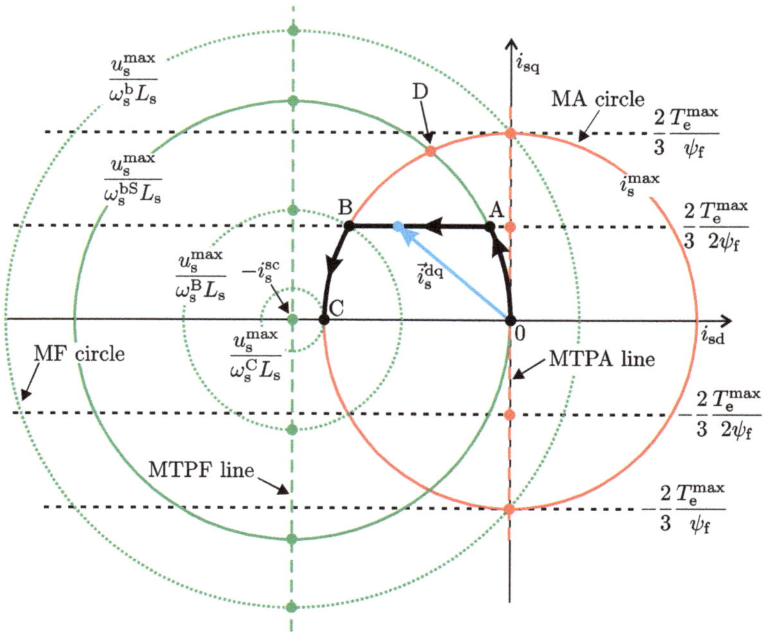

Fig. 7.11 Current locus diagram: operational drive trajectory $0 \rightarrow \omega_s^C$ for non-salient synchronous drives with $i_s^{sc} > i_s^{max}$ and constant stator flux linkage control

Note that the speed ω_s^{bS} exceeds the base speed ω_s^b, which is to be expected, given that operation is not along the MTPA line. For speeds greater than ω_s^{bS}, field weakening operation occurs for the trajectory A $\rightarrow$ B $\rightarrow$ C as discussed in the previous section.

The current-speed diagram given in Fig. 7.12 shows that field weakening operation is required for $\omega_s > \omega_s^{bS}$. Also clearly noticeable from this figure is the torque limitation caused by the need to couple the direct and quadrature currents in order to maintain constant stator flux linkage in the speed range $\omega_s \leq \omega_s^{bS}$. The tutorial given in Sect. 7.6.2 shows a simulation of a drive which utilizes the constant stator flux linkage control strategy.

7.2.4 *Field Weakening for Electrically Excited Non-salient Drive, with Constant Stator Flux and Unity Power Factor Control*

Machines which utilize a field winding have an additional degree of freedom which may be utilized to achieve unity power factor operation. As a consequence, the converter is not required to provide any reactive power. In addition, constant stator

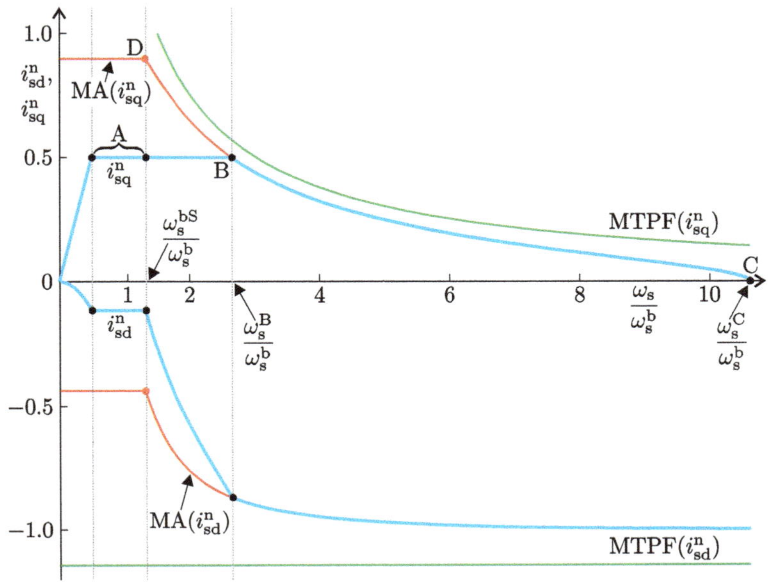

Fig. 7.12 Current-speed diagram: operational drive trajectory $0 \rightarrow \omega_s^C$ for non-salient synchronous drives with $i_s^{sc} > i_s^{max}$ and constant stator flux linkage control

Fig. 7.13 Rotor-oriented control with constant stator flux linkage and unity power factor

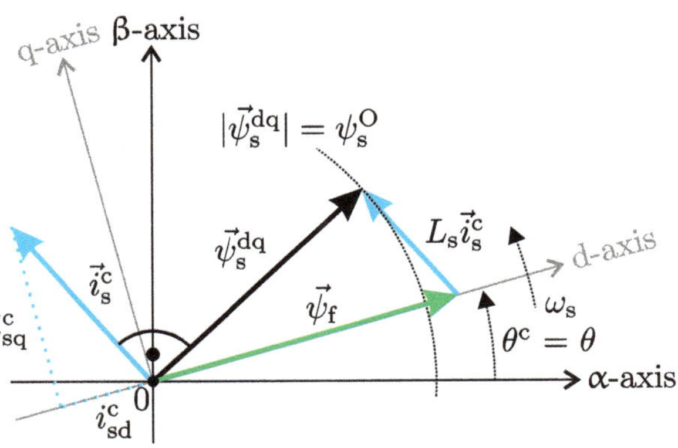

flux linkage operation $\psi_s = \psi_f$ is achieved as discussed in the previous section. The basic control strategy may be deduced by realizing that the voltage $\vec{u}_s^{dq}$ can be approximated as $\vec{u}_s^{dq} \simeq j\omega_s\vec{\psi}_s^{dq}$ given that the voltage drop across the stator resistance is assumed to be low. Consequently, unity power factor is achieved by ensuring that the voltage vector $\vec{u}_s^{dq}$ and current vector $\vec{i}_s^{dq}$ are kept in phase. This condition implies that the current vector should remain orthogonal to the stator flux linkage vector $\vec{\psi}_s^{dq}$ as shown in Fig. 7.13. This again implies that the field flux linkage amplitude ψ_f must be varied for a given user defined value of the quadrature current i_{sq}.

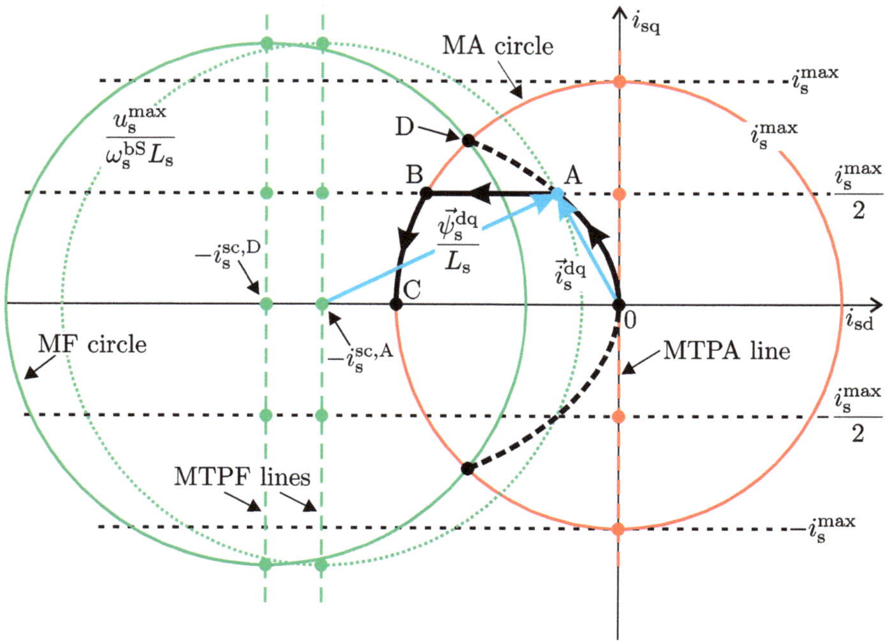

Fig. 7.14 Operational drive trajectory $0 \rightarrow \omega_s^C$ for non-salient synchronous drives with constant stator flux linkage and unity power factor control

The concept of manipulating the field flux linkage ψ_f and therefore the short circuit current i_s^{sc} below the speed ω_s^{bS} of the drive implies that the variable $\kappa = i_s^{sc}/i_s^{max}$ is no longer constant. For the sake of comparison with the previous case, the normalized short circuit current variable κ^O is introduced which is equal to the constant value of the rated short circuit current ratio used above. The flux linkage level ψ_s^O as shown in Fig. 7.13 is chosen to be the same value as assumed in the previous section. Consequently, the defined base speed ω_s^{bS} of the drive remains unchanged.

The process of determining how the field flux linkage ψ_f and the corresponding short circuit current i_s^{sc} must be chosen to maintain unity power factor operation is undertaken with the aid of the current locus diagram given in Fig. 7.14.

Shown in Fig. 7.14 is the trajectory $0 \rightarrow$ A which corresponds to operation below the base speed. This trajectory coincides partly with the dashed locus $0 \rightarrow$ A $\rightarrow$ D which is no longer circular as was the case previously (see Fig. 7.11). The reason for this is that the field flux linkage level must be adjusted as a function of the user defined torque reference value to maintain the stator flux linkage vector $\vec{\psi}_s^{dq}$ orthogonal to the current reference vector $\vec{i}_s^{dq}$ as discussed above.

Note that this objective is reached at the cost of a reduced operating quadrature current range as may be observed by the location of operating point D on the maximum current curve relative to the previous case (see Fig. 7.11). After some goniometric manipulations it may be shown that this point is linked to the normalized current values

$$i_{\mathrm{sd}}^{\mathrm{n,D}} = -\frac{1}{\sqrt{\left(\kappa^{\mathrm{O}}\right)^2 + 1}} \tag{7.22a}$$

$$i_{\mathrm{sq}}^{\mathrm{n,D}} = \frac{\kappa^{\mathrm{O}}}{\sqrt{\left(\kappa^{\mathrm{O}}\right)^2 + 1}} \tag{7.22b}$$

$$i_{\mathrm{s}}^{\mathrm{sc,n,max}} = i_{\mathrm{s}}^{\mathrm{sc,n,D}} = \kappa^{\mathrm{D}} = \sqrt{\left(\kappa^{\mathrm{O}}\right)^2 + 1}. \tag{7.22c}$$

The measure of quadrature current reduction may be quantified by comparing the $i_{\mathrm{sq}}^{\mathrm{n}}$ values according to Eqs. (7.20) and (7.22) where the latter corresponds to stator flux linkage control without power factor control. A careful analysis of Fig. 7.14 gives that the dashed locus $0 \rightarrow \mathrm{A} \rightarrow \mathrm{D}$ may be expressed as

$$i_{\mathrm{sd}}^{\mathrm{n}} = -\frac{\left(i_{\mathrm{sq}}^{\mathrm{n}}\right)^2}{\sqrt{\left(\kappa^{\mathrm{O}}\right)^2 - \left(i_{\mathrm{sq}}^{\mathrm{n}}\right)^2}} \tag{7.23a}$$

$$i_{\mathrm{sq}}^{\mathrm{n}} \propto T^* \tag{7.23b}$$

$$i_{\mathrm{s}}^{\mathrm{sc,n}} = \kappa = \frac{\left(\kappa^{\mathrm{O}}\right)^2}{\sqrt{\left(\kappa^{\mathrm{O}}\right)^2 - \left(i_{\mathrm{sq}}^{\mathrm{n}}\right)^2}}. \tag{7.23c}$$

where $i_{\mathrm{sq}}^{\mathrm{n}}$ corresponds to the user defined normalized current reference. The variable $i_{\mathrm{s}}^{\mathrm{sc,n}}$ represents the normalized short circuit flux linkage (in PM applications referred to as κ) and its value must be increased as the quadrature current reference is increased. The changing short circuit current value is noticeable in Fig. 7.14 by the displacement of the constant stator flux linkage circles which is required to maintain the combination of constant stator flux linkage and unity power factor below the base speed. For operation above the base speed, which is the remaining part of the chosen excitation trajectory $\mathrm{A} \rightarrow \mathrm{B} \rightarrow \mathrm{C}$, field weakening is activated. Due to the changing normalized short circuit current, the relationship for $\mathrm{A} \rightarrow \mathrm{B}$ trajectory slightly deviates from equation set (7.14) and may be expressed as

$$i_{\mathrm{sd}}^{\mathrm{n}} = -\kappa + \sqrt{\left(\frac{\omega_{\mathrm{s}}^{\mathrm{b}}}{\omega_{\mathrm{s}}}\right)^2 \left(1 + \left(\kappa^{\mathrm{O}}\right)^2\right) - \left(i_{\mathrm{sq}}^{\mathrm{n}}\right)^2} \tag{7.24a}$$

$$i_{\mathrm{sq}}^{\mathrm{n}} = \mathrm{const.} \tag{7.24b}$$

For the same reason, trajectory $\mathrm{B} \rightarrow \mathrm{C}$ in this case is based on relationship

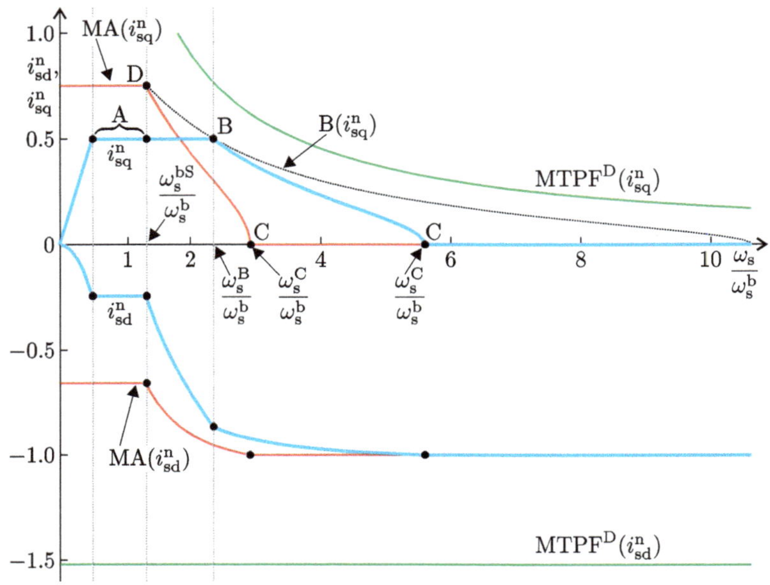

Fig. 7.15 Current-speed diagram for non-salient synchronous drives with constant stator flux linkage and unity power factor control

$$i^{n}_{sd} = \frac{1}{2\kappa}\left[\left(\frac{\omega_s^b}{\omega_s}\right)^2\left(1+\left(\kappa^{O}\right)^2\right)-\left(1+\kappa^2\right)\right] \tag{7.25a}$$

$$i^{n}_{sq} = \sqrt{1-\left(i^{n}_{sd}\right)^2} \tag{7.25b}$$

and also differs from equation set (7.17). As may be seen in Fig. 7.15 and equation set (7.25), the maximum normalized speed of the drive operating under constant stator flux linkage and unity power factor control varies depending on the actual value of normalized short circuit current κ

$$\frac{\omega_s^C}{\omega_s^b} = \sqrt{\frac{1+\left(\kappa^{O}\right)^2}{(1-\kappa)^2}}. \tag{7.26}$$

Note that the linear relationship between torque and quadrature current is no longer maintained with this control strategy given that the field flux linkage ψ_f is a function of the quadrature current. In the approach taken here, the initial field flux linkage ψ_f (for $i_{sq} = 0$) was chosen equal to the value used in the previous section. Consequently, the field flux linkage level ψ_f increases as the quadrature current is increased. Normally, the field flux linkage ψ_f must not exceed its rated value which implies that the rated field flux linkage is realized when the machine is operating under maximum torque conditions, i.e., point D in Fig. 7.14. A consequence of this control approach is that the initial field flux linkage level ψ_f must be chosen lower than the rated value. This will become apparent from the tutorial given in Sect. 7.6.3.

7.3 Control of Salient Synchronous Machines

The operating principles discussed previously for the non-salient machine were primarily aimed at determining an optimum trajectory for the controller current vector $\vec{i}_s^{dq}$ for drive operation over a given shaft speed range. For this purpose, specific operating trajectories such as MTPA and MTPF were introduced, together with transition regions between these two. For the salient machine, similar control strategies can be defined, despite the fact that the torque is determined by the direct and quadrature current as was discussed in Sect. 6.2. Consequently, the process of determining the required current references i_d^c and i_q^c for a rotor-oriented current control algorithm is more sophisticated.

Lines of Constant Torque

A convenient starting point for this analysis is the salient field-oriented model discussed in Sect. 6.2.2. The torque equation (6.20) can be rewritten in a normalized form using the saliency factor χ from Eq. (6.18):

$$T_e^n = \frac{T_e}{\frac{3}{2}\psi_f \, i_s^{max}} = i_{sq}^n \left(1 - \frac{2\chi}{\kappa} i_{sd}^n\right). \tag{7.27}$$

Note that the normalization is undertaken with respect to the maximum torque level of a non-salient machine. The variable χ is greater than zero for $L_{sd} < L_{sq}$, resulting in a maximum Torque $T_e^{n,max} > 1$. Furthermore, the normalized short circuit current κ is assumed to be constant (i.e., excitation is provided by permanent magnets). Equation (7.27) confirms an earlier observation that torque is a function of both currents. Consequently, the lines of constant torque in the salient synchronous current locus diagram of Fig. 7.4 are no longer horizontal lines. The constant torque contours (black dashed lines) shown in the salient synchronous current locus diagram are defined by Eq. (7.27). The analysis in this section concentrates on motor operation in the second quadrant of the current locus diagram for $\chi > 0$. The trajectories for the third quadrant can be found by mirroring the shown trajectories along the d-axis.

Current Limit (Maximum Ampere, MA)

As for the non-salient machine, the MA circle is shown in Fig. 7.16. The corresponding current-speed diagram is shwon in Fig. 7.17.

Maximum Torque per Ampere (MTPA) Line

Maximum torque T_e^{max} is given where the MA circle touches the lines of constant torque in one point. This operating point produces the largest torque which can

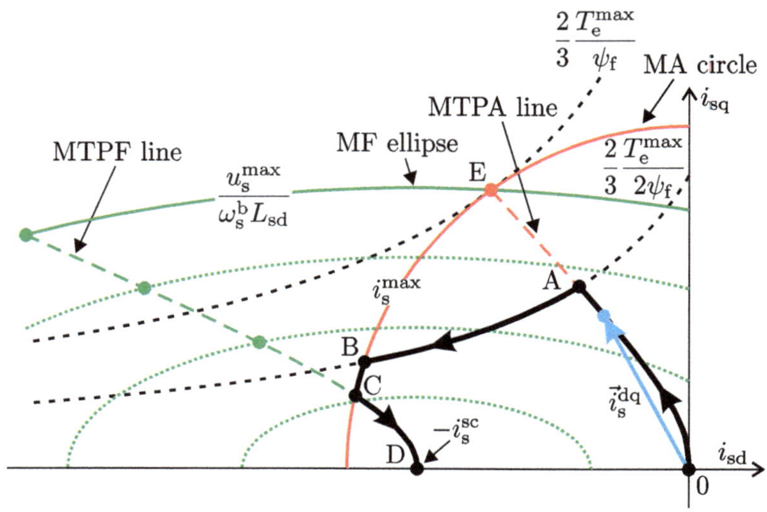

Fig. 7.16 Current locus diagram: Operational drive trajectory $0 \rightarrow \omega_s^C$ for salient synchronous drives with $i_s^{sc} < i_s^{max}$

be realized with the chosen value of i_s^{max}. It is referred to as the *maximum torque point*, which is part of the MTPA trajectory of the salient machine. The entire MTPA trajectory is found by introducing a set of concentric circles centered on the origin with radius $0 \rightarrow i_s^{max}$. For a given i_s value the MTPA point is the point in which the circle touches a line of constant torque in one point:

$$\left(i_{sd}^n\right)^2 + \left(i_{sq}^n\right)^2 = \left(i_s^n\right)^2 \tag{7.28}$$

with $i_s^n = i_s/i_s^{max}$. Subsequent use of Eq. (7.28) with Eq. (7.27) leads to an expression for the torque as function of the normalized direct axis current i_{sd}^n. Differentiation gives the MTPA value of i_{sd}^n as function of i_s^n, which can in turn be used to find the corresponding i_{sq}^n value. The resulting set of coordinates which defines the MTPA trajectory is given in the following equation set:

$$i_{sd}^n = \frac{\kappa}{8\chi} - \sqrt{\frac{\left(i_s^n\right)^2}{2} + \left(\frac{\kappa}{8\chi}\right)^2} \tag{7.29a}$$

$$i_{sq}^n = \sqrt{\left(i_s^n\right)^2 - \left(i_{sd}^n\right)^2} \tag{7.29b}$$

From a control objective it is helpful to re-map the variable i_s^n as function of the torque T_e^n, using Eqs. (7.29) and (7.27). The resulting direct and quadrature reference values are shown in Fig. 7.18 for operation below the base speed. These values can be used as input to the current controller for MTPA.

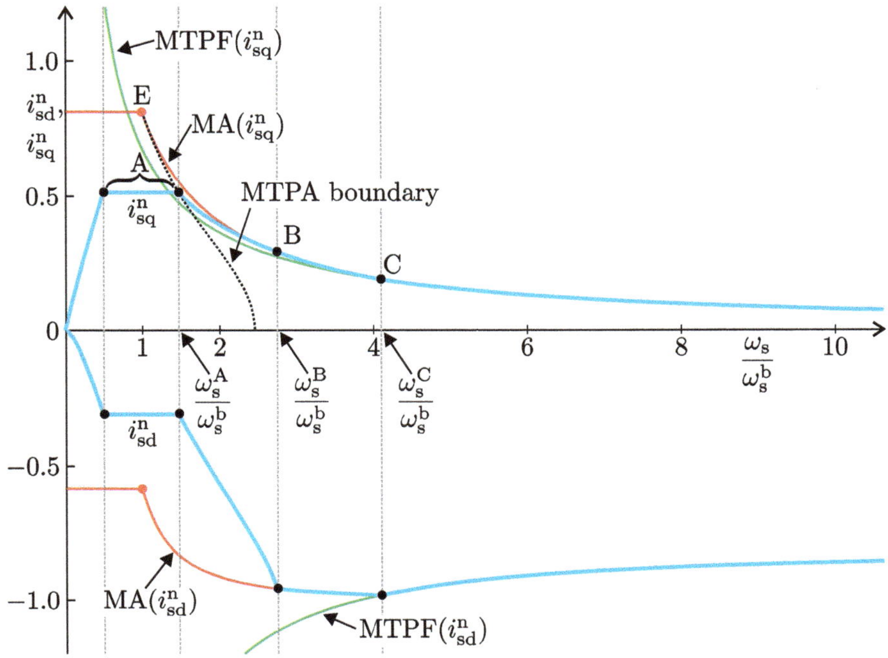

Fig. 7.17 Current-speed diagram: Operational drive trajectory normalized current versus speed plane for salient synchronous drives with $i_s^{sc} < i_s^{max}$

Fig. 7.18 Direct and quadrature current as a function of torque in MTPA mode below base speed for a salient synchronous drive

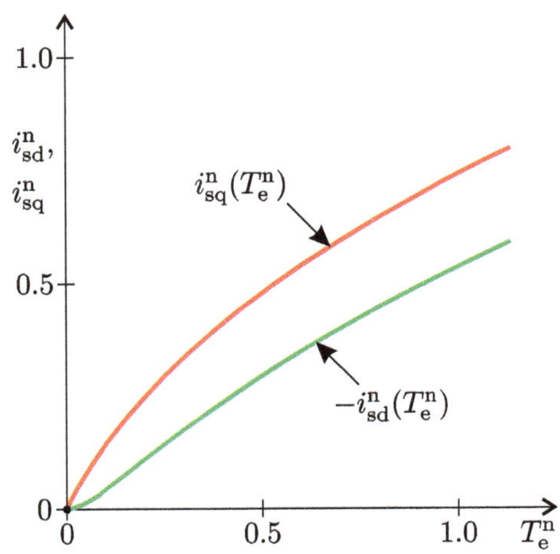

$$(i_{sd}^{n})^2 + \left(\frac{T_e^n}{\left(1 - \frac{2\chi}{\kappa} i_{sd}^n\right)}\right)^2 = (i_s^n)^2 \tag{7.30}$$

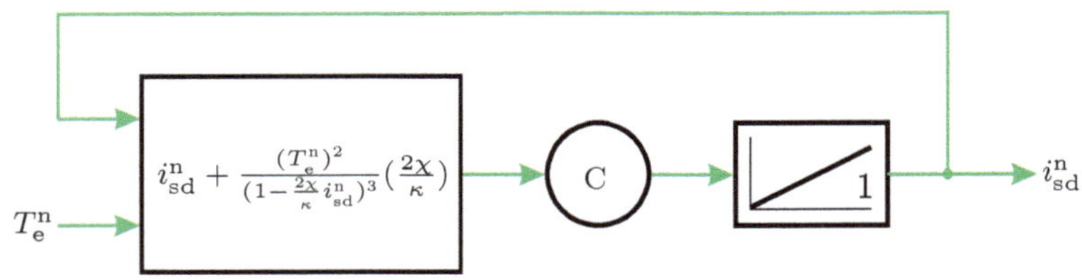

Fig. 7.19 Direct axis reference controller in MTPA mode below base speed for a salient synchronous drive

A convenient novel direct axis reference controller [2] can be found using Eq. (7.30) and considering the derivative with respect to $i_{\rm sd}^{\rm n}$, namely:

$$i_{\rm sd}^{\rm n} + \frac{\left(T_{\rm e}^{\rm n}\right)^2}{\left(1 - \frac{2\chi}{\kappa} i_{\rm sd}^{\rm n}\right)^3} \left(\frac{2\chi}{\kappa}\right) = 0 \tag{7.31}$$

The originality of the solution envisaged here is to consider expression (7.31) as the input of a relaxation circuit that consists of a gain C and integrator of which its output is the required $i_{\rm sd}^{\rm n}$ value as may be observed from Fig. 7.19. The corresponding quadrature reference value can then be found using the output of the direct axis reference controller and Eq. (7.27) (where $i_{\rm sq}^{\rm n}$ must be written explicitly). These values can be used as input to the current controller for MTPA, i.e., for operation below base speed.

Voltage Limit (Maximum Flux Linkage, MF)

The maximum flux linkage (MF) lines can be found in a similar way as done for the non-salient case starting from Eq. (6.16):

$$\left(i_{\rm sd}^{\rm n} + \kappa\right)^2 + \left(i_{\rm sq}^{\rm n}\right)^2 (2\chi + 1)^2 = i_{\rm O}^2 \tag{7.32}$$

The variable $i_{\rm O} = \psi_{\rm s}^{\rm max}/L_{\rm sd} i_{\rm s}^{\rm max}$ is directly linked to the maximum flux linkage $\psi_{\rm s}^{\rm max} = u_{\rm s}^{\rm max}/\omega_{\rm s}$. In case of zero saliency ($\chi = 0$) expression (7.32) reduces to Eq. (7.9), assuming $L_{\rm sd} = L_{\rm s}$. For a salient machine the MF lines are elliptical as may be deduced from expression (7.32). The major axis coincides with the d-axis, while the minor axis is parallel with the q-axis. The intersection of both axes is at the short circuit current point D as shown in Fig. 7.16. As the shaft speed increases, the maximum flux linkage ellipse reduces in size given that the variable $i_{\rm O}$ is inversely proportional to the speed $\omega_{\rm s}$. This implies that the available operating region of the drive is constrained by the area within the ellipse as speed increases. This behavior is the same to the one of non-salient machines. A saliency factor $\chi < 0$ would force operation below the base speed to the first quadrant of the current locus diagram. In this case, operation at high speed is severely compromised, given that the maximum flux linkage circles would need to extend to the first quadrant.

The *base speed* corresponds to the MF ellipse which coincides with the maximum torque point. Operation along the MTPA contour remains viable for shaft speeds up to the base speed. Computation of the base speed may be undertaken by making use of Eq. (7.29), with $i_s^n = 1$ which provides the (normalized) drive saturation coordinates of the current vector $\vec{i}_s^{dq}$. Use of these coordinates with Eq. (7.32) leads to the base speed ω_s^b.

Maximum Torque per Flux Linkage (MTPF) Line

For deriving a field weakening strategy, the MTPF trajectory is used. The MTPF trajectory is given by the operating points in which the MF ellipses and the constant torque curves touch in one point. Expressed differently, they are the points on the MF ellipses which represent maximum torque. A mathematical representation can be found by using Eq. (7.32) with Eq. (7.27) and by differentiation, resulting in an expression for i_{sd}^n as function of i_O. This can in turn be used to find the corresponding i_{sq}^n value. The resulting set of equations which defines the MTPF trajectory is given as follows:

$$i_{sd}^n = -\kappa + (1+2\chi)\frac{\kappa}{8\chi} - \sqrt{\frac{i_O^2}{2} + \left((1+2\chi)\frac{\kappa}{8\chi}\right)^2} \tag{7.33a}$$

$$i_{sq}^n = \frac{1}{2\chi+1}\sqrt{i_O^2 - \left(\kappa + i_{sd}^n\right)^2} \tag{7.33b}$$

The MTPF trajectory is shown in Fig. 7.16 for $\kappa = 0.6$ and $\chi = 0.75$. It is partly within the MA circle, because in this example the short circuit is less than i_s^{max}.

Operating Sequence

In the final part of this section, the control strategy under field weakening condition will be discussed using the excitation scenario introduced for previous examples. The torque is increased from 0 to $0.5\,T_e^{max}$ while the machine accelerates from standstill. This corresponds to the trajectory $0 \rightarrow$ A in the current locus diagram. Operation under MTPA conditions will remain until a shaft speed ω_s^A is reached when the MF contour coincides with point A. Above ω_s^A, constant torque operation is possible by maintaining the current vector $\vec{i}_s^{dq}$ on the maximum flux MF linkage ellipse and on the chosen torque curve A $\rightarrow$ B. This control strategy can be used as long as the currents are within the MA circle and on the right of the MTPF line. This is possible up to speed ω_s^B, when the MA circle is reached. Above ω_s^B, operation proceeds along the MA circle until the MTPF line is reached at point C. For even higher speeds, operation proceeds along the MTPF trajectory C $\rightarrow$ D.

The operating sequence described above can also be shown with the aid of a current-speed diagram as given in Fig. 7.17, where the *blue* curves represent the currents i_{sq}^n and i_{sd}^n. In this figure, the *green* and *red* curves represent the direct and quadrature current curves for operation along the MA contour and the MTPA trajectory. For operation below the base speed, the current references are defined by Fig. 7.18 for the chosen torque reference value. Above the speed ω_s^A, the normalized currents must be found using Eqs. (7.27) and (7.32) until speed ω_s^B is reached. For speeds in excess of ω_s^B, the currents are defined by Eqs. (7.28), with $i_s^n = 1$ and (7.32) until the MTPF line is reached, in which case equation set (7.33) must be used. For an implementation of the control strategies lookup tables are used, which give the required reference current values based on the reference torque and maximum stator flux linkage ψ_s^c. The reason for this is that saturation affects the values of L_{sd} and L_{sq} making it difficult to find analytical expressions. In this case, the lookup tables are usually derived from measurements or from finite element analysis. Nevertheless, the introduction of the equations is useful to give the reader an appreciation of the relationship between variables that control the current reference values.

A tutorial is given in Sect. 7.6.4, which exemplifies and visualizes the control strategies discussed in this section. For this purpose, a current source salient machine model is introduced and used in conjunction with a controller that delivers the appropriate current reference values as a function of shaft speed and a user defined torque reference value.

7.4 Field-Oriented Control Module with a Current-Controlled Synchronous Machine

The process of integrating the controller with the machine is discussed in this section. Use is made of a current source based model of the machine because by using the current source based model concepts can be readily examined without having to consider the dynamics of the current controller. In practice, this approach is realistic given that the time constants linked to the current controllers are small in comparison to those of the machine. Figure 7.20 shows a salient machine model although the approach discussed here is equally applicable to non-salient machines by setting $L_{sd} = L_{sq} = L_s$. The field flux linkage ψ_f may be provided by permanent magnets or a field winding. In the latter case, the synchronous machine control module may be used to regulate the field flux linkage in order to control the power factor as was explained in Sect. 7.2.4.

The main purpose of the SM control module is to generate the required synchronous coordinate reference currents i_{sd}^c and i_{sq}^c based on the user defined torque reference value. A number of field weakening control strategies have been introduced in the previous sections for both salient and non-salient machines which ensure that the machine can operate within the design envelope. Also shown in Fig. 7.20 is a CFO (coordinate calculator for field orientation) module which provides the reference rotor angle θ^c for the synchronous controller and the dq $\rightarrow$ $\alpha\beta$ conversion module which yields the current vector $\vec{i}_s$ used as an input to the machine (in this case) or current controller. For position sensorless applications, i.e., those which do not use a mechanical shaft sensor, use is made of the electrical

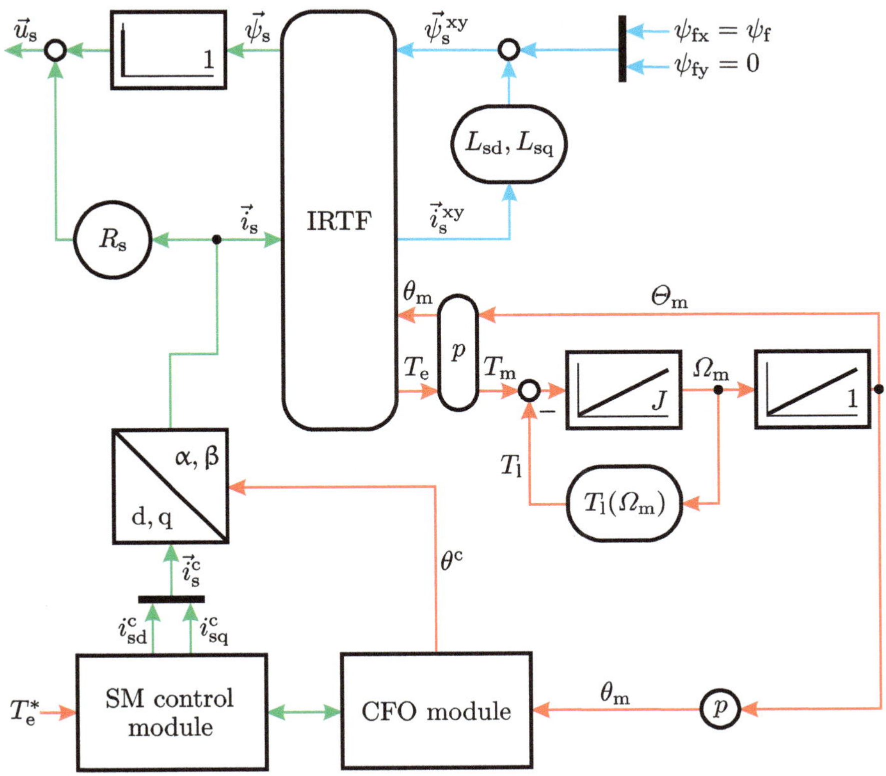

Fig. 7.20 Field-oriented control with current source synchronous machine model

variables, which then serve as inputs to the CFO module. In most cases an encoder is used (as shown) which provides the shaft angle position for the controller.

Interfacing a non-salient machine with a controller is a relatively straightforward task provided that the machine parameters are known. For the salient synchronous machine this process is more laborious given the need to generate lookup tables which is an approach more akin with the switched reluctance drive that will be discussed in Chap. 10.

7.5 Interfacing the Field-Oriented Control Module with a Voltage Source Connected Synchronous Machine

Practical implementation of the current-controlled machine discussed in the previous section calls for the use of a voltage source converter. Consequently, a three-phase current control approach as discussed in Sect. 3.2 must be integrated with the drive. In this section, a three-phase model based current controller is used because these are generally favored in practical drive applications. The generic representation of the current controller in question as shown in Fig. 3.19 is connected to a generalized load which defines the parameters for the discrete model based synchronous current controller which is tied to a flux vector $\vec{\psi}$.

For the synchronous drive, the latter vector must be replaced by the field flux linkage vector $\vec{\psi}_f = \psi_f e^{j\theta}$. Furthermore, the parameters used for the controller need to be redefined given that the load is in this case a synchronous machine. For the purpose of determining the parameters, it is helpful to consider the synchronous direct and quadrature axis salient symbolic models of the machine shown in Fig. 6.15. For the direct axis current controller, calculation of the sampled average voltage reference $U_s{}^*(t_k)$ is carried out with the aid of equation (3.22a) in which the parameters R and L must be replaced by the variables R_s and L_{sd}. A similar approach must be used for the computation of quadrature axis controller sampled average voltage reference $U_s{}^*(t_k)$ where use is made of Eq. (3.22b) in which case the parameters R and L must be replaced by the machine parameters R_s and L_{sq}. Furthermore, this expression shows the presence of a disturbance decoupling term $u_e = \omega_e \psi_e$, which corresponds to the term $\omega_m \psi_f$ in a synchronous drive. Note that in the discussion given above, a salient synchronous model has been purposely entertained given that the non-salient machine model is readily accommodated by choosing the direct and quadrature axis inductances equal.

The resulting drive structure as shown in Fig. 7.21 brings together key concepts such as modulation and current control introduced in earlier chapters of this book. Readily apparent in Fig. 7.21 is the *predictive model based current controller* which uses the current reference values produced by the *SM control module* as discussed in the previous section. The observant reader will note that only the predominant disturbance decoupling term $\omega_m \psi_f$ is shown in the current controller module. The remaining terms $\omega_m L_{sd}$ and $-\omega_m L_{sq}$ are not shown for the sake of readability. However, these terms are included in the accompanying tutorial given in Sect. 7.6.5 which is concerned with a voltage source converter that is connected to a salient synchronous machine operating with a model based current controller. Also shown in Fig. 7.21 is the generic model of the IRTF based salient synchronous machine according to Fig. 6.13. From a simulation perspective, the voltage source converter/modulator structure is often replaced by an alternative module which calculates the requires supply vector directly from the average voltage references generated by the current controller. This approach as outlined in Sect. 3.3.5 reduces the simulation run time because a larger computational step size may be entertained. For the sake of clarity, it is emphasized that the model shown is equally applicable to the non-salient machine provided the appropriate control algorithm is used within the *SM control* and *CFO* modules. Furthermore, the generic machine model according to Fig. 6.3 must be used instead of the salient model, which in effect calls for the introduction of a common stator inductance L_s as discussed previously.

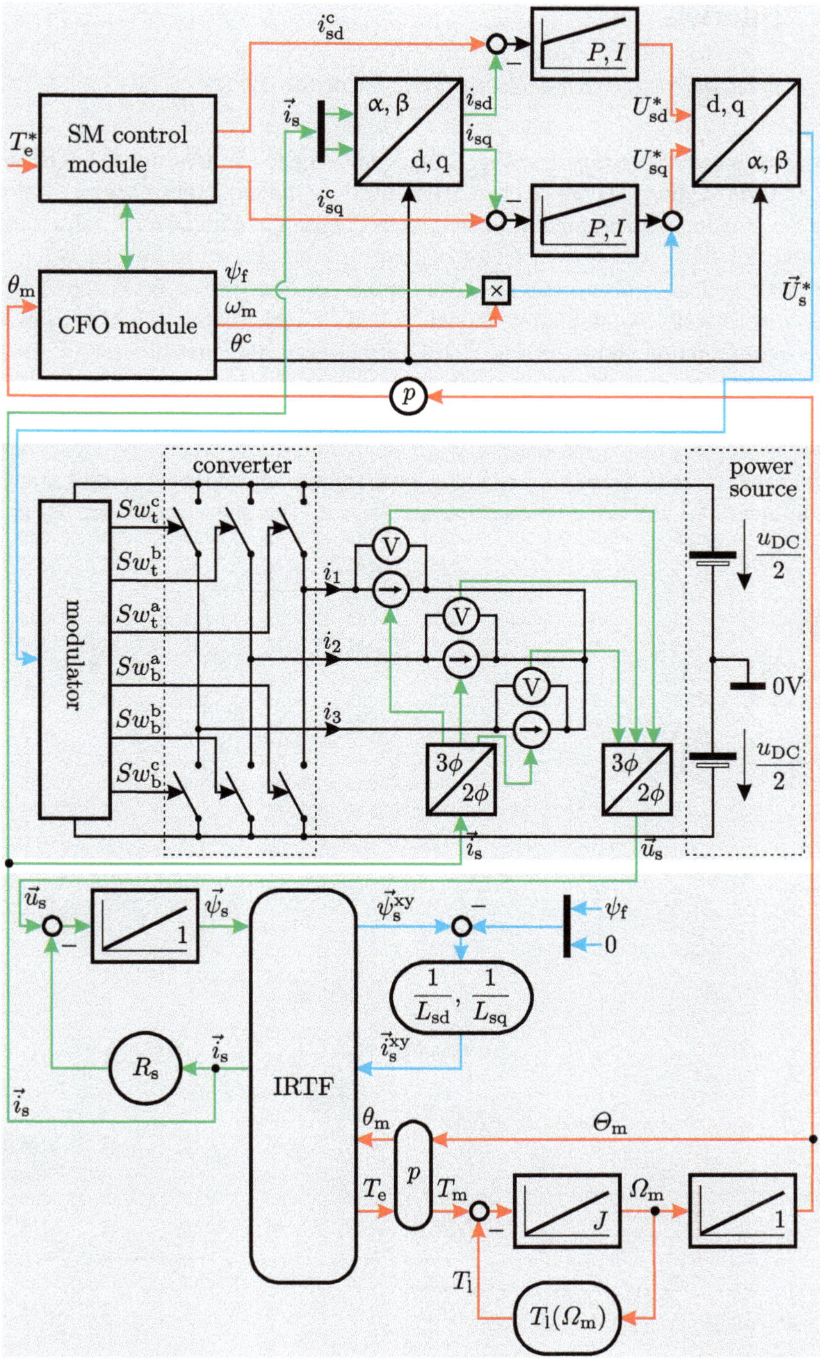

Fig. 7.21 Field-oriented control with voltage source synchronous machine model

7.6 Tutorials

7.6.1 Tutorial 1: Non-salient Synchronous Drive

The drive concept presented in Fig. 7.20 is investigated in this tutorial. The non-salient PMSM discussed in Sect. 6.3.1 is used as the example machine. For the purpose of better visualizing the current locus diagram, the inductance value shown in Table 6.1 is increased by a factor of four and the excitation flux ψ_{f} is halved. The drive voltage and current limit values are set to $u_{\mathrm{s}}^{\max} = 60\,\mathrm{V}$ and $i_{\mathrm{s}}^{\max} = 20\,\mathrm{A}$, respectively. A simulation model should be developed, which satisfies the drive representation shown in Fig. 7.20. Furthermore, the control strategy for the SM controller module should satisfy the approach outlined in Sects. 7.2.1 and 7.2.2, respectively.

An example PLECS simulation is given in Fig. 7.22, where the IRTF based machine and *Field Weakening* module which meets the desired control strategy requirements for the drive in question are shown. The stator resistance R_{s} is set

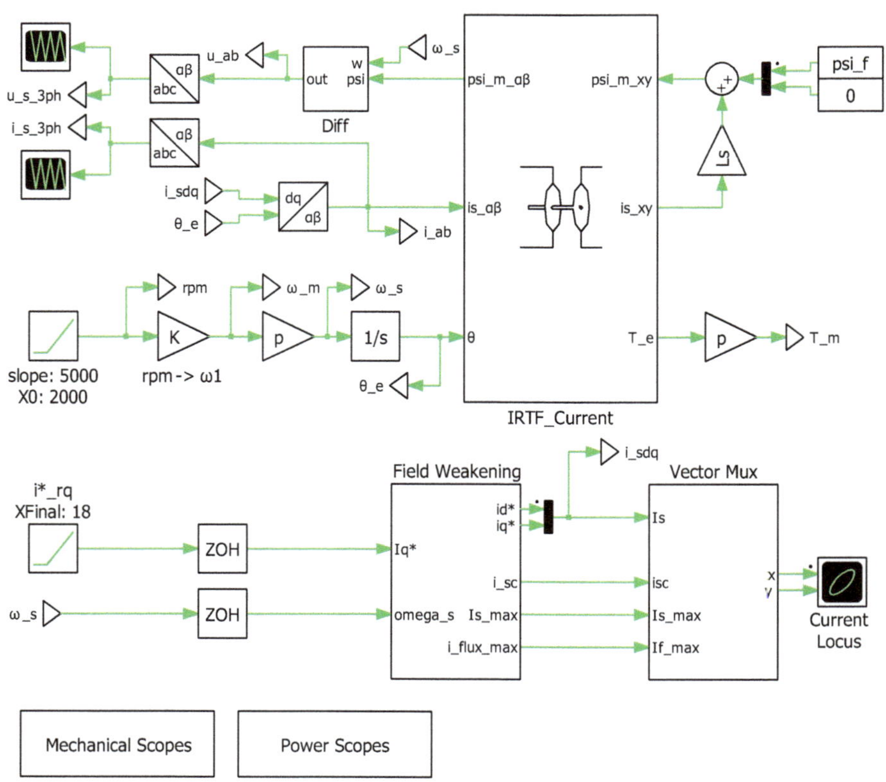

Fig. 7.22 Simulation of non-salient synchronous drive

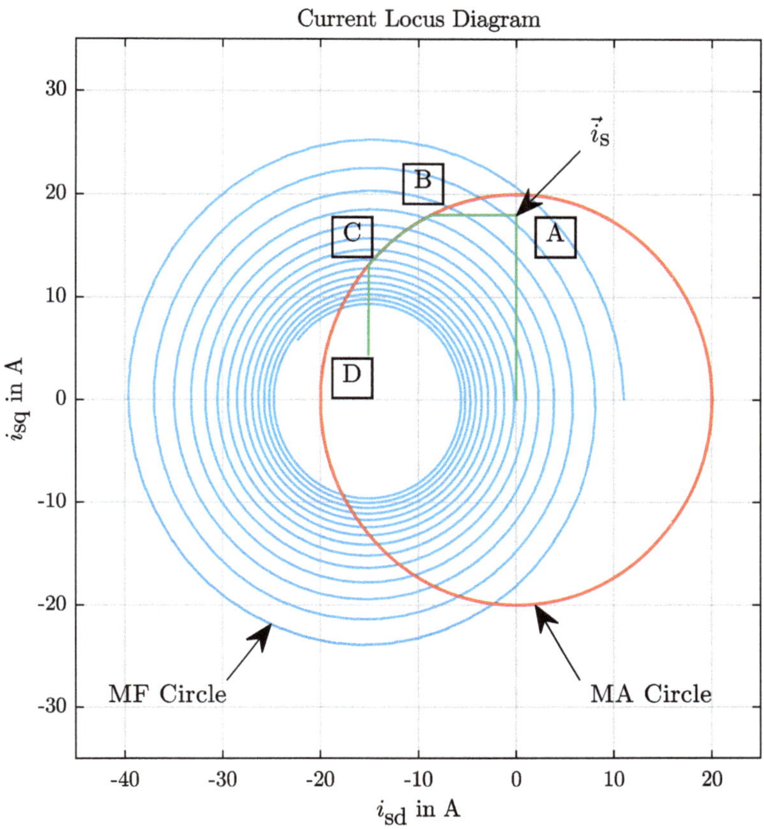

Fig. 7.23 Current locus diagram of a non-salient current source connected PMSM

to zero, given that the voltage drop linked to this parameter is usually low in comparison with the back-EMF term. Moreover, from a didactical perspective it is helpful to observe the stator voltage amplitude without having to consider the contribution caused by the resistance.

The rotor speed of the machine is set by the ramp function, which ramps up from 2000 rpm at $t = 0$ s to 12,000 rpm at $t = 2$ s. The *Vector Mux* subsystem plots the machine current vector $\vec{i}_s^{\,dq}$ (green), as generated by the controller, along with the maximum flux *spiral* (in dashed blue) and the maximum current circle (red). The maximum flux constraint is drawn as a vector with a magnitude of the instantaneous maximum current due to flux limitation ("i_flux_max" in Fig. 7.22), centered at the short circuit current, with an arbitrary angular speed. Since the rotor speed is linearly increasing, the magnitude is linearly decreasing with time, creating a spiral. *Max flux circles* (MF circles) can be obtained by using a constant rotor speed.

The letter notation of Fig. 7.8 is used in Fig. 7.23 to discuss the operation regions of the control. Point A represents operation in the base speed region. The flux is weakened between A and B, however the same torque can still be generated, which is the basic field weakening. Once the current limit is reached, the trajectory over the max current circle is followed between B and C. Once the MTPF line is reached, this line is followed between C and D. Note that the amplitude of the supply voltage is held to $u_{\mathrm{s}}^{\mathrm{max}} = 60\,\mathrm{V}$ as field weakening is enforced when the speed is increased.

The reader is urged to reconsider the problem discussed in this section for the case where the excitation flux ψ_{f} is not halved, which corresponds to the case where the short circuit current point lies outside the maximum current circle (see Fig. 7.8).

7.6.2 *Tutorial 2: Non-salient Synchronous Drive, Constant Stator Flux Operation*

An extension to the previous tutorial is considered here, which is concerned with drive operation under constant stator flux, as discussed in Sect. 7.2.3. The drive concept and parameters are not changed, only a new controller is employed which meets the requirements of maintaining the condition $|\vec{\psi}_{\mathrm{s}}| = \psi_{\mathrm{f}}$ below the base speed $\omega_{\mathrm{m}}^{\mathrm{bS}}$.

Calculate the mechanical shaft speed (in rpm) which corresponds to $\omega_{\mathrm{m}}^{\mathrm{bS}}$ and determine the percentage reduction in torque capability, which must be tolerated for achieving constant flux operation. Furthermore, provide a current locus diagram which shows the maximum flux circle at the new base speed as well as the maximum current circle. Set the simulation model shaft speed to $\omega_{\mathrm{m}}^{\mathrm{bS}}$ and vary the quadrature input current between -20 and $20\,\mathrm{A}$ to achieve a current locus that traverses a trajectory from the second quadrant saturation point to the third quadrant saturation point. Note that these quadrature current values cannot be reached due to max current limitation.

Calculation of the revised base speed $\omega_{\mathrm{m}}^{\mathrm{bS}}$ may be undertaken with the aid of equation (7.21). This gives the new value relative to the base speed found in the previous tutorial. Subsequent analysis using the parameters given shows that the new base speed is equal to $n_{\mathrm{m}}^{\mathrm{bS}} = 1725$ rpm, which is the highest speed under which constant stator flux operation at the desired flux level can be realized for this drive. The maximum achievable torque for this drive may be found by determining the largest quadrature current as defined by Eq. (7.20) (shown in its normalized form). The field flux level remains unchanged. Hence, the torque reduction can be readily calculated using the parameters given. This leads to a maximum quadrature current of $i_{\mathrm{sq}} = 18.8\,\mathrm{A}$, with a corresponding reduction in torque capability of 5.5%.

The simulation model shown in Fig. 7.24 utilizes a *stator flux control module* which has as inputs the shaft speed ($n_{\mathrm{m}}^{\mathrm{bS}} = 1725$ rpm) and input quadrature current variable set to operate in the range of $i_{\mathrm{q}}^{*} = -20 \rightarrow 20\,\mathrm{A}$, as required for this simulation exercise. The current locus diagram given in Fig. 7.25 shows the maximum flux and maximum current circles, together with the current locus trajectory for the chosen excitation scenario.

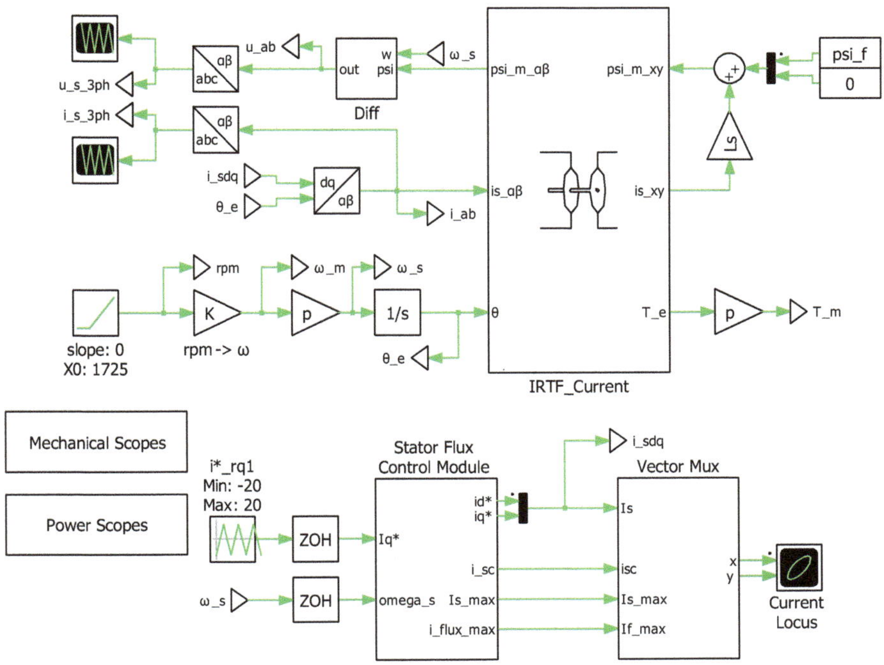

Fig. 7.24 Simulation of non-salient synchronous drive, constant stator flux operation

7.6.3 Tutorial 3: Non-salient Synchronous Drive, Unity Power Factor Operation

The drive concept discussed in the previous section is modified in this tutorial to accommodate unity power factor operation as discussed in Sect. 7.2.4. For this purpose the simulation model is provided with an alternative controller which meets the requirements for this exercise. The field flux is now deemed to be a variable and is chosen in such a manner that the rated flux value of $\psi_{\mathrm{f}} = 0.166\,\mathrm{Wb}$ is realized with the largest allowable quadrature current $i_{\mathrm{sq}}^{\mathrm{max}}$ (while maintaining unity power factor operation).

Calculate this current value and the stator flux level which should be used in the drive. In addition, calculate the corresponding base speed $\omega_{\mathrm{m}}^{\mathrm{bS}}$ using the parameters given in Sect. 7.6.1 and determine the percentage reduction in maximum torque capability of the drive in comparison to operation along the MTPA curve (as discussed in Sect. 7.6.1). Furthermore, provide a current locus diagram which shows the maximum flux circle at the new base speed as well as the maximum current circle. Set the simulation model shaft speed to $\omega_{\mathrm{m}}^{\mathrm{bS}}$ and vary the quadrature input current in the range of $i_{\mathrm{q}}^{*} = -15 \rightarrow 15\,\mathrm{A}$ in order to achieve a current locus that traverses a trajectory from the second quadrant saturation point to the third quadrant saturation point.

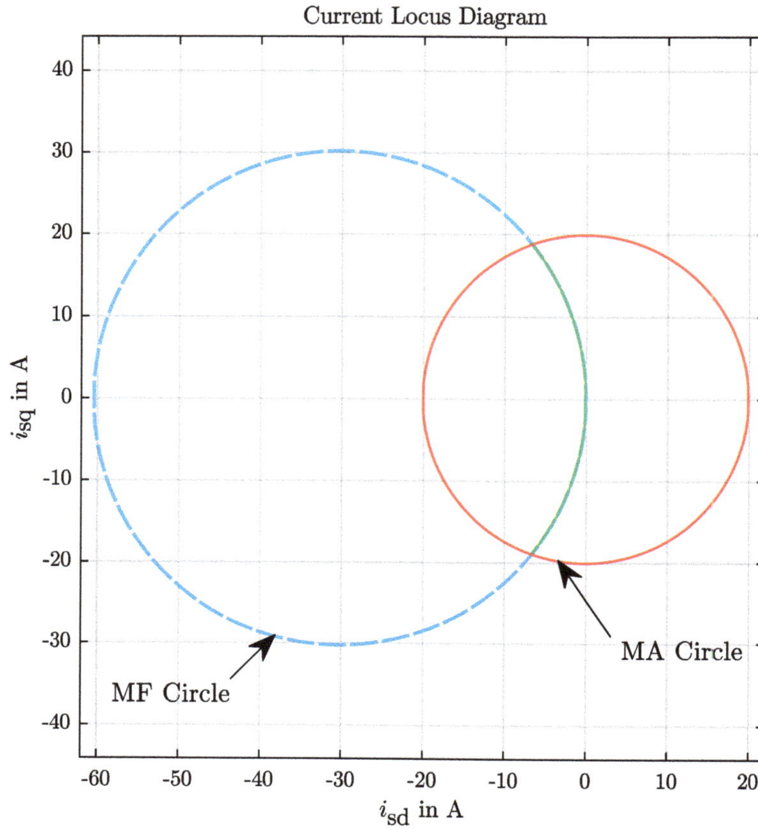

Fig. 7.25 Blondel diagram non-salient current source connected machine, operation with constant stator flux and $n_m^{bS} = 1725$ rpm

Computation of the required field flux level ψ_{fO} for $i_{sq} = 0$ requires access to the variable $\kappa^O = \psi_{fO}/L_s\, i_s^{max}$ and use of the variable $i_s^{sc,max} = \psi_f/L_s\, i_s^{max}$, with $\psi_f = 0.166$ Wb. Subsequent use of Eq. (7.22c) demonstrates that the required flux value is equal to $\psi_{fO} = 0.125$ Wb, which is also the required stator flux value, as discussed in Sect. 7.2.3. The corresponding base speed may be found with the aid of (7.21) with $\kappa = \kappa^O$, which gives $n_m^{bS} = 2291$ rpm. The maximum quadrature current may be calculated to be $(i_{sq})^{max} = 15.0$ A with the aid of equation (7.22b). If this current is used, then the field flux will be at the rated value $\psi_f = 0.166$ Wb as used for operation under MTPA conditions. Hence, the percentage maximum torque reduction is equal to 25%, which must be accepted for operating under zero reactive input power conditions.

The simulation model given in Fig. 7.26 utilizes a *Field Weakening Unity Power Factor* module which allows the drive to operate under unity power factor conditions. The power factor is calculated with the aid of a real and reactive power module and the reader may ascertain that its value remains at the required value

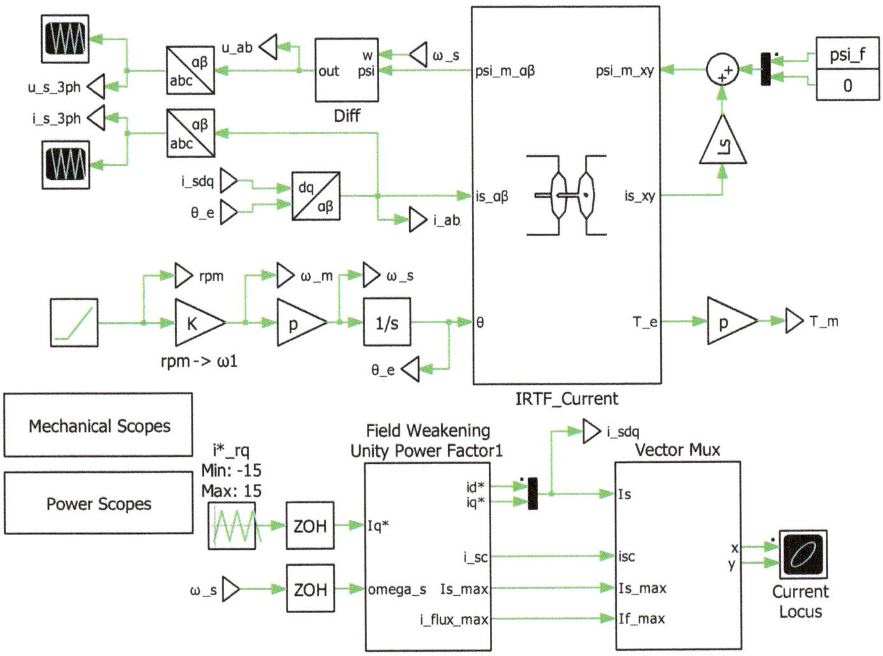

Fig. 7.26 Simulation of non-salient synchronous drive, unity power factor operation

(within an acceptable tolerance band) when the quadrature current reference is varied over the range $i_q^* = -15.0 \rightarrow 15.0\,\text{A}$, as required for this simulation exercise. The shaft speed for this simulation should be set to the base speed value of $n_m^{bS} = 2291\,\text{rpm}$.

Computation of the current trajectory is carried out as discussed in the previous tutorials and in this case variation of the quadrature current takes place with a stator flux level of $\psi_s = 0.125\,\text{Wb}$ and unity power factor as may be observed. The current locus diagram given in Fig. 7.27 shows the discrete current locus as well as the base speed flux circle and maximum current circle.

7.6.4 *Tutorial 4: Salient Synchronous Drive*

The non-salient drive concept discussed in the previous tutorials can be modified to accommodate saliency, as has already been shown in Sect. 6.3.3. For this exercise the direct axis inductance is set to $L_{sd} = 5.5\,\text{mH}$ (which is the value for L_s used in previous cases) and choose a saliency factor of $\chi = 0.75$. Furthermore, the drive parameters u_s^{max}, i_s^{max} are left unchanged. Calculate the mechanical base speed (in *rpm*) of the drive and the corresponding maximum shaft torque. A simulation model is provided, as shown in Fig. 7.28, which utilizes a *Flux Weakening Salient* module,

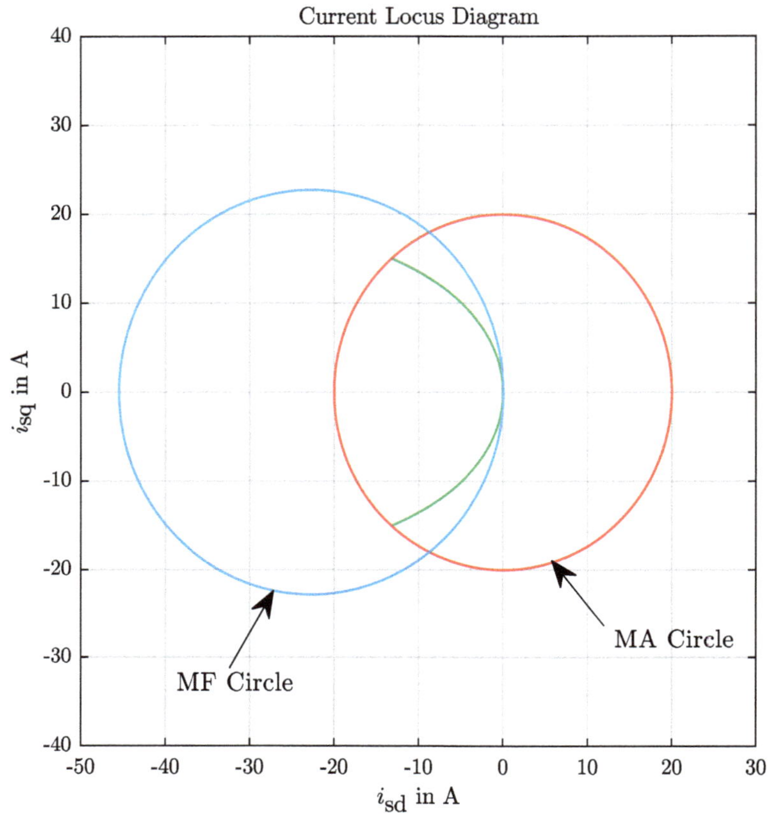

Fig. 7.27 Blondel diagram non-salient current source connected machine, operation with constant stator flux/unity power factor and $n_{\mathrm{m}}^{\mathrm{bS}} = 2291$ rpm

which satisfies the control strategies outlined in Sect. 7.3. The model in question, which is based on the generic diagram given in Fig. 7.20 (with $R_{\mathrm{s}} = 0$), allows the user to set the mechanical shaft speed and reference shaft torque value for the controller module.

At the beginning of the simulation, the mechanical shaft speed is kept at zero, while the shaft torque is varied over the range of $T_{\mathrm{e}}^{*} = -8.0 \rightarrow 4\,\mathrm{Nm}$. Therefore the MTPA operation is *swept* to achieve the related current locus. Then, the mechanical shaft speed is linearly increased so that the controller goes through the modes of basic flux weakening (where the current locus is a "constant torque line"), current limit, and finally the MTPF. The current locus along with maximum flux spiral and the maximum current circle is given in Fig. 7.29.

Computation of the vector $\vec{i}_{\mathrm{s}}^{\mathrm{dq}}$ endpoint coordinates at the drive saturation point may be undertaken with the aid of equation (7.29), with $i_{\mathrm{s}}^{\mathrm{n}} = 1$. Subsequent use of these coordinates with torque expression (7.27) leads to the maximum shaft torque of the drive which was found to be $T_{\mathrm{em}}^{\mathrm{max}} = 9.0\,\mathrm{Nm}$. The same set of coordinates

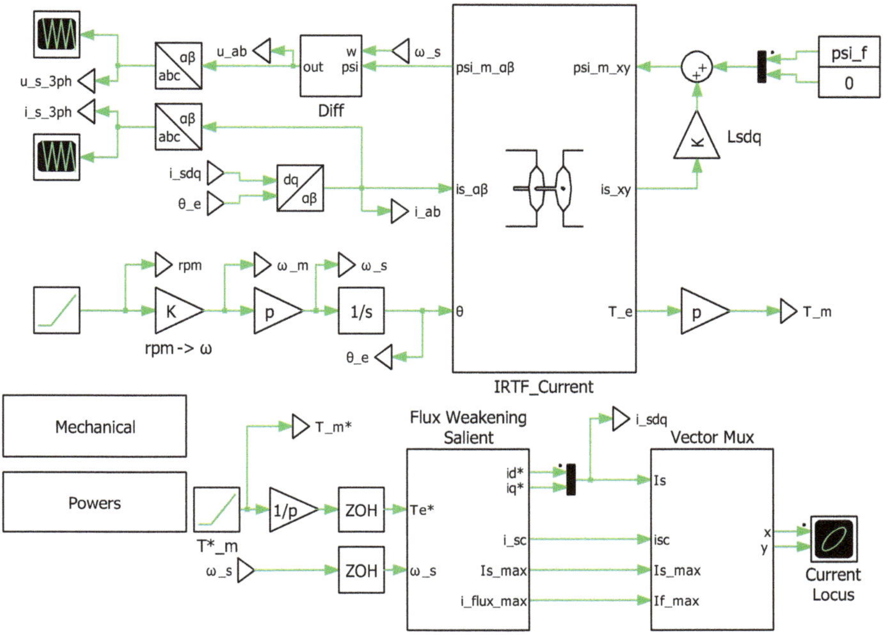

Fig. 7.28 Simulation of salient synchronous drive

may also be used with Eq. (7.32), which yields the variable i_{O} that is linked to the base speed $\omega_{\mathrm{m}}^{\mathrm{b}}$ of the drive. Evaluation of said expression along the suggested lines shows that the base speed (expressed in terms of the mechanical shaft speed) is equal to $n_{\mathrm{m}}^{\mathrm{b}} = 1279$ rpm. Increasing the speed further at $T_{\mathrm{e}}^{*} = 4.0\,\mathrm{Nm}$, the vector $\vec{i}_{\mathrm{s}}^{\mathrm{dq}}$ reaches the maximum current circle at a rotor speed of $n_{\mathrm{m}} = 2479$ rpm.

7.6.5 Tutorial 5: PM Salient Synchronous Drive with Model Based Current Control

The final tutorial in this chapter examines the operation of a voltage source connected machine operating with a model based current controller. The proposed concept as discussed in Sect. 7.5 utilizes a four-pole salient machine as discussed in the previous tutorials. In this exercise the current references produced by the salient current controller are used as inputs $i_{\mathrm{d}}^{\mathrm{c}}$, $i_{\mathrm{q}}^{\mathrm{c}}$ for the current controller as shown in Fig. 7.21. The parameters of the machine are as in the previous section, with exception of the inertia which has been arbitrarily set to $J = 0.005\,\mathrm{kg\,m^2}$. A sampling rate of 10 kHz is assumed for the discrete current controllers.

The input reference torque is set to $T_{\mathrm{e}}^{*} = 6.0\,\mathrm{Nm}$ at $t = 10\,\mathrm{ms}$. A torque reversal of $T_{\mathrm{e}}^{*} = 6.0 \rightarrow -3.0\,\mathrm{Nm}$ is carried out at $t = 300\,\mathrm{ms}$. A first order filter with a

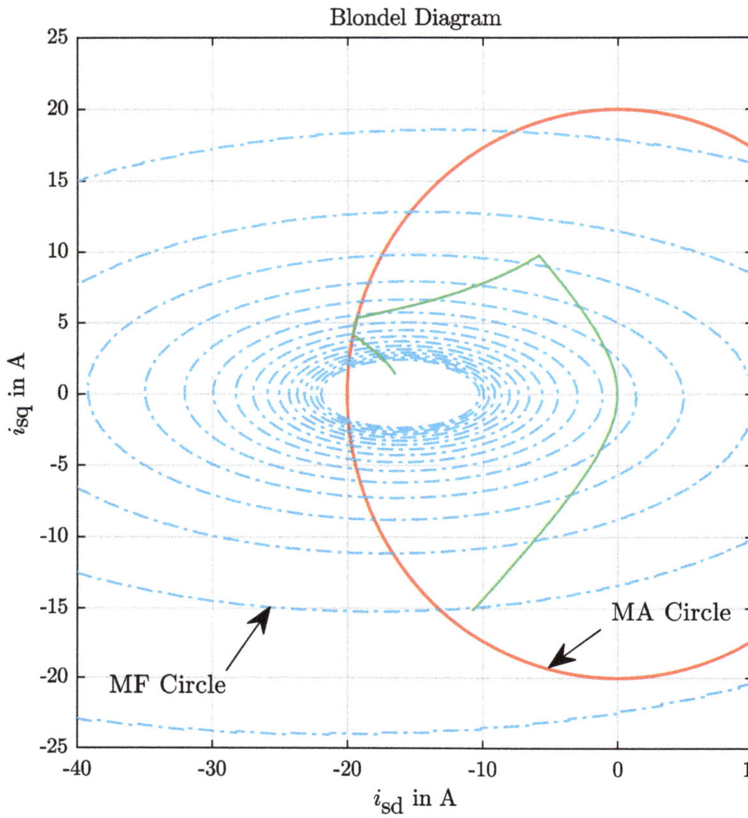

Fig. 7.29 Blondel diagram salient current source connected synchronous machine

time constant of $\tau = 2\,\text{ms}$ is placed between torque reference and synchronous drive controller to limit the torque variations to realistic values. Requesting an instantaneous change in torque is not realistic in any practical drive system, due to limitations imposed by the converter DC bus voltage. Total simulation run time is set to 500 ms. The DC supply is equal to $u_{DC} = 200\,\text{V}$. Both shaft speed and shaft angle are assumed to be available for the controllers. The modulator and converter are not implemented at circuit level, i.e., the switching effects should be excluded (as discussed in Sect. 3.3.5) to better visualize the operation of the drive.

The first task to be undertaken is to calculate the gains for the two current controllers and in addition identify the disturbance decoupling terms which must be introduced. Secondly examine the simulation model of the drive as given in Fig. 7.30 and plot the following results:

- Sampled direct/quadrature reference i_{sd}^{c}, i_{sq}^{c} and *measured* currents i_{sd}, i_{sq}.
- Controller shaft input torque reference T_e^* and *actual* shaft torque T_{em}.

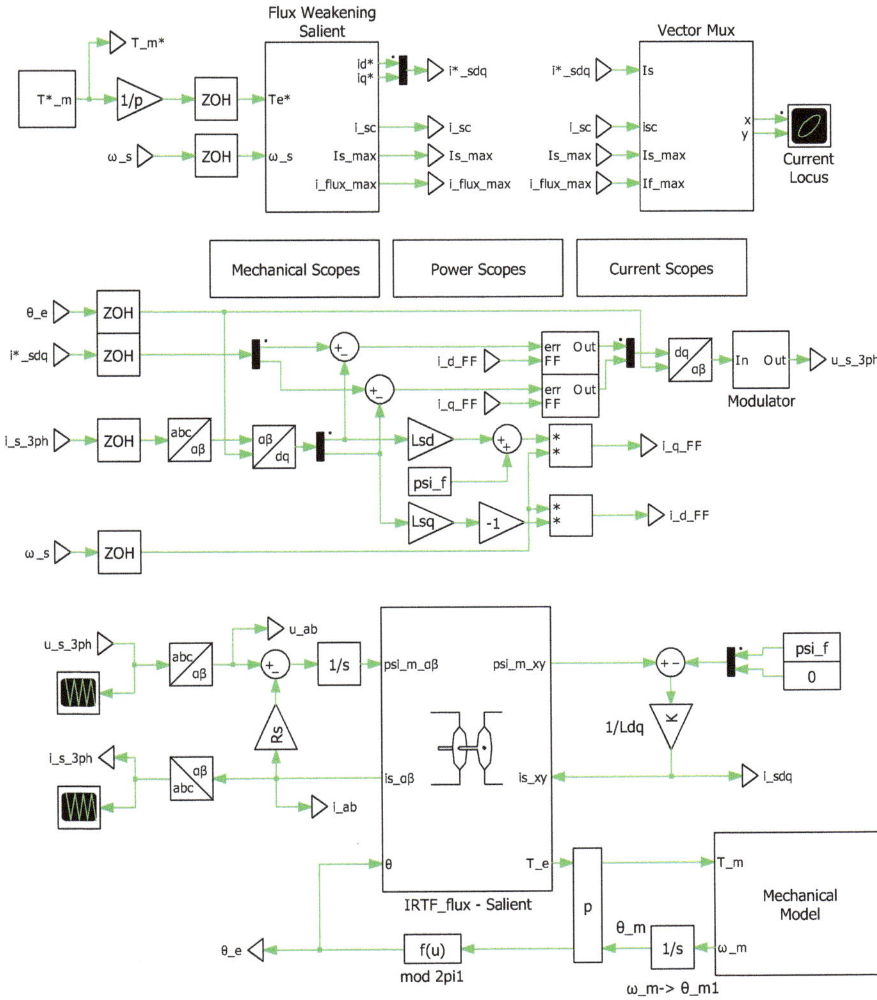

Fig. 7.30 Simulation of salient synchronous drive with voltage source converter

- Shaft speed n_{m} (in rpm and scaled by a factor $1/20$) and stator voltage vector amplitude $|\vec{u}_{\mathrm{s}}|$.

The computation of the current controller gains follows the approach taken for the tutorial shown in Fig. 3.30. For the salient machine, the direct and quadrature inductance calculated with equation set (3.22) differ. Hence, the direct and quadrature proportional gains will reflect the reluctance transformation in use. The integral terms will be equal for both controllers. Said equations also define the disturbance decoupling terms which must be used, keeping in mind the salient nature of the machine.

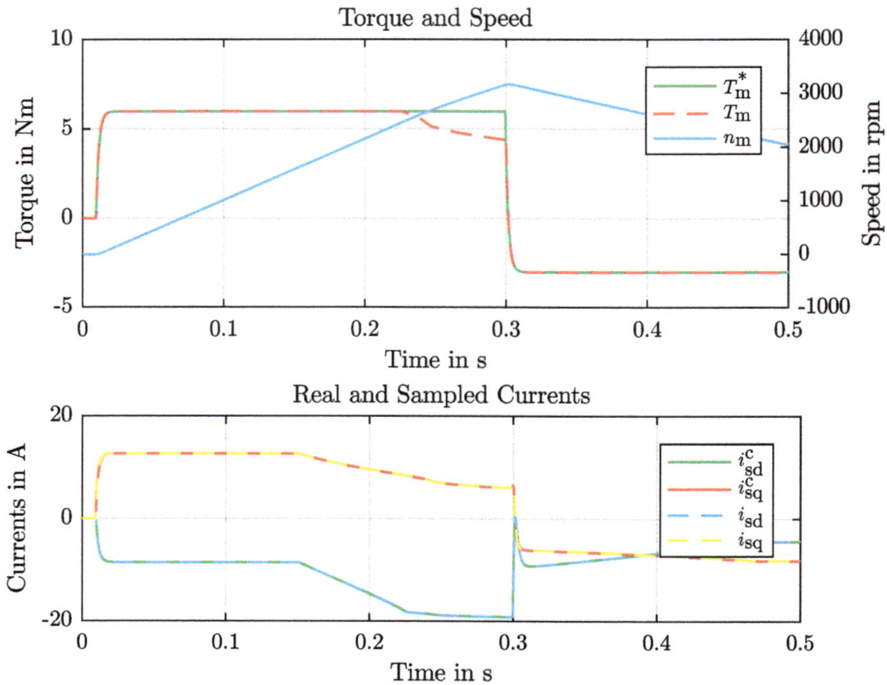

Fig. 7.31 Simulation results for salient synchronous drive with voltage source converter

A torque reference module is shown in Fig. 7.30, which also contains the first order filter used to moderate the torque reference signal supplied to the *salient control module*. The converter limits are set as $u_s^{max} = 60\,\text{V}$, $i_s^{max} = 20\,\text{A}$. Note that the actual supply vector may become larger as the control strategy deploy does not take into account the voltage potential across the stator resistance and the controller reactions to errors.

The results provided in Fig. 7.31 show the operation of the drive in detail. The torque reference step at $t = 10\,\text{ms}$ is provided to the salient control module, which in turn set the required reference direct and quadrature currents. The current controller average voltage per sample values are changed, which in turn change the *measured* direct and quadrature currents. These in turn serve to deliver the required shaft torque. Speed increases linearly as long as the shaft torque remains constant. At approximately $t = 150\,\text{ms}$, the current controller reference values (and therefore the currents in the machine) change because the drive reaches the *drive saturation point*, which occurs at shaft speed $n_m^b = 1600$ rpm in this example. As speed increases beyond the base speed, at approximately $t = 225$ ms, the current controller reference value reaches the current limit. Therefore, the reference torque cannot be supplied any more by the machine. At $t = 300\,\text{ms}$ a reference torque reversal to $T_{em} = -3\,\text{Nm}$ occurs which is matched by an appropriate shaft torque change. The drive example shown was undertaken with a model based current controller. It is left as an exercise for the reader to reconsider this problem for the case where a hysteresis controller as shown in Fig. 3.13 is used with the drive in question.

References

1. Leonhard W (2001). Control of electrical drives, 3rd edn. Springer, Berlin
2. Pulle DWJ (2016) Texas instruments workshop program

Chapter 8
Induction Machine Modeling Concepts

Electrical drives with induction machines remain the dominant market leader in the field. The combination of a robust low cost squirrel-cage machine, high power density converter, and versatile controller yields a highly adaptable drive for wide ranging rugged industrial applications.

In this chapter, induction machine models are developed which will be used in the next chapter of this book, e.g. to control voltage source inverters with field-oriented controls. Initially, a brief review of the induction machine with squirrel-cage rotor is given in terms of a cross-sectional view and simplified symbolic and generic models. A detailed description of the fundamentals can be found in [8, 10, 13], and [14]. As a platform for introducing field-oriented models, first models without leakage inductances are derived showing the essence of torque production of the machine. Central to this chapter is the introduction of a *universal* flux linkage model which allows a three-to-two inductance transformation leading to a simplified IRTF machine model. This universal model is the stepping stone to the *universal field-oriented* (UFO) machine model which gives a basic understanding of the transient behavior of induction machines [4]. Furthermore, this model forms the cornerstone for the development of field-oriented control algorithms. At the end of this chapter, attention is given to single-phase induction machines. These machines are widely used in domestic appliances and as such it is important to have access to dynamic and steady-state models. A set of tutorials is provided which allows the user to interactively explore the concepts presented in this chapter.

8.1 Induction Machine with Squirrel-Cage Rotor

Figure 8.1 shows the cross-section of an induction machine with a so-called *squirrel-cage* rotor. The squirrel cage consists of a set of conductors, i.e. rotor bars, (shown in *red*), which are short-circuited at both ends by a conductive ring. The cage

R. W. De Doncker et al., *Advanced Electrical Drives*, Power Systems,
https://doi.org/10.1007/978-3-030-48977-9_8

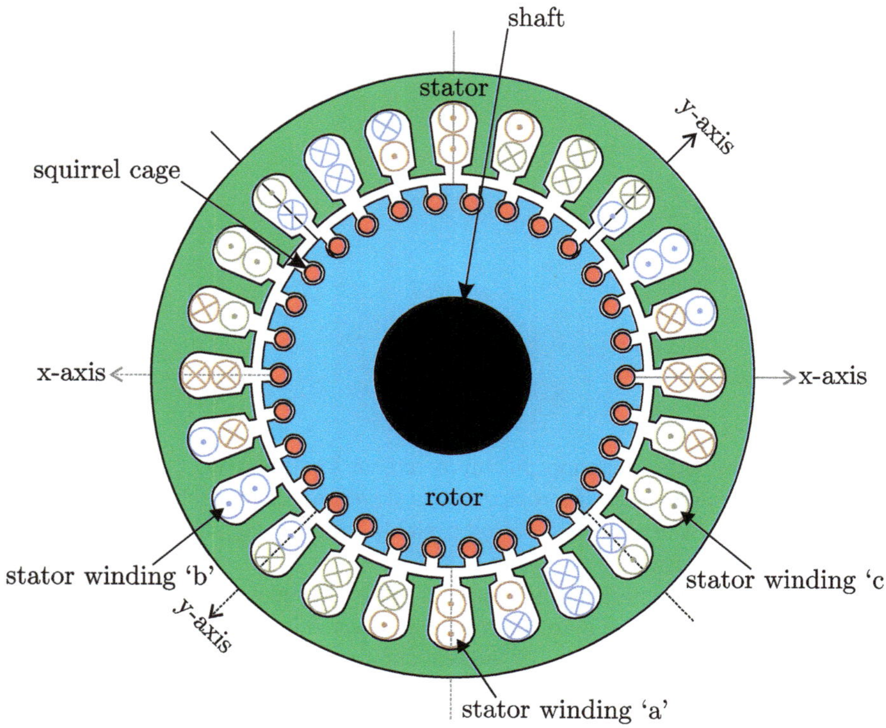

Fig. 8.1 Four-pole induction machine with squirrel-cage rotor, showing a two-layer winding and rotor with rotor bars

is embedded in the rotor lamination as may be observed from Fig. 8.1. A three-phase two-layer winding is housed in the stator of a four-pole machine.

A rotating field created by the stator winding is penetrating the rotor. If the rotor is rotating asynchronously to the stator field (which means it is rotating at a different speed), alternating currents are induced in the squirrel cage. These currents, together with the stator field, are responsible for the torque production of the machine. This is why asynchronous machines are commonly known as *induction machines*. If the stator field and the rotor are rotating synchronously, no currents will be induced and no torque can be produced. Note that, regardless of the rotor speed, the rotor field and stator field are still rotating synchronously with a phase shift. The difference between the speed of the rotor and the fields is compensated by the frequency of the rotor currents, the so-called rotor slip frequency. The asynchronous nature of the machine is the reason why for the control of induction machines, only the position of stator field and rotor field are required and not the absolute rotor position.

Squirrel cage based induction machines are widely used in industry. However, other configurations include *single-phase machines* and *doubly fed induction machines*. Single-phase machines are discussed in Sect. 8.5. Modeling and control of doubly fed induction machines can be found in literature [12]. These machines are found in wind turbines and large pump drives. Such machines make use of a rotor that is provided with a set of three-phase windings which are connected to a stationary converter or a set of resistors via slip-rings. Note, however, that when the rotor side converter is current controlled, the doubly fed machine behaves as a synchronous machine which has a three-phase rotating field winding for its rotor flux excitation [12]. In this case, the term doubly fed induction machine can only relate to the construction of a wound rotor induction machine. From a controls and dynamic performance perspective the term is a misnomer.

8.2 Zero Leakage Inductance Models of Induction Machines

For the purpose of understanding the basic dynamic and steady-state behavior of the induction machine, it is helpful to initially ignore the presence of stator and rotor leakage inductances. Therefore, symbolic and generic models will be introduced first without leakage inductances showing the essence of torque production of the machine. Furthermore, the zero leakage approach serves as an effective platform for introducing field-oriented models.

8.2.1 IRTF Based Model of the Induction Machine

As a first step in the development of dynamic models for induction machines, it is instructive to consider a simplified IRTF based symbolic concept as given in Fig. 8.2. Using an IRTF means to deploy a dual coordinate reference frame which is linked to the stator and to the rotor of the machine.

The mathematical equation set that conforms to the model according to Fig. 8.2 is as follows:

$$\vec{u}_{\mathrm{s}} = R_{\mathrm{s}}\,\vec{i}_{\mathrm{s}} + \frac{\mathrm{d}\vec{\psi}_{\mathrm{m}}}{\mathrm{d}t} \tag{8.1a}$$

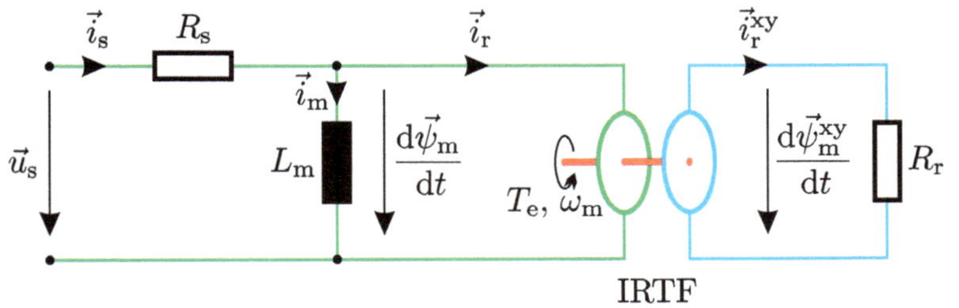

Fig. 8.2 Zero leakage IRTF based induction machine model

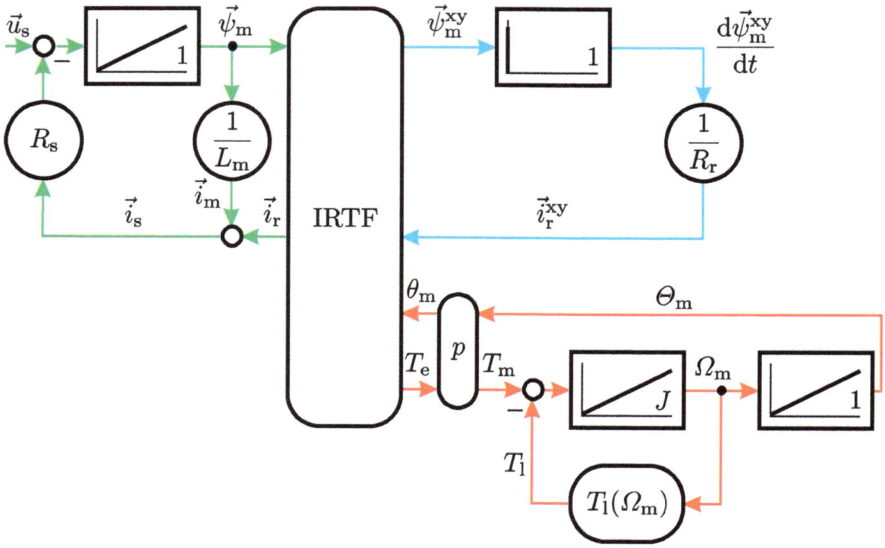

Fig. 8.3 Generic model, zero leakage IRTF based induction motor model

$$\vec{\psi}_m = L_m(\vec{i}_s - \vec{i}_r) \tag{8.1b}$$

$$0 = -R_r\, \vec{i}_r^{xy} + \frac{d\vec{\psi}_m^{xy}}{dt}. \tag{8.1c}$$

Note that the model presented here is very similar to that shown in Fig. 4.10. In the latter case, current excitation was assumed, whereas here voltage excitation is imposed for connecting the machine to a voltage source converter. Furthermore, the magnetizing inductance is shown on the other side of the IRTF which can be done with impunity.

The development of a corresponding generic model representation of this simplified two-pole machine model is readily undertaken with the aid of equation (8.1) and the two IRTF related equations (4.5) for electromagnetic torque calculation and (4.6) as a load model. An implementation example is given in Fig. 8.3. It makes use of a differentiator module which is unavoidable when the leakage inductance has been ignored. Note that, for reasons of numerical stability, the use of differentiators should be treated with care in a simulation environment. The tutorial in Sect. 8.6.1 is based on the generic model presented in this chapter.

8.2.2 Field-Oriented Model

A suitable starting point for the development of a field-oriented model is the IRTF based machine model shown in Fig. 8.2 and the vector diagram given in Fig. 8.4. The

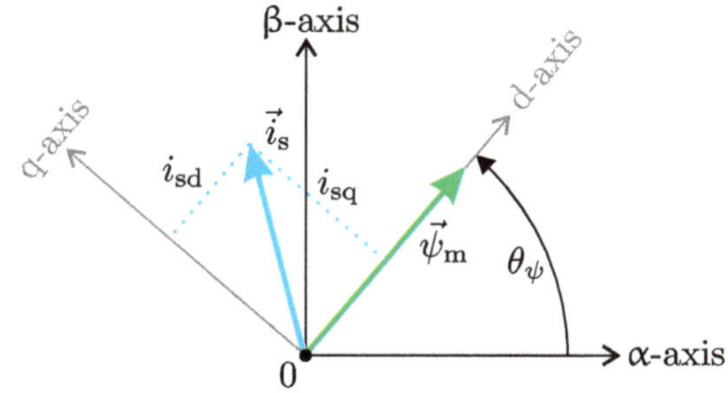

Fig. 8.4 Vector diagram for zero leakage induction machine model

latter diagram shows the stator current and magnetizing flux linkage space vectors $\vec{i}_s$ and $\vec{\psi}_m$, respectively. The latter vector may be represented as $\vec{\psi}_m = \psi_m e^{j\theta_\psi}$, where ψ_m is the amplitude and θ_ψ the angle between the flux linkage vector and real axis of the stationary reference plane.

Stationary based equations are transformed to the dq plane using $\vec{A} = \vec{A}^{dq} e^{j\theta_{\bar{\psi}}}$ with $\omega_s = {}^{d\theta_{\bar{\psi}}}/_{dt}$. The conversion process of rotor-oriented equations to a synchronous reference frame, linked to the flux linkage vector $\vec{\psi}_m$, is achieved by a two-step process. Firstly, the rotor based equations are transformed to a stationary reference frame after which the transformation to a synchronous reference frame can be implemented. Equation set (8.1) may be transformed to the dq plane with the aid of the approach outlined above which leads to

$$\vec{u}_s^{dq} = R_s \vec{i}_s^{dq} + \frac{d\vec{\psi}_m^{dq}}{dt} + j\omega_s \vec{\psi}_m^{dq} \tag{8.2a}$$

$$\frac{\vec{\psi}_m^{dq}}{L_m} = \vec{i}_s^{dq} - \vec{i}_r^{dq} \tag{8.2b}$$

$$\frac{d\vec{\psi}_m^{dq}}{dt} = R_r \vec{i}_r^{dq} - j(\omega_s - \omega_m)\vec{\psi}_m^{dq}. \tag{8.2c}$$

It is noted that in this case some further simplification is possible given that $\vec{\psi}_m^{dq} = \psi_m$, i.e., the vector is real because the direct axis is aligned with this vector. All the remaining vectors have both a real (direct) and an imaginary (quadrature) component, for example, $\vec{i}_s^{dq} = i_{sd} + j i_{sq}$. Further development of equation set (8.2) in terms of grouping the real components leads to

$$u_{sd} = R_s i_{sd} + \frac{d\psi_m}{dt} \tag{8.3a}$$

$$\frac{\psi_m}{L_m} = i_{sd} - i_{rd} \tag{8.3b}$$

$$\frac{d\psi_m}{dt} = R_r i_{rd}. \tag{8.3c}$$

Grouping the imaginary components of equation set (8.2) leads to

$$u_{sq} = R_s\, i_{sq} + \omega_s\, \psi_m \tag{8.4a}$$

$$i_{rq} = i_{sq} \tag{8.4b}$$

$$\omega_s\, \psi_m = R_r\, i_{rq} + \omega_m\, \psi_m. \tag{8.4c}$$

The symbolic direct and quadrature model as given in Fig. 8.5 satisfies equation sets (8.3) and (8.4), respectively.

The corresponding generic current-fed induction machine model with currents i_{sd} and i_{sq} as input variables, which corresponds to the symbolic field-oriented model, is given in Fig. 8.6.

The model according to Figs. 8.5 and 8.6 provides some fundamental insights with respect to the operation of the machine. If, for example, a current i_{sd} is applied to the machine at $t = 0$, the flux linkage ψ_m will assume a steady-state value $\psi_m = L_m\, i_{sd}$ after a transitional period which is governed by the time constant L_m/R_r. Once the flux linkage reaches its steady-state value, the variables $d\psi_m/dt$ and i_{rd} (see Fig. 8.5) will be equal to zero.

The torque T_e is determined by $3/2\psi_m i_{sq}$ as an amplitude invariant vector notation is assumed. These components can be controlled independently as is the case for a separately excited DC machine with a compensation winding. In the latter case, the variables ψ_m and i_{sq} are replaced by the field flux linkage ψ_f and armature current i_a, respectively. A voltage variable $\omega_s\, \psi_m$ as shown in Fig. 8.5 is defined by the sum of the voltages $\omega_m\, \psi_m$ and $R_r i_{sq}$ (see Eq. (8.4c)). The latter is equal to the product of the *slip frequency* $\omega_{sl} = \omega_s - \omega_m$ and flux linkage ψ_m. If a step increase in the current i_{sq} is made, while maintaining a constant i_{sd} value, the torque and slip frequency must also increase stepwise.

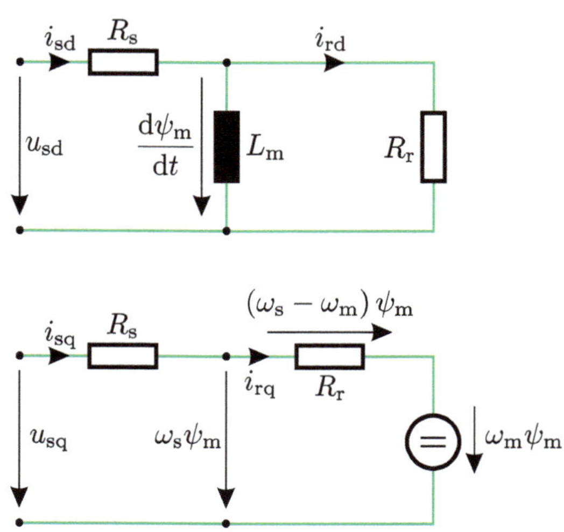

Fig. 8.5 Direct and quadrature field-oriented symbolic model with zero leakage inductance

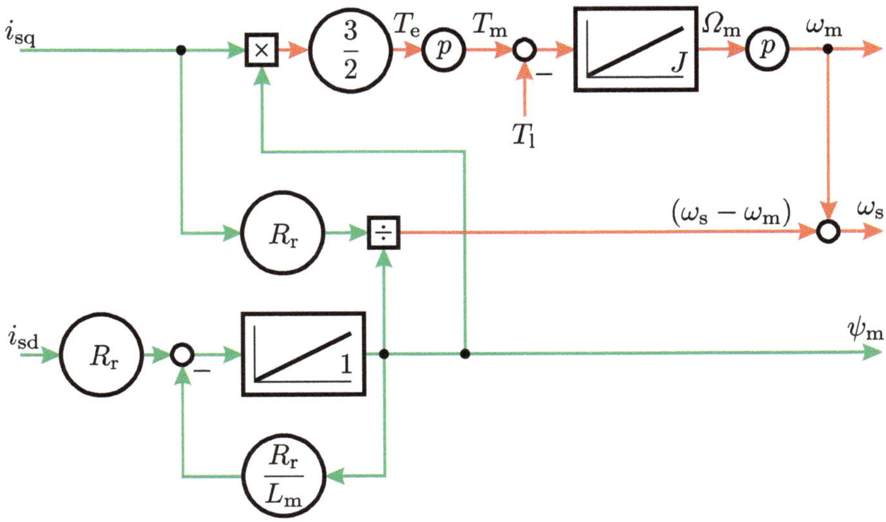

Fig. 8.6 Generic representation of a current source fed field-oriented induction machine model, with zero leakage inductance

Tutorial Results

Figure 8.7 shows a typical transient response for a machine in its current form, without a mechanical load, in terms of the torque T_m, shaft speed ω_m, slip frequency ω_{sl}, and electrical frequency ω_s. In this example, a current $i_{sd} = 4.0\,\text{A}$ is applied at $t = 0$ and a quadrature stator current step $i_{sq} = 0 \rightarrow 4.0\,\text{A}$ is made at $t = 0.8\,\text{s}$. The results shown are obtained with the model given in the tutorial (see Sect. 8.6.4). The reader is referred to Sect. 8.6.4 for further details of the simulation model.

Flux/Current Diagram

Further insight with regard to this type of model can be obtained by considering the flux/current diagram of the machine taken at a particular instance in time. Figure 8.8 shows the flux lines ϕ_d linked with the flux linkage ψ_m. A current distribution in the stator windings and squirrel-cage rotor is shown for both cases. The current and flux distributions tied to the dq-plane rotate at speed ω_s, while the rotor rotates with a shaft speed ω_m.

The direct axis flux distribution (Fig. 8.8a) shows the d-axis which is aligned with the flux linkage vector ψ_m. The current i_{sd} is shown in distributed form on the stator side. Note that the rotor component shows no current, i.e. $i_{rd} = 0$. The quadrature model, see Fig. 8.8b, shows the flux distribution ϕ_d and the stator and rotor current distributions which correspond to i_{sq} and i_{rq}, respectively. It is noted that the two current distributions are in opposition in the actual machine given, while the model assumes the condition $i_{sq} = i_{rq}$. The reason for the supposed discrepancy is linked to the choice of input/output power conventions of the IRTF model. The

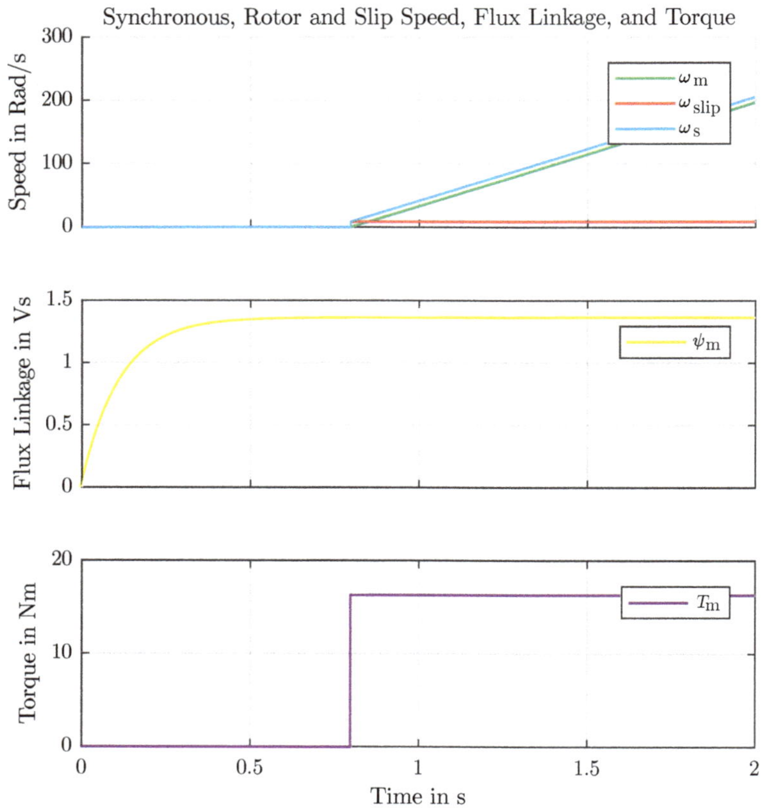

Fig. 8.7 Transient response of field-oriented model with zero leakage inductance

resultant current distribution in the stator is formed by the two stator components shown in Fig. 8.8.

8.3 Machine Models with Leakage Inductances

In practical machines, not all the magnetic flux linkage is fully coupled between the stator windings and the rotor squirrel-cage. The so-called leakage flux paths are present on the stator and rotor side of the machine which in modeling terms are represented by leakage inductances $L_{\sigma s}$ and $L_{\sigma r}$, respectively. In this section, the IRTF and field-oriented modeling approach used in the previous section is extended to accommodate magnetic flux leakage of the machine. In this context, a *universal field-oriented* (UFO) approach will be introduced, which will prove to be instrumental for the development of a field-oriented controller in the next chapter [3].

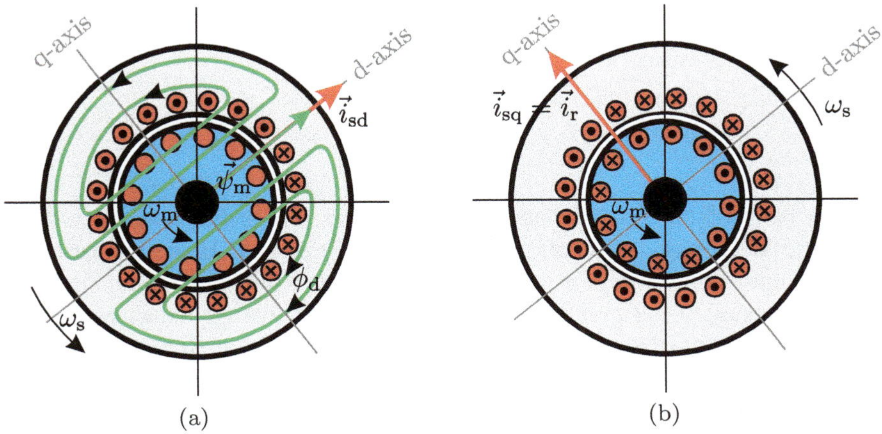

Fig. 8.8 Direct and quadrature flux/current diagram in steady-state. (**a**) d-axis flux/current diagram. (**b**) q-axis current diagram

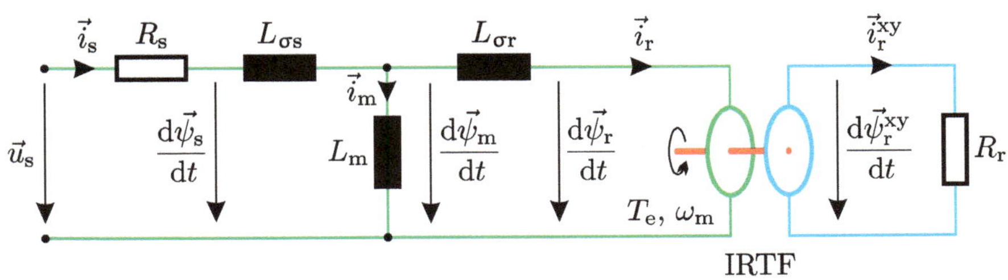

Fig. 8.9 Three inductance, IRTF based induction machine model

8.3.1 Fundamental IRTF Based Model

The simplified model according to Fig. 8.2 is extended to include the rotor and stator based leakage inductances $L_{\sigma r}$ and $L_{\sigma s}$, respectively. The rotor leakage inductance has been conveniently relocated to the stator side of the IRTF module to form a three-element circuit network which consists of the two leakage inductances and the magnetizing inductance L_m. Note that the use of an IRTF module allows the positioning of an inductance to either side without having to change its value. Furthermore, relocating the leakage inductance does not affect the torque T_e, as was shown in Sect. 4.1.

The equation set which corresponds to Fig. 8.9 is as follows:

$$\vec{u}_s = R_s \vec{i}_s + \frac{d\vec{\psi}_s}{dt} \tag{8.5a}$$

$$\vec{\psi}_s = \vec{\psi}_m + L_{\sigma s} \vec{i}_s \tag{8.5b}$$

$$\vec{\psi}_{\mathrm{r}} = \vec{\psi}_{\mathrm{m}} - L_{\sigma\mathrm{r}}\vec{i}_{\mathrm{r}} \tag{8.5c}$$

$$\vec{\psi}_{\mathrm{m}} = L_{\mathrm{m}}\left(\vec{i}_{\mathrm{s}} - \vec{i}_{\mathrm{r}}\right) \tag{8.5d}$$

$$0 = -R_{\mathrm{r}}\vec{i}_{\mathrm{r}}^{\,\mathrm{xy}} + \frac{\mathrm{d}\vec{\psi}_{\mathrm{r}}^{\,\mathrm{xy}}}{\mathrm{d}t}. \tag{8.5e}$$

On the basis of the symbolic model given in Fig. 8.9 and equation set (8.5), a generic IRTF based symbolic model can be developed. However, particular attention must be given in terms of its numerical implementation to avoid algebraic loops. The cause of this problem is the presence of two leakage inductance circuit elements which may be avoided as will become apparent in the following section. Furthermore, it is difficult to determine individual values for these two inductances in a squirrel-cage motor. The reason for this is that these values are usually determined by a *locked rotor* test [14] which yields an estimate for the combined leakage inductance $L_{\sigma\mathrm{s}} + L_{\sigma\mathrm{r}}$. The combined leakage inductance value is then (usually) arbitrarily split evenly to arrive at values for the stator and rotor. As a result, the rotor leakage inductance is measured at line frequency (50 Hz or 60 Hz). Due to rotor deep bar and saturation effects [2, 11], the rotor leakage inductance value can deviate strongly when operating under field-oriented control, i.e. at slip frequency.

8.3.2 Universal IRTF Based Model

The IRTF model according to Fig. 8.9 can be transformed to a so-called *universal* three-inductance configuration which makes use of a transformation coefficient a. By changing the value of this parameter, the user is able to alter the model from, for example, a three- to two-inductance type model where the circuit element which represents the leakage inductance can be located on either side of the equivalent magnetizing inductance component. The term *universal* reflects the flexibility of this new model in terms of being able to change the inductance parameters simultaneously without affecting the no-load or short circuit impedance of the original inductance network. It will be shown in Sect. 8.3.4 that such a transformation capability is of importance for the development of a so-called *universal field-oriented* (UFO) machine model [3].

Parameter Definition for the Universal Model

The aim of this section is therefore to define a set of inductance parameters L_{M}, $L_{\sigma\mathrm{S}}$, and $L_{\sigma\mathrm{R}}$ for a revised symbolic machine model as given in Fig. 8.10 which is able to replace the three-element inductance network of the original model (see Fig. 8.9). The impedance as viewed from either side of the revised inductance network must correspond to the values found in the original inductance network and should not be affected by changes in the arbitrary transformation factor a. In order to achieve this

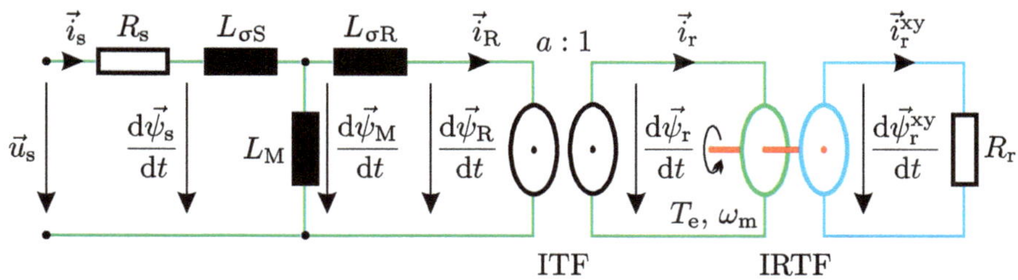

Fig. 8.10 Universal, IRTF based induction machine model, with ITF module

aim, an ideal transformer (ITF) module with arbitrary transformation ratio $a:1$ is introduced in the new model.

The transformation process is initiated by considering equation set (8.5) which is linked to the model given in Fig. 8.9. In particular, it is helpful to consider Eqs. (8.5b) and (8.5d) which may also be written as

$$\vec{\psi}_s = L_s\vec{i}_s - L_m\vec{i}_r - a\,L_m\vec{i}_s + a\,L_m\vec{i}_s \tag{8.6}$$

with $L_s = L_m + L_{\sigma s}$. This expression can then be written as

$$\vec{\psi}_s = \underbrace{(L_s - a\,L_m)}_{L_{\sigma S}}\,\vec{i}_s + \underbrace{a\,L_m}_{L_M}\big(\vec{i}_s - \vec{i}_R\big), \tag{8.7}$$

where the parameters $L_{\sigma S}$ and L_M are introduced, representing a generalized leakage inductance and magnetizing inductance. Furthermore, a scaled rotor current vector $\vec{i}_R$ is introduced in Eq. (8.7) which, along with the scaled rotor flux linkage vector $\vec{\psi}_R$, is defined as

$$\vec{i}_R = \frac{\vec{i}_r}{a} \tag{8.8a}$$

$$\vec{\psi}_R = a\,\vec{\psi}_r. \tag{8.8b}$$

The scaled rotor flux linkage vector $\vec{\psi}_R$ represents the scaled (by the transformation factor a) rotor flux linkage vector $\vec{\psi}_r$. The choice of scaling for $\vec{i}_R$ and $\vec{\psi}_R$ is such that the product of the current and flux linkage vectors as well as the impedance remain unaffected by the scaling. In the *universal* model (see Fig. 8.10) Eq. (8.8) is represented by the ITF module with winding ratio $a:1$. Equations (8.8b) and (8.5c) form the basis for the second part of the proposed model transformation. Use of these two equations to represent the scaled rotor flux linkage vector $\vec{\psi}_R$ gives

$$\vec{\psi}_R = a\,L_m\vec{i}_s - a^2\,L_r\vec{i}_R - a\,L_m\vec{i}_R + a\,L_m\vec{i}_R \tag{8.9}$$

with $L_r = L_m + L_{\sigma r}$. This expression can also be rewritten as

$$\vec{\psi}_R = \underbrace{a\,L_m}_{L_M}\left(\vec{i}_s - \vec{i}_R\right) - \underbrace{\left(a^2\,L_r - a\,L_m\right)}_{L_{\sigma R}}\vec{i}_R, \tag{8.10}$$

where a second leakage inductance parameter $L_{\sigma R}$ is introduced. The resultant flux linkage vector based equation set as given by Eqs. (8.7) and (8.10) can also be written as

$$\vec{\psi}_s = L_{\sigma S}\,\vec{i}_s + L_M\,\vec{i}_M \tag{8.11a}$$

$$\vec{\psi}_R = L_M\,\vec{i}_M - L_{\sigma R}\,\vec{i}_R, \tag{8.11b}$$

where $\vec{i}_M = \vec{i}_s - \vec{i}_R$ represents the scaled magnetizing current vector. The flux linkage equation set (8.11) contains a set of leakage inductances and a magnetizing inductance which are a function of the transformation variable a. This new set on inductances is summarized in equation set (8.12):

$$L_{\sigma S} = L_m\left(\frac{L_s}{L_m} - a\right) \tag{8.12a}$$

$$L_{\sigma R} = a\,L_r\left(a - \frac{L_m}{L_r}\right) \tag{8.12b}$$

$$L_M = a\,L_m. \tag{8.12c}$$

Observation of equation set (8.12) shows that if the transformation variable a is bound by the condition

$$\frac{L_m}{L_r} \leq a \leq \frac{L_s}{L_m}, \tag{8.13}$$

then the leakage inductances $L_{\sigma S}$ and $L_{\sigma R}$ remain greater than or equal to zero. A value of $a = 1$ leads to the original so-called *three-inductance parameter* model shown in Fig. 8.9. A value of $a = {L_s}/{L_m}$ leads to a *two-inductance* model with $L_{\sigma S} = 0$, i.e. the stator leakage inductance is completely eliminated, conversely setting $a = {L_m}/{L_r}$ makes $L_{\sigma R} = 0$, i.e. the rotor leakage inductance is eliminated.

The variation of a as defined by Eq. (8.13) is relatively small, hence it may be helpful to introduce a percentage leakage inductance transformation coefficient Γ_a which is defined over the range $\pm 100\%$, with $\Gamma_a = 0\%$ set to correspond to a universal model with $a = 1$, i.e. no leakage inductance has been eliminated by the transformation in the model. The relationship between Γ_a and a may be written as

$$\text{if } \Gamma_a \geq 0: \qquad a = 1 + \frac{\Gamma_a}{100}\left(\frac{L_\mathrm{s}}{L_\mathrm{m}} - 1\right) \tag{8.14a}$$

$$\text{if } \Gamma_a < 0: \qquad a = 1 + \frac{\Gamma_a}{100}\left(1 - \frac{L_\mathrm{m}}{L_\mathrm{r}}\right). \tag{8.14b}$$

Hence, a value of $\Gamma_a = 100\%$ corresponds to a universal model with $a = {}^{L_\mathrm{s}}/{}_{L_\mathrm{m}}$ and $L_{\sigma\mathrm{S}} = 0$, i.e. the stator leakage inductance is 100% eliminated, conversely setting $\Gamma_a = -100\%$ gives $a = {}^{L_\mathrm{m}}/{}_{L_\mathrm{r}}$ and $L_{\sigma\mathrm{R}} = 0$, i.e. the rotor leakage inductance is 100% eliminated.

Vector Representation of the Universal Model

It is instructive to consider the model transformation from a graphical perspective. This may be achieved by considering the flux linkage/current vector equations (8.5b) and (8.5d) for the original model and representing these in a vector diagram for an arbitrarily chosen set of currents $\vec{i}_\mathrm{s}$ and $\vec{i}_\mathrm{r}$ and a set of inductances L_m, $L_{\sigma\mathrm{s}}$, and $L_{\sigma\mathrm{r}}$.

An example of such a vector plot is shown in Fig. 8.11a. A similar exercise can also be undertaken for the universal model which requires access to the flux linkage/current equations (8.11a) and (8.11b). In this case, the transformation variable must be given a specific value in order to derive the vector plot which corresponds to the current vectors and inductances chosen for the original model. The universal model based vector diagram as given in Fig. 8.11b is shown with a transformation variable of $\Gamma_a = 50\%$. Figure 8.11a and b show an a-axis which is a line defined by the endpoints of the vectors $\vec{\psi}_\mathrm{s}$ and $\vec{\psi}_\mathrm{m}$, respectively. This line is significant because it represents the locus of the vector $\vec{\psi}_\mathrm{M}$ endpoint as function of the transformation variable a [3]. For any user defined value of a the transformation

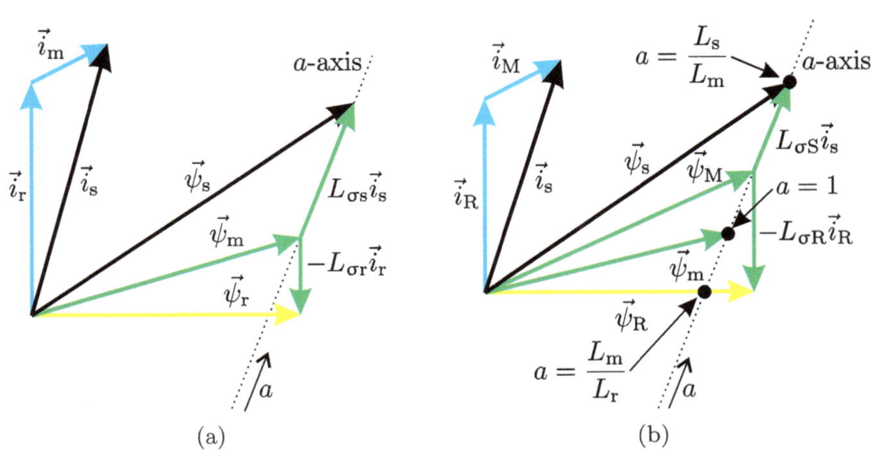

Fig. 8.11 Comparison of vector diagrams for original and universal induction machine model. (**a**) Vector diagram for original model (Fig. 8.9), d-axis aligned with $\vec{\psi}_\mathrm{m}$. (**b**) Vector diagram for universal model with $\Gamma_a = 50\%$, d-axis aligned with $\vec{\psi}_\mathrm{M}$

defines the location of the vector $\vec{\psi}_{\mathrm{M}}$ and corresponding vectors $\vec{\psi}_{\mathrm{R}}$ and $\vec{i}_{\mathrm{R}}$, whereby the latter two vectors are only changed (with respect to the vectors $\vec{\psi}_{\mathrm{r}}$ and $\vec{i}_{\mathrm{r}}$) in terms of their magnitude (see Eq. (8.8)), as may also be observed from Fig. 8.11b.

Symbolic Representation of the Universal Model

The universal model in its present form uses an ITF module to transform the vectors $\vec{\psi}_{\mathrm{R}}$ and $\vec{i}_{\mathrm{R}}$ to their original values $\vec{\psi}_{\mathrm{r}}$ and $\vec{i}_{\mathrm{r}}$. From a modeling perspective it is not strictly necessary to undertake such a transformation provided that the user takes into account the fact that a scaling of these rotor based variables takes place which is dependent on the choice of transformation factor a. The ITF module may be omitted by relocating the IRTF and rotor resistance R_{r} to the primary side of the ITF module. Relocating the IRTF module will not affect the torque but the referred rotor resistance R_{R} must be calculated using

$$R_{\mathrm{R}} = a^2 R_{\mathrm{r}}. \tag{8.15}$$

The resultant universal IRTF based symbolic machine model is shown in Fig. 8.12.

The corresponding equation set for the universal IRTF based model is of the form:

$$\vec{u}_{\mathrm{s}} = R_{\mathrm{s}} \vec{i}_{\mathrm{s}} + \frac{\mathrm{d}\vec{\psi}_{\mathrm{s}}}{\mathrm{d}t} \tag{8.16a}$$

$$\vec{\psi}_{\mathrm{s}} = \vec{\psi}_{\mathrm{M}} + L_{\sigma\mathrm{S}} \vec{i}_{\mathrm{s}} \tag{8.16b}$$

$$\vec{\psi}_{\mathrm{R}} = \vec{\psi}_{\mathrm{M}} - L_{\sigma\mathrm{R}} \vec{i}_{\mathrm{R}} \tag{8.16c}$$

$$\frac{\vec{\psi}_{\mathrm{M}}}{L_{\mathrm{M}}} = \vec{i}_{\mathrm{s}} - \vec{i}_{\mathrm{R}} \tag{8.16d}$$

$$0 = -R_{\mathrm{R}} \vec{i}_{\mathrm{R}}^{\,xy} + \frac{\mathrm{d}\vec{\psi}_{\mathrm{R}}^{\,xy}}{\mathrm{d}t}. \tag{8.16e}$$

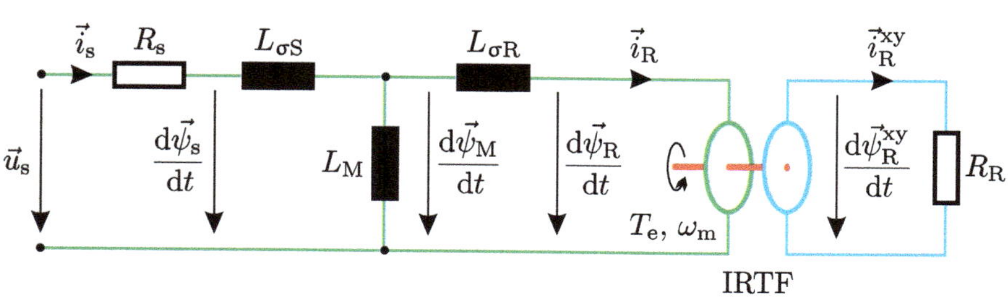

Fig. 8.12 Universal, IRTF based induction machine model

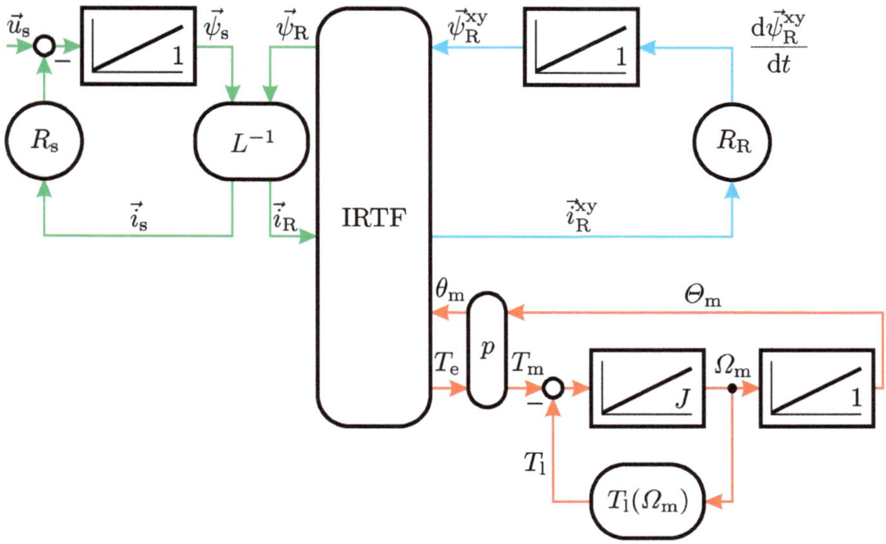

Fig. 8.13 Generic model representation of a universal IRTF based induction machine

The significance of the universal model transformation can be demonstrated by considering the following three values (of which two are chosen at opposite ends of the scale, see Eq. (8.13)) for the transformation variable a, namely:

- $a = \frac{L_m}{L_r}$ ($\Gamma_a = -100\%$): Under these conditions, the three-inductance model shown in Fig. 8.12 is reduced to two inductances $L_{\sigma S}$ and L_M, i.e., $L_{\sigma R} = 0$. Furthermore, the vector $\vec{\psi}_M$ equals $\vec{\psi}_R$ as may also be observed with the aid of Fig. 8.11b. This model will be referred to the *rotor flux based IRTF model*.
- $a = 1$ ($\Gamma_a = 0\%$) The universal model is reduced to the original *three-inductance* parameter model as given in Fig. 8.9. This model is referred to as the *airgap flux based IRTF model*.
- $a = \frac{L_s}{L_m}$ ($\Gamma_a = 100\%$) The universal model according to Fig. 8.12 is reduced to an *alternative* two-inductance model with $L_{\sigma S} = 0$. Furthermore, the vector $\vec{\psi}_M$ is equal to $\vec{\psi}_s$ under these circumstances, as may also be observed with the aid of Fig. 8.11b. This model will be referred to as the *stator flux based IRTF model*.

Generic Representation of the Universal Model

A generic representation of the symbolic model as given in Fig. 8.13 may be developed with the aid of the terminal equations (8.16a) and (8.16e).

The model in question makes use of a generic module identified by the name L^{-1} which represents the matrix $\left[L^{-1}\right]$ defined by Eq. (8.17).

$$\begin{bmatrix}\vec{i}_{\rm s}\\ \vec{i}_{\rm R}\end{bmatrix} = \underbrace{\frac{1}{L_{\rm S}\,L_{\rm R}-(L_{\rm M})^2}\begin{bmatrix}L_{\rm R} & -L_{\rm M}\\ L_{\rm M} & -L_{\rm S}\end{bmatrix}}_{\left[L^{-1}\right]}\begin{bmatrix}\vec{\psi}_{\rm s}\\ \vec{\psi}_{\rm R}\end{bmatrix} \tag{8.17}$$

with $L_{\rm S} = L_{\rm M} + L_{\sigma{\rm S}}$ and $L_{\rm R} = L_{\rm M} + L_{\sigma{\rm R}}$, where $L_{\rm M}$, $L_{\sigma{\rm S}}$, and $L_{\sigma{\rm R}}$ are defined by Eq. (8.12). Expression (8.17) is found with the aid of equation (8.11) in which the vector $\vec{i}_{\rm M}$ is redefined in terms of the current vectors $\vec{i}_{\rm s}$ and $\vec{i}_{\rm R}$. The model according to Fig. 8.13 does not generate the vector $\vec{\psi}_{\rm M}$ explicitly, instead this variable can be calculated using Eq. (8.16c). It is emphasized that this generic model can be used for any transformation factor a within the range defined by Eq. (8.13) without encountering any algebraic loops in the simulation. The reader is reminded of the fact that changes in the transformation variable a affect the matrix L^{-1} terms as well as the variable $R_{\rm R}$. In some cases, it is beneficial to represent Eq. (8.17) in terms of the variables $L_{\rm m}$, $L_{\sigma{\rm s}}$, and $L_{\sigma{\rm r}}$, which represent the inductances of the original model (see Fig. 8.9). Use of Eq. (8.12) with Eq. (8.17) leads to

$$\begin{bmatrix}\vec{i}_{\rm s}\\ \vec{i}_{\rm R}\end{bmatrix} = \underbrace{\frac{1}{\sigma_{\rm u}\,L_{\rm s}}\begin{bmatrix}1 & -\frac{1}{a}\left(\frac{L_{\rm m}}{L_{\rm r}}\right)\\ \frac{1}{a}\left(\frac{L_{\rm m}}{L_{\rm r}}\right) & -\frac{1}{a^2}\left(\frac{L_{\rm s}}{L_{\rm r}}\right)\end{bmatrix}}_{\left[L^{-1}\right]}\begin{bmatrix}\vec{\psi}_{\rm s}\\ \vec{\psi}_{\rm R}\end{bmatrix}, \tag{8.18}$$

where $\sigma_{\rm u} = 1 - L_{\rm m}^2/L_{\rm s}L_{\rm r}$ represents the leakage factor [3, 10, 13], which is a machine characteristic and not a function of the transformation variable a. The tutorial given in Sect. 8.6.2 is directly based on the generic model (see Fig. 8.13) and demonstrates the use of the *universal* transformation concept as discussed in this section.

8.3.2.1 Rotor Flux Based IRTF Model

Use of the transformation variable $a = L_{\rm m}/L_{\rm r}$ ($\Gamma_a = -100\%$) with the model according to Fig. 8.12 leads to the symbolic model shown in Fig. 8.14.

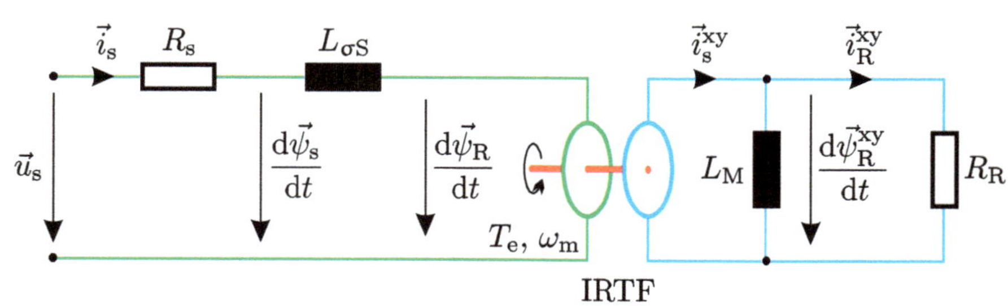

Fig. 8.14 IRTF based induction machine model with $a = L_{\rm m}/L_{\rm r}$

The corresponding equation set for the rotor flux based IRTF based model is of the form

$$\vec{u}_s = R_s \vec{i}_s + \frac{d\vec{\psi}_s}{dt} \tag{8.19a}$$

$$\vec{\psi}_s = \vec{\psi}_R + L_{\sigma S} \vec{i}_s \tag{8.19b}$$

$$\frac{\vec{\psi}_R^{xy}}{L_M} = \vec{i}_s^{xy} - \vec{i}_R^{xy} \tag{8.19c}$$

$$0 = -R_R \vec{i}_R^{xy} + \frac{d\vec{\psi}_R^{xy}}{dt}. \tag{8.19d}$$

The symbolic model according to Fig. 8.14 will be used for representing the *standard* induction machine. The generic dynamic model as given in Fig. 8.15 corresponds to this symbolic model and equation set (8.19). It is emphasized that the rotor flux based IRTF model is able to accommodate dynamic as well as steady-state operation. However, for the latter a phasor type analysis may be more convenient as will be shown in Sect. 8.3.6.

Note that the IRTF model can be readily extended to include (among others) homopolar effects, rotor skin effect, however such models are not included here as these are outside the scope of this book.

Tutorial Results

In Sect. 8.6.7, a simulation example is given which is based on the model described above. In this tutorial, the model is used to examine the line start of a 22 kW four-pole delta connected induction machine. An example of the results obtained with

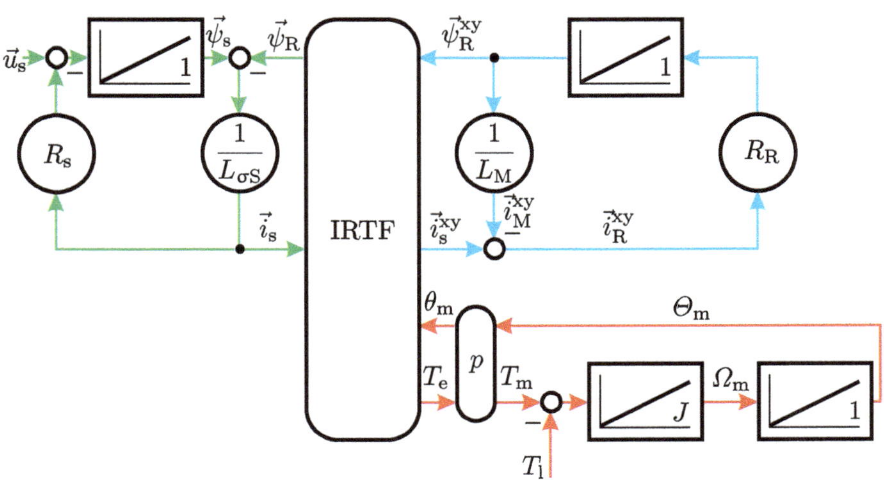

Fig. 8.15 Four-parameter, rotor flux based IRTF induction motor model

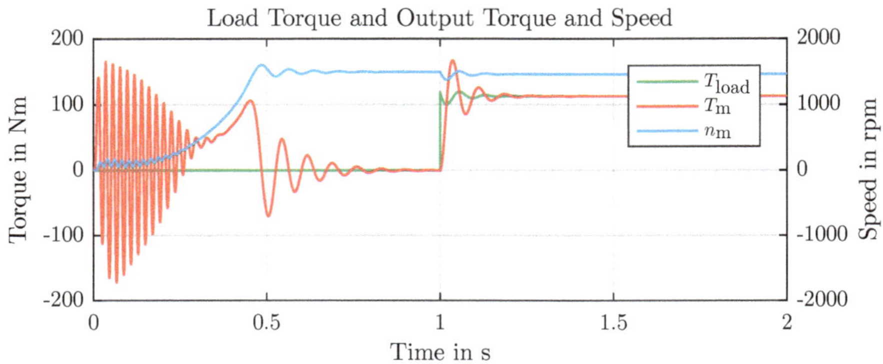

Fig. 8.16 Line start simulation of a 22 kW delta connected machine, showing shaft torque, speed, and line current

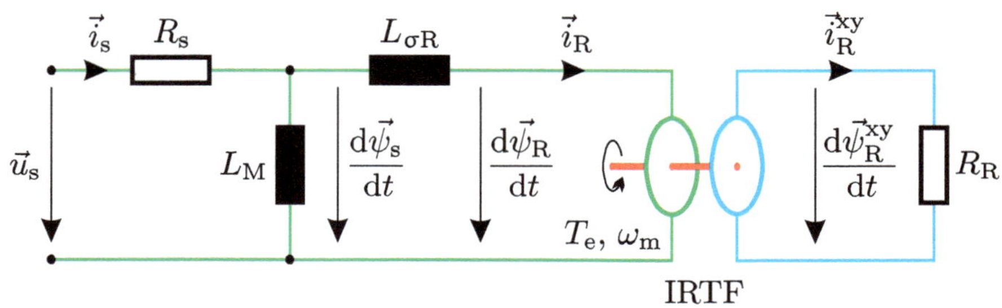

Fig. 8.17 IRTF based induction machine model with $a = {}^{L_s}/_{L_m}$

this simulation model, as indicated in Fig. 8.16, shows the machine shaft torque, shaft speed, and line current over a 2 s start-up sequence, where a load torque step is applied at $t = 1$ s. The reader is referred to the tutorial given in Sect. 8.6.7 for further details and the opportunity to interactively examine this model concept.

8.3.2.2 Stator Flux Based IRTF Model

Setting the transformation variable to $a = {}^{L_s}/_{L_m}$ ($\Gamma_a = 100\%$) reduces the symbolic model given in Fig. 8.12 to the form shown in Fig. 8.17 [15]. This model is instructive because it shows how the rotor current is affected by the series impedance formed by the leakage inductance $L_{\sigma R}$ and rotor resistance R_R.

The corresponding equation set for the stator flux based IRTF model may be found by making use of equation set (8.16) with $L_{\sigma S} = 0$, $L_M = L_s$, and $\vec{\psi}_M = \vec{\psi}_s$, which gives

$$\vec{u}_s = R_s \vec{i}_s + \frac{d\vec{\psi}_s}{dt} \tag{8.20a}$$

$$\vec{\psi}_{\mathrm{s}} = \vec{\psi}_{\mathrm{R}} + L_{\sigma\mathrm{R}}\,\vec{i}_{\mathrm{s}} \tag{8.20b}$$

$$\frac{\vec{\psi}_{\mathrm{s}}}{L_{\mathrm{M}}} = \vec{i}_{\mathrm{s}} - \vec{i}_{\mathrm{R}} \tag{8.20c}$$

$$0 = -R_{\mathrm{R}}\,\vec{i}_{\mathrm{R}}^{\,\mathrm{xy}} + \frac{\mathrm{d}\vec{\psi}_{\mathrm{R}}^{\,\mathrm{xy}}}{\mathrm{d}t}. \tag{8.20d}$$

8.3.3 Universal Stationary Frame Oriented Model

The IRTF based models introduced in Sect. 8.3.2 have components which are linked to vectors in a stationary as well as a shaft-oriented reference frame. To simplify the analysis, a model is derived where all voltage, current, and flux linkage vectors are linked to a *common stationary reference frame*. To realize this aim, the rotor coordinate based Eq. (8.16e) must be converted to stationary coordinates. The general space vector conversion required for this task is of the form $\vec{A} = \vec{A}^{\mathrm{xy}}\mathrm{e}^{\mathrm{j}\theta}$ with $\theta = \omega_{\mathrm{m}}t$. The revised rotor based equation in stationary coordinates is of the form

$$0 = -R_{\mathrm{R}}\,\vec{i}_{\mathrm{R}} + \frac{\mathrm{d}\vec{\psi}_{\mathrm{R}}}{\mathrm{d}t} - \mathrm{j}\omega_{\mathrm{m}}\,\vec{\psi}_{\mathrm{R}}. \tag{8.21}$$

Use of Eq. (8.21) and the stationary frame oriented elements of equation set (8.16) leads to the symbolic model given in Fig. 8.18 which appears in the drives literature with different values for the transformation variable a. Note that the conversion process used here is convenient when dealing with steady-state sinusoidal supplies. In this case, the space vectors can be readily transformed to stationary phasor diagrams.

The corresponding generic model representation as shown in Fig. 8.19 makes use of the generic module L^{-1} introduced in the Sect. 8.3.2. A tutorial which is based on the generic model given here is discussed in Sect. 8.6.3.

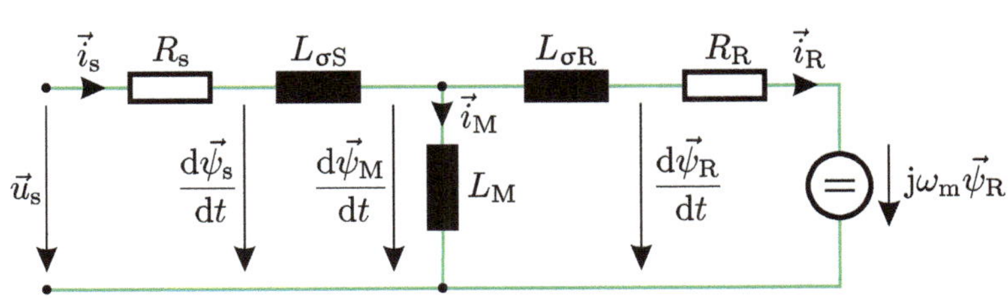

Fig. 8.18 Universal, stationary frame oriented symbolic machine model

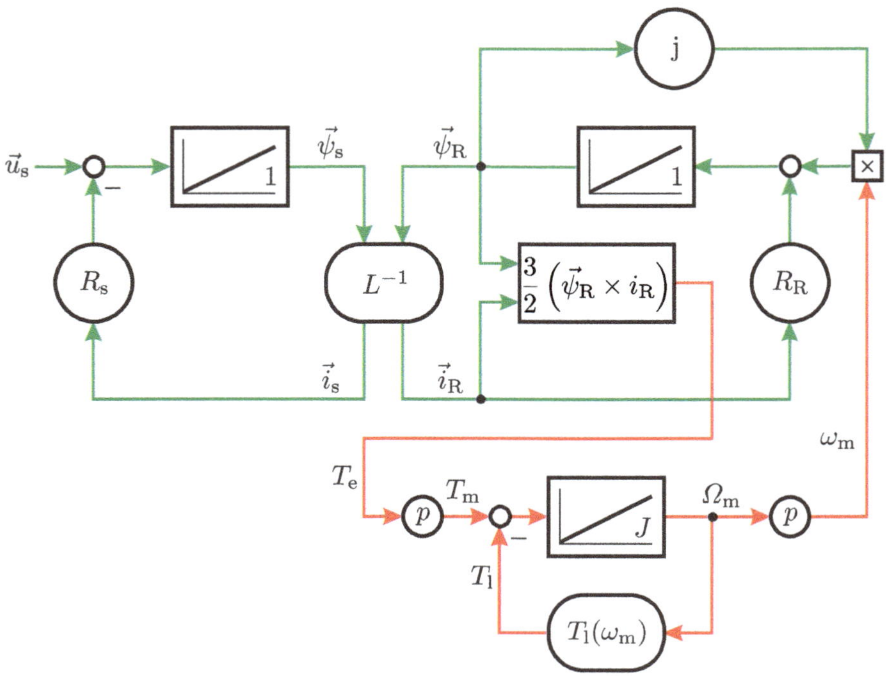

Fig. 8.19 Generic representation, stationary frame oriented induction machine drive model

8.3.4 Universal Field-Oriented (UFO) Model

In the machine models discussed previously, the current, voltage, and flux linkage space vectors were defined with respect to a stationary and/or shaft-oriented reference frame. In this section, a so-called *universal field-oriented* (UFO) transformation is introduced where the stator and rotor based space equations are tied to the flux linkage vector $\vec{\psi}_M$.

This approach combines the advantages of a universal inductance model, as discussed in Sect. 8.3.2, with a so-called field-oriented transformation that leads to synchronous model representation. The development of UFO based models in this section is particularly instructive for the development of field-oriented control concepts in Chap. 9. In this context, the models to be discussed are current-excited because vector-controlled drives often utilize some form of current control as discussed in Chap. 3.

Development of a Symbolic UFO Model

The development of a UFO type model is based on the field-oriented (synchronous) reference frame with a so-called *direct* and *quadrature* axis, i.e., $\vec{x}^{dq} = x_d + j x_q$. The direct axis is aligned with flux linkage vector $\vec{\psi}_M$, hence $\psi_{Md} = \psi_M$ and $\psi_{Mq} = 0$, as can be observed in Fig. 8.20.

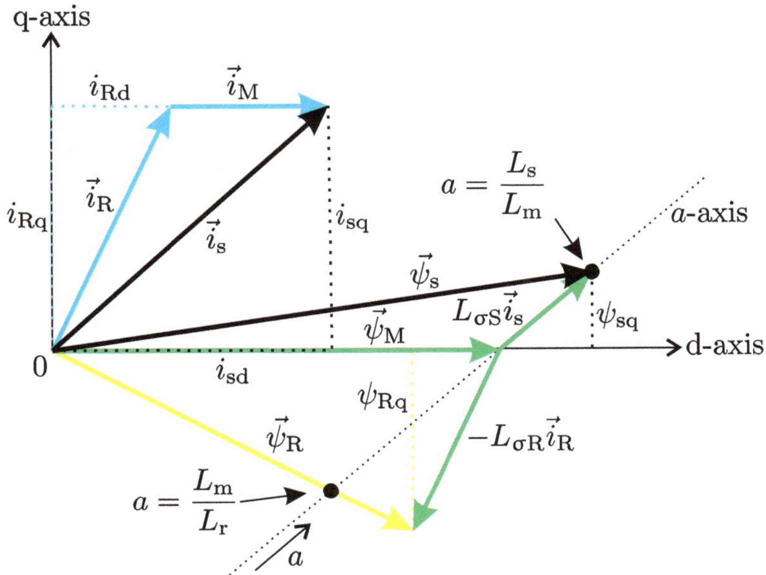

Fig. 8.20 Vector diagram with direct and quadrature axis

The approach required to derive the symbolic and generic model of the UFO model with leakage inductances $L_{\sigma\mathrm{S}}$ and $L_{\sigma\mathrm{R}}$ (see Fig. 8.12) is similar to the method described for the zero leakage case. Consequently, the coordinate transformation process may be undertaken with the aid of equations (8.16) and (8.21). However, in this case the stator and rotor flux linkage space vectors $\vec{\psi}_{\mathrm{s}}$ and $\vec{\psi}_{\mathrm{R}}$ must in the course of this transformation be expressed in terms of the magnetizing vector $\vec{\psi}_{\mathrm{M}}$, given that the d-axis of synchronous reference frame is aligned with this variable (as shown in Fig. 8.20). The equation set for the generalized UFO based model may be written as

$$\vec{u}_{\mathrm{s}}^{\,\mathrm{dq}} = R_{\mathrm{s}}\,\vec{i}_{\mathrm{s}}^{\,\mathrm{dq}} + \frac{\mathrm{d}\vec{\psi}_{\mathrm{s}}^{\,\mathrm{dq}}}{\mathrm{d}t} + \mathrm{j}\omega_{\mathrm{s}}\,\vec{\psi}_{\mathrm{s}}^{\,\mathrm{dq}} \tag{8.22a}$$

$$\vec{\psi}_{\mathrm{s}}^{\,\mathrm{dq}} = \psi_{\mathrm{M}} + L_{\sigma\mathrm{S}}\,\vec{i}_{\mathrm{s}}^{\,\mathrm{dq}} \tag{8.22b}$$

$$\vec{\psi}_{\mathrm{R}}^{\,\mathrm{dq}} = \psi_{\mathrm{M}} - L_{\sigma\mathrm{R}}\,\vec{i}_{\mathrm{R}}^{\,\mathrm{dq}} \tag{8.22c}$$

$$\frac{\psi_{\mathrm{M}}}{L_{\mathrm{M}}} = \vec{i}_{\mathrm{s}}^{\,\mathrm{dq}} - \vec{i}_{\mathrm{R}}^{\,\mathrm{dq}} \tag{8.22d}$$

$$\frac{\mathrm{d}\vec{\psi}_{\mathrm{R}}^{\,\mathrm{dq}}}{\mathrm{d}t} = R_{\mathrm{R}}\,\vec{i}_{\mathrm{R}}^{\,\mathrm{dq}} - \mathrm{j}\,(\omega_{\mathrm{s}} - \omega_{\mathrm{m}})\,\vec{\psi}_{\mathrm{R}}^{\,\mathrm{dq}}. \tag{8.22e}$$

The symbolic direct and quadrature models, as shown in Fig. 8.21, are found by rearranging equation set (8.22) and by grouping the real and imaginary terms. The

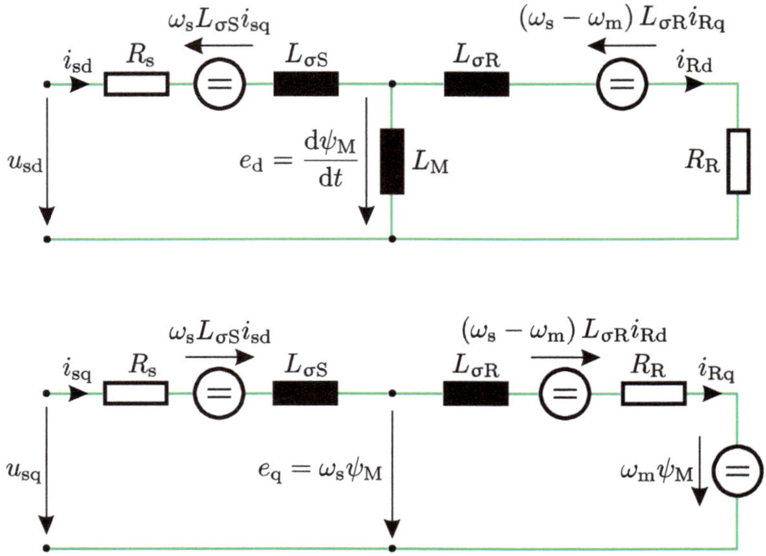

Fig. 8.21 Symbolic UFO model: direct/quadrature axis configuration, with leakage inductance

real terms for the direct axis model are

$$u_{sd} = R_s\, i_{sd} - \omega_s\, L_{\sigma S}\, i_{sq} + L_{\sigma S}\frac{di_{sd}}{dt} + \frac{d\psi_M}{dt} \tag{8.23a}$$

$$\frac{\psi_M}{L_M} = i_{sd} - i_{Rd} \tag{8.23b}$$

$$\frac{d\psi_M}{dt} = L_{\sigma R}\frac{di_{Rd}}{dt} - (\omega_s - \omega_m)\, L_{\sigma R}\, i_{Rq} + R_R\, i_{Rd}. \tag{8.23c}$$

The imaginary terms for the quadrature axis model form the following equation set:

$$u_{sq} = R_s\, i_{sq} + L_{\sigma S}\frac{di_{sq}}{dt} + \omega_s\, L_{\sigma S}\, i_{sd} + e_q \tag{8.24a}$$

$$i_{sq} = i_{Rq} \tag{8.24b}$$

$$e_q = L_{\sigma R}\frac{di_{Rq}}{dt} + (\omega_s - \omega_m)\, L_{\sigma R}\, i_{Rd} + R_R\, i_{Rq} + \omega_m\, \psi_M. \tag{8.24c}$$

The impact of including the leakage inductance components of the machine in the symbolic direct and quadrature models is significant as may be observed by comparing Fig. 8.21 with the zero leakage model in Fig. 8.5. The model complexity increases as the rotor current vector $\vec{i}_r^{dq}$ is no longer perpendicular to the main magnetizing flux vector $\vec{\psi}_m$.

Development of a Generic UFO Model

The development of a generic stator current model which corresponds to Fig. 8.21 may be undertaken with the aid of equations (8.23c) and (8.24c), and replacing the variables i_{Rd} and i_{Rq} with the variables i_{sd}, i_{sq}, and ψ_{M} as defined by Eqs. (8.23b) and (8.24b). Subsequent mathematical manipulation gives

$$\frac{\mathrm{d}}{\mathrm{d}t}\left(\frac{L_{\mathrm{R}}}{L_{\mathrm{M}}}\psi_{\mathrm{M}}\right)+\frac{R_{\mathrm{R}}}{L_{\mathrm{M}}}\psi_{\mathrm{M}}=R_{\mathrm{R}}\,i_{\mathrm{sd}}+L_{\sigma\mathrm{R}}\frac{\mathrm{d}i_{\mathrm{sd}}}{\mathrm{d}t}-\omega_{\mathrm{sl}}\,L_{\sigma\mathrm{R}}\,i_{\mathrm{sq}} \tag{8.25a}$$

$$\omega_{\mathrm{sl}}\left(\frac{L_{\mathrm{R}}}{L_{\mathrm{M}}}\psi_{\mathrm{M}}-L_{\sigma\mathrm{R}}\,i_{\mathrm{sd}}\right)=L_{\sigma\mathrm{R}}\frac{\mathrm{d}i_{\mathrm{sq}}}{\mathrm{d}t}+R_{\mathrm{R}}\,i_{\mathrm{sq}} \tag{8.25b}$$

with $L_{\mathrm{R}}=L_{\sigma\mathrm{R}}+L_{\mathrm{M}}$ and slip frequency $\omega_{\mathrm{sl}}=(\omega_{\mathrm{s}}-\omega_{\mathrm{m}})$ as defined previously. Integrating the first order differential equation (8.25a) generates the flux linkage state variable ψ_{M} from the inputs i_{sd} and i_{sq}. However, to complete this model the slip frequency ω_{sl} must also be expressed in terms of the input and state variables i_{sd}, i_{sq}, and ψ_{M}. The latter can readily be achieved by rewriting Eq. (8.25b) as follows:

$$\omega_{\mathrm{sl}}=\frac{L_{\sigma\mathrm{R}}\frac{\mathrm{d}i_{\mathrm{sq}}}{\mathrm{d}t}+R_{\mathrm{R}}\,i_{\mathrm{sq}}}{\frac{L_{\mathrm{R}}}{L_{\mathrm{M}}}\psi_{\mathrm{M}}-L_{\sigma\mathrm{R}}\,i_{\mathrm{sd}}}. \tag{8.26}$$

Note that the slip frequency is an internal model variable, not a state variable, as it is computed from a division in the time domain (not a differential equation). Equations (8.25) and (8.26) together with the torque equation $T_{\mathrm{e}}=\frac{3}{2}\psi_{\mathrm{M}}\,i_{\mathrm{sq}}$ define the complete generic direct/quadrature model of the UFO machine concept shown in Fig. 8.22. Some indication with respect to its functioning is possible at this stage by considering this model without rotor leakage, in which case the parameter $L_{\sigma\mathrm{R}}$ should be set to zero. The model according to Fig. 8.22 is particularly useful for considering two special cases where the d-axis and corresponding $\vec{\psi}_{\mathrm{M}}$ are aligned with either the rotor flux linkage vector $\vec{\psi}_{\mathrm{R}}$ or stator flux linkage vector $\vec{\psi}_{\mathrm{s}}$. Both cases will be discussed in the next two subsections.

8.3.4.1 Rotor Flux Oriented Model

If a transformation value of $a={}^{L_{\mathrm{m}}}/_{L_{\mathrm{r}}}$ is used in the generalized UFO model, a d-axis alignment with the rotor flux linkage vector $\vec{\psi}_{\mathrm{R}}$ takes place as may be observed from Fig. 8.23. This vector diagram shows the spatial orientation of the current/flux linkage vectors for rotor flux oriented models. As such, this figure is a specific case of the more generalized case shown in Fig. 8.20.

With the present choice of transformation value, the symbolic direct and quadrature models shown in Fig. 8.21 convert to the form indicated in Fig. 8.24. The reason for this is that the leakage inductance $L_{\sigma\mathrm{R}}$ will be zero for the selected a value (see Eq. (8.12b)).

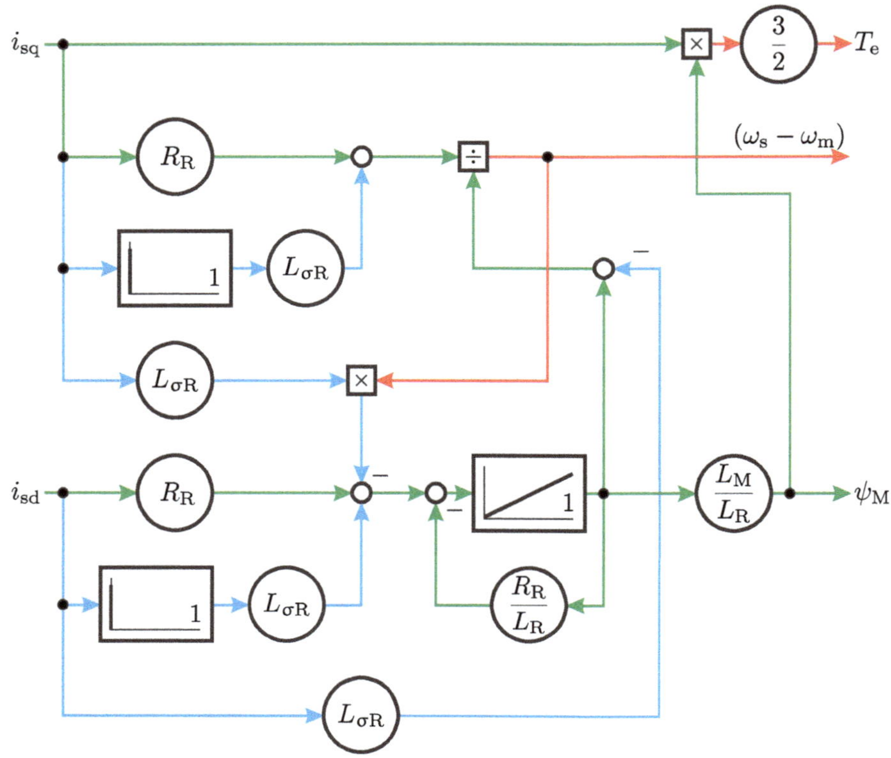

Fig. 8.22 Generic direct and quadrature UFO based, stator current source induction machine model

The implications for the generic model which is linked with the choice of dq coordinates may be observed with the aid of Fig. 8.25. This figure is based on Fig. 8.22 where all $L_{\sigma R}$ related terms are omitted as $L_{\sigma R} = 0$ for rotor flux orientation ($a = L_m/L_r$). Under these circumstances, the model reduces to the form given by Fig. 8.6 (where here, due to the choice of a, R_R replaces R_r and L_R replaces L_m), which represents the zero rotor leakage model. The latter is perhaps not surprising given that the model assumes current excitation, which implies, as may be observed from the symbolic model in Fig. 8.24, that only the voltages e_d and e_q are affected by the parameters and variables to the *right* of these diagrams. An important benefit of a rotor-oriented flux model is that there is a complete decoupling between the direct and quadrature currents. This implies that a change in torque may be undertaken by changing the quadrature current only as may be observed from Fig. 8.6. Likewise, a change in direct current will only affect the flux linkage magnitude $\psi_M = \psi_R$. Given the simplicity of these models, they were used first by the inventors of field-oriented control in a time when no digital implementation of the algorithm was possible [1, 5]. A model representation with a

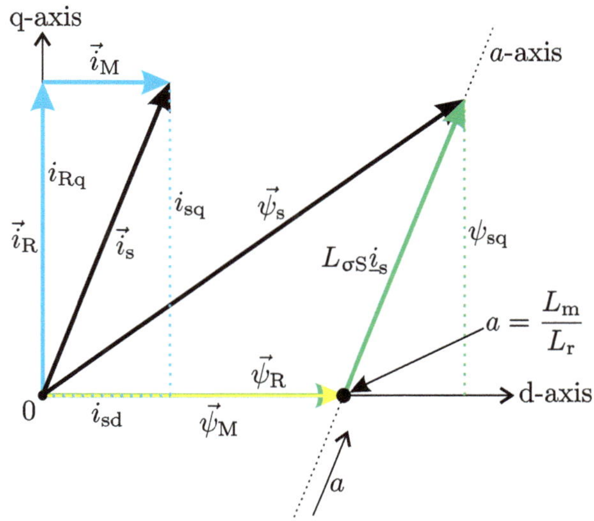

Fig. 8.23 Rotor flux oriented vector diagram

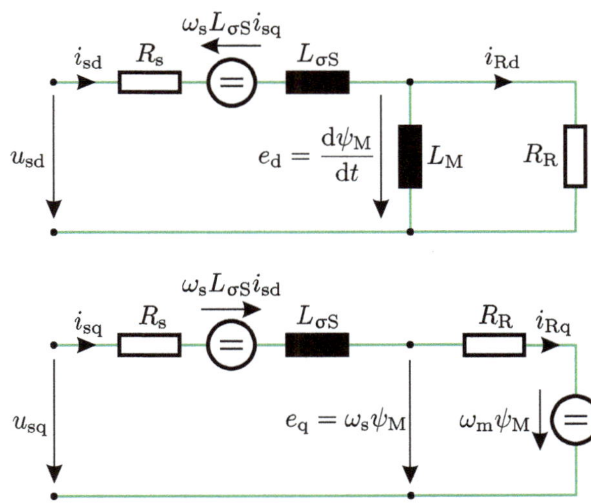

Fig. 8.24 Symbolic rotor flux oriented model: direct and quadrature axis topologies

single leakage inductance $L_{\sigma S}$ on the stator side of the magnetizing inductance L_M is often used.

Tutorial Results

The tutorial given in Sect. 8.6.5 is based on the UFO generic model concept shown in Fig. 8.22. The current excitation and parameters R_r and L_m used here, to derive the results shown in Fig. 8.26, are the same as those used to derive the zero leakage results given in Fig. 8.7. However, the model has been extended to accommodate the leakage inductance parameters $L_{\sigma s}$ and $L_{\sigma r}$ and the transformation variable a is introduced. Variable a may be varied to allow the model to be used in any UFO reference. Here, the model is set to a rotor flux oriented reference frame, and the

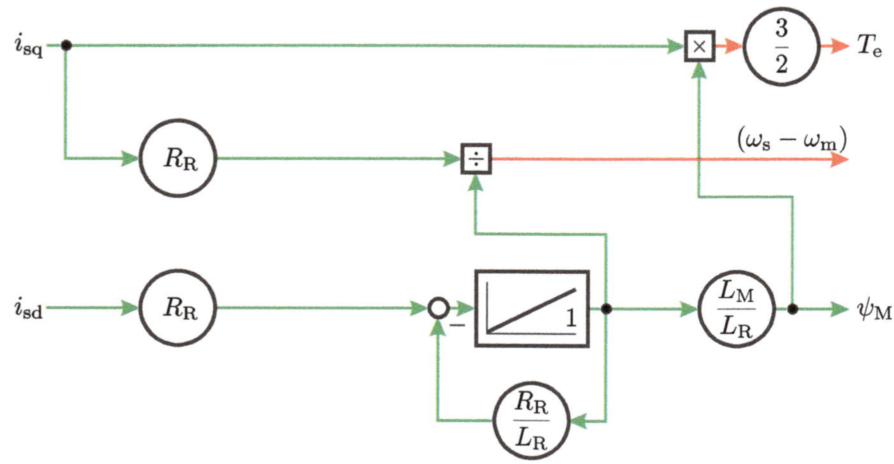

Fig. 8.25 Block diagram for rotor flux based, current source induction machine model

transient results shown in Fig. 8.26 underline the fact that the direct and quadrature models are decoupled, given that a change in the torque does not affect the flux linkage ψ_M.

A qualitative comparison of the results shown in Figs. 8.7 and 8.26 demonstrates that the results are similar. This implies that the use of a rotor flux oriented model leads to a transient response which matches that of a machine without leakage inductance. It is precisely the combination of using direct/quadrature current excitation and setting the leakage inductance parameter $L_{\sigma R}$ to zero which leads to a decoupled model that may be readily used for realizing field-oriented control in a rotor flux oriented reference frame.

8.3.4.2 Stator Flux Oriented Model

If the transformation variable is set to $a = L_s/L_m$, the dq symbolic model is reduced from the configuration shown in Fig. 8.21 to that indicated in Fig. 8.27. The reason for this is that the leakage inductance $L_{\sigma S}$ will be zero for this a value (see Eq. (8.12b)).

The corresponding vector diagram for this model representation as shown in Fig. 8.28 highlights the fact that the flux linkage vector $\vec{\psi}_M$ (and therefore the d-axis) is now aligned with the vector $\vec{\psi}_s$. A further observation of Fig. 8.28 shows that the magnitudes of the vectors $\vec{\psi}_R$ and $\vec{i}_R$ are changed. However, their spatial orientation with respect to the vectors $\vec{\psi}_r$ and $\vec{i}_r$ remains unaffected. In contrast to the rotor flux oriented symbolic models, the direct and quadrature models exhibit a degree of cross-coupling which is dictated by the leakage inductance $L_{\sigma R}$.

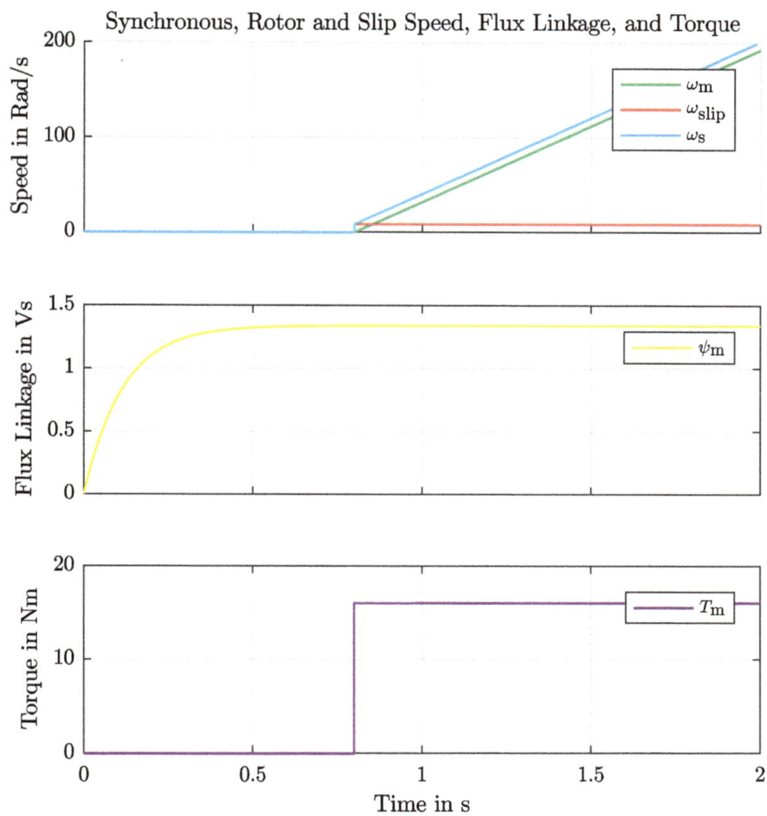

Fig. 8.26 Transient response of rotor flux oriented UFO model

Tutorial Results

The tutorial (see Sect. 8.6.5) used in the previous section for the presentation of machine transient results may be conveniently used in this section, provided that the transformation value is set to $a = L_s/L_m$. With this choice of transformation variable, the UFO based simulation model will operate with a stator flux oriented reference frame. The current excitation and machine parameters used to derive the results given in Fig. 8.26 are also applied to the numerical results given in Fig. 8.29.

A qualitative comparison between the results shown in Figs. 8.29 and 8.26 demonstrates that they are markedly different. For example, a Dirac type response occurs in the slip frequency ω_{sl} which in turn appears in the electrical frequency ω_s. The cause of this phenomenon is the step change in the quadrature current which leads to an instantaneous change of the flux linkage vector $\vec{\psi}_s$ with respect to the vector $\vec{\psi}_R$. Note that due to the leakage inductances such a step change of current is not possible in practice as it would require an infinitely high voltage pulse. In addition, a step in the slip frequency occurs which leads to a change in the flux linkage $\psi_M = \vec{\psi}_s$. The results given in Fig. 8.29 confirm the presence of coupling

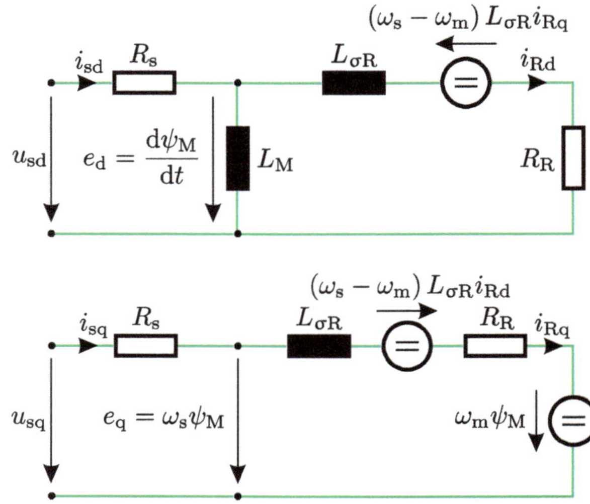

Fig. 8.27 Symbolic stator flux oriented model: direct/quadrature axis configurations

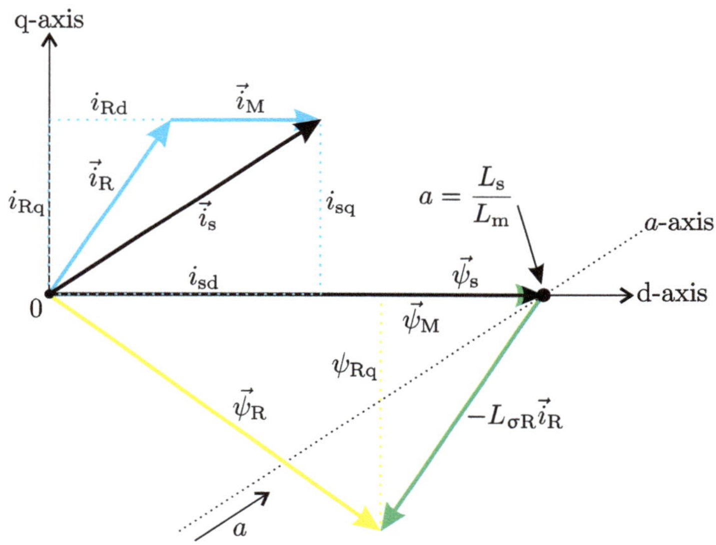

Fig. 8.28 Stator flux oriented vector diagram

between the direct and quadrature models and the impact of a torque step on the instantaneous slip frequency. Inverting the model in question for the purpose of developing stator flux oriented control is therefore a more demanding task as will become apparent in Chap. 9.

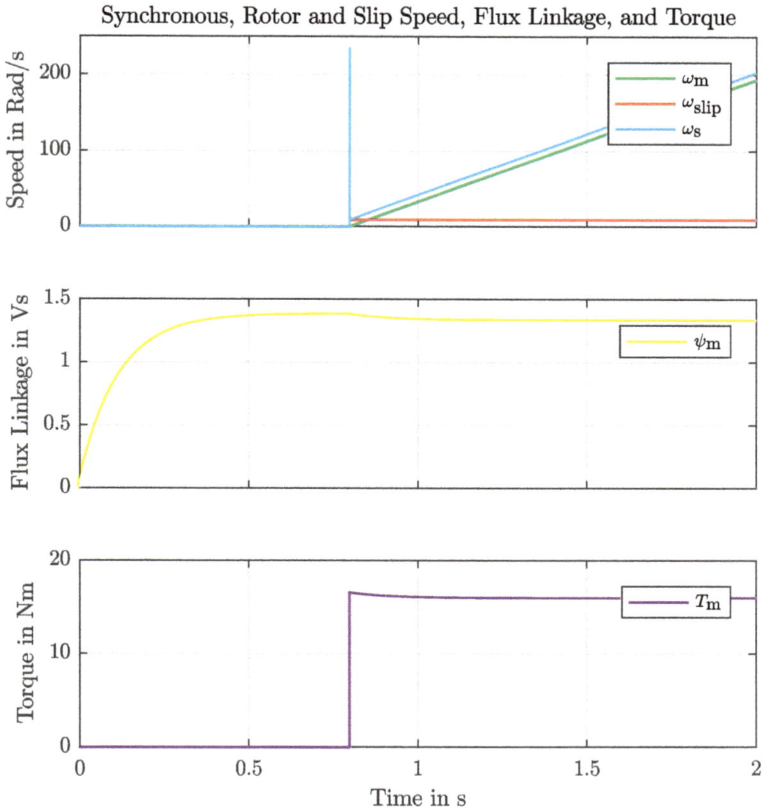

Fig. 8.29 Transient response of stator flux oriented UFO model

8.3.5 Synchronous Frame Oriented Heyland Diagram

As mentioned earlier, most drives assume some type of direct/quadrature axis current control, where the flux linkage ψ_{M} is maintained at a constant value. It is therefore of interest to consider the quasi-stationary stator current $\vec{i}_{\mathrm{s}}^{\,\mathrm{dq}}$ locus under varying slip conditions and constant flux linkage ψ_{M}. This analysis may be undertaken by considering the *steady-state* form of Eqs. (8.23c) and (8.24c) which upon elimination of $\omega_{\mathrm{sl}} = (\omega_{\mathrm{s}} - \omega_{\mathrm{m}})$ and after some mathematical manipulation leads to

$$\left(i_{\mathrm{sd}} - \frac{\psi_{\mathrm{M}}}{2}\left(\frac{1}{L_{\sigma\mathrm{R}}} + \frac{2}{L_{\mathrm{M}}}\right)\right)^2 + i_{\mathrm{sq}}^2 = \frac{\psi_{\mathrm{M}}^2}{4L_{\sigma\mathrm{R}}^2}. \tag{8.27}$$

This expression represents a circle in the complex dq plane with its center at coordinates $(\psi_{\mathrm{M}}/2(1/L_{\sigma\mathrm{R}} + 2/L_{\mathrm{M}}), 0)$ and radius $\psi_{\mathrm{M}}/2L_{\sigma\mathrm{R}}$ as indicated in Fig. 8.30a.

Fig. 8.30 Current locus $\vec{i}_s^{dq}$ for UFO model and two transformation values. (**a**) $L_m/L_r < a \leq L_s/L_m$. (**b**) $a = L_m/L_r$

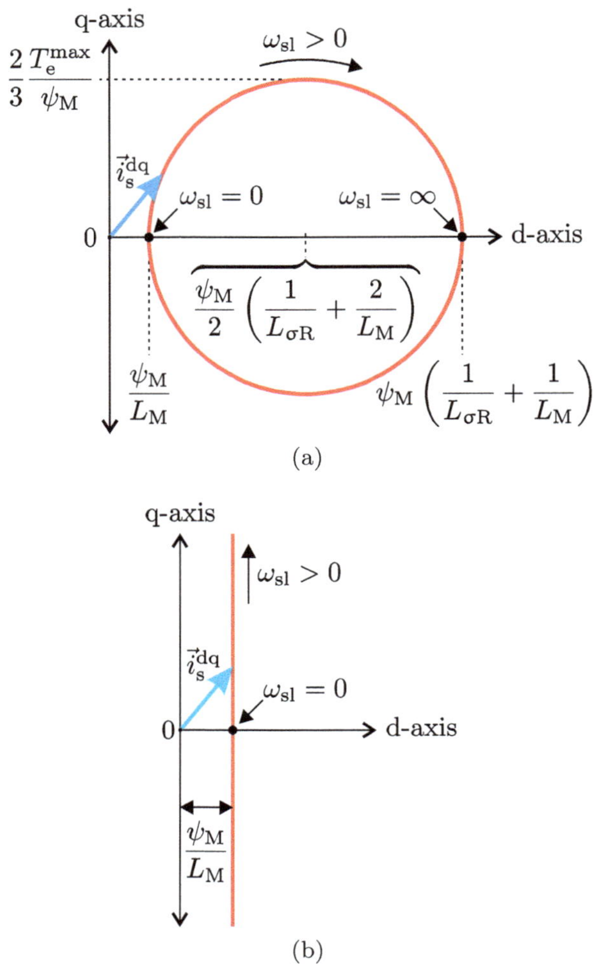

This type of current locus diagram, known as a *Heyland diagram* [6, 7, 9], shows the interaction between the stator current vector $\vec{i}_s$ (in this case represented in synchronous coordinates), the torque T_e, and the slip frequency ω_{sl}. The circle shows that a given ψ_M value corresponds to a maximum quadrature current and maximum torque T_e^{max} value. In practice, this maximum quadrature current value is normally outside the rated value, given that $L_{\sigma R} \ll L_M$. The influence of the transformation variable is clearly apparent in Fig. 8.30b given the dependency of the latter on the parameters $L_{\sigma R}$ and L_M (see Eq. (8.12)). When this transformation ratio is changed from $L_s/L_m \rightarrow L_m/L_r$ the radius increases to infinity, given the variation of the leakage inductance from $L_{\sigma R} \rightarrow 0$. This implies that the current locus at $a = L_m/L_r$ is reduced to a straight line as shown in Fig. 8.30b.

8.3.6 Steady-State Analysis of Voltage Source Connected Induction Machines

The steady-state characteristics of the induction machine are studied with the aid of Fig. 8.14, where it is assumed that the stator is connected to a three-phase sinusoidal supply which is represented by the space vector $\vec{u}_{\rm s} = \hat{u}_{\rm s}\,{\rm e}^{{\rm j}\omega_{\rm s}t}$. The basic characteristics of the machine are deemed to be the Heyland diagram of the p pole pair machine and torque/speed curve. For the development of this type of model, it is helpful to redefine the relevant space vector equations in terms of phasors. The general steady-state relationship for three-phase balanced systems between a space vector $\vec{x}$ and phasor $\underline{x}$ is of the form $\vec{x} = \underline{x}{\rm e}^{{\rm j}\omega_{\rm s}t}$ [14]. Given this relationship, the supply voltage phasor may be expressed as $\underline{u}_{\rm s} = \hat{u}_{\rm s}$. Furthermore, the phasor representation of the rotor flux linkage on the stator and rotor side of the IRTF may be written as

$$\vec{\psi}_{\rm R} = \underline{\psi}_{\rm R}\,{\rm e}^{{\rm j}\omega_{\rm s}t} \tag{8.28a}$$

$$\vec{\psi}_{\rm R}^{\rm xy} = \underline{\psi}_{\rm R}\,{\rm e}^{{\rm j}(\omega_{\rm s}t-\theta)}, \tag{8.28b}$$

where θ is equal to $\omega_{\rm m}t$ (constant speed operation). Use of Eqs. (8.19) and (8.28) leads to the following phasor based equation set for the machine

$$\underline{u}_{\rm s} - \underline{e}_{\rm R} = R_{\rm s}\,\underline{i}_{\rm s} + {\rm j}\omega_{\rm s}\,L_{\sigma{\rm S}}\,\underline{i}_{\rm s} \tag{8.29a}$$

$$\underline{e}_{\rm R} = {\rm j}\omega_{\rm s}\,\underline{\psi}_{\rm R} \tag{8.29b}$$

$$\underline{e}_{\rm R}^{\rm xy} = {\rm j}\,(\omega_{\rm s} - \omega_{\rm m})\,\underline{\psi}_{\rm R} \tag{8.29c}$$

$$\underline{i}_{\rm R}^{\rm xy} = \frac{\underline{e}_{\rm R}^{\rm xy}}{R_{\rm R}} \tag{8.29d}$$

$$\underline{\psi}_{\rm R}^{\rm xy} = L_{\rm M}\left(\underline{i}_{\rm s}^{\rm xy} - \underline{i}_{\rm R}^{\rm xy}\right). \tag{8.29e}$$

Elimination of the flux linkage phasor from Eqs. (8.29b) and (8.29c) leads to an expression for the air-gap EMF on the stator and rotor side of the IRTF, namely

$$\underline{e}_{\rm R}^{\rm xy} = \underline{e}_{\rm R}\,s, \tag{8.30}$$

where s is known as the slip of the machine and is given by

$$s = 1 - \frac{\omega_{\rm m}}{\omega_{\rm s}}. \tag{8.31}$$

The slip according to Eq. (8.31) is simply the ratio between the rotor rotational frequency $\omega_{\rm sl} = \omega_{\rm s} - \omega_{\rm m}$ (as apparent on the rotor side of the IRTF) and the stator rotational frequency $\omega_{\rm s}$. The Heyland diagram for this model is found by making

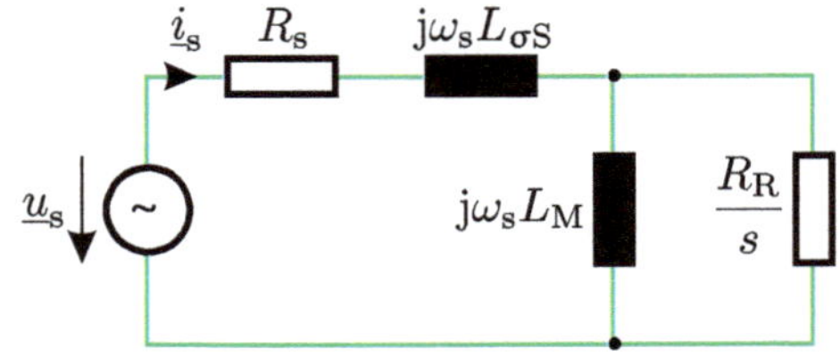

Fig. 8.31 Equivalent circuit of an induction machine with $L_{\sigma S}$, R_s and L_M, and voltage source (steady-state version of dynamic model in Fig. 8.14)

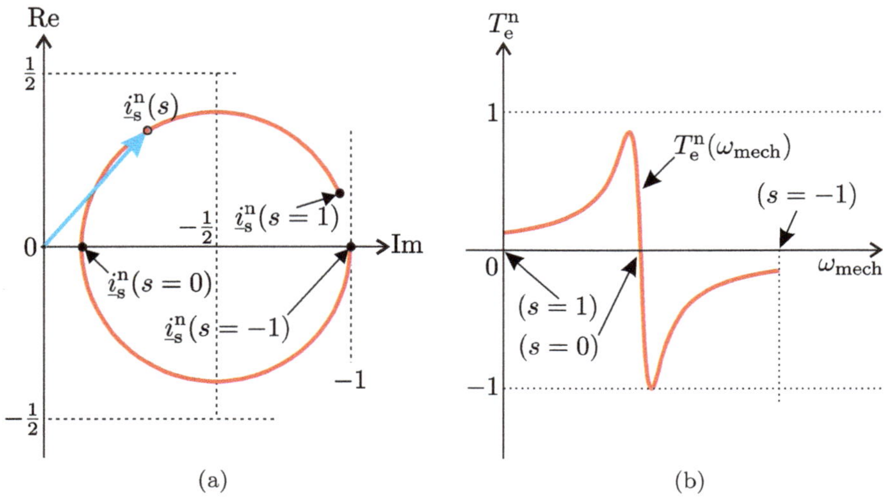

Fig. 8.32 Steady-state characteristics of voltage source connected induction machine, complete model according to Fig. 8.31. (**a**) Normalized stator current phasor. (**b**) Torque versus speed

use of Eq. (8.29) which leads to the following expression for the phasor based stator current:

$$\underline{i}_s = \frac{\hat{u}_s \left(\frac{R_R}{s} + j\omega_s L_M\right)}{j\omega_s L_M \frac{R_R}{s} + (R_s + j\omega_s L_{\sigma S}) \left(\frac{R_R}{s} + j\omega_s L_M\right)}. \tag{8.32}$$

The equivalent circuit which corresponds with Eq. (8.32) is shown in Fig. 8.31.

A normalization of expression (8.32) is convenient, which is of the form $\underline{i}_s^n = \underline{i}_s/(\hat{u}_s/\omega_s L_{\sigma S})$. An example of an Heyland diagram as calculated using Eq. (8.32) (in normalized form) is given in Fig. 8.32a, for a slip range of $-1 \leq s \leq 1$ using parameters of the 22 kW machine discussed in the tutorial linked to this part of the book (see Sect. 8.6.8).

The corresponding torque T_e of this machine may be calculated as

$$T_e = \frac{3}{2}\frac{1}{\omega_s} \underbrace{\left(\Re\left\{\underline{u}_s \underline{i}_s^*\right\} - R_s \, \underline{i}_s \, \underline{i}_s^*\right)}_{p_{\text{air-gap}}}. \tag{8.33}$$

Equation (8.33) underlines the fact that the torque can be calculated on the basis of the *air-gap power* $p_{\text{air-gap}}$, i.e., the power which *crosses* the air-gap to the rotor (see Eq. (8.36)). A normalization of Eq. (8.33) (as introduced in [14]) of the form $T_e^n = T_e/\hat{T}_e$, with $\hat{T}_e = \hat{u}_s^2/2\omega_s^2 L_{\sigma S}$, leads, with the aid of equation (8.32), to the torque speed curve given in Fig. 8.32b.

8.4 Parameter Identification and Estimates

The direct and quadrature stator flux model as given by Fig. 8.27 is also useful in terms of obtaining estimates for the parameters L_S, $L_{\sigma R}$, and R_R on the basis of the measured stator resistance R_S, measured no-load stator current, and given nameplate data. Furthermore, the approach given here provides a guideline for the flux linkage magnitudes to be used for vector control. The first part of this section outlines the calculation steps which leads to these parameters. Under no-load steady-state conditions, the stator flux oriented model is greatly simplified given that $i_{sq} = 0$ and $\omega_m = \omega_s$. This means that the d-axis and q-axis voltages are given by $u_{sd} = R_s i_{sd}$ and $u_{sq} = \omega_s \vec{\psi}_s$, respectively. The no-load stator current (vector amplitude) $i_s^{\text{noload}} = i_{sd}$. Consequently, the stator flux linkage can be calculated according to

$$\psi_s = \frac{1}{\omega_s}\sqrt{u_s^2 - \left(R_s i_s^{\text{noload}}\right)^2}, \tag{8.34}$$

where u_s represents the applied rated stator voltage (vector amplitude, amplitude invariant). The self inductance $L_s = L_M$ is, according to Fig. 8.27, of the form

$$L_s = \frac{\psi_s}{i_s^{\text{noload}}}, \tag{8.35}$$

where ψ_s is calculated using Eq. (8.34). An estimate for the rotor resistance R_R is found by considering Fig. 8.27 under steady-state rated torque conditions. Hence, we assume the machine is operating at its nameplate rated speed value ω_m^{nom} (electrical rated shaft speed in rad/s). The rated electrical torque T_e^n is then estimated by calculating the so-called *air-gap power* $T_e\omega_s$, which is equal to the terminal input power minus the stator copper losses which gives

$$T_e^{\text{nom}} = \frac{3}{2}\frac{1}{\omega_s}\left(\underbrace{u_s\, i_s^{\text{nom}}\cos(\rho_s)}_{p_{\text{in}}} - R_s\left(i_s^{\text{nom}}\right)^2\right), \tag{8.36}$$

where ω_s represents the electrical stator frequency $2\pi f$, with frequency f in Hz. The variable i_s^{nom} represents the rated stator current vector amplitude, while ρ_s represents the angle between the stator voltage vector and stator current vector (this parameter comes directly from the nameplate data, i.e., power factor). On the basis of the rated torque we can calculate the rated quadrature current $i_{\mathrm{sq}}^{\mathrm{nom}}$ using Eqs. (8.34) and (8.36) as

$$i_{\mathrm{sq}}^{\mathrm{nom}} = \frac{2}{3}\frac{T_{\mathrm{e}}^{\mathrm{n}}}{\psi_{\mathrm{s}}}. \tag{8.37}$$

From the quadrature model, shown in Fig. 8.27, we can derive, under steady-state rated conditions, the voltage equation $e_{\mathrm{sq}} = (\omega_{\mathrm{s}} - \omega_{\mathrm{m}})L_{\sigma\mathrm{R}}\, i_{\mathrm{sd}} + R_{\mathrm{R}}\, i_{\mathrm{sq}} + \omega_{\mathrm{m}}\, \psi_{\mathrm{s}}$, which under the assumption of $L_{\sigma\mathrm{R}}\, i_{\mathrm{sd}} \ll \psi_{\mathrm{s}}$ reduces to

$$R_{\mathrm{R}} \simeq \frac{\left(\omega_{\mathrm{s}} - \omega_{\mathrm{m}}^{\mathrm{nom}}\right)\psi_{\mathrm{s}}}{i_{\mathrm{sq}}^{\mathrm{nom}}}. \tag{8.38}$$

The remaining parameter, being the leakage inductance $L_{\sigma\mathrm{R}}$, comes from Fig. 8.30a (with $a = L_{\mathrm{s}}/L_{\mathrm{m}}$). This figure shows that the direct axis current value i_{sd} increases as the torque increases. The Heyland diagram circle is a direct function of the leakage. It may be shown that we can approximate the direct current i_{sd} versus i_{sq} function as

$$i_{\mathrm{sd}} \simeq \underbrace{\frac{\psi_{\mathrm{s}}}{L_{\mathrm{s}}}}_{i_{\mathrm{s}}^{\mathrm{noload}}} + L_{\sigma\mathrm{R}}\frac{i_{\mathrm{sq}}^2}{\psi_{\mathrm{s}}}. \tag{8.39}$$

Equation (8.39), when used under rated conditions, i.e., $i_{\mathrm{sd}} = i_{\mathrm{sd}}^{\mathrm{nom}}$ and $i_{\mathrm{sq}} = i_{\mathrm{sq}}^{\mathrm{nom}}$, gives us a direct estimate for the leakage inductance, namely

$$L_{\sigma\mathrm{R}} \simeq \left(i_{\mathrm{sd}}^{\mathrm{nom}} - i_{\mathrm{s}}^{\mathrm{noload}}\right)\frac{\psi_{\mathrm{s}}}{\left(i_{\mathrm{sq}}^{\mathrm{nom}}\right)^2}, \tag{8.40}$$

where $i_{\mathrm{sd}}^{\mathrm{nom}}$ is calculated using the stator current $i_{\mathrm{s}}^{\mathrm{nom}}$ and Eq. (8.37) as

$$i_{\mathrm{sd}}^{\mathrm{nom}} = \sqrt{\left(i_{\mathrm{s}}^{\mathrm{nom}}\right)^2 - \left(i_{\mathrm{sq}}^{\mathrm{nom}}\right)^2}. \tag{8.41}$$

The conversion to a model with $a = 1$ is based on the assumption of equal rotor and stator leakage inductances, i.e., $L_{\sigma\mathrm{s}} = L_{\sigma\mathrm{r}}$ and $L_{\mathrm{s}} = L_{\mathrm{r}}$. Use of Eq. (8.12b), with $a = L_{\mathrm{s}}/L_{\mathrm{m}}$, gives

$$L_{\mathrm{m}} = \sqrt{\frac{L_{\mathrm{s}}^3}{L_{\sigma\mathrm{R}} + L_{\mathrm{s}}}} \tag{8.42a}$$

$$L_{\sigma\mathrm{s}} = L_{\mathrm{s}} - L_{\mathrm{m}} \tag{8.42b}$$

$$R_{\mathrm{r}} = \left(\frac{L_{\mathrm{m}}}{L_{\mathrm{s}}}\right)^2 R_{\mathrm{R}}. \tag{8.42c}$$

The stator flux linkage magnitude ψ_{s} as given by Eq. (8.34) and rated currents $i_{\mathrm{sd}}^{\mathrm{nom}}$ and $i_{\mathrm{sq}}^{\mathrm{nom}}$ provide the basis for calculating a value for the flux linkage ψ_{M} and use of Eq. (8.16b), namely

$$\psi_{Md} = \psi_{\mathrm{s}} - L_{\sigma\mathrm{S}}\, i_{\mathrm{sd}}^{\mathrm{nom}} \tag{8.43a}$$

$$\psi_{Mq} = -L_{\sigma\mathrm{S}}\, i_{\mathrm{sq}}^{\mathrm{nom}} \tag{8.43b}$$

which leads to the required flux linkage magnitude ψ_{M} for any given transformation value a.

$$\psi_{\mathrm{M}} = \sqrt{\psi_{\mathrm{Md}}^2 + \psi_{\mathrm{Md}}^2} \tag{8.44}$$

The tutorial at the end of this chapter gives an example of using this approach for calculating the rated flux linkage values and motor parameters for a given machine.

8.5 Single-Phase Induction Machines

In this section, we will consider the use of the IRTF based concept for representing single-phase, squirrel-cage based induction machines. The machine in question is provided with two orthogonal-oriented stator windings referred to as the *run winding* (main winding) and *auxiliary winding*, respectively, as indicated in Fig. 8.33. The number of winding turns per phase may in this case differ, hence the phase resistance of the run winding and auxiliary winding are defined as R_{aux} and R_{run}, respectively. Furthermore, a factor k_{aux} is introduced which represents the auxiliary-to-run winding turns ratio. A single-phase sinusoidal supply is assumed which is directly connected to the run winding.

Types of Single-Phase Induction Machines

There are various types of single-phase induction machines. For the so-called *split-coil* machines, the auxiliary winding is also connected directly to the supply source. For *capacitor start* type machines, a capacitor C is placed in series with the auxiliary winding and the supply (see Fig. 8.33) and a shaft speed operated switch (also shown in Fig. 8.33) disconnects the auxiliary winding when the speed reaches a predetermined operating speed. The so-called *capacitor-run* machines, first proposed by Steinmetz, are provided with a fixed capacitor between supply and winding which removes the need for a switch. A combination of phase winding topologies described above is found in the industry, hence it is of interest to present a generalized modeling approach in this section which will allow the user to examine the dynamic and steady-state behavior of such machines.

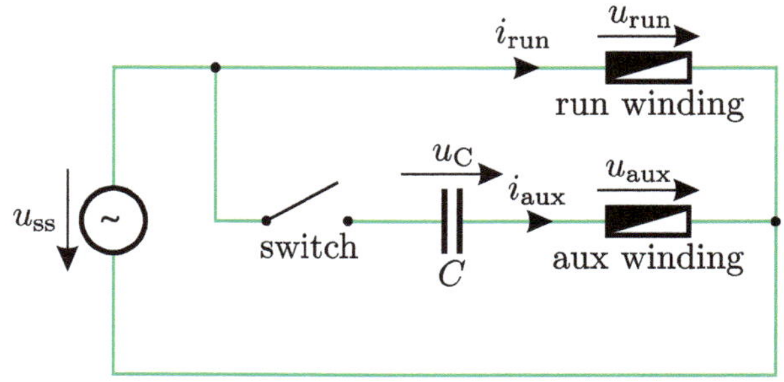

Fig. 8.33 Generalized wiring diagram of single-phase motor

Dynamic Model of Single-Phase Induction Machine

A convenient starting point for this modeling process is the rotor flux based machine model discussed in Sect. 8.3.2.1. The use of a space vector model is particularly helpful given that this is in fact a two-phase representation of the induction machine. On the stator side of the IRTF components and variables linked to the α-axis may, for example, be assigned to the auxiliary winding as is the case in this section. With this choice of axis assignment, the run winding parameters and variables are tied to the β-space vector axis hence the supply voltage and current space vectors $\vec{u}_{ss}$ and $\vec{i}_{ss}$ may be written as

$$\vec{u}_{ss} = u_{aux} + j\, u_{run} \tag{8.45a}$$

$$\vec{i}_{ss} = i_{aux} + j\, i_{run}, \tag{8.45b}$$

where u_{aux}, u_{run}, i_{aux}, and i_{run} are given in Fig. 8.33. For the generalized model assumed here, a capacitor C is connected in series with the auxiliary winding in which case the winding voltage u_{aux} may be found using

$$u_{aux} = u_{ss} - \frac{1}{C}\int i_{aux}\, dt, \tag{8.46}$$

where u_{ss} represents the supply voltage to the machine which is arbitrarily defined as

$$u_{ss} = \hat{u}_{ss} \cos \omega_s t. \tag{8.47}$$

The development of a symbolic model for single-phase induction machines is readily initiated by considering the rotor flux based IRTF model according to Fig. 8.14. The parameters $L_{\sigma S}$, L_M, and R_R used to model standard three-phase machines as discussed in Sect. 8.3.2.1 are based on the stator referred parameters $L_{\sigma s}$, $L_{\sigma s}$, L_m, and R_r. Here, the number of turns on the run and auxiliary winding may differ and a choice needs to be made in terms of referring said parameters to a specific winding. The run winding is assigned as the reference winding and the machine parameters are referred to this winding. The resulting symbolic model for

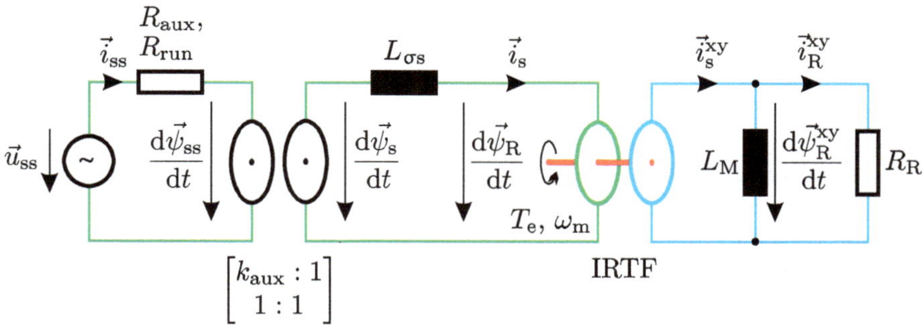

Fig. 8.34 IRTF/ITF based symbolic model for single-phase induction machine

the single-phase machine as given in Fig. 8.34 has resemblance with the rotor flux based machine model for the three-phase machine from Fig. 8.14 which was the starting point of the modeling process.

However, in this case an *asymmetric* ITF module is introduced in order to accommodate the difference in run/auxiliary winding turns. The latter is achieved by introducing an ITF module where the α-axis (assigned to the auxiliary winding) winding ratio is set to $k_{aux} : 1$, while the β-axis winding ratio is set to unity. The ITF equation set is therefore of the form

$$i_{s\alpha} = k_{aux}\, i_{aux} \tag{8.48a}$$

$$i_{s\beta} = i_{run} \tag{8.48b}$$

while the corresponding ITF flux linkages may be defined as

$$\psi_{aux} = k_{aux}\, \psi_{s\alpha} \tag{8.49a}$$

$$\psi_{run} = \psi_{s\beta}, \tag{8.49b}$$

where the variables ψ_{aux} and ψ_{run} are introduced which are linked with the space vector $\vec{\psi}_{ss} = \psi_{aux} + j\psi_{run}$. The equation set which corresponds to Fig. 8.34 is of the form

$$u_{aux} = R_{aux}\, i_{aux} + \frac{d\psi_{aux}}{dt} \tag{8.50a}$$

$$u_{run} = R_{run}\, i_{run} + \frac{d\psi_{run}}{dt} \tag{8.50b}$$

$$\vec{\psi}_s = \vec{\psi}_R + L_{\sigma s}\, \vec{i}_s \tag{8.50c}$$

$$\frac{\vec{\psi}_R^{xy}}{L_M} = \vec{i}_s^{xy} - \vec{i}_R^{xy} \tag{8.50d}$$

$$\vec{e}_{\mathrm{R}}^{\mathrm{xy}} = \frac{\mathrm{d}\vec{\psi}_{\mathrm{R}}^{\mathrm{xy}}}{\mathrm{d}t} \tag{8.50e}$$

$$\vec{e}_{\mathrm{R}}^{\mathrm{xy}} = R_{\mathrm{R}}\,\vec{i}_{\mathrm{R}}^{\mathrm{xy}}, \tag{8.50f}$$

where the parameters R_{aux} and R_{run} represent the resistance of the auxiliary and run winding, respectively. Note that the single-phase symbolic model and corresponding equation set reduce to the rotor flux oriented model from Sect. 8.3.2.1 in case the auxiliary and run windings are identical in terms of the winding configuration. Under these circumstances, the phase resistances will be equal and the winding ratio k_{aux} will be equal to 1, in which case the ITF module can be omitted from the symbolic diagram.

The generic model as given by Fig. 8.35 which corresponds to the symbolic model shown in Fig. 8.34 and equation set (8.50) shows the introduction of scalar variables on the primary side of the (asymmetric) ITF module.

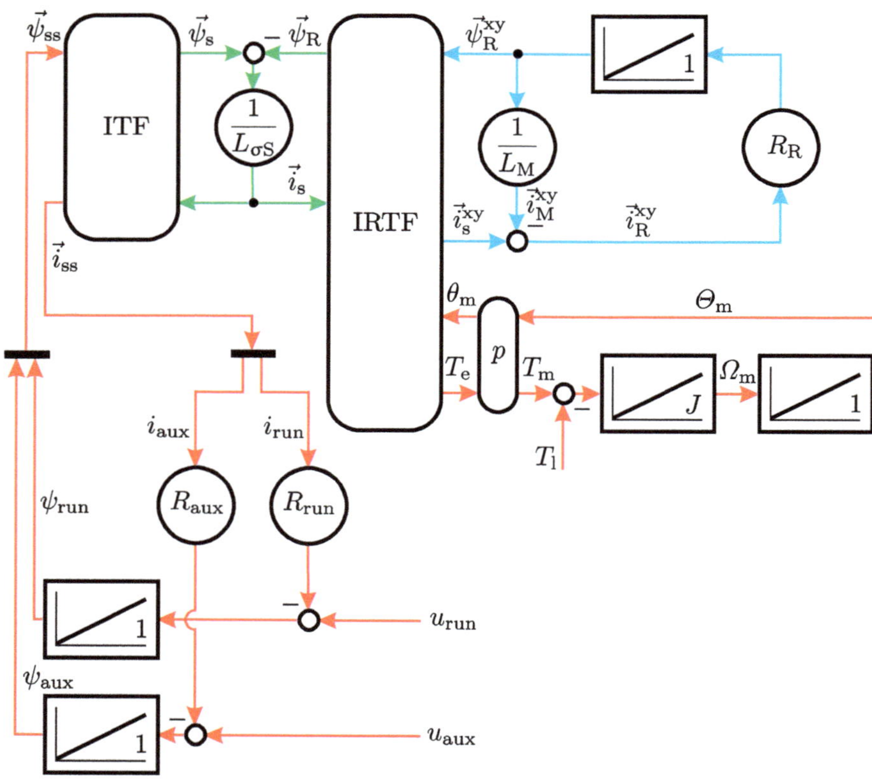

Fig. 8.35 Generic model for single-phase induction machines

Tutorial Results
In Sect. 8.6.9, a simulation example is given which is based on the model described above. In this tutorial, the model is used to examine the line start of a 150 W four-pole, capacitor-run type single-phase induction machine. For this purpose, the model voltage input u_{run} is connected to a 50 Hz sinusoidal voltage source as defined by Eq. (8.47). Furthermore, the model voltage u_{aux} is derived with the aid of equation (8.46) which may be implemented with the aid of an additional integrator module and gain module with gain $1/C$. An example of the results obtained with this simulation model, as indicated in Fig. 8.36, shows the machine shaft torque and shaft speed during a 2 s start-up sequence where a load torque step is applied at $t = 1.5$ s. The reader is referred to the tutorial given in Sect. 8.6.9 for further details and the opportunity to interactively examine this model concept.

It is emphasized that the capacitor-run machine model can be readily adapted to different single-phase machine concepts. For the purpose of modeling a capacitor start machine, a speed dependent switch may be introduced which purposely sets the auxiliary winding resistance R_{aux} to a substantially larger (in comparison with the auxiliary winding resistance) value. For the purpose of modeling a split-phase induction machine, the capacitor gain module $1/C$ may be simply set to zero in which case both voltage inputs of the model are connected to the same sinusoidal voltage source. Modeling of, for example, a TRIAC controlled single-phase machine can be undertaken by the introduction of circuit based CASPOC elements such as a TRIAC and a circuit/block conversion module (see Fig. 3.28).

8.5.1 *Steady-State Analysis of Capacitor-Run Single-Phase Induction Machines*

A phasor based analysis of the generalized single-phase model is considered here to determine the steady-state run and auxiliary peak (or RMS) currents, average torque, and torque ripple as function of the shaft speed for capacitor-run single-phase induction machines.

Capacitor as Integral Part of the Machine
For the purpose of this analysis, it is helpful to consider the capacitor in series with the auxiliary winding as an integral part of the machine, i.e., instead of defining the stator voltage space vector $\vec{u}_{\mathrm{ss}}$ as

$$\vec{u}_{\mathrm{ss}} = u_{\mathrm{aux}} + \mathrm{j}\, u_{\mathrm{run}}$$

as in Eq. (8.45) we redefine it as

$$\vec{u}_{\mathrm{ss}} = u_{\mathrm{ss}\alpha} + \mathrm{j}\, u_{\mathrm{ss}\beta}, \tag{8.51}$$

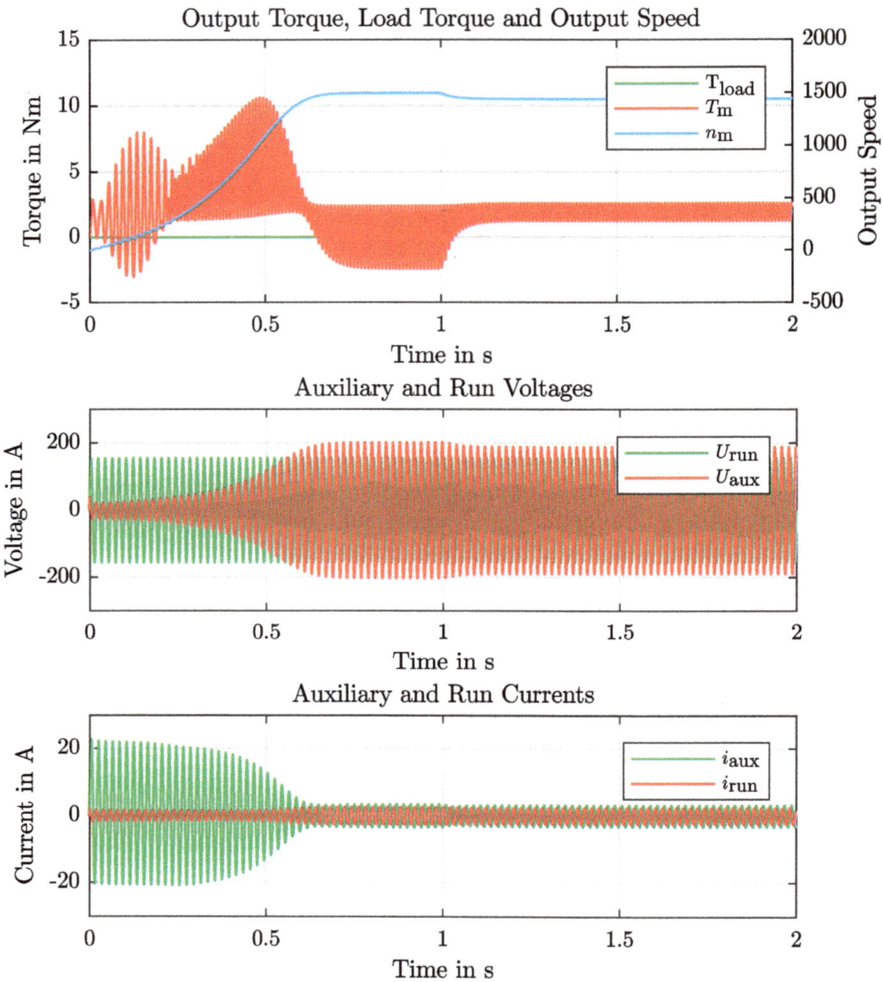

Fig. 8.36 Line start example of a capacitor-run single-phase induction machine

where $u_{ss\alpha}$ is now the voltage drop over the capacitor in series with the auxiliary winding while $u_{ss\beta}$ remains the resulting voltage drop over the run winding, compare Fig. 8.33. Due to the parallel connection to the single-phase supply voltage, both $u_{ss\alpha}$ and $u_{ss\beta}$ equal u_{ss} (with u_{ss} defined in Eq. (8.47)). The orientation of the space vector components with the auxiliary winding in the α-direction and the run winding in the β-direction remains unchanged.

For $u_{ss\alpha}$, Eq. (8.50a) is substituted into Eq. (8.46) which leads to

$$u_{ss\alpha} = u_{aux} + \frac{1}{C} \int i_{aux} \, dt$$

$$\Leftrightarrow u_{ss\alpha} = R_{aux}\, i_{aux} + \frac{1}{C}\int i_{aux}\,dt + \frac{d\psi_{aux}}{dt}, \tag{8.52}$$

where u_{ss} is defined by Eq. (8.47). The run winding is directly connected to the supply source and hence $u_{ss\beta}$ is formed by

$$u_{ss\beta} = u_{run}$$
$$\Leftrightarrow u_{ss\beta} = R_{run}\, i_{run} + \frac{d\psi_{run}}{dt}. \tag{8.53}$$

Phasor Notation of Voltage Equations

In steady-state, the auxiliary and run voltages $u_{ss\alpha}$ and $u_{ss\beta}$ are sinusoidal functions of time. A sinusoidal quantity is represented by the expression $u = \hat{u}\cos(\omega_s t + \rho)$ and can be expressed as $u = \Re\{\underline{u}\, e^{j\omega_s t}\}$ where the phasor $\underline{u} = \hat{u}\, e^{j\rho}$ is introduced. In phasor notation and matrix format, Eqs. (8.52) and (8.53) can therefore be rewritten as

$$\begin{bmatrix} \underline{u}_{ss\alpha} \\ \underline{u}_{ss\beta} \end{bmatrix} = \begin{bmatrix} \underline{Z}_{aux} & 0 \\ 0 & R_{run} \end{bmatrix} \begin{bmatrix} \underline{i}_{run} \\ \underline{i}_{aux} \end{bmatrix} + j\omega_s \begin{bmatrix} \underline{\psi}_{aux} \\ \underline{\psi}_{run} \end{bmatrix}, \tag{8.54}$$

where the impedance $\underline{Z}_{aux} = R_{aux} + {}^{1}/_{j\omega_{s3}}$.

Equation set (8.54) is now referred to the secondary side of the ITF module shown in Fig. 8.34. This is achieved by multiplying Eq. (8.54) by a factor ${}^{1}/_{k_{aux}}$ which gives

$$\begin{bmatrix} \underline{u}_{s\alpha} \\ \underline{u}_{s\beta} \end{bmatrix} = \underbrace{\begin{bmatrix} \underline{Z}_{s\alpha} & 0 \\ 0 & R_{s\beta} \end{bmatrix}}_{\underline{Z}_{sp}} \begin{bmatrix} \underline{i}_{s\alpha} \\ \underline{i}_{s\beta} \end{bmatrix} + j\omega_s \begin{bmatrix} \underline{\psi}_{s\alpha} \\ \underline{\psi}_{s\beta} \end{bmatrix} \tag{8.55}$$

with

$$\underline{u}_{s\alpha} = \frac{\underline{u}_s}{k_{aux}}$$
$$\underline{u}_{s\beta} = \underline{u}_s$$
$$\underline{i}_{s\alpha} = k_{aux}\, \underline{i}_{aux}$$
$$\underline{i}_{s\beta} = \underline{i}_{run}$$
$$\underline{\psi}_{s\alpha} = \frac{\underline{\psi}_{aux}}{k_{aux}}$$
$$\underline{\psi}_{s\beta} = \underline{\psi}_{run}$$

and

$$\underline{Z}_{s\alpha} = \frac{\underline{Z}_{aux}}{k_{aux}^2}$$

$$R_{s\beta} = R_{run}.$$

Positive and Negative Sequence Phasors

With the aid of Euler's formula, the time domain quantities $u_{s\alpha}$ and $u_{s\beta}$ can be expressed by their phasor from the above Eq. (8.55) as

$$u_{s\alpha} = \frac{1}{2}\left(\underline{u}_{s\alpha}\, e^{j\omega_s t} + \underline{u}_{s\alpha}{}^{*} e^{-j\omega_s t}\right) \tag{8.56a}$$

$$u_{s\beta} = \frac{1}{2}\left(\underline{u}_{s\beta}\, e^{j\omega_s t} + \underline{u}_{s\beta}{}^{*} e^{-j\omega_s t}\right). \tag{8.56b}$$

An observation of Eq. (8.56) shows that the two sinusoidal scalar quantities $u_{s\alpha}$ and $u_{s\beta}$ may be represented by two counter-rotating space vectors and two phasors $\underline{u}_{s\alpha}$ and $\underline{u}_{s\beta}$.

The two time domain quantities $u_{s\alpha}$ and $u_{s\beta}$, according to the symbolic and generic diagram of the machine, represent the space vector $\vec{u}_s = u_{s\alpha} + ju_{s\beta}$. Use of Eq. (8.56) with this expression and grouping the phasors linked to the terms $e^{j\omega_s t}$ and $e^{-j\omega_s t}$ gives

$$\vec{u}_s = \underbrace{\left(\frac{1}{2}\underline{u}_{s\alpha} + j\frac{1}{2}\underline{u}_{s\beta}\right)}_{\underline{u}_{s+}} e^{j\omega_s t} + \underbrace{\left(\frac{1}{2}\underline{u}_{s\alpha}{}^{*} + j\frac{1}{2}\underline{u}_{s\beta}{}^{*}\right)}_{\underline{u}_{s-}{}^{*}} e^{-j\omega_s t}. \tag{8.57}$$

The resultant equation as represented by expression (8.58) shows that the space vector $\vec{u}_s$ may also be represented by two counter-rotating space vectors which in turn are linked to two phasors $\underline{u}_{s+}$ and $\underline{u}_{s-}$ that are tied to the rotation direction. The subscript notation +, − underlines the fact that these phasors are referred to as the so-called *positive* and *negative sequence phasors*,

$$\vec{u}_s = \underline{u}_{s+} e^{j\omega_s t} + \underline{u}_{s-}{}^{*} e^{-j\omega_s t}. \tag{8.58}$$

Observation of Eq. (8.57) shows that the relationship between the phasors $\underline{u}_{s\alpha}$ and $\underline{u}_{s\beta}$ and phasors $\underline{u}_{s+}$ and $\underline{u}_{s-}$ may be expressed in terms of a matrix based expression (8.59),

$$\begin{bmatrix} \underline{u}_{s+} \\ \underline{u}_{s-} \end{bmatrix} = \underbrace{\frac{1}{2}\begin{bmatrix} 1 & j \\ 1 & -j \end{bmatrix}}_{A} \begin{bmatrix} \underline{u}_{s\alpha} \\ \underline{u}_{s\beta} \end{bmatrix}, \tag{8.59}$$

where A is referred to as the transformation matrix. For three-phase machines operating under symmetrical conditions, the two phasors $\underline{u}_{s\alpha}$ and $\underline{u}_{s\beta}$ will be orthogonal and symmetric. If, for example, the two phasors are set to $\underline{u}_{s\alpha} = \hat{u}_s$ and $\underline{u}_{s\beta} = -j\hat{u}_s$, respectively, Eq. (8.58) is reduced to $\vec{u}_s = \hat{u}_s e^{j\omega_s t}$, i.e., the negative sequence phasor $\underline{u}_{s-}$ will then be zero.

The positive and negative phasor variables may be transformed to the α and β-phasor as

$$\begin{bmatrix} \underline{u}_{s\alpha} \\ \underline{u}_{s\beta} \end{bmatrix} = \underbrace{\begin{bmatrix} 1 & 1 \\ -j & j \end{bmatrix}}_{A^{-1}} \begin{bmatrix} \underline{u}_{s+} \\ \underline{u}_{s-} \end{bmatrix}. \tag{8.60}$$

This transform is the inverse of equation (8.59). The matrix A^{-1} is the inverse matrix of the matrix A.

Positive and Negative Sequence Model

The transformation process described above for the two scalar variables $u_{s\alpha}$ and $u_{s\beta}$ can be equally applied to other machine variables such as the flux linkage and current. Consequently, the conversion process in question may be used to develop a so-called positive and negative sequence phasor model of the machine, with inputs $\underline{u}_{s+}$ and $\underline{u}_{s-}$.

The transformation to positive and negative sequence based phasors is achieved by multiplying both sides of Eq. (8.55) with the transformation matrix A as given in Eq. (8.59). The resulting equation is

$$\begin{bmatrix} \underline{u}_{s+} \\ \underline{u}_{s-} \end{bmatrix} = \underbrace{\frac{1}{2}\begin{bmatrix} \underline{Z}_{s\alpha} + R_{s\beta} & \underline{Z}_{s\alpha} - R_{s\beta} \\ \underline{Z}_{s\alpha} - R_{s\beta} & \underline{Z}_{s\alpha} + R_{s\beta} \end{bmatrix}}_{\underline{Z}_{pn} = A\,\underline{Z}_{sp}\,A^{-1}} \begin{bmatrix} \underline{i}_{s+} \\ \underline{i}_{s-} \end{bmatrix} + j\omega_s \begin{bmatrix} \underline{\psi}_{s+} \\ \underline{\psi}_{s-} \end{bmatrix}. \tag{8.61}$$

In this expression, the terms $\underline{\psi}_{s+}$ and $\underline{\psi}_{s-}$ need to be considered in detail. This is achieved by rewriting Eq. (8.50c) in terms of positive and negative sequence phasors. The resultant equation set, which has been multiplied by factor $j\omega_s$ to simplify the ensuing analysis, becomes

$$j\omega_s \underline{\psi}_{s+} = j\omega_s L_\sigma \underline{i}_{s+} + j\omega_s \underline{\psi}_{R+} \tag{8.62a}$$

$$j\omega_s \underline{\psi}_{s-} = j\omega_s L_\sigma \underline{i}_{s-} + j\omega_s \underline{\psi}_{R-}, \tag{8.62b}$$

where the phasors $j\omega_s \underline{\psi}_{R+}$ and $j\omega_s \underline{\psi}_{R-}$ appear. These may be further developed by making use of expression (8.50e) in which the space vectors are shown in rotating coordinates. For the conversion to positive and negative sequence phasors using the approach outlined above, it is helpful to reconsider the relationship between stationary and rotating space vectors, namely $\vec{A} = \vec{A}^{xy} e^{j\theta}$ where θ represents the shaft angle which, at constant speed, may also be written as $\theta = \omega_m t$. The positive/negative phasors linked to expression (8.50e) may after some mathematical handling be written as

$$\underline{e}_{R+} = \left(1 - \frac{\omega_m}{\omega_s}\right) j\omega_s \underline{\psi}_{R+} \tag{8.63a}$$

$$\underline{e}_{R-} = \left(1 + \frac{\omega_m}{\omega_s}\right) j\omega_s \underline{\psi}_{R-}. \tag{8.63b}$$

Some simplification of equation set (8.63) may be realized by introducing the *slip* s as defined by expression (8.31). Note that the slip relationship was defined for a unidirectional set of space vectors. In the case here, two counter-rotating space vectors are used, hence we will arbitrarily assume that the slip is determined with respect to the positive rotating vectors. The left-hand side of equation set (8.63) may also be developed further with the aid of equations (8.50c), (8.50d), and (8.50f) and converting the latter to positive/negative phasors. Subsequent manipulation of Eq. (8.63) using the approach outlined and substitution of the phasor based equation into Eq. (8.62) gives

$$j\omega_s \underline{\psi}_{s+} = \underline{i}_{s+} \overbrace{\left(j\omega_s L_{\sigma S} + \frac{j\omega_s L_M \frac{R_R}{s}}{\frac{R_R}{s} + j\omega_s L_M} \right)}^{Z_{s+}} \tag{8.64a}$$

$$j\omega_s \underline{\psi}_{s-} = \underline{i}_{s-} \underbrace{\left(j\omega_s L_{\sigma S} + \frac{j\omega_s L_M \frac{R_R}{2-s}}{\frac{R_R}{2-s} + j\omega_s L_M} \right)}_{Z_{s-}}, \tag{8.64b}$$

where Z_{s+} and Z_{s-} represent a positive and negative sequence impedance network which consists of the leakage inductance in series with a parallel network formed by a slip-dependent rotor resistance and magnetizing inductance. Note that the positive sequence impedance Z_{s+} is in fact the network shown in Fig. 8.31 for the steady-state of the three-phase induction machine (without the resistance R_s), while the negative sequence is similar but utilizes a different slip-dependent rotor resistance.

The positive and negative sequence currents are obtained by substituting the flux linkage expression (8.64) into the voltage equation (8.61),

$$\begin{bmatrix} \underline{u}_{s+} \\ \underline{u}_{s-} \end{bmatrix} = \frac{1}{2} \begin{bmatrix} \underline{Z}_{s\alpha} + R_{s\beta} + 2Z_{s+} & \underline{Z}_{s\alpha} - R_{s\beta} \\ \underline{Z}_{s\alpha} - R_{s\beta} & \underline{Z}_{s\alpha} + R_{s\beta} + 2Z_{s-} \end{bmatrix} \begin{bmatrix} \underline{i}_{s+} \\ \underline{i}_{s-} \end{bmatrix}. \tag{8.65}$$

Hence, on the basis of a given steady-state sinusoidal voltage excitation u_s, the voltage phasors $\underline{u}_{s+}$ and $\underline{u}_{s-}$ may be found using Eq. (8.57). In turn, these phasors are used as an input for expression (8.65) from which the currents $\underline{i}_{s+}$ and $\underline{i}_{s-}$ for a given shaft speed may be derived. These current phasors can with the aid of equation (8.60) (for currents instead of voltages) be used to obtain the run and auxiliary currents (in phasor form). For the purpose of developing a steady-state equivalent model it is helpful to rewrite Eq. (8.65) as

$$\begin{aligned}\underline{u}_{s+} &= (R_{run} + Z_{s+})\,\underline{i}_{s+} \\ &\quad + \left\{\frac{1}{2}\left(\frac{R_{aux}}{k_{aux}^2} - R_{run}\right) + \frac{1}{j2\omega_s C k_{aux}^2}\right\}(\underline{i}_{s+} + \underline{i}_{s-})\end{aligned} \tag{8.66a}$$

$$\begin{aligned}\underline{u}_{s-} &= (R_{run} + Z_{s-})\,\underline{i}_{s-} \\ &\quad + \left\{\frac{1}{2}\left(\frac{R_{aux}}{k_{aux}^2} - R_{run}\right) + \frac{1}{j2\omega_s C k_{aux}^2}\right\}(\underline{i}_{s+} + \underline{i}_{s-}),\end{aligned} \tag{8.66b}$$

where the auxiliary and run winding variables are reintroduced as defined for the dynamic model in order to enhance the readability of the resultant expression. A model representation as introduced in Fig. 8.37 is consistent with mathematical expression (8.66). Readily apparent in this model are the two impedances Z_{s+} and Z_{s-} as defined in Eq. (8.64).

Torque Calculation in Steady-State

We will now consider the calculation of the average torque and peak amplitude of the pulsating torque under steady-state conditions using the model given in Fig. 8.37. A suitable starting point for this analysis is the general torque expression in phasor notation

$$T_e = \Im\left\{\underline{\psi}_s^{\,*}\,\underline{i}_s\right\}. \tag{8.67}$$

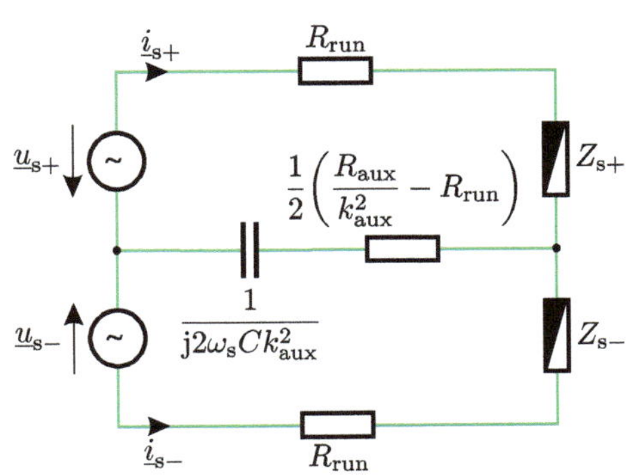

Fig. 8.37 Equivalent steady-state circuit of a single-phase induction machine with auxiliary capacitor

This expression may be rewritten in terms of the positive and negative sequence flux linkage and current phasors $\psi_{s\pm}$ and $\underline{i}_{s\pm}$ by making use of Eq. (8.58) in which the voltage variables must be replaced by the appropriate space vectors and phasors. Subsequent mathematical manipulation using this approach gives

$$\begin{aligned} T_e = & \underbrace{\Im\left\{\underline{\psi}_{s+}{}^{*}\,\underline{i}_{s+}\right\} + \Im\left\{\underline{\psi}_{s-}\,\underline{i}_{s-}{}^{*}\right\}}_{T_e^{av}} \\ & + \underbrace{\Im\left\{\underline{\psi}_{s-}\,\underline{i}_{s+}\,e^{j2\omega_s t}\right\} + \Im\left\{\underline{\psi}_{s+}{}^{*}\,\underline{i}_{s-}{}^{*}\,e^{-j2\omega_s t}\right\}}_{T_e^{ripple}}. \end{aligned} \tag{8.68}$$

Observation of Eq. (8.68) shows that the torque expression has an average (non-time-dependent) component T_e^{av} that may be found with the aid of the positive and negative current phasors. Equation (8.68) also shows a sinusoidal torque ripple component T_e^{ripple} which has a frequency that is double the voltage supply excitation frequency ω_s. The torque T_e can be rewritten as

$$T_e = T_e^{av} + T_e^{R} \sin(2\omega_s t) + T_e^{X} \cos(2\omega_s t) \tag{8.69}$$

with

$$\begin{aligned} T_e^{R} &= \Re\left\{\underline{\psi}_{s-}\,\underline{i}_{s+}\right\} - \Re\left\{\underline{\psi}_{s+}{}^{*}\,\underline{i}_{s-}{}^{*}\right\} \\ T_e^{X} &= \Im\left\{\underline{\psi}_{s-}\,\underline{i}_{s+}\right\} + \Im\left\{\underline{\psi}_{s+}{}^{*}\,\underline{i}_{s-}{}^{*}\right\}. \end{aligned}$$

The peak amplitude $\hat{T}_e^{ripple}$ of this torque ripple component is

$$\hat{T}_e^{ripple} = \sqrt{(T_e^{R})^2 + (T_e^{X})^2}. \tag{8.70}$$

Tutorial Results

A numerical example of a steady-state analysis for the single-phase capacitor-run induction machine used for the dynamic model example is given in Fig. 8.38. The results represent the average torque, peak torque ripple, peak auxiliary, and peak run currents as function of the ratio between shaft speed and synchronous speed $p\,\omega_s$ over a slip range from $s = 0$ to $s = 2$.

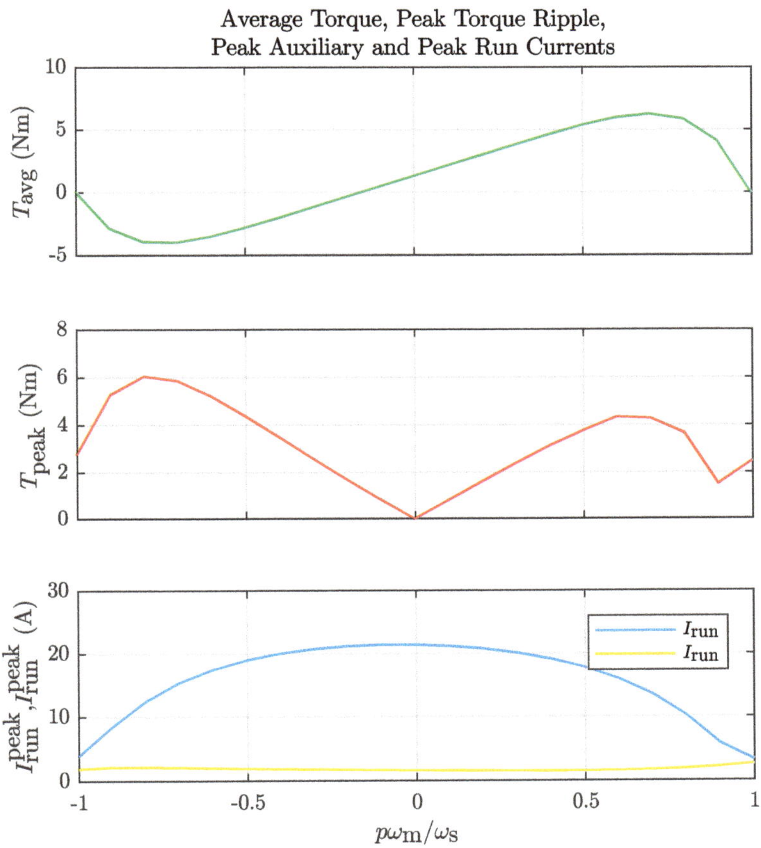

Fig. 8.38 Steady-state analysis example of a capacitor-run single-phase induction machine

8.6 Tutorials

8.6.1 Tutorial 1: Simplified Induction Machine Model

In this tutorial, a simulation based on the simplified generic model as shown in Fig. 8.3 is considered. A rotating flux linkage vector $\vec{\psi}_m$ with angular frequency $\omega_s = 100\pi$ rad/s and amplitude $\psi_m = 1.25$ Vs is generated as an input to the model. The flux linkage vector is used instead of a voltage vector to emphasize the fact that it is the flux linkage which plays a key role with respect to the operation of the machine. The machine is connected to a mechanical load with a quadratic torque speed curve which can be manipulated inside the *Mechanical Model* subsystem. The machine has an inertia of $J = 0.001\,\text{kg m}^2$, a stator resistance of $R_s = 6.9\,\Omega$, a rotor resistance of $R_r = 3.0\,\Omega$, and a magnetizing inductance of $L_m = 340.9$ mH.

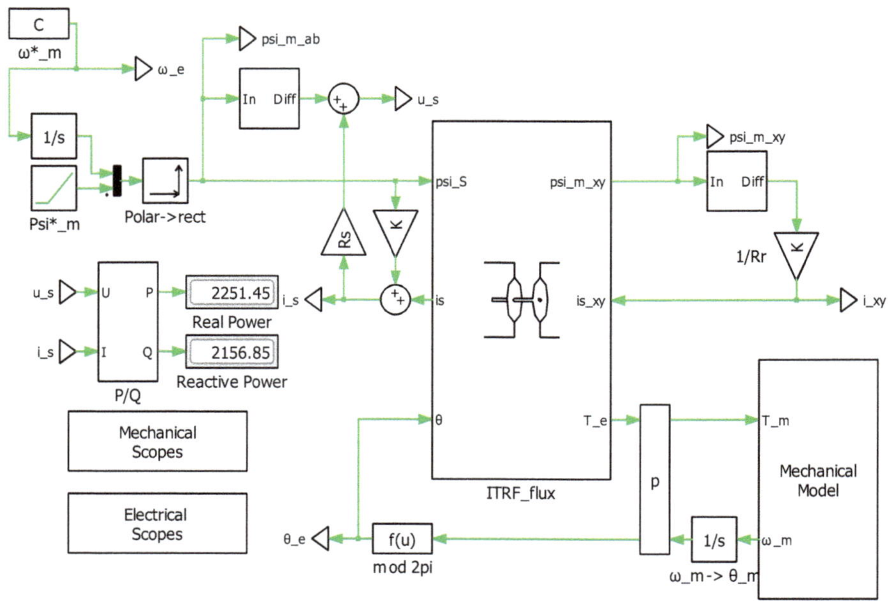

Fig. 8.39 Simulation of an induction machine using the simplified model

The aim of this example is to examine the steady-state operation of a simplified induction machine under different loads. Therefore the readers are encouraged to change the mechanical model parameters and observe the effects on the operation of the machine. The simulation model is given in Fig. 8.39.

Note that a derivative is used in this model to calculate the rotor current $i_{\mathrm{r}}^{\mathrm{xy}}$. Although this example is provided here to help readers understand the topic better, it is recommended to avoid differentiators whenever possible. Differentiators may cause numeric instabilities or significantly increased simulation times and should be used with caution, especially when step responses are applied to the model.

8.6.2 *Tutorial 2: Universal Induction Machine Model*

A simulation that represents a universal IRTF based induction machine model is developed in this tutorial. The aim is to develop a simulation which reflects the generic model given in Fig. 8.13. Furthermore, the user should be able to vary the transformation variable a so that the impact of this transformation can be seen on a vector diagram of the type given in Fig. 8.11b. To achieve this objective, the flux vectors $\vec{\psi}_{\mathrm{s}}$, $\vec{\psi}_{\mathrm{M}}$, and $\vec{\psi}_{\mathrm{R}}$ should be transformed to a coordinate reference frame that is tied to the flux vector $\vec{\psi}_{\mathrm{m}}$, as given in Fig. 8.11a. In addition to the

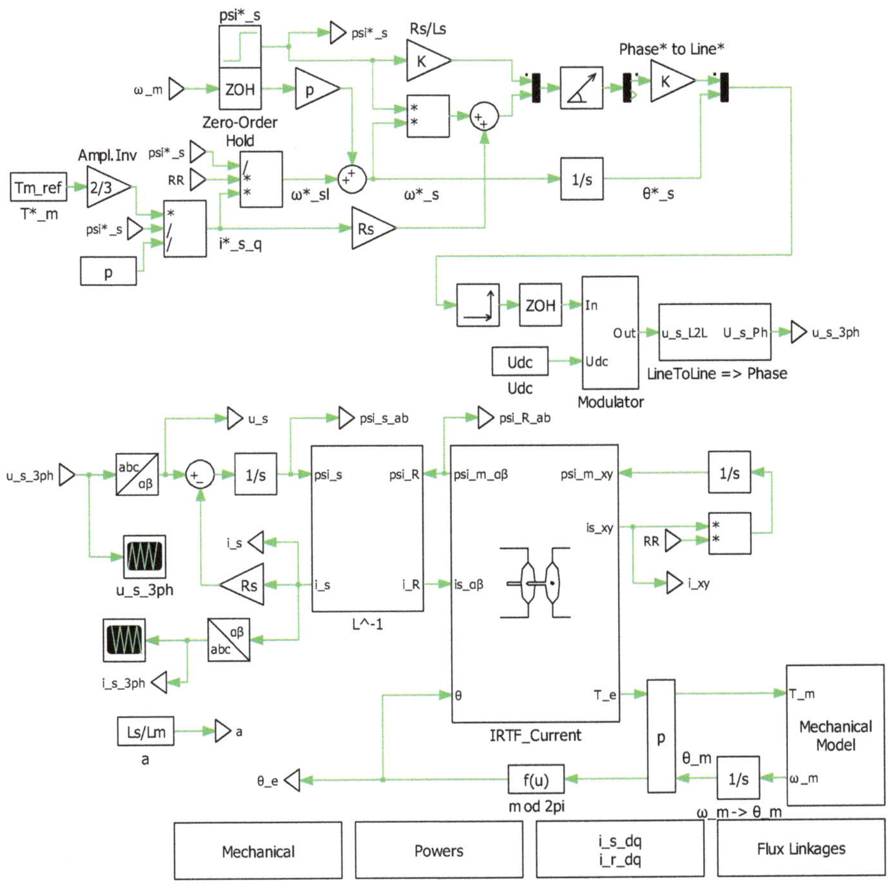

Fig. 8.40 Simulation of an induction machine, using the universal model

parameters given in previous tutorial, a stator inductance of $L_s = 346.9\,\text{mH}$ and rotor inductance of $L_r = 346.9\,\text{mH}$ are used. With this choice of parameters, the limits of the transformation variable a are $0.9827 \leq a \leq 1.0177$, according to Eq. (8.13).

A PLECS model is given in Fig. 8.40. The inverse inductance matrix L^{-1} is implemented as a subsystem. The transformed inductance values, the inverse inductance matrix, and the related current calculations are carried out in this subsystem. The transformation variable a is given as a constant as can be changed at the beginning of each simulation. The transformed rotor resistance R_R has also been implemented as a subsystem, where the gain is equal to $R_R = a^2 R_r$, as in Eq. (8.15). The flux linkage vectors $\vec{\psi}_s$, $\vec{\psi}_M$, $\vec{\psi}_R$ are transformed to a synchronous reference frame linked to the untransformed airgap flux linkage vector $\vec{\psi}_m$. This vector is calculated using Eq. (8.5b), with $L_{\sigma s} = L_s - L_m$. The instantaneous angle θ_m of this vector $\vec{\psi}_m$ is used for these transformations.

The flux linkage vectors $\vec{\psi}_{\mathrm{S}}$, $\vec{\psi}_{\mathrm{M}}$, $\vec{\psi}_{\mathrm{R}}$ as well as the torque and speed of the machine are given in Fig. 8.41. It is important to note that the output torque and speed, as well as stator voltages and currents are not affected by the changes in the parameter a, as expected. Finally, it is instructive to vary the load torque setting and observe the changes to the flux vectors and node values of the simulation model.

8.6.3 Tutorial 3: Universal Stationary Frame Oriented Induction Machine Model

The implementation of a stationary oriented model of the induction machine as discussed in Sect. 8.3.3 is explored in this tutorial. The aim is to develop a simulation model which is based on the generic structure given in Fig. 8.19. The machine is connected to a three-phase 220 V, 50 Hz sinusoidal source. The machine inductances and resistances, as well as input voltage and load characteristics given in the previous tutorial are used. The effect changing the transformation factor a on the shaft torque and speed for a given load torque is examined under steady-state conditions.

The PLECS simulation of a stationary oriented induction machine model is given in Fig. 8.42. If the simulation is executed for different transformation values a, the user may ascertain that the torque and speed readings remain unaffected. Changes to the load torque setting, realized by way of the load module torque setting, give the user the ability to examine, for example, the steady-state torque speed curve of this machine. Note that the results of this simulation also match the results of the previous simulation, given in Fig. 8.41.

8.6.4 Tutorial 4: Current-Controlled Zero Leakage Flux Oriented Machine Model

A zero leakage, UFO based induction machine model is developed in this tutorial, as discussed in Sect. 8.2.2. The generic structure shown in Fig. 8.6 is used to create the simulation model. The machine inductance and resistance parameters given in the previous tutorials are used also in this tutorial. The machine is operated under no-load conditions. A direct and quadrature current signals are used as inputs, where the direct axis is set to $i_{\mathrm{d}} = 4\,\mathrm{A}$ at the start of the simulation. A step in the quadrature current from $0 \rightarrow 4\ A$ is used at $t = 0.8\,\mathrm{s}$. The aim is to build a model of the machine and examine the transient responses of the shaft torque T_{e}, flux linkage ψ_{m}, and angular speed variables ω_{m}, ω_{sl}, ω_{s}.

A PLECS model, given in Fig. 8.43, was used to derive the results shown in Fig. 8.44. It is worth noting that the step change in direct current results in a first order delayed response in flux linkage ψ_{m}, whereas a step change in quadrature

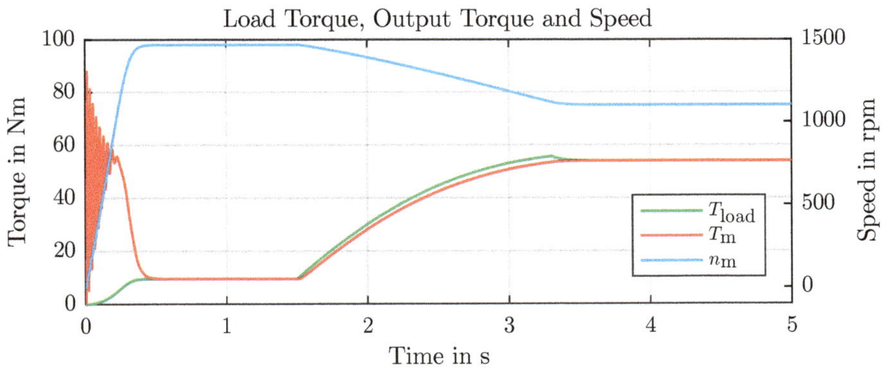

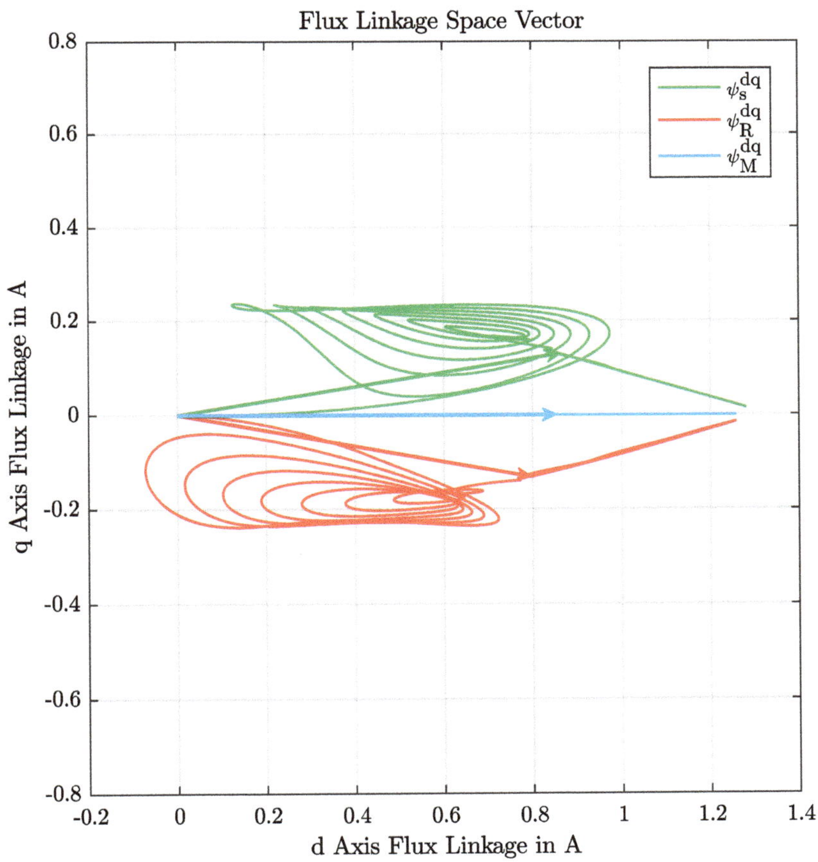

Fig. 8.41 Simulation results of the universal induction machine model ($a = 1$)

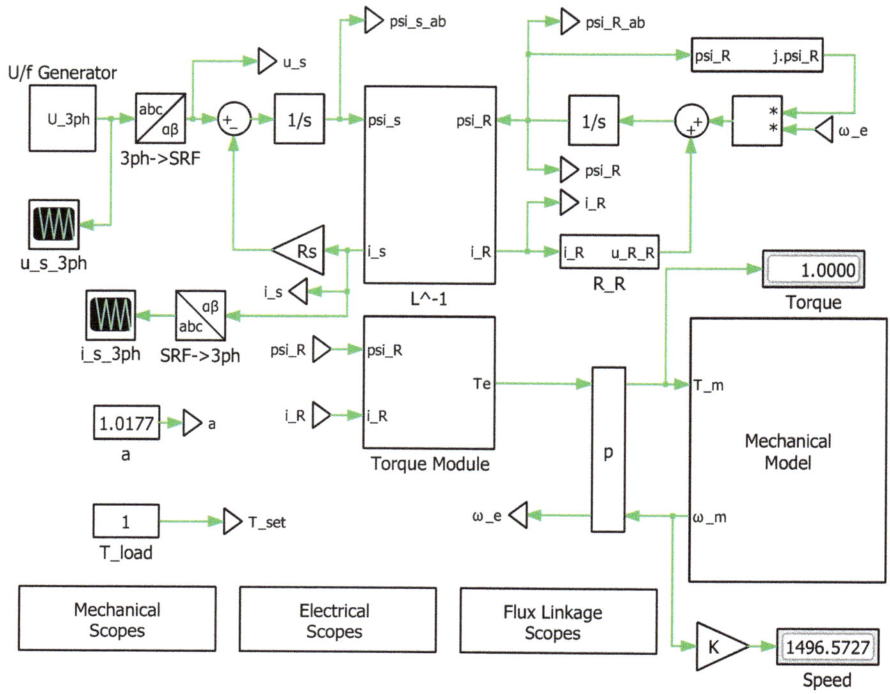

Fig. 8.42 Simulation of universal model with stationary oriented reference frame

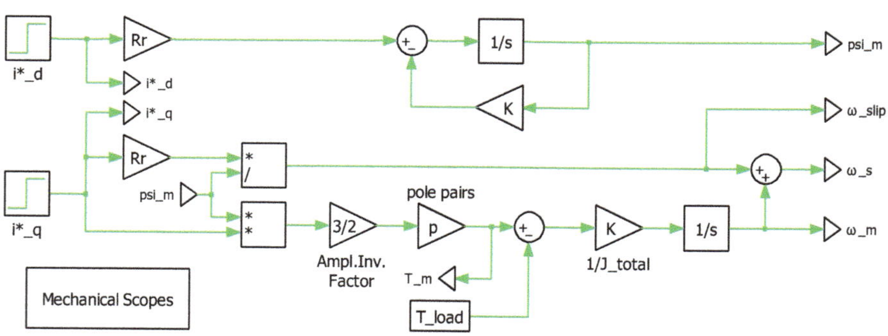

Fig. 8.43 Simulation of current controlled, flux oriented model with zero leakage

current results in an instantaneous step change in slip frequency as well as torque. The machine shaft speed increases linearly as the result of the constant total torque. Since the slip frequency is constant, the stator frequency also increases linearly.

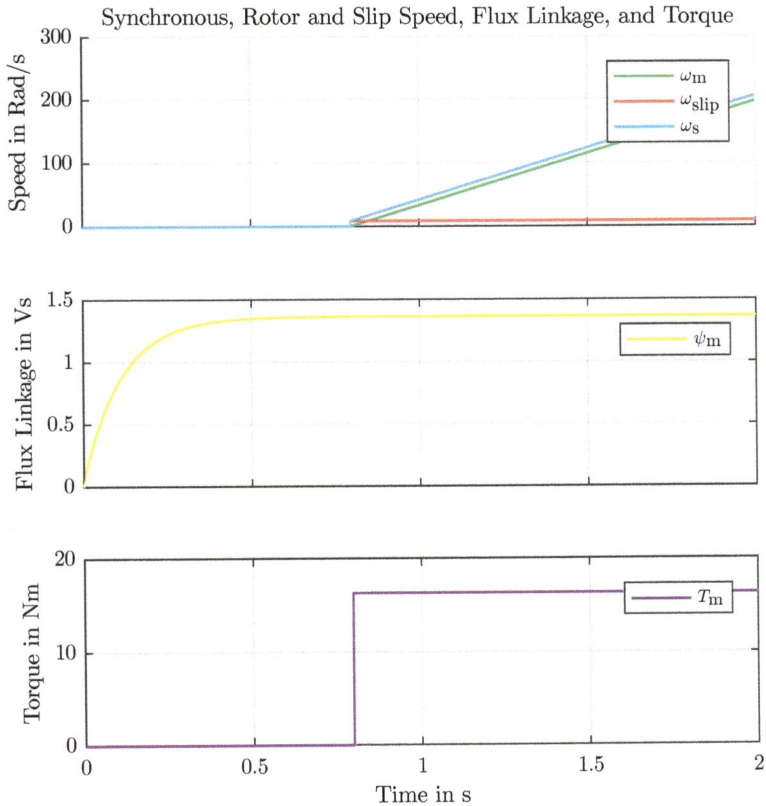

Fig. 8.44 Simulation results of a current controlled, flux oriented model with zero leakage inductances

8.6.5 *Tutorial 5: Current Controlled Universal Field-Oriented (UFO) Model*

In this tutorial a generalized UFO model with leakage inductances is developed which is based on the generic structure shown in Fig. 8.22. The aim is to consider the transient responses of the model as in the previous chapter for different transformation ratios a. The simulation parameters for this simulation are not changed compared to the previous tutorials of this chapter. The simulation model, as shown in Fig. 8.45, complies with the generic model of the machine. Setting the transformation variable a to either side of its range of $0.9827 \leq a \leq 1.0177$ leads to a rotor or stator flux oriented model, with corresponding transient results given by Figs. 8.26 and 8.29, respectively. Note that *ramp* inputs instead of (unrealistic) step change inputs are used for direct and quadrature current references to avoid unrealistically high values at the outputs of the *differentiator* modules.

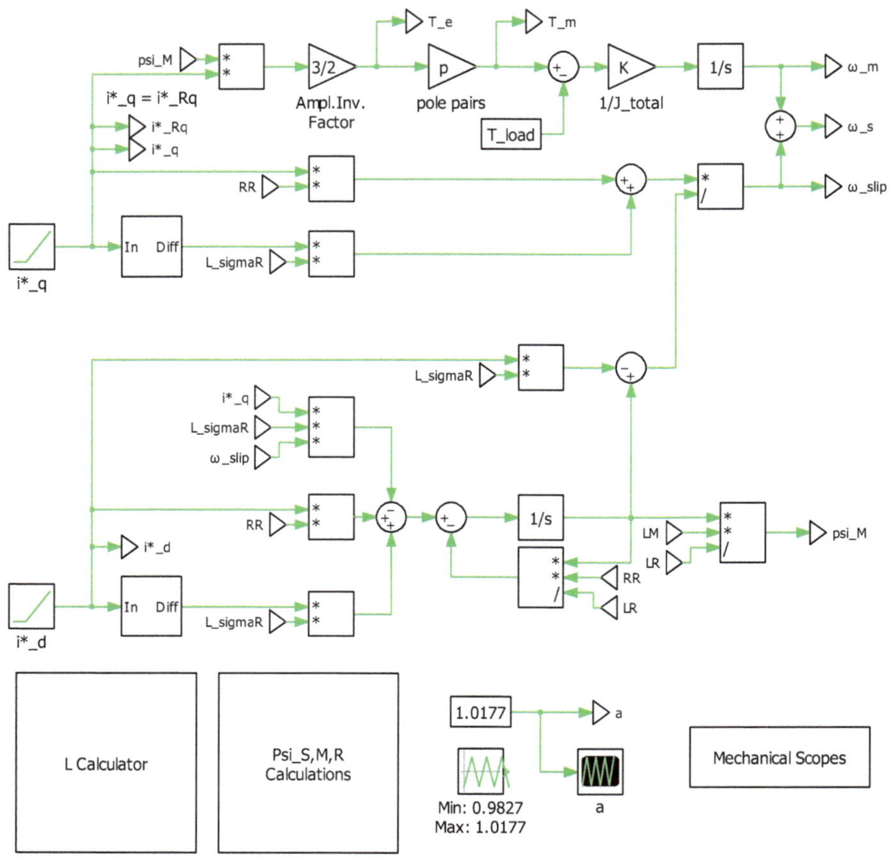

Fig. 8.45 Simulation of a current controlled, flux oriented model with leakage inductances

8.6.6 *Tutorial 6: Parameter Estimation Using Name Plate Data and Known Stator Resistance*

The name plate data given in Table 8.1 corresponds to a *delta* connected motor. In addition to these data, the measured RMS no-load line current is given as $I_1^{\text{noload}} = 8.41\,\text{A}$, and the resistance between two terminals with the machine at standstill is measured to be $R_l = 0.40\,\Omega$.

On the basis of the approach set out in Sect. 8.4, a MATLAB m-file is built to calculate the parameters L_s, $L_{\sigma s}$, $L_{\sigma r}$, and R_r, where it may be assumed that the leakage inductances are equal. Furthermore, the rated stator flux $\vec{\psi}_s$, magnetizing flux ψ_M, and the parameters $L_{\sigma S}$, L_M, R_R for a rotor flux field-oriented model are calculated.

Table 8.1 Name plate data

Parameters	Value
Rated power ($P_{\mathrm{out}}^{\mathrm{nom}}$)	17,332 W
Rated shaft speed ($n_{\mathrm{m}}^{\mathrm{nom}}$)	1465 rpm
Rated RMS line voltage ($U_{\mathrm{l}}^{\mathrm{nom}}$)	415 V
Rated RMS line current ($I_{\mathrm{l}}^{\mathrm{nom}}$)	30 A
Supply frequency (f_{s})	50 Hz
Rated power factor ($\mathrm{pf}^{\mathrm{nom}}$)	0.80
Number of pole pairs (p)	2

Table 8.2 Estimated machine parameters

Parameters	Value
Stator inductance (L_{s})	272 mH
Magnetizing inductance (L_{m})	260 mH
Leakage inductance ($L_{\sigma\mathrm{s}}$)	12 mH
Leakage inductance ($L_{\sigma\mathrm{r}}$)	12 mH
Rotor resistance (R_{r})	0.54 Ω
Stator resistance (R_{s})	0.6 Ω

A possible approach to this problem may be initiated by examining the measured resistance R_{l} between two terminals of the machine. For a delta connected machine, the equivalent resistance *seen* by two terminals is equal to resistance of one phase with the other two series connected phases in parallel. Hence, the stator resistance is equal to $R_{\mathrm{s}} = {}^{3}/{}_{2}R_{\mathrm{l}}$.

The peak rated phase current is equal to $i_{\mathrm{D}}^{\mathrm{nom}} = \sqrt{2}/\sqrt{3} \cdot I_{\mathrm{l}}^{\mathrm{nom}}$, whereas the magnitude of the current and voltage space vectors defined in Eq. (8.34) are equal to $u_{\mathrm{s}} = \sqrt{2} \cdot U_{\mathrm{l}}^{\mathrm{nom}}$ and $i_{\mathrm{s}}^{\mathrm{no\ load}} = \sqrt{2}/\sqrt{3} \cdot I_{\mathrm{l}}^{\mathrm{noload}}$, since the machine is delta connected. The stator flux linkage vector amplitude $\vec{\psi}_{\mathrm{s}}$ under no-load conditions and inductance L_{s} can now be calculated, given that $\omega_{\mathrm{s}} = 2\pi f_{\mathrm{s}}$. Computation of the parameters L_{s}, $L_{\sigma\mathrm{s}}$, $L_{\sigma\mathrm{r}}$, R_{r}, with the assumption of $L_{\sigma\mathrm{s}} = L_{\sigma\mathrm{r}}$, proceeds using the approach set out in Sect. 8.4 which leads to the data shown in Table 8.2.

It is *crucial* to note that the no-load flux linkage vector and rated flux linkage vector are assumed equal in magnitude throughout the calculations in Sect. 8.4. These values may be assumed equal for some quick calculation. However, the small difference of a few percent becomes relevant when a simulation is set up. If this difference is not taken into account, torque and slip frequency values for the rated operating point from the simulation do not match the calculations. Rated flux linkage vector magnitude is calculated using the rated current and rated power factor information.

The second part of this tutorial requires the conversion of the parameters given in Table 8.2 to a revised set of universal model parameters as defined by Eqs. (8.12) and (8.15) with $a = {}^{L_{\mathrm{m}}}/{}_{L_{\mathrm{r}}}$. The new set of parameters as given in Table 8.3 correspond to a rotor flux oriented UFO model which in turn is used in conjunction with Eqs. (8.43) and (8.44) to calculate the required (rated) magnetizing flux amplitude $\psi_M^{nom} = \psi_R^{nom}$. This flux linkage value together with the earlier found stator flux linkage value and measured stator resistance are also shown in Table 8.3.

Table 8.3 Estimated machine parameters and flux linkage values for a rotor flux oriented model with $a = L_m/L_r$

Parameters	Value
Magnetizing inductance (L_M)	248.5 mH
Leakage inductance ($L_{\sigma S}$)	23.6 mH
Leakage inductance ($L_{\sigma R}$)	0 mH
Rotor resistance (R_R)	0.494 Ω
Stator resistance (R_s)	0.600 Ω
Rated stator flux linkage (ψ_s^{nom})	1.83 Vs
Rated rotor flux linkage (ψ_R^{nom})	1.67 Vs

```
% The machine is delta connected
VsR      = 415     % [V] RMS line to line voltage
IsR      = 30      % [A] rated RMS line current
IsN      = 8.41    % [A] noload RMS line current
Rlin     = 0.40    % [Ohm] line-to-line resistance
J_total = 100e-3   % [kgm^2] Total inertia

% Nameplate data
PoutR    = 17332   % [W] rated output power
nR       = 1465    % [1/min] rated shaft speed
fs       = 50      % [Hz] supply frequency
pfR      = 0.80    % [1] rated power factor
p        = 2       % [1] pole-pair number

% Calculating
ws       = 2*pi*fs      % [rad/s] electrical freq.
wm       = nR*p*2*pi/60 % [rad/s] rated electrical shaft freq.
wsl      = ws-wm        % [rad/s] rated electrical shaft freq.
Te       = PoutR/wm     % [Nm] rated electrical torque per pole
Tm       = Te*2         % [Nm] rated torque
Rs       = 3/2*Rlin     % [Ohm] stator resistance
isR      = sqrt(2)*IsR/sqrt(3)   % [A] rated i_s ampl.
isN      = sqrt(2)*IsN/sqrt(3)   % [A] noload i_s ampl.
usR      = sqrt(2)*VsR           % [V] supply voltage ampl.

% Calculate remaining data
psi_s    = sqrt(usR^2-(isN*Rs)^2)/ws % [Vs] no load stator flux linkage
Ls       = psi_s/isN                 % [H] self inductance stator
psi_s_R = sqrt((usR-isR*Rs*0.8)^2+(isR*Rs*0.6)^2)/ws
                                 % [Vs] rated stator flux
isqR     = Te/psi_s_R*2/3        % [A] rated isq
isdR     = sqrt(isR^2-isqR^2)    % [A] rated isd
is_o_R   = psi_s_R / Ls          % [A] rated "magnetizing current"
                                 % the current flowing through L_M
                                 % in fig 8.21 - Symbolic UFO Model
ir_d     = isdR - is_o_R         % [A] rated rotor d-axis current
RR       = wsl*psi_s_R*isqR / (ir_d^2+isqR^2 )
                                 % [Ohm] rotor resistance, a=Ls/Lm
Lrsig    = ir_d/isqR*RR/wsl      % [H] leakage inductance, a=Ls/Lm

% Calculate parameters for a=1 assume Lsigr=Lsigs
Lr       = Ls                       % [H] rotor inductance
Lm       = sqrt(Ls^3/(Lrsig+Ls))    % [H] magnetizing inductance
Lsigs    = Ls-Lm                    % [H] stator leakage inductance
```

```
Rr        = RR /(Ls/Lm)^2            % [Ohm] rotor resistance
psi_r_R_vec = psi_s_R - isdR*L_Ssig - 1i*isqR*L_Ssig*2 -ir_d*L_Ssig;
                   % [Vs] rotor flux linkage vector (complex number)
psi_r_R     = abs(psi_r_R_vec);
                   % [Vs] rotor flux linkage vector magnitude

% Parameters and flux for any 'a': example rotor flux orientation
a = Lm/Lr      % rotor  field oriented
%a = Ls/Lm       % stator field oriented
L_Ssig = Lm*(Ls/Lm-a);
L_M = a*Lm
R_R = a^2*Rr
psiMd = psi_s-L_Ssig*isdR
psiMq = -L_Ssig*isqR
psi_M = sqrt(psiMd^2+psiMq^2) % [Vs] rated magnetizing flux
```

Listing 8.1 Matlab file for calculation

8.6.7 Tutorial 7: Grid Connected Induction Machine

A dynamic model of a three-phase induction machine is considered in this tutorial. Such a model can be implemented by making use of the generic machine model shown in Fig. 8.15. The set of parameters as given in Table 8.3 are employed. The machine is initially operated without a mechanical load. At time $t = 1$ s, a 113 Nm (rated) torque load step is applied to the shaft. The schematics of a PLECS simulation for this tutorial is given in Fig. 8.46, while the simulation results are given

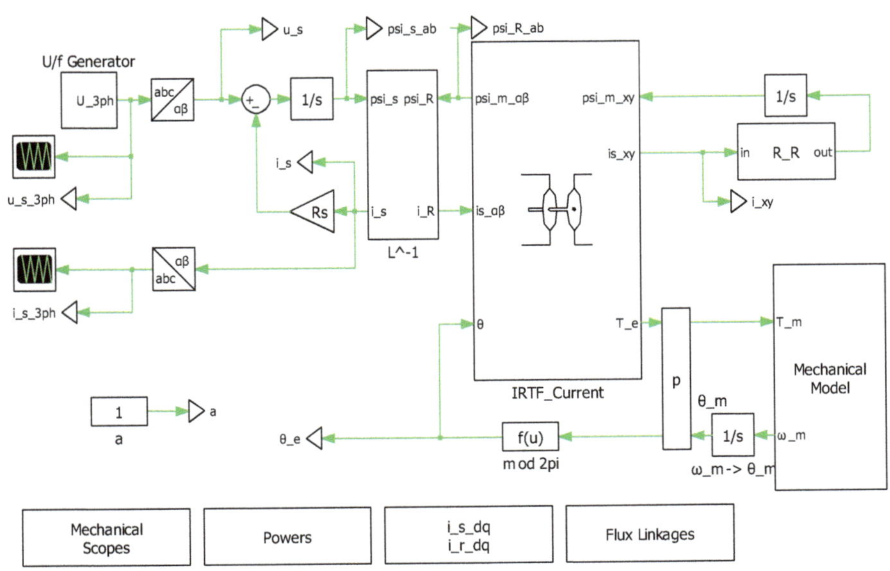

Fig. 8.46 Simulation of grid connected induction machine

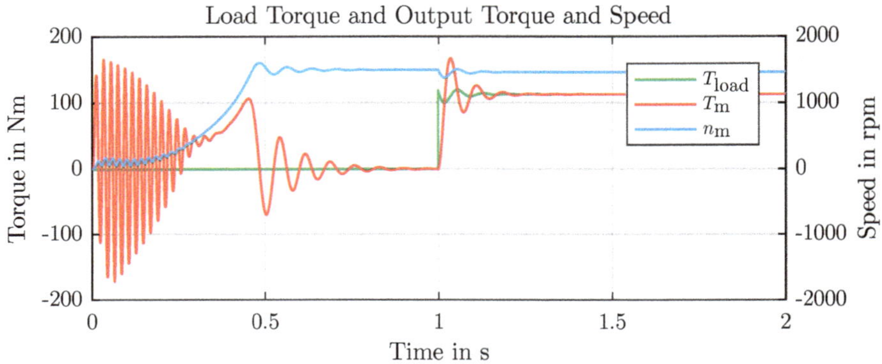

Fig. 8.47 Simulation results of grid connected induction machine

in Fig. 8.47. Note that the output torque oscillates when the machine is connected to the network at the beginning of the simulation, as well as after the load step is applied.

8.6.8 *Tutorial 8: Steady-State Characteristics, Grid Connected Induction Machine*

The steady-state characteristics of T_e, I_l, $\vec{\psi}_s$, P_{out} of the induction machine is discussed in this tutorial. To obtain these characteristics, the simulation model of the previous tutorial is employed. The *Mechanical Model* subsystem is manipulated to keep the shaft speed at desired values independent of the torque generated by the machine. By varying the shaft speed between $0 \rightarrow 3000$ rpm slowly compared to the time constants of the machine, the steady-state characteristics of the machine can be obtained. The mechanical torque T_e, RMS line current I_l, flux linkage $|\underline{\psi}_s|$, and output power P_{out} as function of the shaft speed are given in Fig. 8.48.

8.6.9 *Tutorial 9: Grid Connected Single-Phase Induction Machine*

This tutorial is concerned with the implementation of a dynamic simulation model of a single-phase, capacitor-run induction machine connected to a 110 V(RMS), 50 Hz supply. For this purpose a machine prototype with parameters as given in Table 8.4 will be considered in terms of its dynamic behavior when subjected to line start

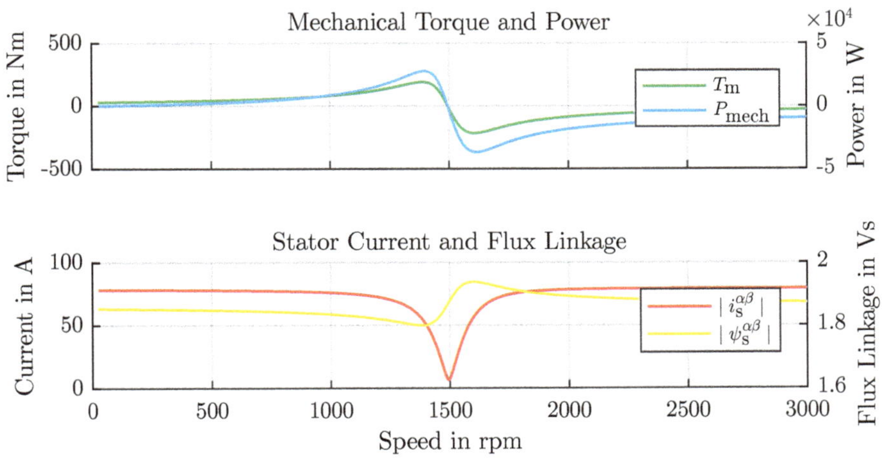

Fig. 8.48 Steady-state characteristics of the 17.3 kW induction machine

Table 8.4 Single-phase, capacitor-run machine parameters

Parameters	Value
Leakage inductance ($L_{\sigma s}$)	12.8 mH
Stator resistance (run winding) (R_{run})	2.02 Ω
Stator resistance (aux winding) (R_{aux})	7.14 Ω
Rotor resistance (R_R)	3.87 Ω
Magnetizing inductance (L_{M})	171.8 mH
Asymmetric ITF winding ratio (k_{aux})	1.18
Capacitor size (C)	31.0 μF
Inertia (J)	1.46×10^{-2} kg m^2
Pole pairs (p)	2
Initial rotor speed ($\omega_{\text{m}}^{\text{o}}$)	0 rad/s

conditions. Of interest is to examine the instantaneous torque, shaft speed, and phase currents during the run-up sequence of the motor. Central to the approach given here are the concepts discussed in Sect. 8.5 which, as may be observed, provides a more general approach to the topic of modeling single-phase machines. This tutorial is restricted to a capacitor-run machine, but the model may be readily adapted to split-coil and capacitor start type machines. In the latter case, the auxiliary winding resistance must be controlled in a manner outlined in the previous tutorial.

A suitable starting point for the proposed analysis is the generic model of the single-phase machine as shown in Fig. 8.35 in which the auxiliary winding must (in this case) be connected to the run winding via a capacitor C. Given this approach, an additional set of generic modules must be provided to implement equation (8.46).

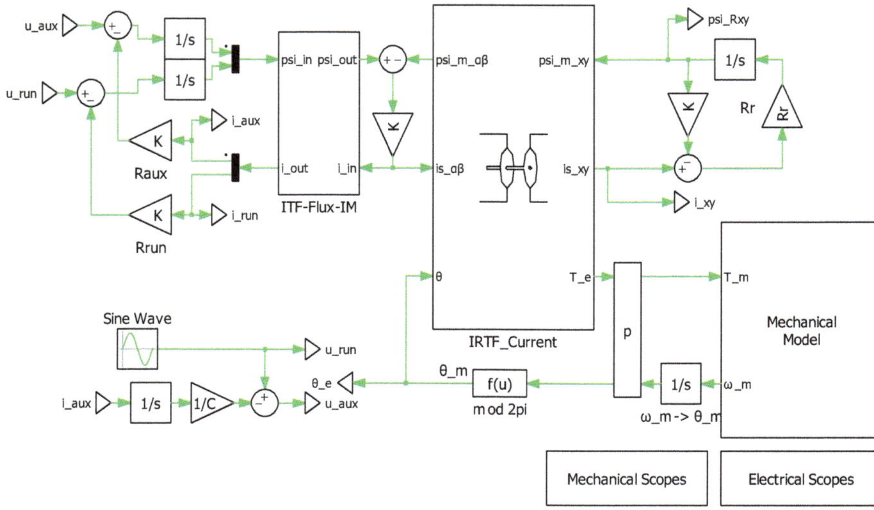

Fig. 8.49 Simulation of single-phase capacitor-run motor

The model as given in Fig. 8.49 shows the ITF and IRTF modules as present in the generic model. A load step of 2 Nm is applied to the machine at $t = 1$ s to also observe the step response as well as behavior under load.

The simulation results that show instantaneous torque and shaft speed, as well as auxiliary and run voltages and currents are given in Fig. 8.50. It can be observed in these results that the torque has a rather large pulsating component, in addition to an average torque component, which (under steady-state conditions) corresponds to the applied load torque value of 2 Nm.

It is also instructive to examine the interaction between the rotor flux $\vec{\psi}_R^{xy}$ and stator current vector $\vec{i}_s^{xy}$ (in rotor coordinates) of the machine, given that these define the instantaneous torque in the machine. The vector plot as given in Fig. 8.51 shows that the relationship between flux and current vectors is by no means constant over the course of the simulation.

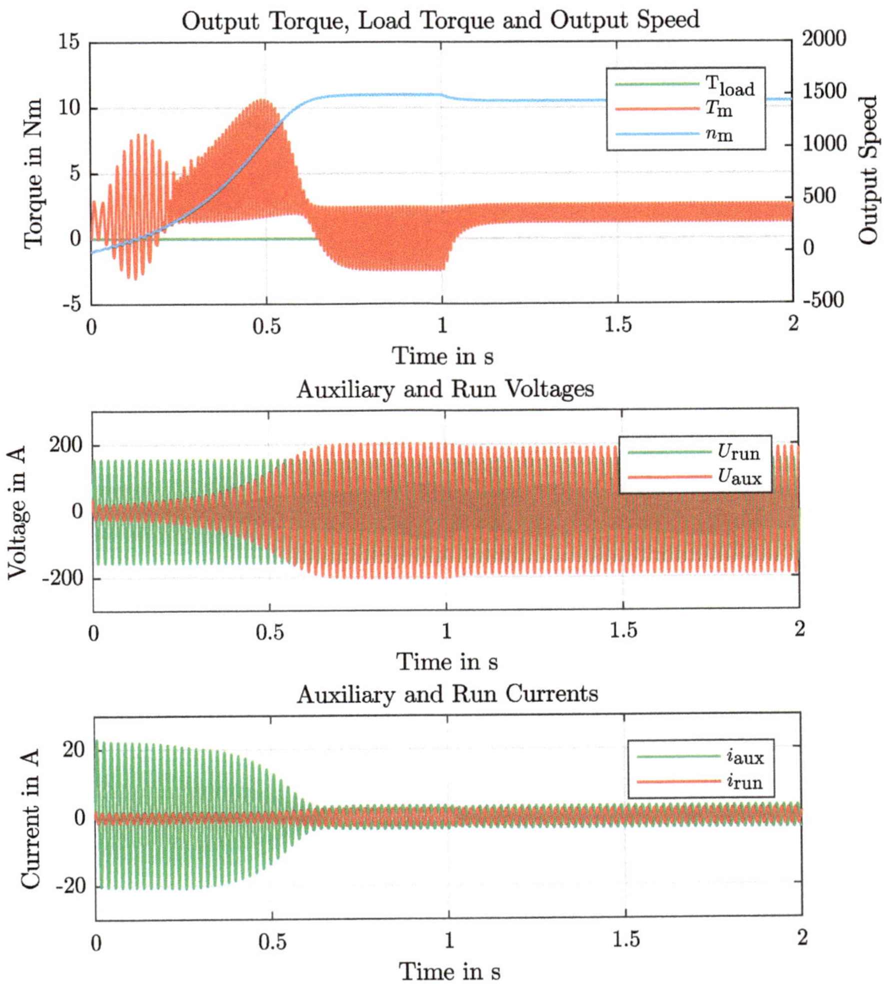

Fig. 8.50 Simulation results of a line start of single-phase capacitor-run induction machine

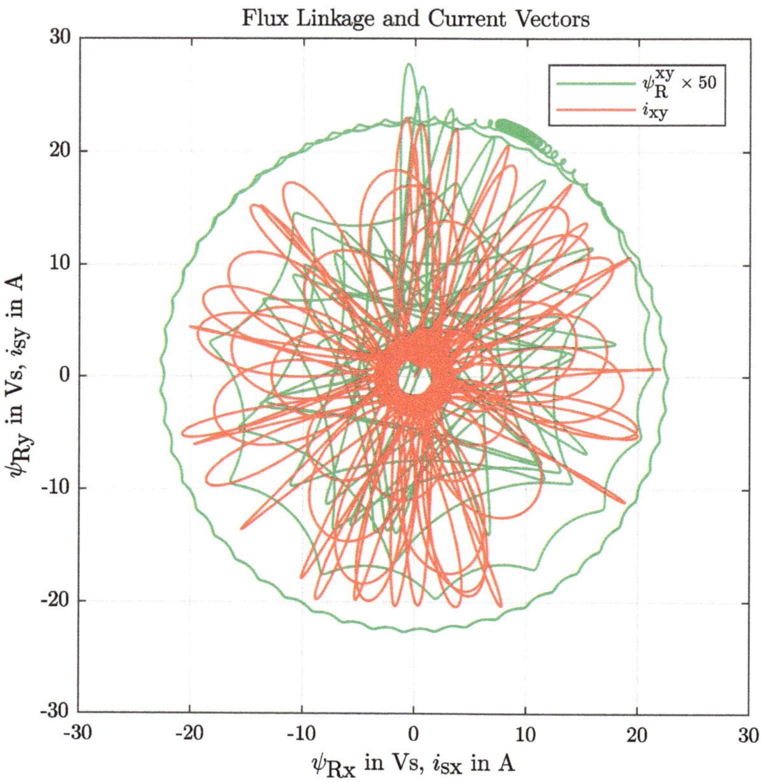

Fig. 8.51 Line start of single-phase capacitor-run induction machine, vector plot $\vec{\psi}_R^{xy} \cdot 50$ (*green trace*), and $\vec{i}_s^{xy}$ (*red trace*)

References

1. Blaschke F (1972) The principle of field orientation as applied to the new transvektor closed-loop control system for rotating-field machines. Siemens Rev 39(5):217–219
2. De Doncker R (1992) Field-oriented controllers with rotor deep bar compensation circuits [induction machines]. IEEE Trans Ind Appl 28(5):1062–1071. https://doi.org/10.1109/28.158830
3. De Doncker R, Novotny D (1994) The universal field oriented controller. IEEE Trans Ind Appl 30(1):92–100. https://doi.org/10.1109/28.273626
4. Dorf RC, Bishop RH (2007) Modern control systems, 11th edn. Prentice Hall, Upper Saddle River
5. Hasse K (1969) Zur dynamik drehzahlgeregelter antriebe mit stromrichtergespeisten asynchron-kurzschlussläufermaschinen. PhD Thesis, TH Darmstadt
6. Heyland A (1894) Ein graphisches Verfahren zur Vorausberechnung von Transformatoren und Mehrphasenmotoren. ETZ 15(11):561–564
7. Heyland A (1906) A graphical treatment of the induction motor. McGraw, New York
8. Hughes A (2006) Electric motors and drives: fundamentals, types and applications, 3rd edn. Newnes, Oxford
9. Jordan H, Weis M (1969) Asynchronmaschinen. Vieweg, Braunschweig
10. Leonhard W (2001). Control of electrical drives, 3rd edn. Springer, Berlin

11. Lorenz R, Novotny D (1990) Saturation effects in field-oriented induction machines. IEEE Trans Ind Appl 26(2):283–289. https://doi.org/10.1109/28.54254
12. Muller S, Deicke M, De Doncker R (2002) Doubly fed induction generator systems for wind turbines. IEEE Ind Appl Mag. 8(3):26–33. https://doi.org/10.1109/2943.999610
13. Novotny DW, Lipo TA (1996) Vector control and dynamics of AC drives (Monographs in electrical and electronic engineering). Oxford University Press, Oxford
14. Veltman A, Pulle DWJ, De Doncker R (2007) Fundamentals of electrical drives. Spinger, Berlin
15. Xu X, De Doncker R, Novotny D (1988) A stator flux oriented induction machine drive. In: PESC'88 record 19th annual IEEE power electronics specialists conference, 1988, vol 2, pp 870–876. https://doi.org/10.1109/PESC.1988.18219

Chapter 9
Control of Induction Machine Drives

In this chapter, attention is given to the control concepts that can be used to achieve independent torque and flux linkage control of induction machines over a wide speed range. Following the *machine model inversion principles*, the machine models introduced in the previous chapter will again be used for deriving suitable controller structures.

First, classical *V/f* control concepts with very limited dynamic performance capabilities will be discussed to illustrate that these simple controllers, which are based on steady-state models, can become unstable. Next, field-oriented control techniques will be discussed which build on the fundamental model given in Sect. 4.2.2. A key role is assigned to the presentation of a *universal field-oriented* (UFO) model which is used in conjunction with a current source induction machine model. At a later stage, this model is replaced with a voltage source induction model which is connected to a synchronous current-controlled converter. Finally, attention is given to the operational aspects of the drive which takes into account the maximum supply voltage and current constraints. In this context field weakening techniques are also discussed. A set of tutorials is again provided to further familiarize the reader with the concepts presented in this chapter.

9.1 Voltage-to-Frequency (V/f) Control

In low-performance adjustable speed drives the issue of controlling torque or speed of a mechanical load connected to an induction motor is not critical in terms of required dynamic performance. For these applications, the classical *voltage-to-frequency* (V/f) control can be used resulting in a very simple drive control which can easily be implemented in analog hardware. It was therefore extensively used in inverter controllers before digital hardware was available. The key concepts of the

R. W. De Doncker et al., *Advanced Electrical Drives*, Power Systems,
https://doi.org/10.1007/978-3-030-48977-9_9

V/f control are based on the steady-state model of the induction machine and are presented in the following.

The drive structure in question does not utilize closed loop current control as discussed in Chap. 3. Instead, only the stator voltage $\hat{u}_s$ and the stator frequency ω_s are controlled in a way that

$$\psi_s = \text{const.} \tag{9.1}$$

The lack of a current control in V/f control is a key issue which distinguishes this low-dynamic drive concept from, for example, the dynamic field-oriented control concepts to be discussed in the subsequent section of this chapter. Furthermore, it will be shown in this section that the omission of current control can in some applications lead to instability. The ensuing discussion is undertaken for a two-pole machine to facilitate the readability of the figures and equations.

Steady-State Equations for V/f Control

Prior to examining the algorithms and generic models which may be invoked for this type of control it is helpful to consider the steady-state relationship which exists between the torque T_e, flux linkage ψ_s, and angular frequency ω_s.

For this purpose, the stator flux oriented model, discussed in Sect. 8.3.4.2, is introduced because it is advantageous given the simplicity of the control algorithm that can be achieved. Under these conditions, the variables ψ_M and L_M shown in Fig. 8.27 may be written as ψ_s and L_s respectively. The equation set which corresponds to the stator flux model viewed under steady-state conditions may be written as

$$u_{sd} = R_s\, i_{sd} \tag{9.2a}$$

$$u_{sq} = R_s\, i_{sq} + \omega_s\, \psi_s \tag{9.2b}$$

$$\omega_s\, \psi_s = \omega_{sl}\, L_{\sigma R}\, i_{Rd} + R_R\, i_{sq} + \omega_m\, \psi_s \tag{9.2c}$$

$$i_{sd} = \frac{\psi_s}{L_s} + \underbrace{\frac{\omega_{sl}\, L_{\sigma R}\, i_{sq}}{R_R}}_{i_{Rd}} \tag{9.2d}$$

$$i_{sq} = \frac{2}{3}\frac{T_e}{\psi_s}. \tag{9.2e}$$

Elimination of the variable i_{Rd} from expression (9.2c) and expression (9.2d) gives

$$L_{\sigma R}\, i_{sq} = \frac{\psi_s}{\frac{\hat{\omega}_{sl}}{\omega_{sl}} + \frac{\omega_{sl}}{\hat{\omega}_{sl}}} \tag{9.3}$$

with $\hat{\omega}_{sl} = {}^{R_R}/_{L_{\sigma R}}$ and $\omega_{sl} = \omega_s - \omega_m$. Use of the latter equation with expression (9.2e) leads to the following torque equation known as the *Kloss formula*:

$$T_e = \frac{2\hat{T}_e}{\frac{\hat{\omega}_{sl}}{\omega_{sl}} + \frac{\omega_{sl}}{\hat{\omega}_{sl}}} \tag{9.4}$$

with

$$\hat{T}_e = \frac{3}{2}\frac{\psi_s^2}{2L_{\sigma R}} \tag{9.5}$$

represents the *pullout torque* and $\hat{\omega}_{sl}$ the *pullout slip frequency* at which the pullout torque occurs. Use of Eq. (9.4), with a given reference stator frequency $\omega_s = \omega_{s1}^*$, leads to the torque versus shaft speed curve shown as the blue curve in Fig. 9.1. When increasing the stator frequency at constant stator flux linkage $\hat{\psi}_s$, as given in Eq. (9.1), another curve shown in Fig. 9.1 arises for stator frequency reference values ω_{s2}^* with $\omega_{s1}^* < \omega_{s2}^*$. Note, by maintaining constant flux linkage ψ_s the amplitude of the pullout torque $\hat{T}_e$ is constant as can be seen from Eq. (9.5).

A typical load torque/shaft speed characteristic $T_l(\omega_m)$ is also shown in Fig. 9.1 to show the *steady-state* operating point of the drive, which is found when the load torque is equal to the torque produced by the machine.

In the following two subsections V/f controllers for speed and torque control will be discussed.

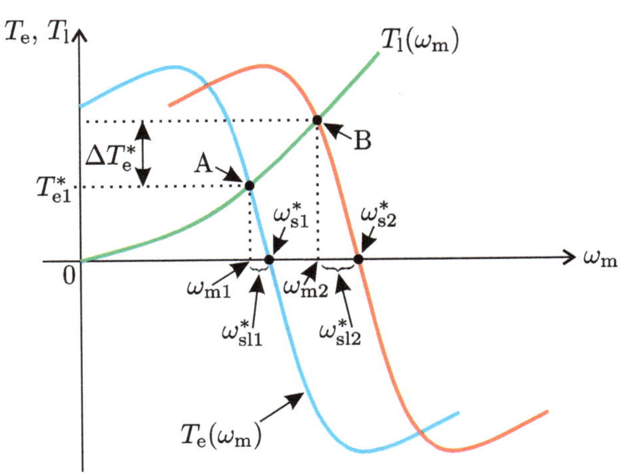

Fig. 9.1 Classical voltage-to-frequency (V/f) drive concept

9.1.1 Simple V/f Speed Controller

This control approach is based on the use of the reference ω_s^* and strives to maintain, without consideration of the stator resistance, a constant stator flux linkage value. The required supply voltage amplitude $\hat{u}_s$ and phase angle ρ_s are determined with the aid of equation set (9.2). Input to the controller is the stator frequency ω_s in which case the phase angle is given as $\rho_s = \omega_s t$. The required voltage amplitude may be written as $\hat{u}_s = \sqrt{u_{sd}^2 + u_{sq}^2}$. In most cases the leakage term in Eq. (9.2d) can be ignored and the voltage u_{sd} can be written as

$$u_{sd} \simeq \psi_s^* \frac{R_s}{L_s} \tag{9.6a}$$

$$u_{sq} \simeq \omega_s^* \, \psi_s^*. \tag{9.6b}$$

The quadrature voltage component u_{sq} shown in Eq. (9.6b) is derived from expression (9.2b) with the implicit assumption that the term $\omega_s \, \psi_s$ is usually larger than the term $R_s \, i_{sq}$. Note that in most applications the reference stator flux linkage matches the nominal stator flux linkage to avoid saturation. By neglecting the stator resistance, the term $\frac{u_{sq}}{\omega_s}$ is held constant, which is why this type of control is referred to as a V/f control. The generic representation of the simple V/f speed controller is given in Fig. 9.2.

If, for example, the speed reference is set to ω_{s1}^*, the steady-state shaft speed will settle to ω_{m1} for the assumed load torque versus speed characteristic. A frequency shift from $\omega_{s1} \rightarrow \omega_{s3}$ in effect translates the torque speed curves as may be observed from Fig. 9.1. Note that this type of controller does not require a shaft speed sensor, which is advantageous for applications where such a device is either unpractical (for robustness reasons) or not possible (in case the load-machine combination excludes the possibility of fitting a sensor). The real machine speed differs from the reference speed by the slip frequency since the latter is not calculated by the controller.

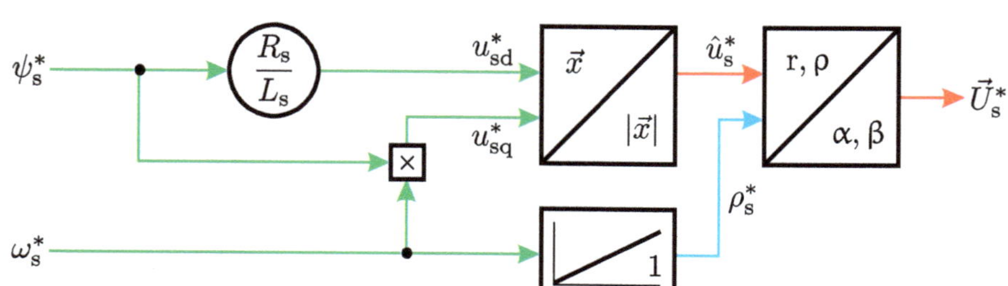

Fig. 9.2 V/f control structure with stator frequency reference input

9.1.1.1 Tutorial Results

An example of a drive which utilizes the control concept shown in Fig. 9.2 is given in tutorial (see Sect. 9.7.2). It uses the same 22 kW machine as in the previous chapter. An example of the results achieved with this simulation is given in Fig. 9.3, showing the stator flux linkage amplitude $|\vec{\psi}_{\rm s}|$, shaft speed, and shaft torque as function of time. The flux linkage reference was set to $\psi_{\rm s}^* = 2.29$ Wb which is the rated stator flux linkage value for this machine. The reference stator frequency, represented in terms of a reference speed, was set to $n_{\rm m}^* = 1465$ rpm. A quadratic load torque versus shaft speed function is introduced which provides rated torque $T_{\rm l}^{\rm nom} = 120$ Nm at rated speed $n_{\rm m}^{\rm nom} = 1465$ rpm. An observation of the results given in Fig. 9.3 shows that the steady-state shaft speed and stator flux linkage amplitude are in reasonable agreement with the corresponding control input values. The steady-state shaft speed will be lower than the controller reference value given that the difference between the two speed settings is proportional to the (load dependent) slip frequency. Further details of the simulation are provided in tutorial (see Sect. 9.7.1).

Note that the controller in its present form has no provision for ensuring that the stator currents are kept within specified limits. In a practical drive this must be undertaken given that changes to the stator frequency reference can lead to

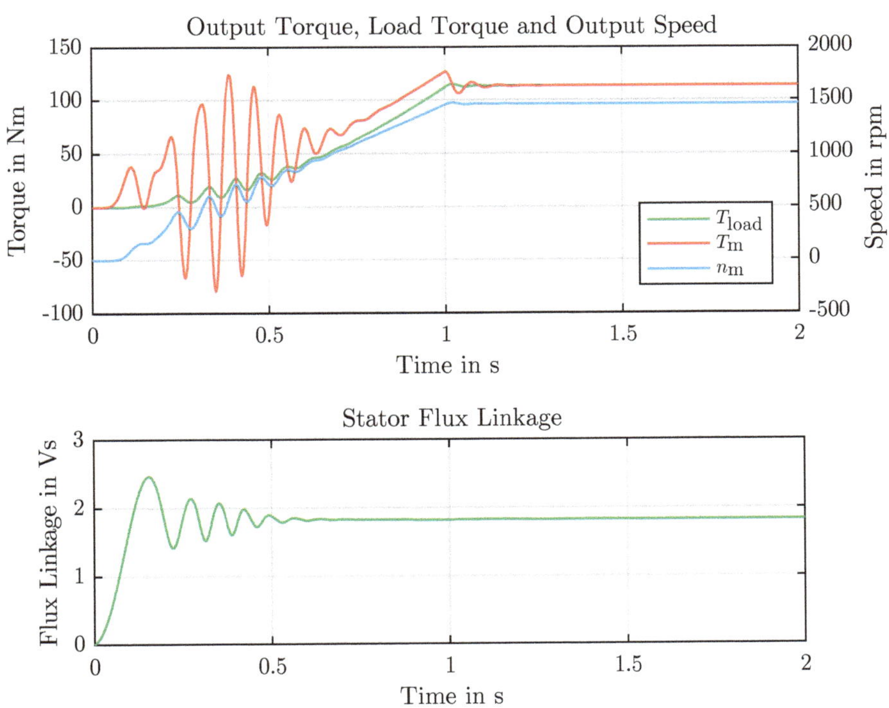

Fig. 9.3 22 kW induction machine with simplified V/f based drive controller

large current perturbations. This can be shown using the model given in Fig. 5.5. Observation of this model shows that a frequency ω_s step change will lead to a step change in u_{sq}, which in turn causes a change in current i_{sq} that is dictated among others by machine operating conditions, i.e., the shaft speed in use. To avoid large current changes, it is therefore prudent to limit the rate of change of the stator frequency reference input to the controller.

In the second part of the tutorial given in Sect. 9.7.1 an operating example is discussed where the stator frequency is ramped up with a constant rate of $d\omega_s^*/dt = 10\,\text{rad/s}^2$. This case is of interest because drives based on the V/f controller concept discussed in this section may exhibit instability [1, 9, 13] under certain conditions. To ascertain this phenomenon, a bipolar rectangular shaped load torque of 0.2 Nm amplitude and frequency of 4 Hz is introduced in the tutorial given in Sect. 9.7.2. A numerical example showing the behavior of the drive during a 20 s ramp up interval with the machine initially at standstill is given in Fig. 9.4. The result shown was undertaken with a rotor resistance which is 10% less than the nominal value calculated for this machine on the basis of name plate data (see Sect. 8.6.6), to better demonstrate this resonance phenomenon.

Observation of the results shows that drive instability occurs at shaft speed of approximately 580 rpm, which has severe effects on the stator currents and

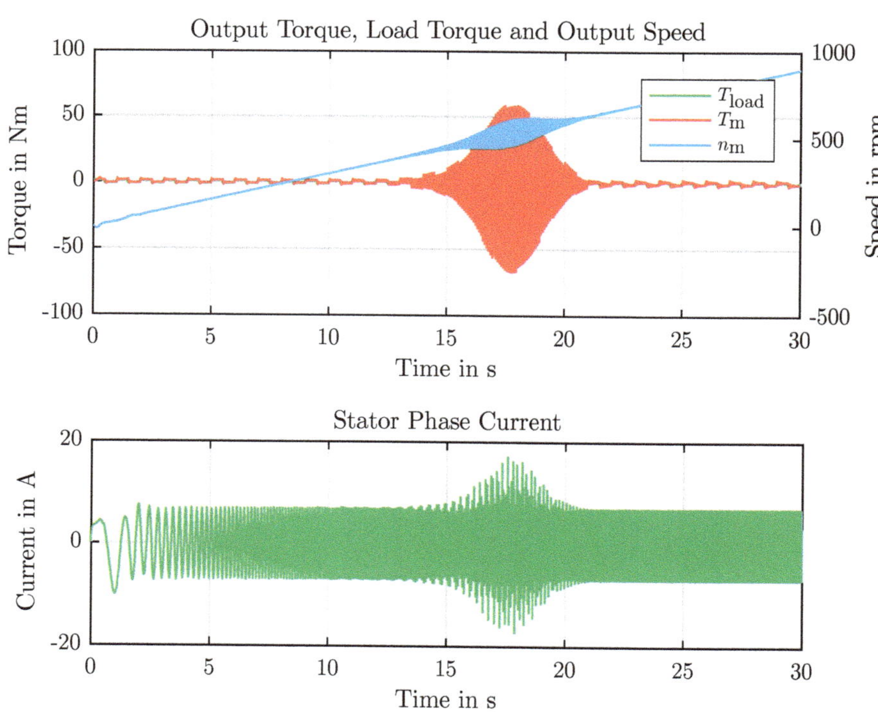

Fig. 9.4 22 kW induction machine with simplified V/f based drive controller exhibiting instability

torque. Note that the instability is substantially reduced when the machine operates with the nominal rotor resistance. Nevertheless, the results demonstrate that induction machines operated open loop at low and medium frequencies may exhibit instabilities. Manufacturers of V/f type drives usually deploy counter measures to alleviate operating speeds where instability can occur. It is also emphasized that the instability depends on a particular combination of machine parameters and operating conditions as is apparent from the example given in Fig. 9.4. Note that running the simulation without the load disturbance torque leads to a result where the instability effect is not triggered. In practice, however, torque harmonics produced by the (slotted stator winding) induction machine also trigger these instabilities.

9.1.2 V/f Torque Controller with Shaft Speed Sensor

For drives which have access to the estimated or measured shaft speed ω_{m} a V/f torque control can be implemented. It uses a torque reference T_{e}^{*} and flux linkage reference ψ_{s}^{*} as inputs. The derivation of the generic model of the V/f torque controller is similar to that used in the previous subsection. The required voltage amplitude $\hat{u}_{\mathrm{s}} = \sqrt{u_{\mathrm{sd}}^2 + u_{\mathrm{sq}}^2}$ is again derived using the stator flux oriented model discussed in Sect. 8.3.4.2. In this case, the stator frequency control input is replaced by the torque reference T_{e}^{*}. This implies that the stator frequency ω_{s}^{*} must now be calculated with the aid of the estimated or measured shaft speed ω_{m}. The stator flux ψ_{s}^{*} control input is unchanged and calculation of the direct axis voltage is undertaken with the aid of equation set (9.6). Computation of the stator frequency reference may be undertaken using $\omega_{\mathrm{s}}^{*} = \omega_{\mathrm{sl}}^{*} + \omega_{\mathrm{m}}$, where the slip frequency variable ω_{sl}^{*} is introduced. The latter may be found using Eq. (9.2c) in which the term containing the leakage inductance $L_{\sigma\mathrm{r}}$ is assumed to be small in comparison to the other two terms. Under this condition the slip frequency may be expressed as

$$\omega_{\mathrm{sl}}^{*} \simeq \frac{R_{\mathrm{R}}\, i_{\mathrm{sq}}^{*}}{\psi_{\mathrm{s}}^{*}}. \tag{9.7}$$

The quadrature current i_{sq} variable which appears in expression (9.7) may be found using Eq. (9.2e) with $\psi_{\mathrm{s}} = \psi_{\mathrm{s}}^{*}$ and $T_{\mathrm{e}} = T_{\mathrm{e}}^{*}$. Computation of the outstanding quadrature voltage variable u_{sq} is readily undertaken using Eq. (9.2b). The resulting generic model which complies with Eqs. (9.2) and (9.7) under the conditions given above is shown in Fig. 9.5. It includes a module which determines the magnitude of the vector $\vec{u}_{\mathrm{s}}^{dq}$ and a polar-to-Cartesian conversion module to calculate the reference stator voltage vector from the variable $\hat{u}_{\mathrm{s}}^{*}$ and ρ_{s}^{*}. The angle ρ_{s}^{*} is found through integration of the calculated stator frequency reference ω_{s}^{*}.

For shaft speeds $\omega_{\mathrm{m}} \gg \omega_{\mathrm{sl}}^{\mathrm{nom}}$, where $\omega_{\mathrm{sl}}^{\mathrm{nom}}$ represents the rated slip frequency, the voltage amplitude $\hat{u}_{\mathrm{s}}^{*}$ is largely determined by the back-EMF $e_{\mathrm{sq}}^{*} = \psi_{\mathrm{s}}^{*}\,\omega_{\mathrm{s}}^{*}$. At low stator frequencies the controller terms u_{sd}^{*} and $R_{\mathrm{s}}\, i_{\mathrm{sq}}^{*}$ become more prevalent

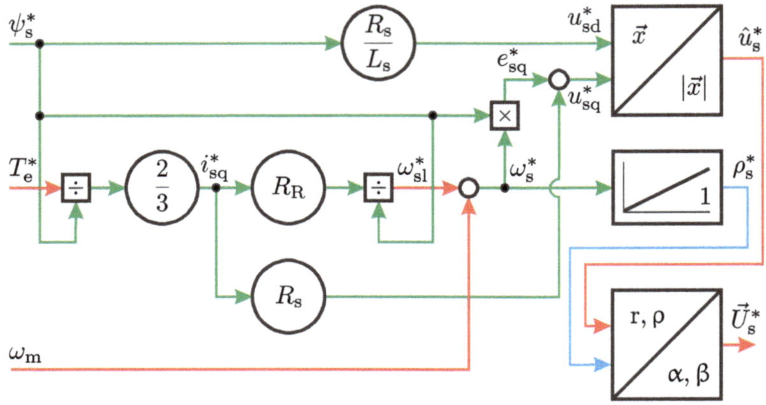

Fig. 9.5 V/f control structure with shaft speed input

in comparison to the term e^*_{sq} which causes the function $\hat{u}^*_{\text{s}}(\omega^*_{\text{s}})$ to deviate from the $e^*_{\text{sq}}(\omega^*_{\text{s}})$ characteristic in this operating region. The latter operating region is referred to as *low speed voltage boost* [8] and in effect compensates the voltage drop across the stator resistance.

In point A in Fig. 9.1, the machine operates with a shaft speed of ω_{m1} while delivering a torque $T_{\text{e}} = T_1$, which in turn is assumed to be equal to the reference torque T^*_{e1} of the control unit. The task of the controller is in this case reduced to determining (for a given value of the flux linkage reference ψ^*_{s}) the required slip frequency reference ω^*_{sl1}, which together with the measured shaft speed produces the required stator flux linkage frequency ω^*_{s1}. If a reference torque step ΔT^*_{e} is applied, the controller will determine the new reference slip frequency ω^*_{sl2} which leads to the revised stator frequency ω^*_{s2} as shown in Fig. 9.1. The increased machine (steady-state) shaft torque accelerates the machine until the new steady-state operating point B is reached. In this point the reference stator frequency is ω^*_{s3}.

9.1.2.1 Tutorial Results

In tutorial (see Sect. 9.7.2) an example of this type of drive is given which makes use of the 22 kW machine used earlier. An example of the results achieved with this simulation is given in Fig. 9.6, which shows the stator flux linkage amplitude $|\vec{\psi}_{\text{s}}|$, shaft speed, and shaft torque as function of time. The flux linkage reference was set to $\psi^*_{\text{s}} = 1.868\,\text{Wb}$, which is the rated stator flux linkage value for this machine. The reference shaft torque for the controller was set to $T^*_{\text{e}} = 113\,\text{Nm}$. An observation of the results given in Fig. 9.6 shows that the steady-state shaft torque and stator flux linkage amplitude are in reasonable agreement with the above-mentioned control input values. Further details of the simulation are provided in tutorial (see Sect. 9.7.2). Note that the V/f torque controller did not decouple torque

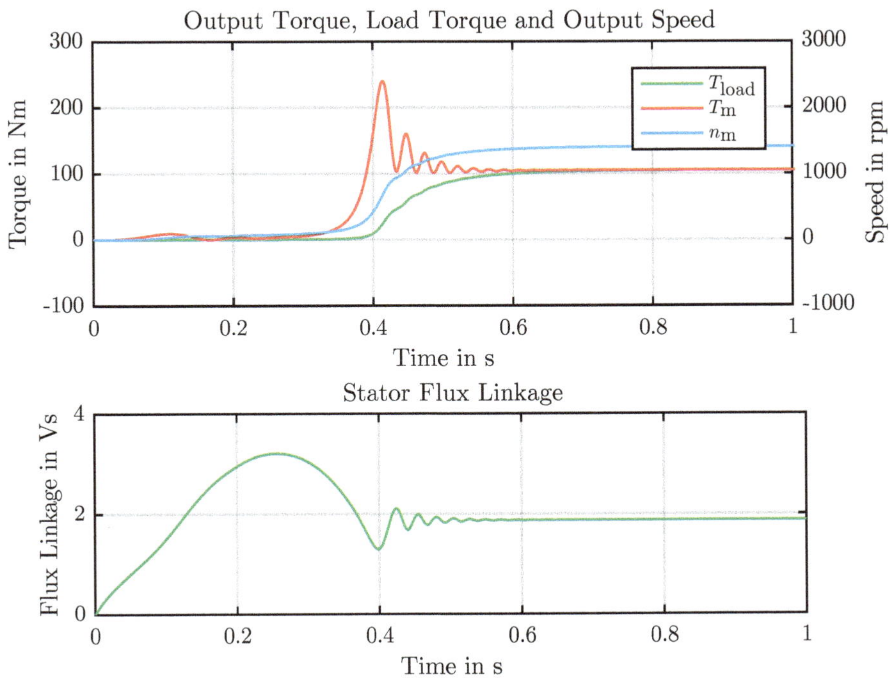

Fig. 9.6 Delta connected 22 kW machine with V/f based drive controller and shaft speed sensor

and flux linkage under transient conditions. Oscillations still occur as magnetic and mechanical energy are exchanged during transients.

9.2 Field-Oriented Control

The advent of affordable high-performance digital signal processors has enabled the introduction of field-oriented control for a plethora of applications involving the use induction machines.

Field-oriented control allows independent control of torque and flux linkage under transient conditions. To provide this decoupling, the design of the controller uses the model inversion approach introduced in the previous chapters for the DC and synchronous machine. The drive structure as given in Fig. 9.7 shows the presence of a vector control module which generates a set of reference signals for use with an induction machine operating under current control. The current control methods outlined in Chap. 3 are directly transferable to the theory and application examples to be discussed in this section.

First, attention is given to the underlying control principles which in turn lead to an in depth treatment of various vector control module implementations which make

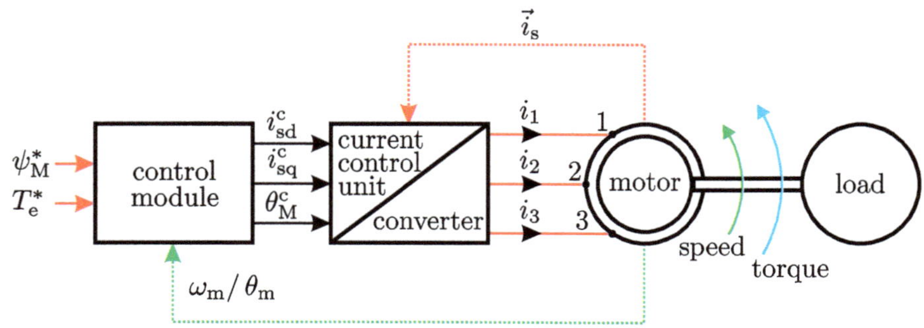

Fig. 9.7 Field-oriented control drive structure

use of the so-called *direct* or *indirect* field-oriented control of which the principles are developed by K. Hasse [7] and F. Blaschke [2]. In the sequel of this section the integration of the vector control module with the current-controlled induction machine is examined using the current control techniques hitherto presented in Chap. 3. Tutorial examples related to this subject matter are given at the end of this chapter.

9.2.1 *Controller Principle*

9.2.1.1 Torque Control

It is instructive to consider first the rotor flux oriented symbolic and generic model as given in Fig. 8.24. The direct and quadrature machine models play an important role in terms of understanding the basic principles by which to achieve decoupled magnetizing flux linkage and precise torque control. Observation of the direct axis model shows that control of the direct axis current i_{sd} provides direct control of the magnetizing flux linkage ψ_M. The time constant by which these flux linkage variations can occur is L_M/R_R. A constant i_{sd} value will therefore result in a constant rotor flux linkage magnitude and constant (zero) e_{sd} value. Torque control is governed by the quadrature stator current component i_{sq}, given that the latter can be varied fast (rate of rise only limited by the leakage inductances) without affecting the flux linkage ψ_M. For example, a torque step with constant magnetizing flux linkage ψ_M value is realized by changing the i_{sq} value in accordance with equation

$$i_{sq} = \frac{2}{3}\frac{T_e}{\psi_M}. \tag{9.8}$$

Hence, the required i_{sq} value corresponds to a specified torque and flux linkage value.

Observation of the quadrature model shown in Fig. 8.24 shows that this current change corresponds to a voltage change across the rotor resistance R_R. This voltage change must be equal to the product of the slip frequency $\omega_{sl} = \omega_s - \omega_m$ and magnetizing flux linkage ψ_M. Since flux linkage and shaft speed cannot change instantaneously, a step increase in the slip frequency is needed. The latter corresponds to an increase in the induced rotor voltage which in turn gives a larger rotor current i_{rq} equal to the new stator quadrature current. Hence, the only variable (besides i_{sq} and i_{rq}) which can and must be changed instantaneously is the stator frequency ω_s. Furthermore, the required stator frequency change must be precise as to ensure that the condition according to Eq. (9.9) is always satisfied.

$$(\omega_s - \omega_m)\, \psi_M = R_R\, i_{sq}. \tag{9.9}$$

This expression follows directly upon application of Kirchhoff's voltage law to the quadrature model shown in Fig. 8.5.

From the previous section it is clear that a current controller is required which is able to deliver the required direct and quadrature currents to the machine. These may in turn be used as inputs to a three-phase current-controlled converter as discussed in Chap. 3.

9.2.1.2 Calculation of Field Orientation (CFO)

As mentioned before, torque control can be achieved by controlling the direct and quadrature currents separately. It is obvious, that the controllers reference dq coordinate system must be aligned with the synchronous dq coordinate system formed by the vectors $\vec{\psi}_M$ and $\vec{e}_M = \mathrm{j}\,\omega_s\,\vec{\psi}_M$. Both systems are rotated by the angle θ_M and θ_M^c, respectively. There are two methods how the controller can determine the angle θ_M^c, namely:

- *Direct field-oriented control (DFO)* which estimates the angle θ_M^c by making use of the flux linkage vector $\vec{\psi}_M$ or voltage vector $\vec{e}_M$. This may be achieved by using sensors in the machine or an observer that uses measured electrical quantities at the terminals of the machine. This so-called *position sensorless* (better *encoderless*) control approach generally refers to operation without mechanical position sensors.
- *Indirect field-oriented control (IFO)* which makes use of mechanical sensors which measure the shaft angle Θ_m (position encoder) or integrate the measured shaft speed Ω_m (tachogenerator). The flux position $\vec{\psi}_M$ position is estimated by adding the slip angle (derived from the machine model) to this shaft angle.

There may however be a misalignment between the controller reference and motor reference coordinate systems as indicated in Fig. 9.8. This misalignment

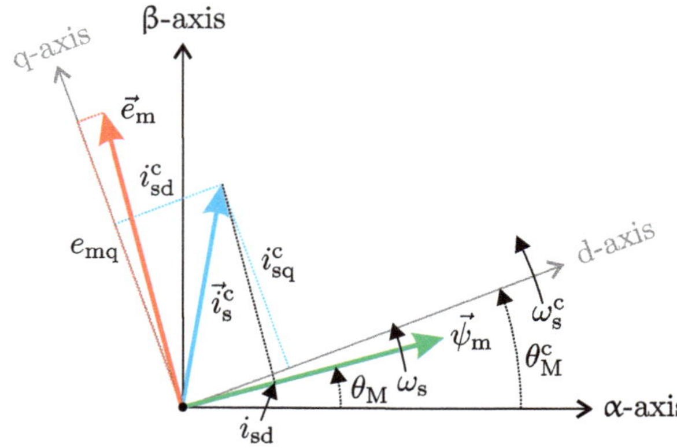

Fig. 9.8 dq-space vector diagram for motor and controller under detuned conditions

causes an error $\Delta\theta_M = \theta_M^c - \theta_M$ between the reference and actual current values which in turn will affect the torque and magnetizing flux linkage level of the machine. In the case shown, the machine will in steady state produce a lower magnetizing flux linkage level, namely $\psi_M = L_M i_{sd}$ than the required value $\psi_M^* = L_M i_{sd}^c$ given that $i_{sd} < i_{sd}^c$. In addition, the actual i_{sq} is slightly higher than the reference value, however, in this case, the torque of the machine will be lower than T_e^* given that the magnetization level of the machine is lower ($T_e = \frac{3}{2} i_{sq}\psi_M$). The coordinate system misalignment, or the so-called detuning, should be reduced to a minimum.

9.2.2 *Controller Structure*

In the previous section the basic controller principles were presented based on a model without leakage inductances. In this section, the leakage inductances are included. The basic *universal field-oriented* UFO model as discussed in Sect. 8.3.4 is used for the development of a generalized controller concept.

The concept is shown in Fig. 9.9. It consists of two basic modules, the *UFO control module*, which decouples torque and flux linkage, and the calculation of the flux orientation *CFO control module*.

The UFO module generates the required reference currents i_{sd}^c and i_{sq}^c which corresponds to the user defined torque reference T_e^* and flux linkage reference ψ_M^* inputs.

The CFO module calculates the d-axis grid reference angle θ_M^c for aligning the field-oriented control coordinate system and (in some applications) estimates the flux linkage magnitude ψ_M. Inputs to the CFO module are, among others, the measured stator voltages $\vec{u}_s$ and stator currents $\vec{i}_s$. In addition, the input vector $\vec{\psi}_{sense}$ is shown, which represents the flux linkage vector, which is measured with the aid of sensors located inside the machine. The remaining inputs ω_m and θ_m represent the measured angle or speed using position encoders or tachogenerators. The reader

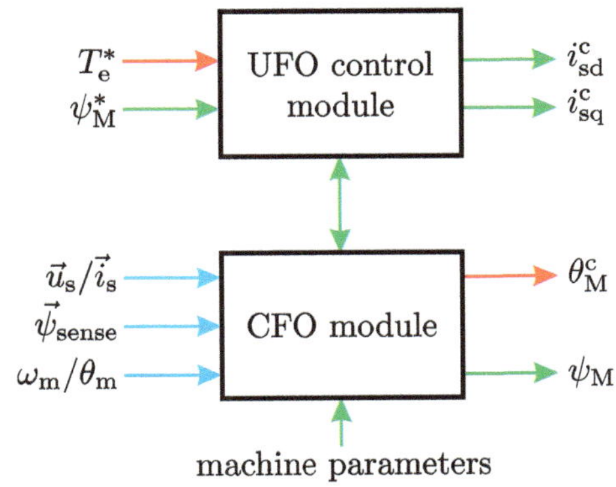

Fig. 9.9 UFO controller structure

should be keenly aware of the fact that not all input variables of the CFO are required at any one time. The choice of inputs used depends on the control strategy envisaged for a particular drive implementation.

It is noted that invariably a number of estimated machine parameters are required for the controller. The number and choice of parameters are again strongly influenced by the choice of (direct and indirect) field orientation. Last but not least, the required drive performance and the sensitivity of the latter to de-tuning effects caused by parameter variations, as a result of temperature variations, magnetic saturation, or the like within the machine will strongly influence the type of controller to be implemented.

9.2.3 UFO Module Structure

A generic representation of the UFO module can be developed by inverting the UFO based *dq* module given in Fig. 8.22 [4, 10]. The inverse model can be obtained by making use of Eqs. (8.25b) and (8.26), leading to the generic controller model given in Fig. 9.10.

A comparison between Figs. 9.10 and 8.22 shows the commonality between the two models. Also apparent from the controller generic diagram is the significance of the inductance parameter $L_{\sigma R}$. Setting this variable to zero, i.e., choosing rotor flux oriented control, will significantly reduce the overall controller complexity. This may be directly observed from Fig. 9.10, given that the *blue* signal paths will then become non-operative.

The diversity of overall controller implementations precludes the possibility of considering all feasible configurations of the modules shown in Fig. 9.9. Instead, a number of realistic controller configuration will be discussed in the following subsections, which includes both *direct* and *indirect* based field-oriented concepts. The merits of the various configurations, their performance, operating range, and

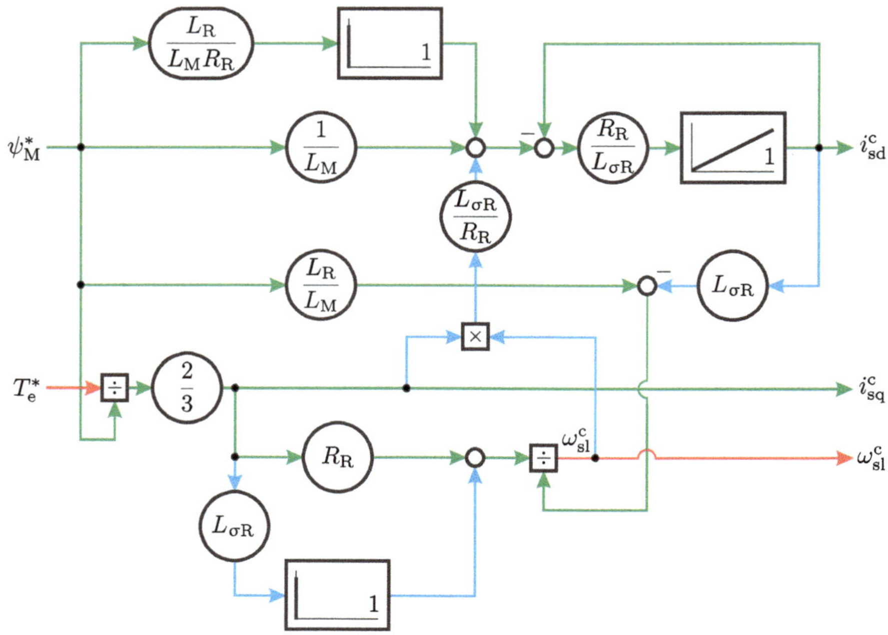

Fig. 9.10 UFO decoupling control structure

sensitivity to detuning will be presented. Tutorials at the end of this chapter will allow the reader to gain additional insight by way of simulations.

9.2.4 *IFO Using Measured Shaft Speed or Shaft Angle*

In fact, the implementation of a CFO module with indirect field orientation (IFO) requires the relative shaft position as an input. If shaft speed is measured, the following equation can be used:

$$\theta^c_M = \int \left(\omega^c_{sl} + \omega_m\right) \, dt \tag{9.10}$$

where ω_{sl} is the rotor slip frequency. This slip frequency can be approximated by the estimated value ω^c_{sl}, given by Eq. (8.26), which is already computed in the UFO decoupler, shown in Fig. 9.10. The implementation of the CFO module with measured shaft speed is given in Fig. 9.11.

Note that measurement and offset error in the shaft speed, in particular at low speeds, may significantly detune the controller, given the fact that typically $\omega_{sl} \ll \omega_m$.

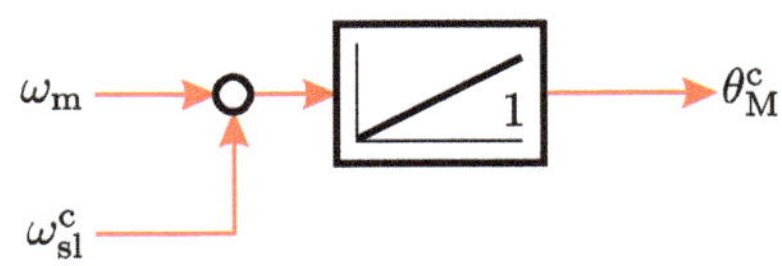

Fig. 9.11 CFO module: IFO using measured shaft speed

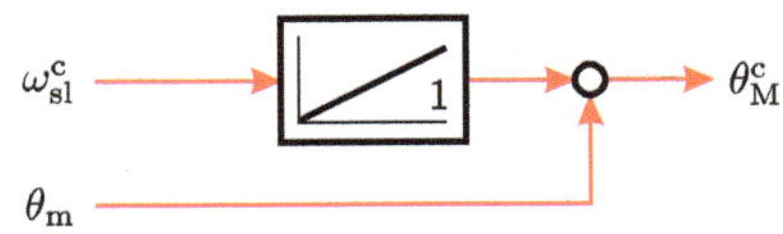

Fig. 9.12 CFO module: IFO with measured shaft angle

The IFO concept can be readily adapted when using an incremental shaft encoder instead of a shaft speed tachometer. Use of a digital encoder is not sensitive to offset errors and can be easily integrated in a digital controller. The CFO module for this approach makes use of the fact that the last term $\int \omega_{\rm m}\,{\rm d}t$ represents the shaft angle $\theta_{\rm m}$, which is in this case the input variable. The revised generic implementation of the CFO modules with measured shaft speed is shown in Fig. 9.12.

For both implementations with measured shaft speed and shaft position, the transformation variable a may be set to any value within the range of $L_{\rm m}/L_{\rm r} \leq a \leq L_{\rm s}/L_{\rm m}$. As can be seen from Fig. 9.10, a step in the quadrature current, which corresponds to a step in torque, leads to a step on the inputs of the differentiator. In practice, this is avoided by introducing a rate limiter for $i_{\rm sq}^{*}$. Otherwise a Dirac impulse would lead a step in the slip frequency $\omega_{\rm sl}$. In terms of parameters dependency, IFO requires accurate estimates for the parameters $L_{\rm M}$ and $R_{\rm R}$ as well as the leakage inductance $L_{\sigma{\rm R}}$. The latter may change under varying load conditions, depending on the type of squirrel-cage rotor in use. Also, the rotor resistance is particularly sensitive to temperature variations.

One disadvantage of the approaches is the need to add a sensor to the shaft, which reduces the overall robustness of the drive, i.e., addition of a signal cable between motor and controller. Furthermore, most "off-the-shelf" machines are provided with one drive shaft (used to connect the mechanical load), which further exacerbates the problem of fitting an encoder. On the other hand, the IFO controller with measured shaft angle can function properly at standstill (zero frequency) because no integrators are used to detect flux linkage position. As the flux reference vector can be chosen freely, most IFO controllers use the rotor flux as the reference vector ($a = L_{\rm m}/L_{\rm r}$). This leads to a simpler controller structure as all rotor side leakage inductances are set to zero. Note, however that the dependency on machine parameters remains. Nevertheless, the rotor flux indirect field-oriented control was widely used at the start of IFO controller developments, when the computational power of microprocessors was limited. Only with the avenue of digital signal processors (DSPs) the implementation of UFO algorithms became practical.

9.2.5 DFO with Air-Gap Flux Sensors

Direct field orientation (DFO) may be realized by measuring the flux density inside the air-gap by way of Hall effect sensors. Figure 9.13 shows the location of two sensors on the real and imaginary axis of a stator based complex plane [2].

A scalar-to-vector module is used to combine the two scalar flux linkage values $\psi_{\mathrm{m}\alpha}$ and $\psi_{\mathrm{m}\beta}$ into a single vector $\vec{\psi}_{\mathrm{m}} = \psi_{\mathrm{m}\alpha} + \mathrm{j}\psi_{\mathrm{m}\beta}$. The choice of reference frame for the field-oriented UFO controller is directly linked to the CFO module, which in this case must generate the required output vector $\vec{\psi}_{\mathrm{M}}$ on the basis of the measured vector $\vec{\psi}_{\mathrm{m}}$ and stator current vector $\vec{i}_{\mathrm{s}}$. The relationship between these vectors may be found with the aid of Eq. (8.11), which leads to

$$\vec{\psi}_{\mathrm{M}} = \vec{\psi}_{\mathrm{m}} - (L_{\sigma\mathrm{S}} - L_{\sigma\mathrm{s}})\,\vec{i}_{\mathrm{s}}. \tag{9.11}$$

Also shown in Fig. 9.13 are the generic modules linked with this equation and the Cartesian to Polar conversion module which generates the reference angle $\theta_{\mathrm{M}}^{\mathrm{c}}$. It is instructive to consider the implications of choosing a specific transformation ratio. For example, use of a transformation value of $a = 1$ in the CFO unit gives an output vector $\vec{\psi}_{\mathrm{M}} = \vec{\psi}_{\mathrm{m}}$ ($L_{\sigma\mathrm{s}} = L_{\sigma\mathrm{S}}$ in Eq. (9.11)). This implies that the stator leakage inductance $L_{\sigma\mathrm{s}}$ or stator current vector $\vec{i}_{\mathrm{s}}$ is not required for this unit,under

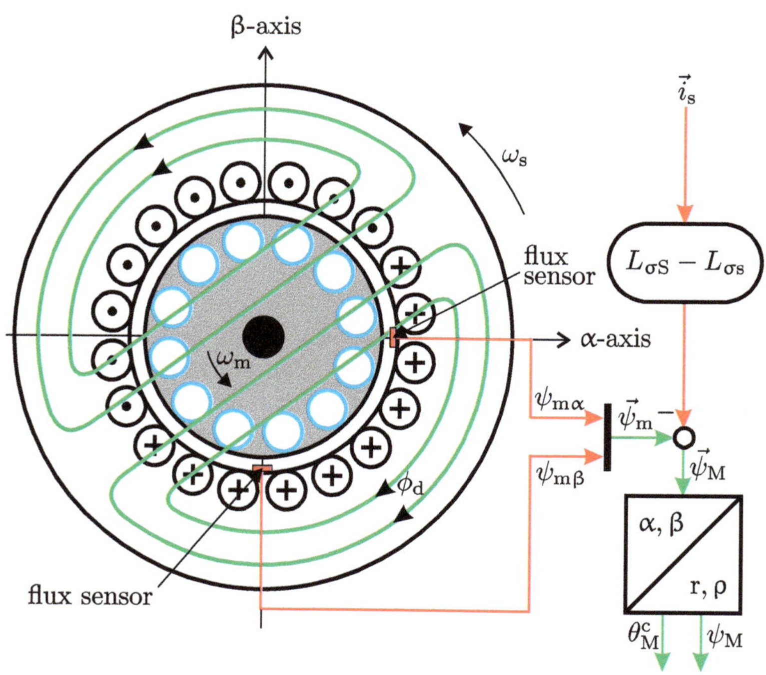

Fig. 9.13 CFO module: DFO with air-gap flux sensors

these circumstances. Hence, the reference angle calculation is devoid of parameter dependency. However, the UFO decoupling module, as shown in Fig. 9.10, does require the (non-zero) leakage inductance $L_{\sigma r}$ in the controller as it uses the air-gap flux oriented controller ($a = 1$). Hence, its complexity is increased when compared with a rotor flux oriented controller $L_{\sigma r} = 0$. Note that this flux observer also provides the magnitude of the air-gap flux. Feeding this signal back to a PI feedback regulator, the air-gap flux oriented controller would have zero steady-state error on the flux command. Ignoring sensor errors, it can be proven that DFO controllers with flux feedback have no steady-state error on torque [10, 11]. Hence, operating DFO in the flux reference frame which is provided directly by the flux sensors should be a must when (steady-state) detuning has to be avoided. Furthermore, controlling the air-gap flux linkage is however an attractive option given that the machine design is usually based on a specific air-gap flux density value. Consequently, maintaining this level under varying load conditions prevents saturation and makes optimum use of the machine.

A rotor flux oriented controller may also be used, in which case the CFO module (as defined by Eq. (9.11)) and UFO module must be used with a transformation variable of $a = L_m/L_r$. However, in this case, the observer would depend on machine model parameters, leading to flux and position errors. These errors cannot be eliminated in steady-state by a regulator, as they are in the feedback path of the regulator. Note in this context that any arbitrarily chosen reference frame may be used provided that the transformation factor is varied within its range in both modules.

Hall effect sensors are often used in laboratories for the air-gap flux density measurement. However, use of these air-gap sensors confines the user to a non-standard machine and severely limits the maximum temperature range of the drive (typically less than 120 °C). Hence, the issue of sensor reliability and drive robustness needs to be addressed when pursuing this control approach. Nevertheless, Hall sensors have been used to detect the flux position in low-power drives, such as drives for hard disks that require precise speed control.

9.2.6 DFO with Sensor Coils

Avoiding the temperature limitations of Hall sensors, sense coils, located in the stator windings on the stationary real and imaginary axis, as shown in Fig. 9.14, can be used to provide the stator flux linkage vector by integration of the induced voltage. Alternatively, taps on stator windings can provide the same information [5, 11].

The induced sensor voltages $e_{s\alpha}$ and $e_{s\beta}$ must be integrated to obtain the estimated stator flux linkage vector $\vec{\psi}_s$. This task is performed in the CFO module which also generates a reference vector $\vec{\psi}_M$ for the UFO controller and making use of Eq. (9.12).

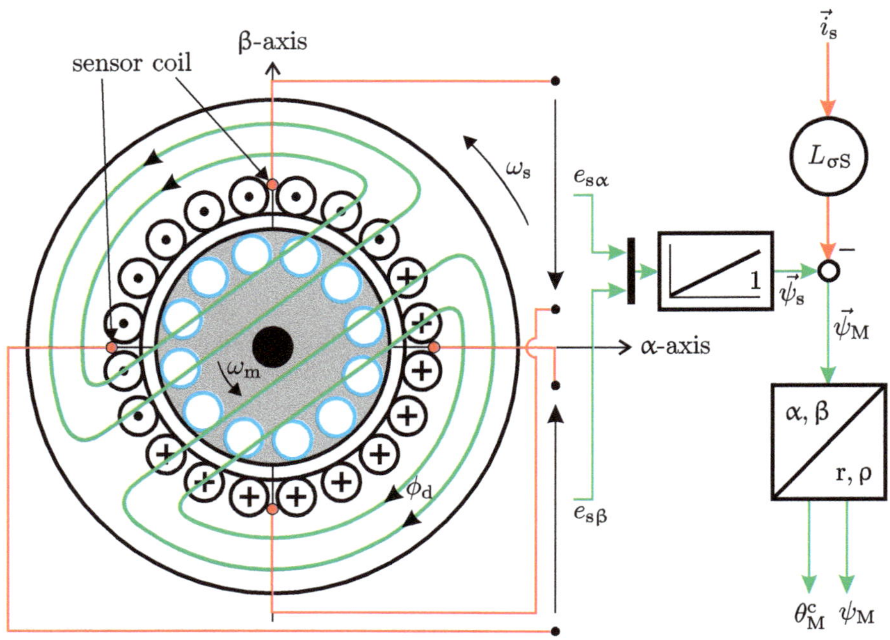

Fig. 9.14 CFO module: DFO with sensor coils

$$\vec{\psi}_M = \vec{\psi}_s - L_{\sigma S}\,\vec{i}_s \tag{9.12}$$

where $\vec{i}_s$ represents the measured current and $L_{\sigma S}$ is the estimated leakage inductance as defined by Eq. (8.12a). The latter is a function of the transformation variable a, which must be given the same value for both UFO and CFO modules. Setting this value to $a = {}^{L_s}/_{L_m}$ corresponds to a value of $L_{\sigma s} = 0$, hence $\vec{\psi}_M = \vec{\psi}_s$ (see Eq. (9.12)) which is beneficial in this case because it avoids the need to measure the stator currents. This type of approach requires the insertion of sensor coils in the machine, which precludes its use for standard "off-the-shelf" machines. Furthermore, the need to integrate the sensor voltages makes this method unsuitable for applications where the frequency of the rotating magnetic field becomes zero.

9.2.7 *DFO with Voltage and Current Transducers*

An alternative approach to deriving the required flux linkage vector $\vec{\psi}_M$ is to consider the machine from the stator terminal side [6, 14, 15]. This approach makes use of the three measured inverter voltages and Eq. (8.16a) which may be rewritten as

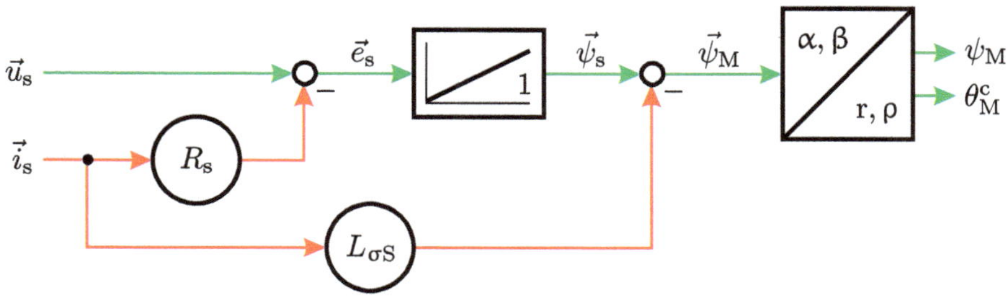

Fig. 9.15 CFO module: DFO with voltage and current transducers

$$\frac{d\hat{\vec{\psi}}_s}{dt} = \underbrace{\vec{u}_s - R_s \vec{i}_s}_{\vec{e}_s} \tag{9.13}$$

in which $\hat{\vec{\psi}}_s$ represents the estimated stator flux linkage vector. Equation (9.13) shows that this method requires access to the stator voltage vector $\vec{u}_s$ and requires knowledge of the stator resistance and the current vector $\vec{i}_s$ as derived by measurement of the stator currents. The *universal* flux linkage vector $\vec{\psi}_M$ may be found using Eq. (9.12) once integration of the *EMF* vector $\vec{e}_s$ has been undertaken. The observer structure (in space vector form) as determined by Eqs. (9.13) and (9.12) is given in Fig. 9.15.

Practical implementation of the DFO observer structure requires access to measured signals on the converter output side which must be linked to the controller module. This may lead to electrical noise problems between these two modules. In addition, an analog or digital integrator with high oversampling rate must be used to directly process the inverter PWM voltage waveforms. Note that synchronized sampling of the stator voltage, synchronized to the PWM modulator, is not feasible given that at the sampling time intervals the zero vector is usually active. The remaining part of the observer (after the integrator) may be implemented either in discrete time or via an analog circuit. However, it is important to note that the observer makes use of the parameter R_s as mentioned above. This parameter is (among others) subject to temperature variations in the motor. This will lead to observer errors which are of particular concern at lower speeds where the converter voltage magnitude approaches the magnitude of the term $R_s \vec{i}_s$. This implies that observers of this type may need to estimate the stator resistance variation by way of measuring or estimating the stator winding temperature [12]. In addition, similar to DFO with stator sensor coils, this method does not function in applications with zero stator frequency.

9.2.8 DFO with Current and Shaft Speed Transducers

This current based observer requires access to the measured currents and the measured or estimated shaft speed. The observer structure is based on equation set (8.19) which together with the IRTF module gives the generic observer structure shown in Fig. 9.16.

A complex representation of the observer structure as shown in Fig. 9.16 highlights the presence of the rotor resistance parameter which is subject to large variations as a result of temperature changes in the rotor.

The observer structure is identical to that used on the rotor side of the IRTF rotor flux based generic motor model shown in Fig. 8.15. Consequently, differences between the observer and actual motor rotor resistance R_{R}, variations in L_{M} due to saturation effects as well as errors between measured and actual shaft speed may affect the estimated *universal* flux linkage vector $\vec{\psi}_{\mathrm{M}}$, in particular at low slip frequencies where the integrator gain will be high. The major advantage of this observer is that it avoids the need to measure the converter voltages. Its use is particularly advantageous in the low shaft speed region, high slip area (low stator frequencies), where errors in the shaft speed or rotor resistance will not contribute significantly to an error in the flux linkage estimate.

Computation of the required *universal* flux linkage vector $\vec{\psi}_{\mathrm{M}}$ may be undertaken with the aid of equation set (8.16) which leads to the following expression:

$$\vec{\psi}_{\mathrm{M}} = \vec{\psi}_{\mathrm{R}} + \underbrace{\left(L_{\sigma\mathrm{S}}(a = \frac{L_{\mathrm{m}}}{L_{\mathrm{r}}}) - L_{\sigma\mathrm{S}}(a) \right)}_{\Delta L} \vec{i}_{\mathrm{s}} \tag{9.14}$$

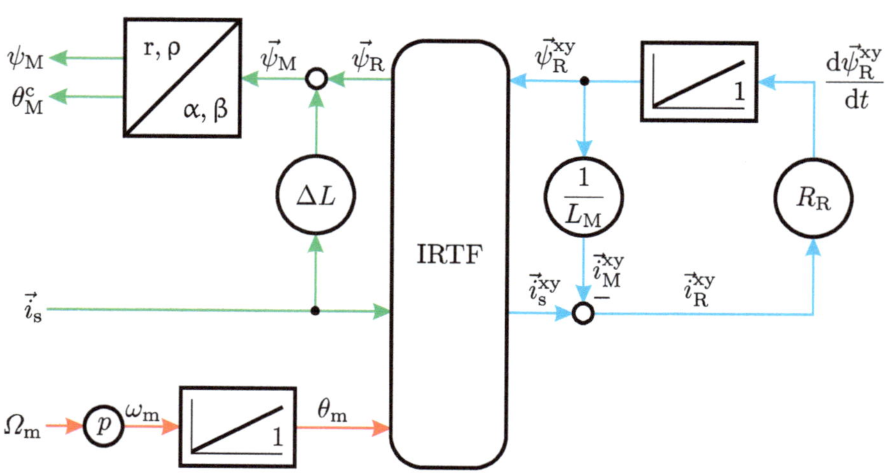

Fig. 9.16 CFO module: DFO with current and shaft speed transducers

in which the gain $\Delta L = (L_{\sigma S}(a = L_m/L_r) - L_{\sigma S}(a))$ is introduced. With the addition of this gain module (see Fig. 9.16) the user is able to select the required transformation value a which corresponds to the flux linkage vector $\vec{\psi}_M$ of the observer.

In some cases, the observer in question is used in conjunction with the previous voltage/current based observer. The advantage of this approach is that at low shaft speeds use can be made of the current based observer, whereas at higher speeds a voltage based observer is used. This approach uses each observer in the speed range for which it is most suited. Care must be taken to ensure a seamless transition from current/shaft speed to voltage/current observer and vice versa.

9.3 Operational Drive Boundaries for Rotor Flux Oriented Control

The approach used to visualize the control laws and operating limits imposed by the maximum voltage and current values for the synchronous and DC drives may be readily extended to the field-oriented current-controlled induction machine drive. As with the previous cases a synchronous $\underline{dq}$ complex plane is used of which the direct axis d is tied to the flux linkage vector $\vec{\psi}_R$, given that rotor flux oriented control is considered here. This approach was widely used first due to the simplicity of the controller structure, which is achieved as a result of a natural (simple) decoupling of the direct and quadrature currents (see Fig. 9.10).

9.3.1 *Steady-State Equations*

The machine equation set tied to the direct and quadrature model shown in Fig. 8.24 is reconsidered. The operational drives limitations are derived for quasi-steady-state, which leads to

$$\vec{u}_s^{dq} = R_s \vec{i}_s^{dq} + j\,\omega_s\,\vec{\psi}_s^{dq} \tag{9.15a}$$

$$\vec{\psi}_s^{dq} = \psi_R + L_{\sigma S}\,\vec{i}_s^{dq} \tag{9.15b}$$

$$\vec{T}_e = \frac{3}{2}\psi_R \times \vec{i}_s^{dq} = \frac{3}{2}\psi_R\, i_{sq} \tag{9.15c}$$

for the voltage, flux linkage, and torque equation, respectively. For quasi-steady-state operation, the flux linkage ψ_R is given as $\psi_R = L_M\, i_{sd}$, which underlines the fact that the direct axis current defines the magnetization in the machine. This is decidedly different when compared to the DC and synchronous machines drives discussed in the previous chapters.

9.3.2 General Machine Operation Below and Above Base Speed

In the *base speed range* (which is yet to be defined) the rotor flux linkage is normally held at its rated value henceforth defined as ψ_R^{max}, which in turn corresponds to a maximum direct axis current value $i_{sd}^{max} = \psi_R^{max}/L_M$. Maintaining maximum flux linkage in the machine is beneficial because it gives the user the flexibility to set the torque between zero and maximum value without the need to adjust the flux linkage level ψ_R, which is governed by the large time constant L_M/R_R (as can be seen in Fig. 8.24). If the flux linkage is held constant, any torque step will only be dictated by the time needed to change the quadrature current i_{sq} (only limited by voltage across $L_{\sigma S}$).

For operation above the base speed the voltage limit is reached. The flux linkage ψ_R needs to be changed and the *field weakening* region is entered. This allows to extend the operating speed of the drive at the cost of reduced dynamic performance.

9.3.3 Lines of Constant Torque

It is helpful to rewrite the torque as

$$T_e = \frac{3}{2} L_M \, i_{sd} \, i_{sq} \tag{9.16}$$

with $\psi_R = L_M \, i_{sd}$. Contours of constant torque according to Eq. (9.16) are given in the operational drive limits diagram for induction machines shown in Fig. 9.17.

9.3.4 Current Limit (Maximum Ampere, MA)

This diagram also shows the maximum current MA circle with radius i_s^{max} introduced earlier. The region within the circle represents the operating region of the drive in terms of the maximum current constraints imposed by the machine, the main supply, or by power electronic converters

$$|\vec{i}_s^{dq}| \leq i_s^{max}. \tag{9.17}$$

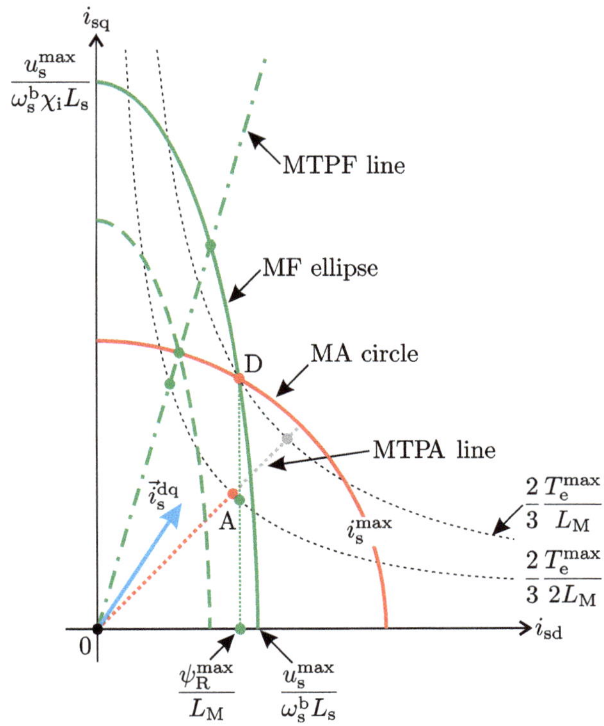

Fig. 9.17 Operational drive limits for rotor flux oriented induction machine drive

9.3.5 Maximum Rotor Flux Constraint

For the induction machine drive only part of the MA region is accessible given the maximum rotor flux constraint imposed by the machine, which is shown in Fig. 9.17 by way of the vertical line that passes through operating point O and D where i_{sd} equals its maximum value i_{sd}^{max} with

$$i_{sd}^{max} = \frac{\psi_R^{max}}{L_M}. \tag{9.18}$$

9.3.6 Voltage Limit (Maximum Flux Linkage, MF)

Figure 9.17 also shows a set of elliptical contours which are tied to the maximum voltage constraint u_s^{max}. The introduction of these maximum flux linkage (MF) contours is based on the use of Eq. (9.15a) and the underlying assumption that the dominant term in this expression is formed by the induced voltage. With this assumption in mind, the stator flux linkage variable ψ_s^{max} is introduced (as was also done for previous drive topologies), which according to Eq. (9.15a) with $|\vec{u}_s^{dq}| = u_s^{max}$ and $\vec{u}_s^{dq} \cong j\,\omega_s\,\vec{\psi}_s^{dq}$, can be written as

$$\psi_s^{max} \cong \frac{u_s^{max}}{\omega_s}. \tag{9.19}$$

From an operational perspective the variable ψ_s^{max} represents the maximum stator flux linkage value that may be realized by the converter for a given speed ω_s and voltage u_s^{max}. Use of $|\vec{\psi}_s^{dq}| = \psi_s^{max}$ with Eq. (9.15b) and equating the real and imaginary components of $\vec{i}_s^{dq}$ gives

$$i_{sd}^2 + \left(\chi_i\, i_{sq}\right)^2 = \left(\frac{\psi_s^{max}}{L_s}\right)^2 \tag{9.20}$$

where

$$\chi_i = \frac{L_{\sigma S}}{L_s} \tag{9.21}$$

represents the ratio between leakage inductance $L_{\sigma S}$ and stator inductance. Observation of Eq. (9.20) shows that this expression is indeed an ellipse with its origin at coordinates (0, 0). It intersects the direct and quadrature axis at $\pm u_s^{max}/\omega_s L_s$ and $\pm u_s^{max}/\chi_i \omega_s L_s$, respectively, as may be observed from Fig. 9.17. Note that the value of χ_i used in Fig. 9.17 was purposely chosen high for didactic reasons, i.e., to better illustrate the various operating trajectories. The elliptical MF contours will change with frequency ω_s and stator voltage (as was the case with previous drive topologies).

9.3.7 *Drive Saturation Point and Base Speed*

A base operating point referred to as the *drive saturation point* or *corner point* may be identified. The base speed of the drive corresponds with the ellipse that intersects the maximum current MA circle and maximum rotor flux linkage (MF) lines, i.e., at point D. The base speed may with the aid of Eqs. (9.20) and (9.19) and $i_{sd}^2 + i_{sq}^2 = (i_s^{max})^2$ be written as

$$\omega_{\mathrm{s}}^{\mathrm{b}} = \left(\frac{u_{\mathrm{s}}^{\max}}{L_{\mathrm{s}}\, i_{\mathrm{s}}^{\max}}\right) \frac{1}{\sqrt{\chi_{\mathrm{i}}^{2} + \kappa_{\mathrm{i}}^{2}\left(1 - \chi_{\mathrm{i}}^{2}\right)}} \tag{9.22}$$

where the variable κ_{i} is introduced, which is defined as

$$\kappa_{\mathrm{i}} = \frac{\psi_{\mathrm{R}}^{\max}}{L_{\mathrm{M}} i_{\mathrm{s}}^{\max}}. \tag{9.23}$$

9.3.8 Maximum Torque per Flux Linkage (MTPF) Line

The *maximum torque per flux linkage* MTPF line for this type of machine may be found by considering the torque contours according to Eq. (9.16) and the elliptical maximum flux linkage (MF) contours (see Eq. (9.20)). For a given torque T_{e}, which corresponds to a hyperbola in Fig. 9.17, an operating frequency will occur where the ellipse will no longer intersect said curve but merely touches the T_{e} hyperbola in a single point. The direct and quadrature currents for this operation along the MTPF line are represented by the condition

$$i_{\mathrm{sd}} = \frac{u_{\mathrm{s}}^{\max}}{\sqrt{2}\omega_{\mathrm{s}}\, L_{\mathrm{s}}} \tag{9.24a}$$

$$i_{\mathrm{sq}} = \frac{i_{\mathrm{sd}}}{\chi_{\mathrm{i}}} \tag{9.24b}$$

which may be derived by use of Eq. (9.16) and differentiation of the latter with respect to the torque after substitution of Eq. (9.20).

Operation along the MTPF line signifies drive operation above the base speed with the highest possible torque level for a given stator flux linkage value. In a practical sense, operation along this line above the base speed is restricted by the need to adhere to the maximum current constraint. This implies that operation along the MTPF line is possible for that part of the trajectory which is within the maximum current MA circle.

9.3.9 *Maximum Torque per Ampere (MTPA) Line*

The *maximum torque per ampere* MTPA line as discussed for previous drive topologies is not shown in Fig. 9.17. This line which may be found by use of Eq. (9.16) and differentiation of the latter with respect to the torque after substitution of the expression $i_{sd}^2 + i_{sq}^2 = (i_s^{max})^2$ (which represents the MA circle) represents the control law

$$i_{sd} = 0 \rightarrow \frac{\psi_R^{max}}{L_M} \tag{9.25a}$$

$$i_{sq} = \pm i_{sd} \tag{9.25b}$$

which signifies an operating trajectory where the highest torque is obtained with the lowest stator current value. For induction machines, operation under MTPA conditions is normally not done, given the need to simultaneously vary the flux linkage level ψ_R and quadrature current, which significantly degrades the dynamic performance of the drive as mentioned earlier. Furthermore, the maximum torque achievable under MTPA conditions is constrained by the maximum allowable direct axis current value i_{sd}^{max}.

9.4 Field Weakening for Rotor Flux Oriented IM Drives

Field weakening for field-oriented induction machine drives is readily achieved by reducing the direct axis current component i_{sd} and manipulating the quadrature current i_{sq} according to a chosen control strategy. The current locus and corresponding dq current versus speed characteristics will be examined in this section, which assumes a rotor flux oriented control structure. The current space vector locus as given in Fig. 9.18 shows three specific operating trajectories which exemplify three different operating regions, namely base speed range, basic field weakening, and maximum torque per flux linkage operation.

9.4.1 *d-Axis Current in Base Speed Range, $0 \leq \omega_s \leq \omega_s^b$ ($\psi_R^{max}/L_M \rightarrow$ A)*

Below the base speed ω_s^b, the d-axis current i_{sd} is set to its rated value, i.e.,

$$i_{sd} = i_{sd}^R = \frac{\psi_R^{max}}{L_M}. \tag{9.26}$$

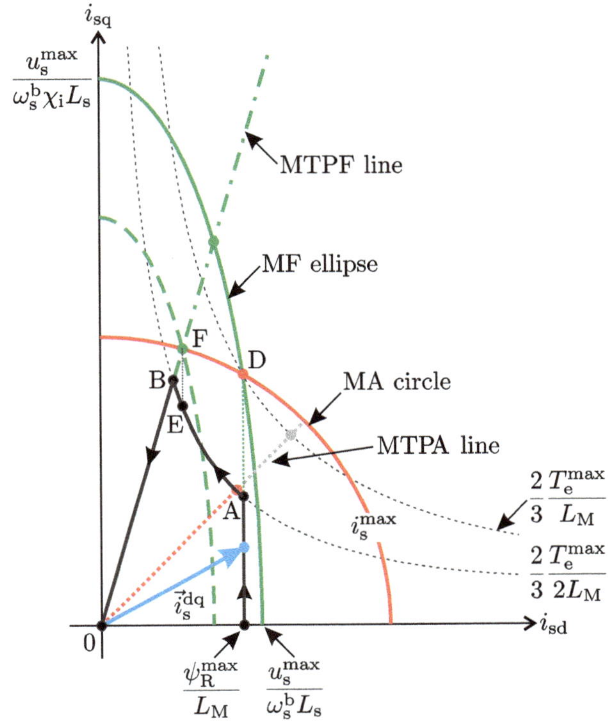

Fig. 9.18 Current locus diagram: Operational drive limits for rotor flux oriented induction machine drive

Note that in point A the torque reaches 50% of its maximum base speed value, which is assumed here as an example.

9.4.2 *d*-Axis Current in Basic Field Weakening, $\omega_s^b \leq \omega_s \leq \omega_s^B$ (A → B)

As speed increases, the maximum flux linkage ellipse decreases in size. The base speed ω_s^b is reached when the ellipse intersects the MA circle at point D. The available torque then quickly approaches zero if i_{sd} is kept constant at its rated value and when the speed is increased further. This occurs at speed ω_s^l when the MF ellipses intersect with the d-axis at $i_s = i_{sd} = \psi_R^{max}/L_M$. With the aid of Eqs. (9.20) and (9.23) ω_s^l is derived as

$$\omega_s^l = \left(\frac{u_s^{max}}{L_s\, i_s^{max}}\right)\frac{1}{\kappa_i}. \tag{9.27}$$

ω_s^l is approximately equal to the base speed ω_s^b, given that the condition $\chi_i \ll \kappa_i$ normally holds. Hence, for operation above base speed, field weakening is applied by reducing the direct current i_{sd}. Strongest field weakening is required where the MA circle intersects the MF ellipse. For the same reason as above, i_{sd} is used for all torque values at one particular speed. To derive the expression for i_{sd} in field weakening, the MA expression $(i_s^{max})^2 = (i_{sd})^2 + (i_{sq})^2$ is used for i_{sq} in the MF equation (9.20). Solving for i_{sd} leads to the expression

$$i_{sd} = \sqrt{\frac{\left(\frac{\psi_s^{max}}{L_s}\right)^2 - \chi_i^2\,(i_s^{max})^2}{1-\chi_i^2}}. \tag{9.28}$$

In normalized form:

$$i_{sd}^n = \sqrt{\frac{\left(i_o^b \frac{\omega_s^b}{\omega_s}\right)^2 - \chi_i^2}{1-\chi_i^2}} \tag{9.29}$$

with $i_o^b = u_s^{max}/\omega_s^b\, L_s\, i_s^{max}$.

Some simplification of expression (9.29) is possible by taking into account that the condition $\chi_i \ll 1$ usually holds (which gives an elongated ellipse along the major axis) in which case the above expression may be reduced to

$$i_{sd}^n \simeq i_o^b \left(\frac{\omega_s^b}{\omega_s}\right). \tag{9.30}$$

9.4.3 d-Axis Current along MTPF Line, $\omega_s^B \leq \omega_s$ (B → (0, 0))

The d-axis current i_{sd} needs to remain greater than the direct current which corresponds to operation along the MTPF line. Therefore, eventually the torque command dependent speed ω_s^B (compare Fig. 9.19) is reached. Above this speed, operation occurs along the MTPF line with i_{sd} being defined by Eq. (9.24a). The maximum available torque decreases accordingly.

9.4.4 q-Axis Current

With the analysis undertaken thus far the direct axis current is fully defined. What remains to be defined is the quadrature current versus speed relationship. A constraint is that the dq current stays within the envelope formed by the maximum current circle and MTPF line as may be observed from current locus diagram in Fig. 9.18.

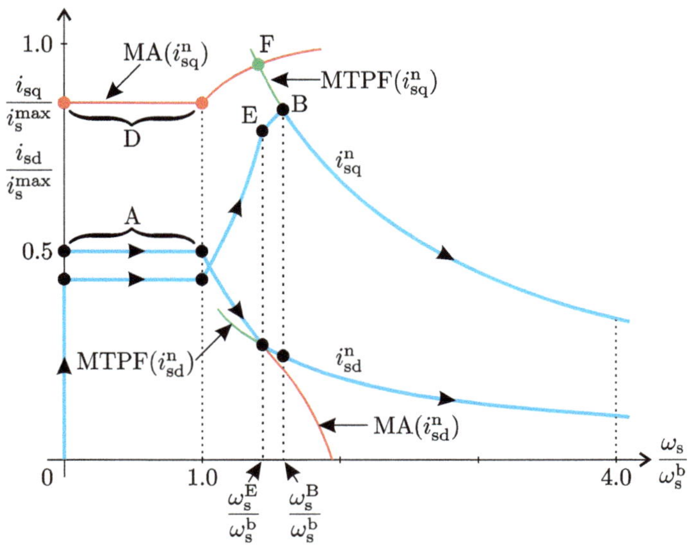

Fig. 9.19 Current-speed diagram for rotor flux oriented induction machine drive

One option is to maintain the user defined torque level beyond the base speed (a control directive also pursued for the synchronous drive) in which case the current is defined as

$$i_{sq} = \frac{2}{3} \frac{T_e}{L_M \, i_{sd}} \tag{9.31}$$

where i_{sd} defined by Eq. (9.26) for the base speed range, Eq. (9.28) during basic field weakening, and Eq. (9.24a) along the MTPF trajectory.

9.4.5 Discussion of Trajectories

In the following, the above derived rules will be briefly discussed on the trajectories in the current locus diagram in Fig. 9.18 and the current-speed diagram in Fig. 9.19.

The first trajectory, from point $\psi_R^{max}/L_M \rightarrow$ A, corresponds to a 50% maximum torque step applied when the machine is at standstill. This operation trajectory is, as required, on the rated i_{sd}^{R} line and occurs in the basic speed range.

The trajectory A to B lies in the basic field weakening range. The torque is kept constant, while the dq current is altered following the voltage limit. Figures 9.18 and 9.19 show that limited operation under constant torque conditions is feasible in the field weakening mode with the chosen control strategy.

At point B the constant torque can no longer be maintained and decreases while the dq current follows the MTPF trajectory.

This statement is underlined by the tutorial given in Sect. 9.7.5 which examines field weakening according to the control principles set out above. The controller in question is used in conjunction with a current source based, induction machine model operating with rotor flux oriented field control.

9.4.6 *Alternative Field Weakening Strategy*

An often used alternative field weakening strategy follows the approach taken for the separately excited machine, where the armature current i_{a} is held constant for an extended period after the base operating speed has been reached (see Sect. 5.2.2). For the IM drive this condition implies maintaining i_{sq} constant for operation in excess of $\omega_{\mathrm{s}}^{\mathrm{b}}$, in which case

$$i_{\mathrm{sd}}^{\mathrm{n}} \simeq i_{\mathrm{o}}^{\mathrm{b}} \left(\frac{\omega_{\mathrm{s}}^{\mathrm{b}}}{\omega_{\mathrm{s}}}\right) \tag{9.32}$$

according to Eq. (9.30) and

$$i_{\mathrm{sq}}^{\mathrm{n}} = \text{constant}. \tag{9.33}$$

The current values according to Eqs. (9.32) and (9.33) are maintained in this so-called constant power mode of operation provided these variables are within the currents which correspond to the MTPF envelope (see equation set (9.24) and Fig. 9.19).

9.5 Interfacing the Field-Oriented Controller with a Current-Controlled Induction Machine

Prior to discussing field-oriented control with a voltage source based machine it is instructive to consider the use of a current based IRTF model. This approach is helpful because it allows the user to examine the interaction between vector control module and induction machine without the need to consider the vagaries of the current controller. The proposed drive concept as given in Fig. 9.20 shows the vector control module as represented by the UFO control module (see Fig. 9.10) and the CFO module. The latter may, as was discussed in the previous sections, use a range of inputs for determining the instantaneous angle θ_{M}^{*} of the flux linkage vector $\vec{\psi}_{\mathrm{M}}$ of the machine. For the sake of simplicity an IFO approach has been arbitrarily selected which makes use of the measured rotor angle θ of the machine (see Sect. 9.2.4), i.e., use is made of a encoder connected to the motor shaft.

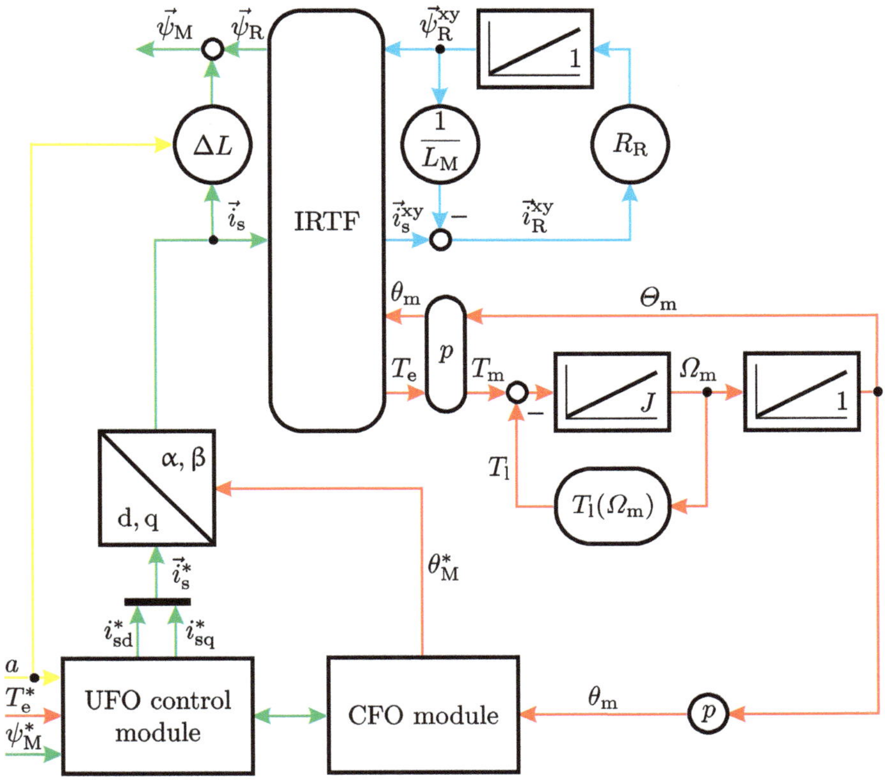

Fig. 9.20 Field-oriented control with current source induction machine model

For the representation of the machine the rotor flux based model as discussed in Sect. 8.3.2.1 is used. However, in this case the model is used with the stator current $\vec{i}_s$ as an input vector instead of the voltage vector $\vec{u}_s$. Observation of equation set (8.19) shows that the corresponding generic model is reduced to the structure given in Fig. 9.20. Note that any of the models given in Chap. 8 could equally have been used to represent the machine. The machine model shown in Fig. 9.20 makes use of a gain module ΔL, as introduced earlier (see Eq. (9.14)), which is used in conjunction with the current $\vec{i}_s$ and flux linkage vector $\vec{\psi}_R$ to generate the flux linkage vector $\vec{\psi}_M$. With the introduction of a gain module ΔL the user is able to examine the impact of changing the transformation variable a on the flux linkage vector $\vec{\psi}_M$ in the machine. More specifically, the amplitude of the flux linkage vector $\vec{\psi}_M$ should match the flux linkage reference ψ_M^* for any chosen value of a within the range $L_m/L_r \leq a \leq L_s/L_m$. Furthermore, the instantaneous torque T_e generated by the machine should match the reference torque T_e^* used by the UFO controller.

9.5.1 Tutorial Results

The tutorial given in Sect. 9.7.3 is a direct embodiment of the drive concept shown in Fig. 9.20. The 22 kW delta connected machine presented earlier in this chapter (see Fig. 9.6) is used again with the flux linkage reference ψ_{M}^{*} is set to 1.83 Wb. A rated shaft torque step from $T_{\mathrm{e}}^{*} = 0 \rightarrow 113\,\mathrm{Nm}$ is applied at $t = 0.5\,\mathrm{s}$ and the torque reference is reversed at $t = 1.5\,\mathrm{s}$. No external load is connected to the machine. The results given in Fig. 9.21 show the shaft torque, shaft speed, and flux linkage ψ_{M},

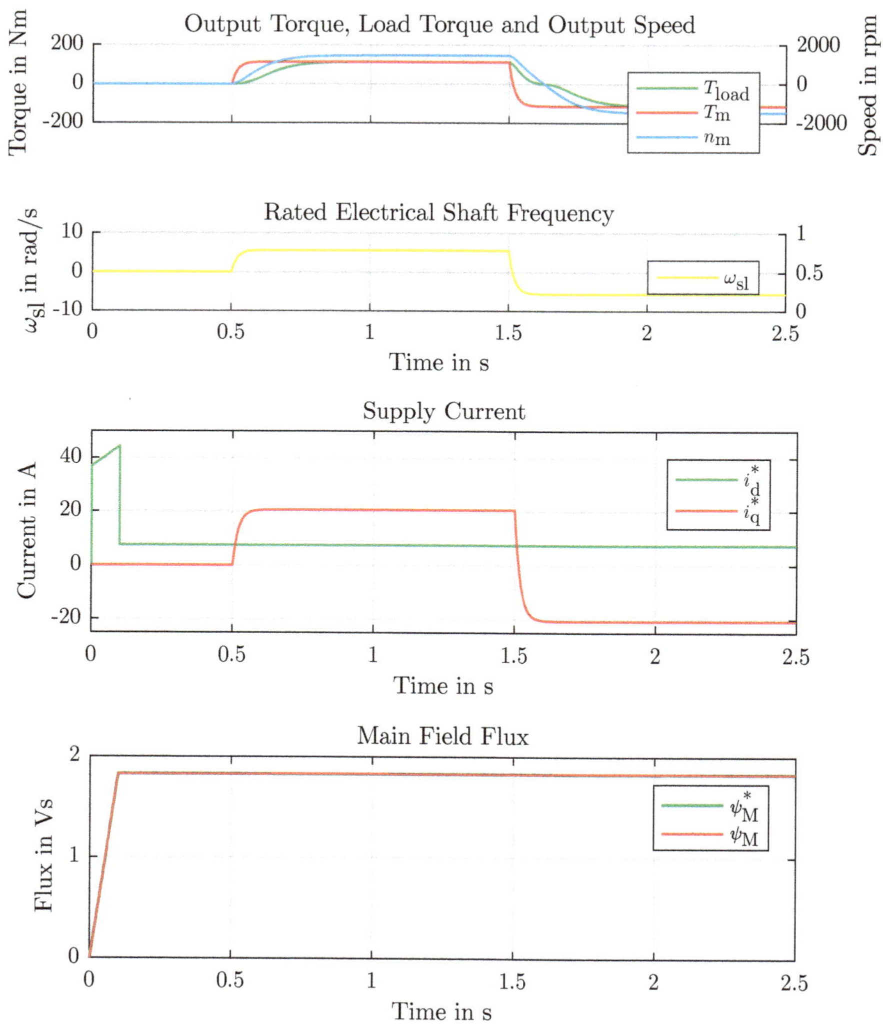

Fig. 9.21 Delta connected 22 kW machine with UFO based drive controller and shaft encoder in rotor flux orientation

which in this case represents the stator flux linkage given that the variable a has been set to allow the vector controller to operate under stator flux oriented control.

A comparison with the results given in Fig. 9.21 and those obtained with the classical control approach (see Fig. 9.6) shows the remarkable degree of control response and stability that can be achieved using field-oriented control. Note that changing the transformation variable a will not change the results shown, as may be deduced by considering the simulation results given in Sect. 9.7.3. Some minor changes are apparent due to the presence of differentiators in the forward path of the controller. Furthermore, the torque and flux linkage step were moderated through the use of a low pass filter with a time constant of 2 ms and 10 ms, respectively. Without these filters unrealistically high currents would appear in the simulation and correspondingly, in the actual machine. The reader is referred to the tutorial given in Sect. 9.7.3 for further details regarding the simulation model and the opportunity to interactive explore the drive concept presented.

9.6 Interfacing the Field-Oriented Controller with a Voltage-Source-Connected Induction Machine

The practical implementation of the current-controlled machine discussed in the previous section requires the use of a voltage source converter. Consequently, a three-phase current control approach as discussed in Sect. 3.2 must be integrated with the drive. In this section, a three-phase model based current controller is used, because these are generally favored in practical drive applications, among others for acoustic noise reasons as was mentioned in Sect. 3.2.2. The generic representation of the current controller in question, as shown in Fig. 3.19 is connected to a generalized load which define the parameters for the discrete model based synchronous current controllers which is tied to the vector $\vec{\psi}_{\mathrm{e}}$. For the induction machine drive, the latter vector must be replaced by the flux linkage vector $\vec{\psi}_{\mathrm{M}} = \psi_{\mathrm{M}}\,\mathrm{e}^{\mathrm{j}\theta^*_{\mathrm{M}}}$, where the angle θ^*_{M} is generated by the CFO module. Furthermore, the parameters used for the controller need to be redefined, given that the *load* is in this case an induction machine. For the purpose of determining the parameters it is helpful to consider the asynchronous rotor flux oriented direct and quadrature axis symbolic models of the machine shown in Fig. 8.24. For the direct axis current controller calculation of the sampled average voltage reference $U^*_{\mathrm{d}}(t_k)$ is carried out with the aid of Eq. (3.22a), in which the parameters R and L must be replaced by the variables R_{s} and $L_{\sigma\mathrm{s}}$ given the nature of the *load* in use. A similar approach must be used for the computation of quadrature axis controller sampled average voltage reference $U^*_{\mathrm{q}}(t_k)$, where Eq. (3.22b) is used, in which case the parameters R and L must be replaced by the machine parameters R_{s} and $L_{\sigma\mathrm{s}}$. Furthermore, this expression shows the presence of a disturbance decoupling term $u_{\mathrm{e}} = \psi_{\mathrm{e}}\omega_{\mathrm{e}}$, which corresponds to the term $\psi_{\mathrm{M}}\omega_{\mathrm{s}}$ in an asynchronous drive.

The resultant drive structure as shown in Fig. 9.22 brings together key concepts, such a modulation and current control introduced in earlier chapters of this book.

Readily apparent in Fig. 9.22 is the model based current controller which uses current reference values produced by the *UFO Control module* as discussed in the previous section. The observant reader will note that a disturbance decoupling term $\psi_M\, \omega_m$ is shown in the current controller module, whereas in theory the frequency ω_s should have been used. In the example given here the shaft speed was deemed to be accessible and it is therefore prudent to utilize this variable, given the need to otherwise obtain an estimate for the slip frequency. This means that the error in the feed-forward term needs to be generated by the current controller integrator, which results in a slight loss of dynamic performance. The remaining terms $\omega_s\, L_{\sigma s} i_{sd}$ and $-\omega_s\, L_{\sigma s} i_{sq}$ are not used in this example, given that their contribution is typically small in comparison with the term e_q. Also shown in Fig. 9.22 is the generic model of the IRTF based rotor flux based induction machine according to Fig. 8.15. From a simulation perspective the voltage source converter/modulator structure is often replaced by an alternative module which calculates the required supply vector directly from the average voltage references generated by the current controller. This approach, as outlined in Sect. 3.3.5, reduces the simulation run time and a larger computational step size may be selected.

9.6.1 Tutorial Results

In the accompanying tutorial given in Sect. 9.7.4 the reader is able to examine in detail an implementation example which involves the 22 kW four-pole delta connected machine used in earlier tutorials in this book. Shown in Fig. 9.23 are some of the results obtained with the simulation model given in Sect. 3.3.5. In this instance the UFO controller was set to operate under stator flux oriented control ($a = {}^{L_s}/_{L_m}$) with a flux linkage reference setting of $\psi_M = 1.83$ Wb which corresponds to the rated stator flux linkage value of the machine in use. A rated torque step was applied at $t = 500$ ms and subsequently reversed at $t = 1500$ ms. The timing of the torque step was purposely delayed to ensure that the stator flux linkage level in the machine was at it steady-state value prior to introducing a quadrature current change. Note that in many applications it is practical to ensure that the machine is kept at its fully magnetized state, when the drive is set to standby. This implies that the drive is able to respond directly if needed. This is particularly relevant in, for example, hoist applications.

The results given in Fig. 9.23 confirm that the stator flux linkage is held constant at its reference value, while the torque is varied in accordance with the user defined reference torque sequence. Note that a low pass filter with 2 ms time constant has been introduced to moderate the torque reference, and its effect is clearly noticeable in Fig. 9.23c. More extensive results with the simulation model operating with a rotor flux oriented UFO controller are given in Sect. 3.3.5.

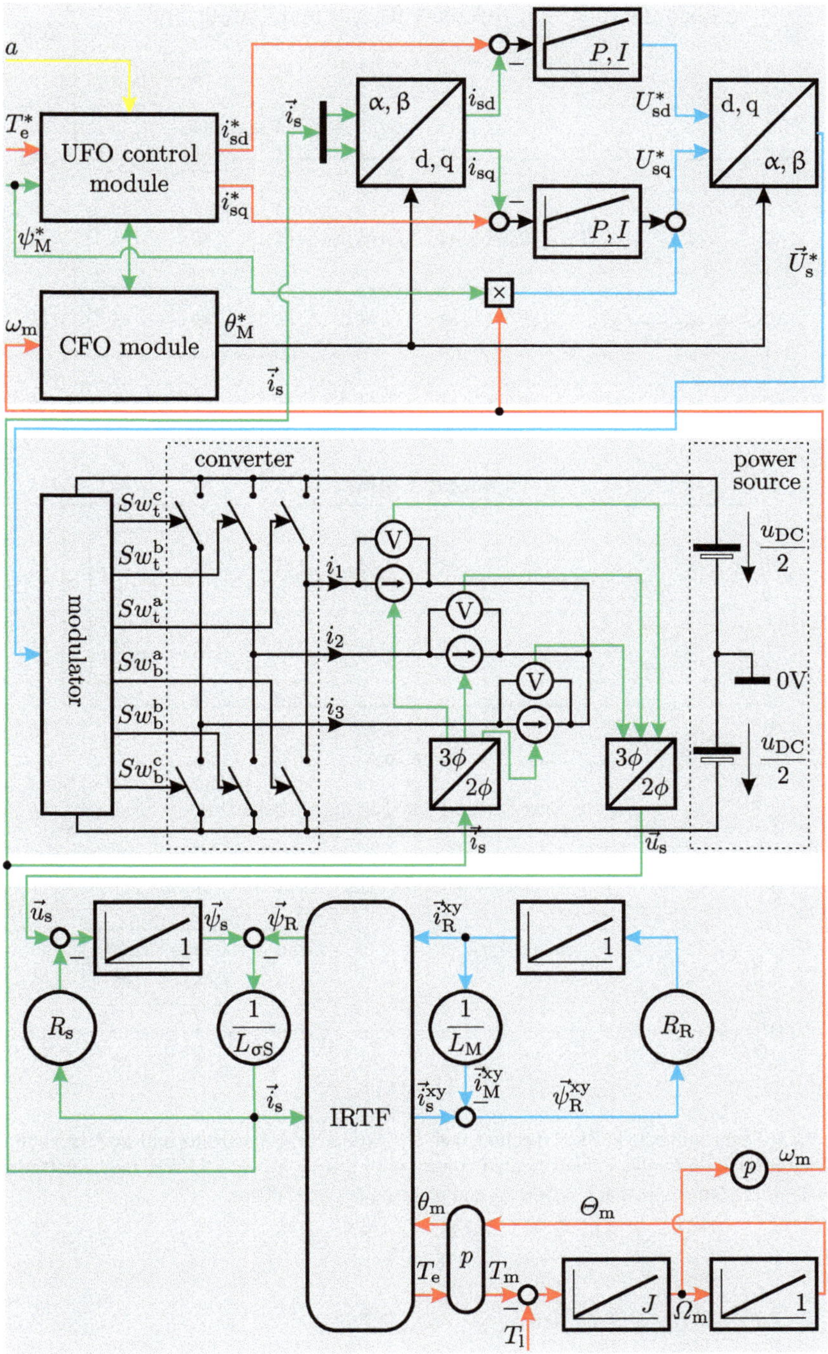

Fig. 9.22 Field-oriented control with voltage source induction machine model and model based current control

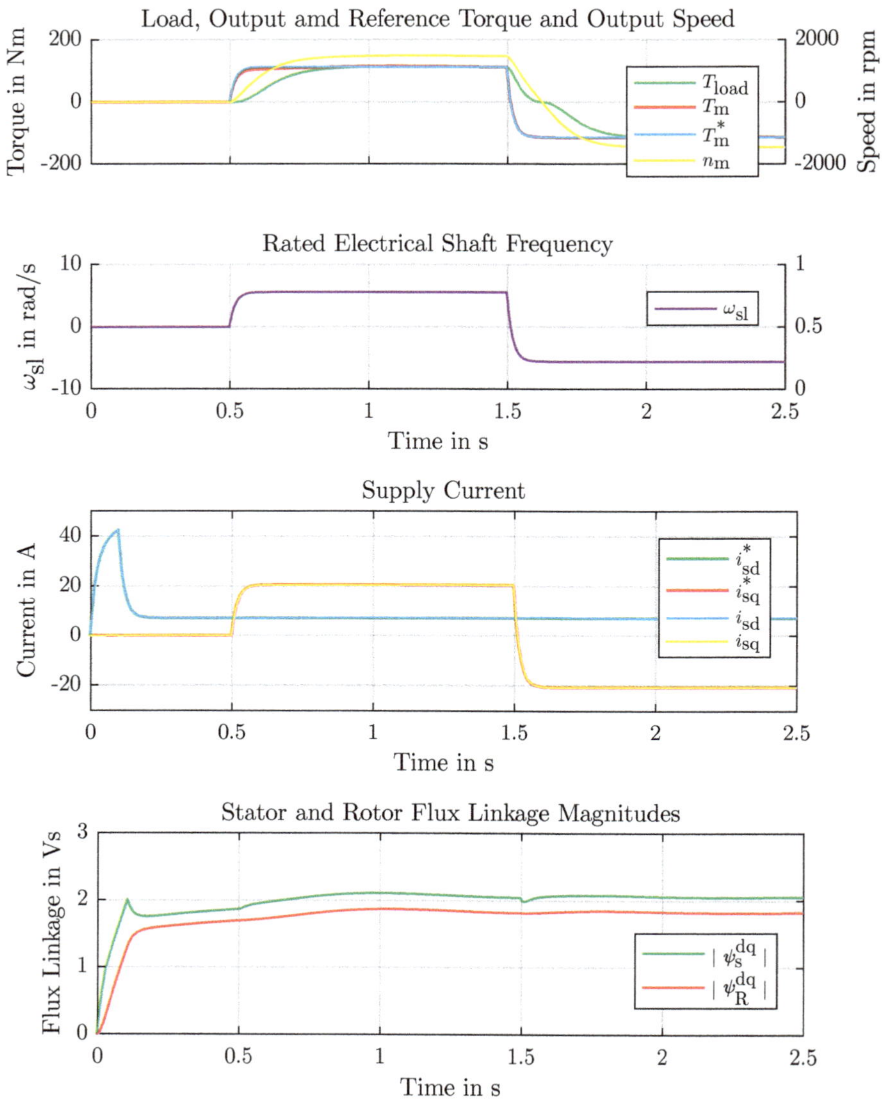

Fig. 9.23 Delta connected 22 kW machine with UFO based drive controller and shaft tacho meter. UFO controller set to stator flux oriented control with $\psi_M = \psi_s = 1.83$ Wb, rated shaft torque step: $0 \rightarrow 113$ Nm applied at $t = 500$ ms and reversed at $t = 1500$ ms

9.7 Tutorials

9.7.1 *Tutorial 1: Simplified V/f Drive*

The purpose of this tutorial is to develop a V/f drive model which utilizes the generic controller structure discussed in Sect. 9.1.1. A 17.3 kW, four pole, delta connected induction machine as presented in tutorial (see Sect. 8.6.7) is used with a converter. A load module is used which generates a quadratic torque versus speed characteristic. This module is configured to generate a load torque of 113 Nm at a shaft speed of $n_{\mathrm{m}} = 1465$ rpm. To simplify the analysis, the PWM converter is replaced by an *average three-phase modulator*, as discussed in tutorial (see Sect. 3.3.5) which avoids the need to model the switching behavior of the converter. Inputs to the modulator are the voltages $U^*_{\mathrm{s}\alpha}$, $U^*_{\mathrm{s}\beta}$, sampled at $T_{\mathrm{s}} = 1$ ms. The DC bus voltage for the converter is set to $u_{\mathrm{DC}} = 1200$ V.

The simulation model given in Fig. 9.24 shows the rotor flux based model as presented in Fig. 8.46. Parameters for this rotor flux oriented model have been introduced in Table 8.3. The structure of the controller module shown in Fig. 9.24

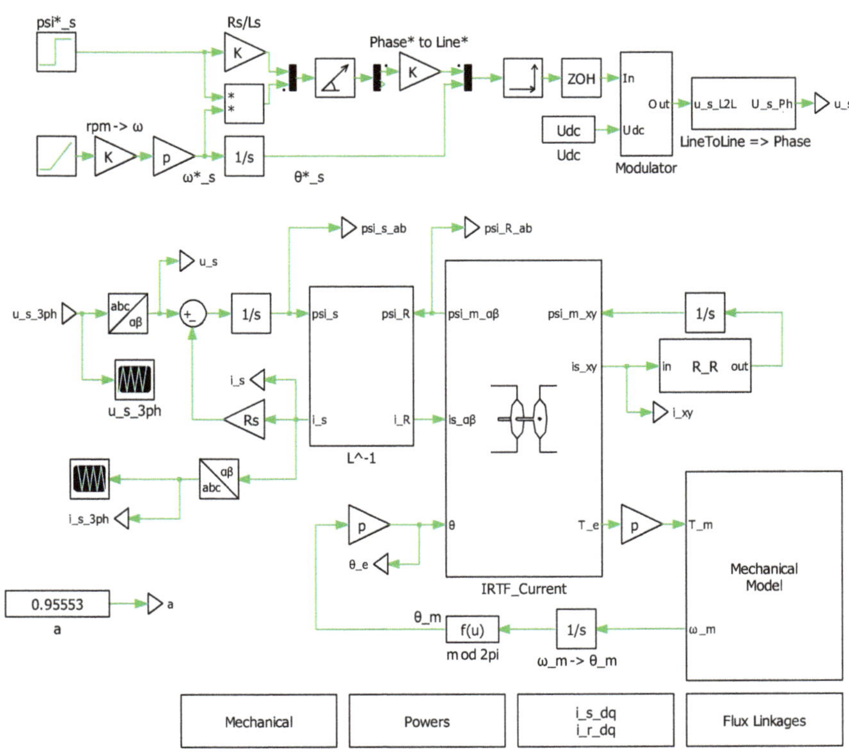

Fig. 9.24 Model of simplified V/f drive

is consistent with the generic model shown in Fig. 9.2. A gain module with gain R_s/L_s is used in the controller, of which the parameters may be obtained from Table 8.3, where the stator inductance is given as $L_s = L_{\sigma s} + L_M$. Note that the parameters given are for a delta connected machine, whereas the voltage inputs for the modulator unit must be expressed for a *star* connected configuration. This requirement is accommodated in the controller by the inclusion of a *Phase* to Line** conversion module which is positioned after the polar to Cartesian conversion unit that generates the voltage vector $\vec{U}_s^*$.

Inputs to the controller are the stator flux linkage reference value, which is set to the rated value of $\vec{\psi}_s^* = 1.83\,\text{Vs}$, and the stator flux frequency. The latter is expressed as $\omega_s^* = {}^{30}/_{\pi}\, p\, n_s^*$ where n_s^* represents the four-pole ($p = 2$) synchronous speed in rpm. Its value has been initially set to the rated speed of the machine. It can be observed in the simulation results that the stator flux linkage reaches the set value after some large transients. The steady-state shaft speed is however lower than the reference value, which is expected given that the controller sets the synchronous speed of the rotating flux linkage vector and not the actual shaft speed. Hence, the difference between these values is the slip frequency which corresponds to the torque generated by the machine. The observed simulaton results are shown in Fig. 9.25.

The second part of this tutorial aims at identifying potential operating regions where instabilities may occur. For this purpose, the *disturbance* load module is used,

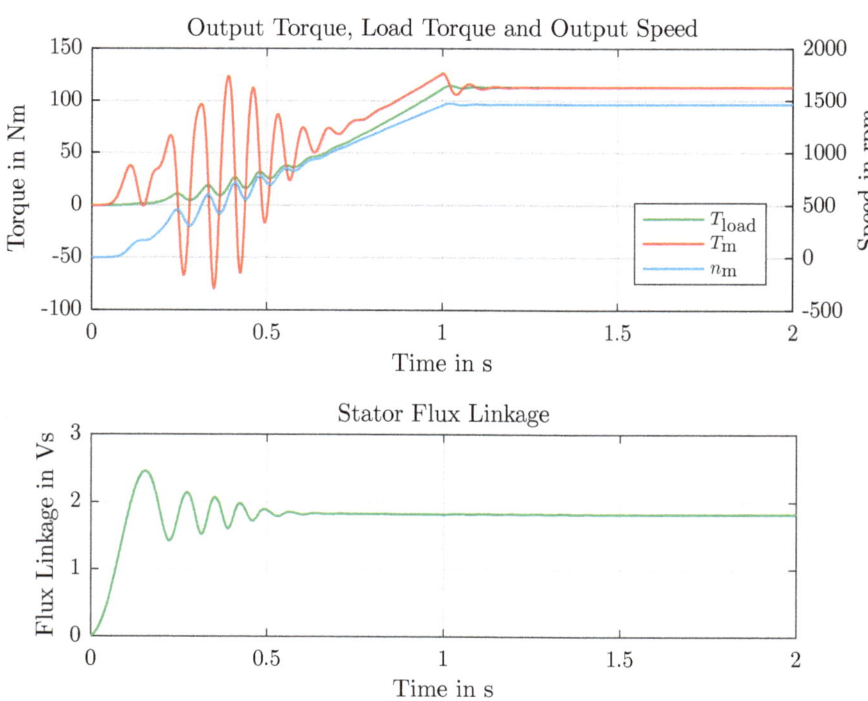

Fig. 9.25 Simulation results of simplified V/f drive

which is a 1 Hz bipolar, 1 Nm square wave oscillator. The introduction of this unit serves to trigger any resonances which may be present over the selected speed range. A very slow ramped synchronous speed is given as reference to the controller with a range of $0 \rightarrow 900$ rpm at a slope of 30 rpm/s, for which the simulation time is increased to 30 s. The results obtained with this speed sweep are shown in Fig. 9.4.

9.7.2 Tutorial 2: V/f Drive with Shaft Speed Sensor

This tutorial examines a V/f based drive with measured or estimated shaft speed feedback. For this purpose, the simulation generated for the previous tutorial is modified to accommodate a controller as shown in Fig. 9.5. Inputs to the control module are the shaft torque reference T_{e}^{*} and stator flux linkage reference ψ_{s}^{*}. The machine configuration, converter and DC supply configuration remain unchanged compared to the previous tutorial.

The simulation, as given in Fig. 9.26, shows the revised controller structure which is in accordance with the generic model discussed in Sect. 9.1.2. An example of the results obtained with this simulation model is given in Fig. 9.6, where the shaft torque reference and stator flux linkage reference were taken to be $T_{\mathrm{e}}^{*} = 113$ Nm and $\psi_{\mathrm{s}}^{*} = 1.868$ Vs, respectively. Examination of the results obtained with the model shows that the corresponding steady-state shaft torque and stator flux linkage are equal to $T_{\mathrm{e}} = 105.1$ Nm and $\vec{\psi}_{\mathrm{s}}^{*} = 1.871$ Vs, respectively. The deviation between reference and actual shaft torque is attributed to the approximation used (Eq. (9.7)) to derive the controller slip frequency reference value, which ignores the leakage inductance $L_{\sigma\mathrm{R}}$. More extensive slip frequency models can however be used, but this is outside the scope of this tutorial.

9.7.3 Tutorial 3: Universal Field-Oriented (UFO) Control with a Current Source Based Machine Model and Known Shaft Angle

A numerical implementation of UFO control is carried out in this tutorial. The same 17.3 kW, four-pole machine is used as in the previous chapters. The machine parameters are given in Table 8.3. A simulation model of the UFO controller is developed as defined by the generic model given in Fig. 9.10. Input to the UFO controller are the variables ψ_{M}, which in this case is set to $\psi_{\mathrm{M}} = 1.83$ Vs. A torque reference T_{e}^{*} is configured to provide a rated shaft torque step from $T_{\mathrm{e}}^{*} = 0 \rightarrow 113$ Nm at $t = 500$ ms and it is reversed at $t = 1500$ ms. A quadratic load as defined in the previous tutorials is connected to the machine.

A shaft encoder is present which provides the electrical shaft angle θ and this variable is used in conjunction with a CFO module as given in Fig. 9.12. Output of

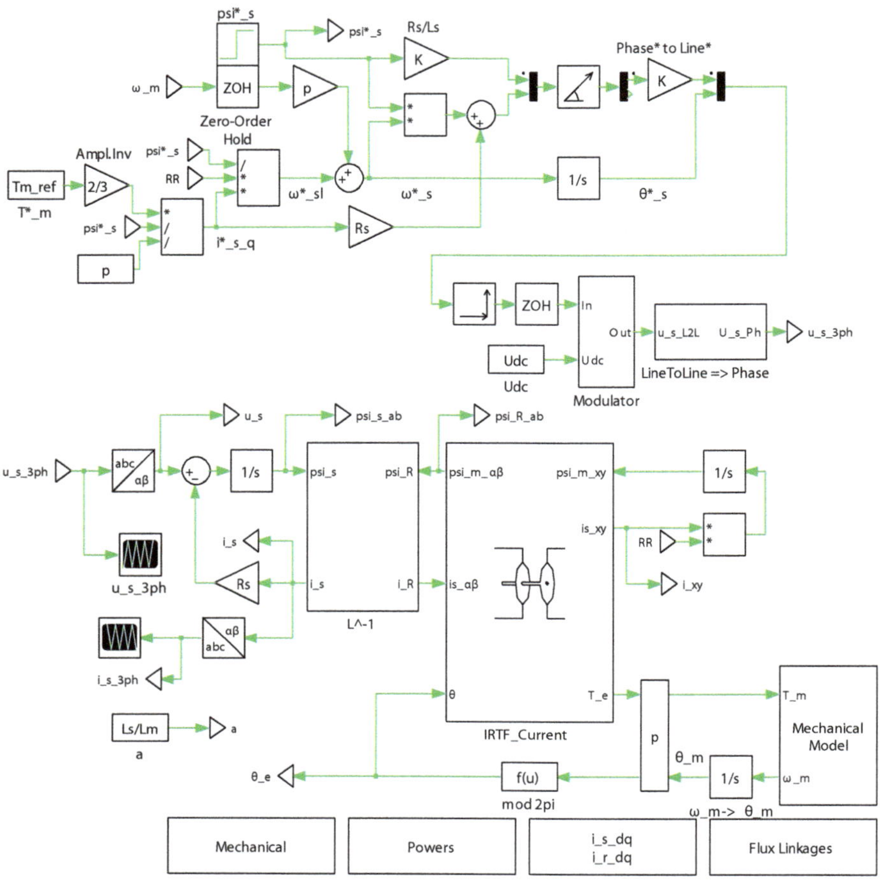

Fig. 9.26 Simulation of V/f drive with shaft speed sensor

the CFO module, which must be interfaced with the UFO module, is the reference angle θ_{M}^{*} as required for achieving field-oriented control.

A four parameter, IRTF based machine model as given by Fig. 8.15 is used. In this case, the stator current $\vec{i}_{\mathrm{s}}$ is taken to be the input to the model, instead of the voltage vector $\vec{u}_{\mathrm{s}}$ given that the UFO module generates the reference current vector $\vec{i}_{\mathrm{s}}^{dq}$ which must be transformed to stationary coordinates by using the reference angle θ_{M}^{*}. The use of a current source model is helpful because it simplifies the model and provides the ability to visualize the operation of the drive without the complexities of the current controller.

The simulation model as shown in Fig. 9.27 satisfies the requirements for this tutorial. Clearly shown are the UFO and CFO module which generate the required variables i_{sd}^{*}, i_{sq}^{*}, and θ_{M}^{*}. The torque reference T_{e}^{*} and flux linkage reference ψ_{M}^{*} can be set in the given subsystems, while the transformation value a can be set in the *simulation parameters* within the range $L_{\mathrm{m}}/L_{\mathrm{r}} \leq a \leq L_{\mathrm{s}}/L_{\mathrm{m}}$ (see Eq. (8.13)). A

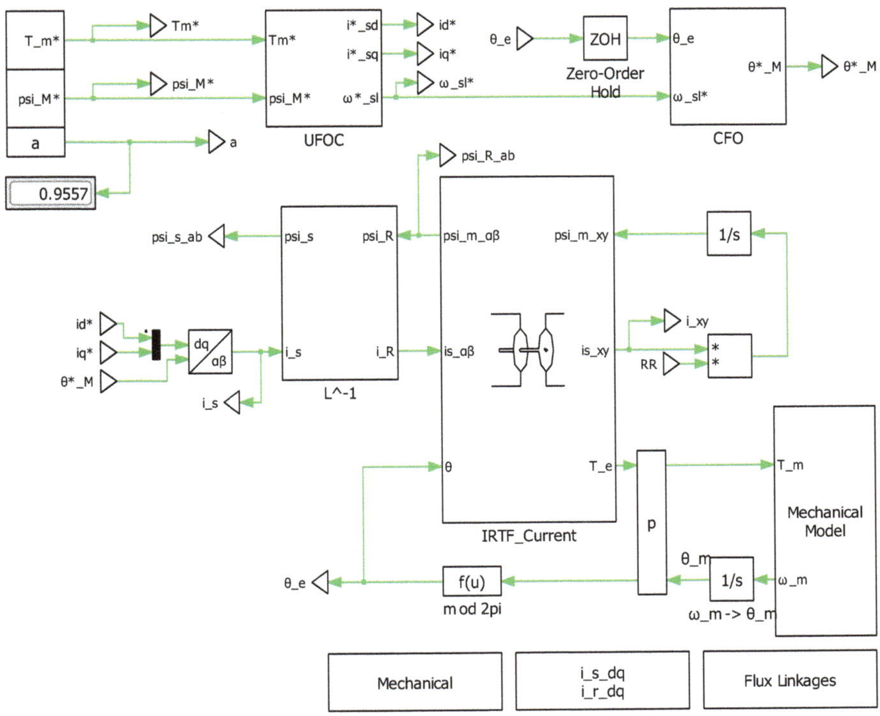

Fig. 9.27 Simulation of UFO controller with current source based machine model

first order filter is used to moderate the reference torque and the time constant for this filter is taken to be 20 ms.

A set of scope modules is provided to show the results obtained with this simulation in Fig. 9.28. These scope outputs are generated for transformation variable $a = L_m/L_r$, i.e., UFO controller set to rotor flux oriented control. Under these conditions the variable ψ_M represents the variable ψ_R. The simulation results for $a = L_s/L_m$, given in Fig. 9.29, show the use of the UFO controller with stator flux oriented control. In this case, the variable ψ_M represents the variable ψ_s.

Note that for both settings the required torque and flux linkage reference values are realized by the UFO controller. Similarly, any intermediate value for the transformation variable a may be chosen by the user. Note that for stator flux oriented control the slip frequency shows a Dirac type function. This is caused by the fact that a torque step under stator flux oriented control is generated by a near instantaneous displacement of the stator flux linkage vector relative to the rotor flux linkage vector. In practice, however, changes in torque should be *rate limited*, for example, to limit mechanical stress or to prevent exciting mechanical eigenfrequencies.

Finally, it may be observed for both cases that a change in torque can be done independently of the flux linkage ψ_M, which is maintained at the required level

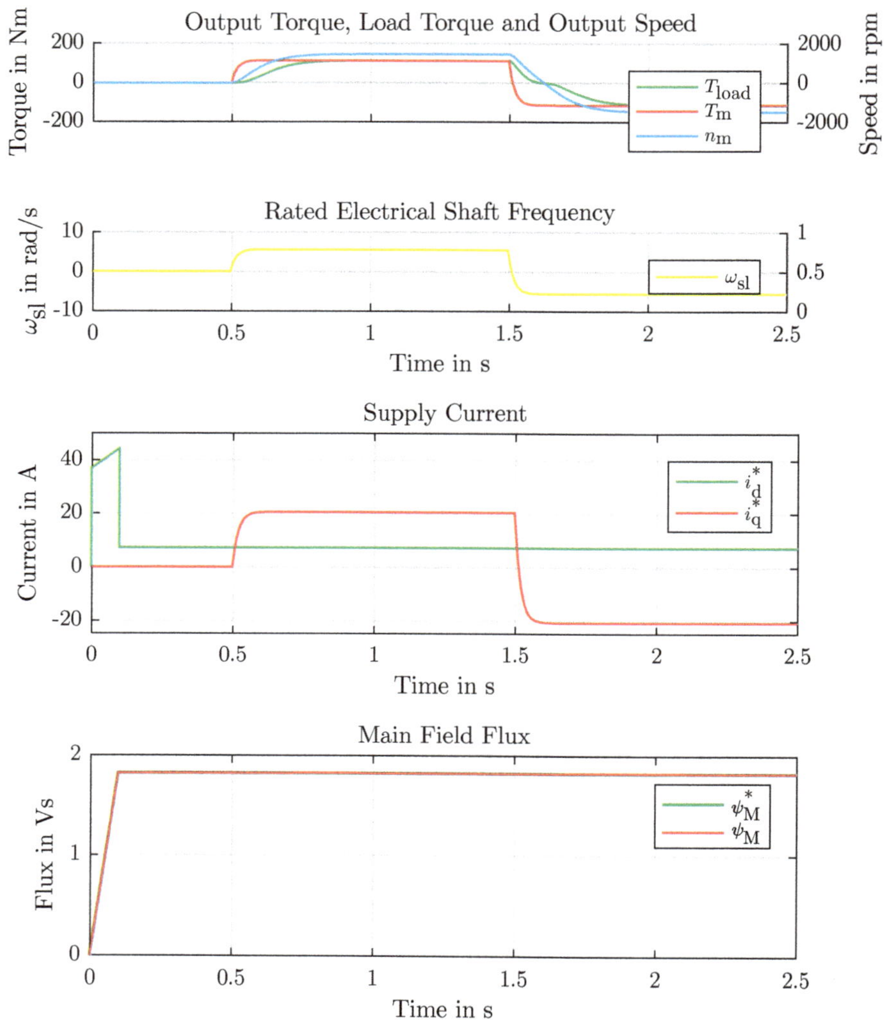

Fig. 9.28 Simulation results for a UFO controller with a current source based machine model, using rotor flux oriented control, $a = 0.955$

when torque transients occur. When applying step functions to the torque reference, some minor perturbations around the desired value occur for stator flux oriented control, due to the use of non-ideal differentiators in the UFO controller. In practice, detuning effects, i.e., discrepancies between model and actual machine parameters will cause some coupling between the flux linkage and torque variables. The user is encouraged to detune the UFO parameters to investigate detuning effects [3].

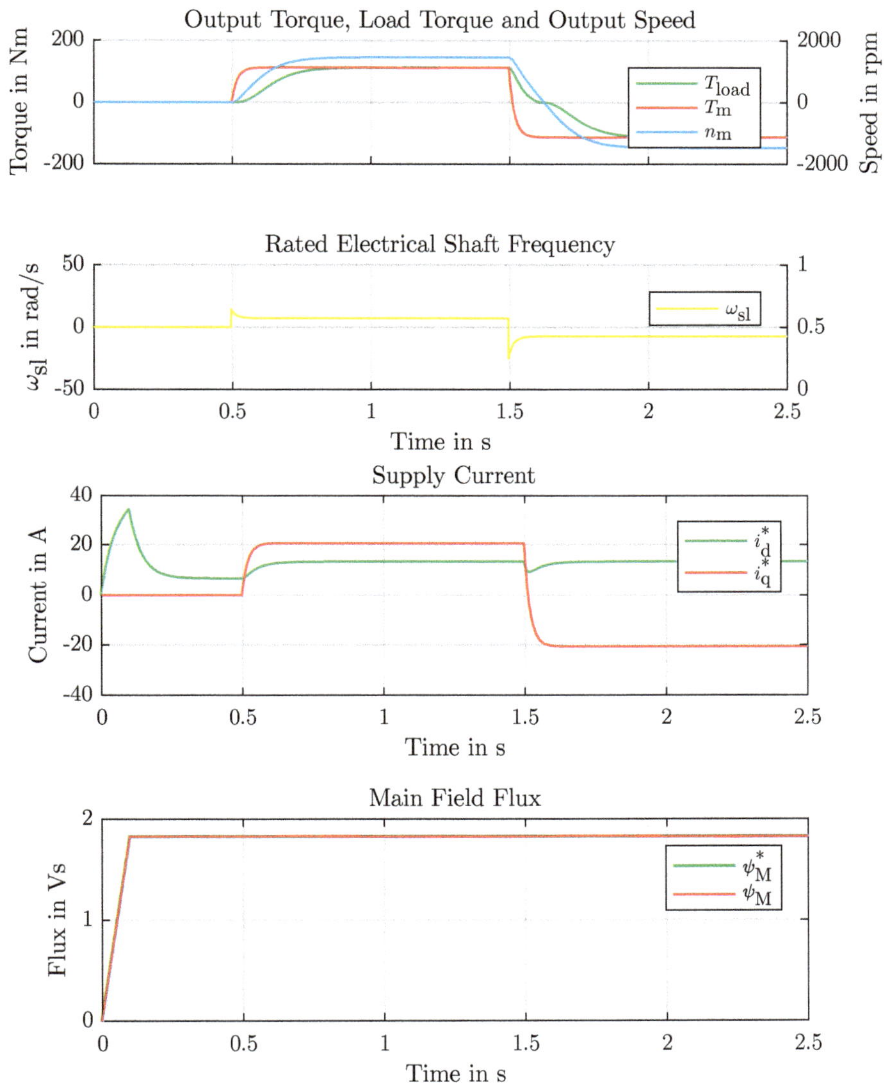

Fig. 9.29 Simulation results for a UFO controller with a current source based machine model, using stator flux oriented control, $a = 1.046$

9.7.4 *Tutorial 4: Induction Machine Drive with UFO Controller and Model Based Current Control*

This tutorial examines the operation of a delta connected four-pole induction machine operating with a model based current controller and voltage source converter. The proposed concept as discussed in Sect. 9.6 utilizes a four-pole induction machine as used in tutorial (see Sect. 9.7.2). In this exercise, the current commands produced by the UFO module are used as inputs $i_{\rm sd}^*$ and $i_{\rm sq}^*$ for the current controller as in Fig. 9.22. The machine parameters are given in Table 8.3. A sampling rate of 10 kHz is assumed for the discrete current controllers.

For this example, the input electromagnetic reference torque is set to $T_{\rm e}^* = 113$ Nm at $t = 500$ ms and this is to be followed by a torque reversal $T_{\rm e}^* = 113.0 \rightarrow -113.0$ Nm at $t = 1500$ ms. A first order filter with a time constant of $\tau = 20$ ms is positioned between the torque reference and synchronous drive controller, in order to limit the torque variations to realistic values. Requesting an instantaneous change in torque is not realistic in any practical drive system, due to limitations imposed by the converter DC bus-voltage. Similarly, a first order filter with a time constant of $\tau = 20$ ms is used to moderate the flux linkage reference value $\psi_{\rm M}$ as used for the UFO controller. Total simulation run time is to be set to 2500 ms, while the DC supply is equal to $u_{\rm DC} = 800$ V. In this simulation the modulator and converter are not to be implemented at circuit level to better visualize the operation of the drive (Fig. 9.31).

The first task is to calculate the gains for the two current controllers and in addition identify the disturbance decoupling terms which must be introduced. Secondly, the simulation model of the drive as given in Fig. 9.30 should be examined, and the following plots should be obtained.

- Sampled reference and *measured* currents $i_{\rm sd}^*$, $i_{\rm sq}^*$, $i_{\rm sd}$, and $i_{\rm sq}$.
- Torque reference $T_{\rm e}^*$, *actual* torque $T_{\rm e}$, and mechanical shaft speed $\omega_{\rm m}$.
- Stator flux linkage $|\vec{\psi}_{\rm s}|$ and rotor flux linkage $|\vec{\psi}_{\rm R}|$ amplitudes.

The computation of the current controller gains and identification of the relevant feed-forward term follows the approach outlined in Sect. 9.6. The two PI controllers shown are of the anti-windup type as discussed in Sect. 3.3.2, which implies that the output limits must be specified. A convenient choice is to set the value to the length of the active converter vector, namely $\pm u_{\rm DC}$ on the grounds that the output voltage vector can never exceed the limits of the hexagon formed by the active converter vectors. The q-axis PI controller is provided with a feed-forward signal $\omega_{\rm m}\,\psi_{\rm M}$ as discussed in Sect. 9.6.

Note that a delta connected machine is in use, which explains the presence of the star/delta and delta/star space vector modules in the simulation model. These could be avoided, by simply assuming an equivalent star connected machine model (which is achieved by dividing the relevant machine parameters by three). In this case, preference is given to using the delta connected machine, as it is instructive

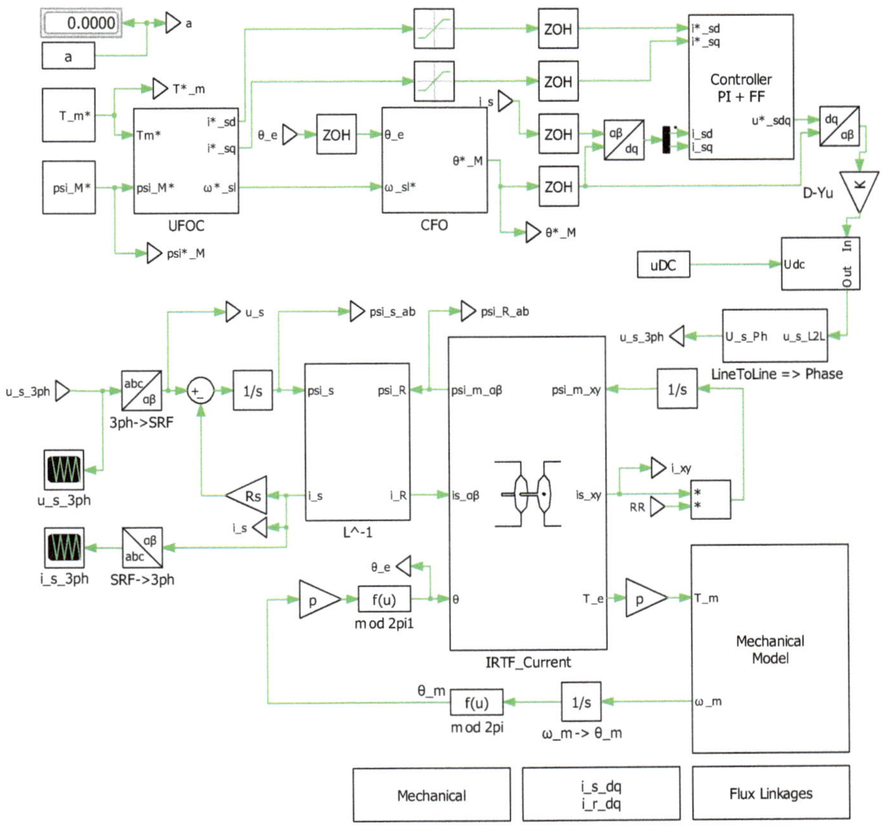

Fig. 9.30 Simulation of a UFO controller with a current source based machine model and model based current control

to show how the voltage converter references must be configured to accommodate converter saturation.

The simulation results, given in Fig. 9.31, show operation of the drive under rotor flux oriented control. It can be observed in the simulation results that the current controller is able to maintain the direct and quadrature currents at the value dictated by the reference values.

Note that in the simulation results given in Fig. 9.31, during the startup sequence, the direct axis current reference reaches beyond the maximum rated current value of 30 A. If the UFO controller outputs are limited to this value, which should be the case in real life applications, the time required to fully magnetize the machine increases. An example simulation is run, where a torque step was imposed at $t = 0.2$ s. At this instance, the machine is not fully magnetized. Consequently, an error will occur between the reference and actual torque (Fig. 9.32).

These results emphasize the need to ensure that the machine magnetization level is held at the reference value during standby operation, in the event that a potential

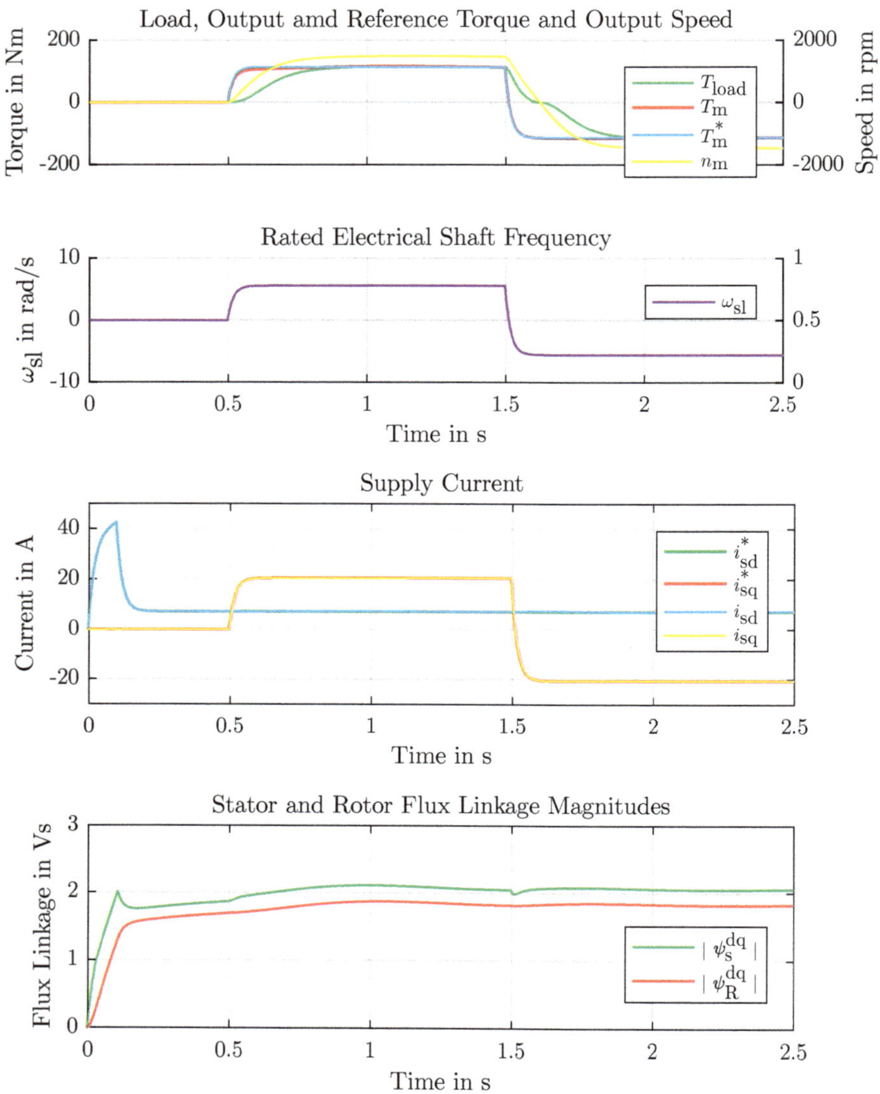

Fig. 9.31 Simulation results of an induction machine drive with a voltage source converter and UFO controller

full torque response may be required. If the machine is demagnetized, a substantial delay must be incurred to allow the machine to reach the flux linkage reference, before a torque step can be applied.

The drive example shown here was undertaken with a model based current controller. It is left as an exercise for the reader to reconsider this problem with a hysteresis controller as shown in Fig. 3.13.

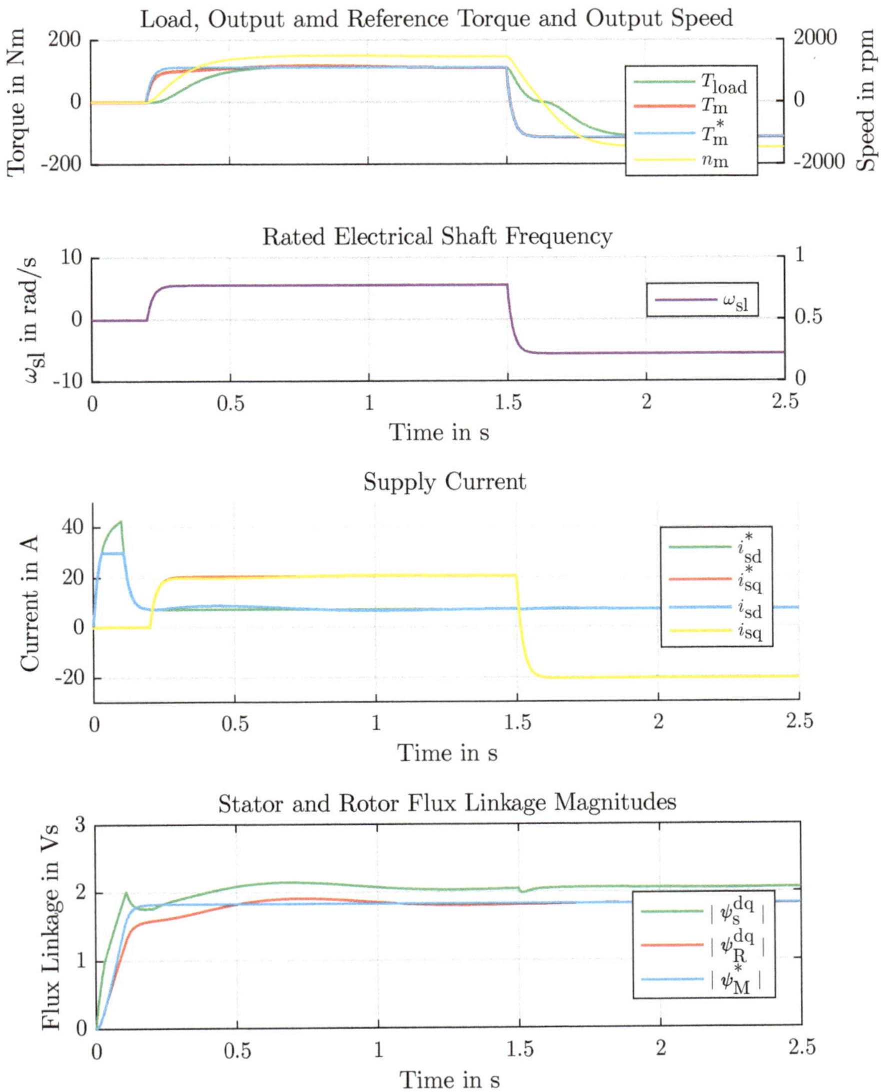

Fig. 9.32 Simulation results of an induction machine drive with a voltage source converter and UFO controller with current limiters

9.7.5 *Tutorial 5: Rotor Flux Oriented Induction Machine Drive with UFO Controller and Field Weakening Controller*

This tutorial examines the use of a field weakening controller which operates according to the control principles set out in Sect. 9.4. Said controller is used together with a UFO controller operating under rotor flux oriented control. The direct and quadrature current outputs are to be used directly with a current source rotor flux based induction machine model as discussed in tutorial (see Sect. 9.7.3).

The rated current for the delta connected remains at $i_s^{max} = 30\,A$, while the maximum supply voltage setting is arbitrarily set to $u_s^{max} = 1000\,V$. A simulation model is developed with a *field weakening controller* which must be provided with a torque reference and electrical stator frequency ω_s. The induction machine model is connected to a load model which runs on a predefined speed trajectory independent of the generated torque. In terms of operational requirements, the tutorial is divided in two parts.

The first part deals with operation of the drive under partial and full load conditions, whereby drive operation is to be examined in two phases, namely:

- Operation with rated rotor flux linkage reference value of $\psi_R = 1.83\,Vs$, while the electrical frequency is set to the base speed of the drive. Under these conditions the shaft torque reference is varied from $T_e^* = 30 \rightarrow 113\,Nm$.
- Operation with maximum torque setting $T_e^* = 113\,Nm$, while the synchronous shaft speed is to be varied over the range $n_s^* = 0 \rightarrow 3000\,rpm$.

The second part of the tutorial is concerned with partial load operation where the torque reference is held at $T_e^* = 80\,Nm$, while the reference flux linkage is set to $\psi_R = 1.83\,Vs$. Under these conditions the synchronous shaft speed is to be varied over the range $n_s^* = 0 \rightarrow 3000\,rpm$. The objective of the tutorial is to plot the trajectory of the synchronous current vector $\vec{i}_s^{dq}$ in a synchronous IM control diagram together with the MA circle and MF ellipses, which correspond to operation at the base speed and maximum synchronous shaft speed of $n_s^{max} = 3072\,rpm$, respectively. In addition, the user is encouraged to carefully examine the simulation model to ascertain whether or not the operating conditions as prescribed are indeed executed by the controllers.

9.7.5.1 Part I

The simulation model, shown in Fig. 9.33, complies with the requirements of this tutorial. Readily identifiable in Fig. 9.33 is the four parameter IRTF model which complies with the generic model shown in Fig. 8.15, where the vector $\vec{i}_s$ and shaft speed are taken as inputs for the model. Computation of the model parameters $L_{\sigma s}$, L_M, and R_R is undertaken with the aid of Eqs. (8.12) and (8.15) and Table 8.2, taking into account the new leakage parameters. The revised four parameter model data of the machine for this tutorial is given in Table 9.1.

The electrical frequency ω_s for the four-pole machine model is found by making use of the slip frequency ω_{sl} value generated by the UFO controller and the shaft speed ω_m, which in turn is obtained via the mechanical load module that defines the synchronous shaft speed n_s^*. Inputs to the UFO controller are the rotor flux linkage ψ_R^* and electrical torque T_e^*, which are derived from the field weakening module

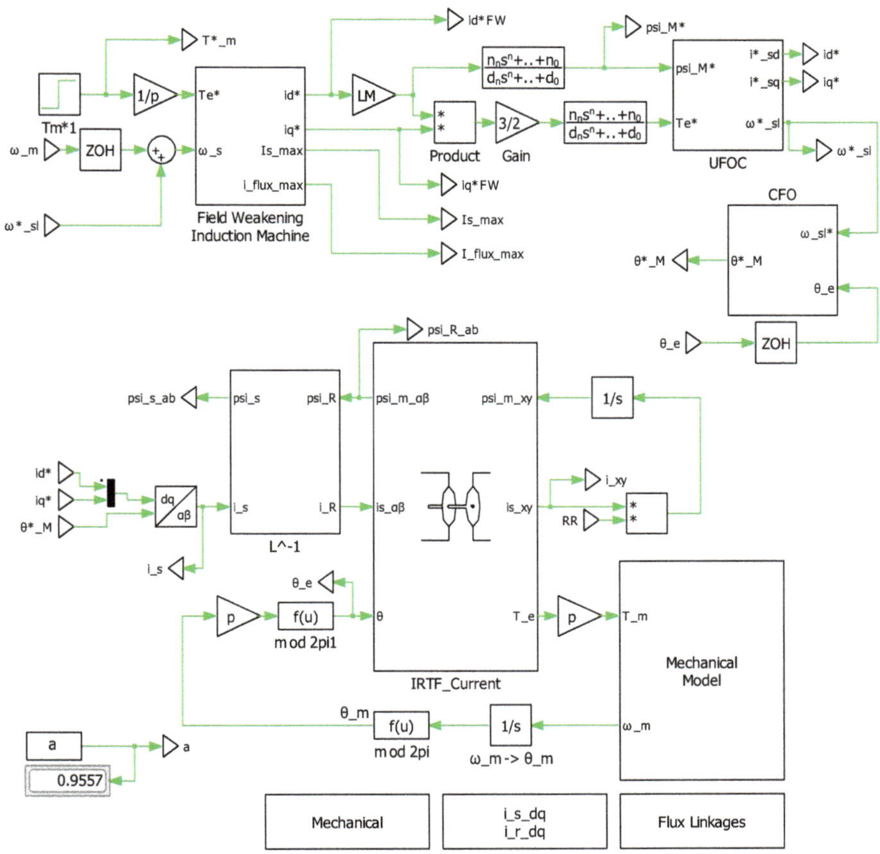

Fig. 9.33 Simulation of an induction machine drive with rotor flux oriented UFO controller and field weakening controller

Table 9.1 Machine parameters and flux linkage values for the rotor flux oriented model, as used in this tutorial

Parameters	Value
Magnetizing inductance (L_M)	248.5 mH
Leakage inductance ($L_{\sigma s}$)	23.6 mH
Leakage inductance ($L_{\sigma r}$)	0 mH
Rotor resistance (R_R)	0.494 Ω
Stator resistance (R_s)	0.600 Ω
Rated stator flux linkage (ψ_s^{nom})	1.83 Vs
Rated rotor flux linkage (ψ_R^{nom})	1.67 Vs

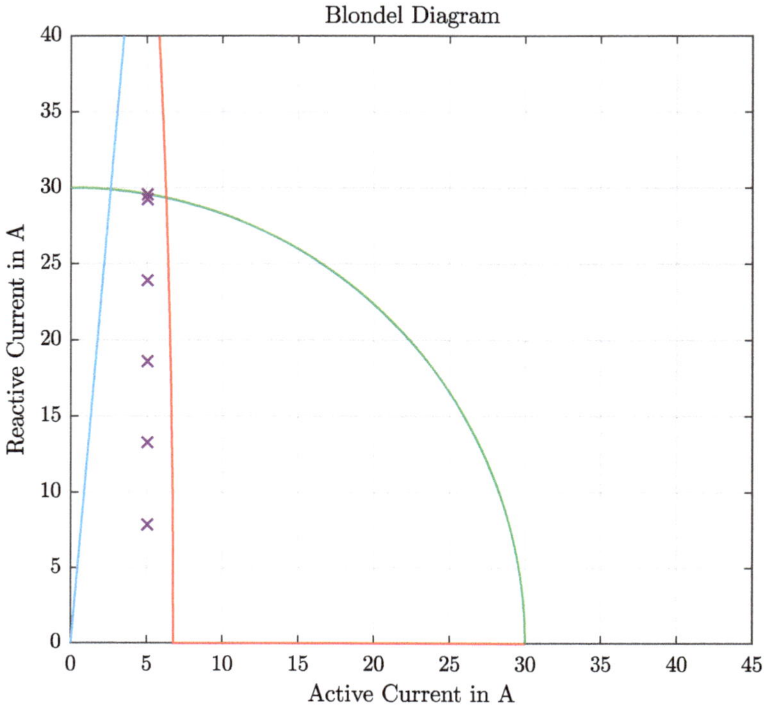

Fig. 9.34 Simulation results for induction machine drive with rotor flux oriented UFO controller and field weakening controller

which provides (among others) the direct and quadrature current reference i_{d}^* and i_{q}^*. Use of these two variables gives $T_{\mathrm{e}}^* = L_{\mathrm{M}} i_{\mathrm{d}}^* i_{\mathrm{q}}^*$ and $\psi_{\mathrm{R}}^* = L_{\mathrm{M}} i_{\mathrm{d}}^*$ as required for the UFO controller, which must also be provided with the relevant data given in Table 9.1. In this tutorial rotor flux oriented control is envisaged which implies that the UFO control input a must be set to $a = {}^{L_{\mathrm{m}}}/_{L_{\mathrm{r}}} = 0.9125$. In addition, the UFO controller must be provided with the five parameter model data as given by Table 8.2, where use must be made of the revised leakage parameters as given at the beginning of this tutorial. The UFO controller used in this tutorial also houses a CFO module which generates the rotor angle θ_{M}^* as required for implementation of indirect rotor flux oriented control, in which use is made of the (known) shaft speed.

A step module is also visible in Fig. 9.33, which provides the shaft torque reference value T_{e}^* that is used indirectly by the field weakening module. This module also provides additional data needed to generate the IM synchronous control diagram, which enables the user to examine the MA circle, MF flux linkage ellipse, and current space vector $\vec{i}_{\mathrm{s}}^{\,dq}$ generated by the UFO controller for a given set of user settings. To accommodate the operational requirements for the first part of the tutorial the base synchronous shaft speed of $n_s^b = 1465$ rpm must be selected in the mechanical load module. Said base speed may be found by making use of Eq. (9.22), Table 9.1, and the drive parameters defined for this tutorial. The torque reference was varied over the range: $T_e^* = 30 \rightarrow 130$ Nm with an incremental step of 20 Nm. The current vector $\vec{i}_s^{dq}$ generated by the UFO controller was recorded for each torque setting and used to plot the some of the results given in Fig. 9.34. More specifically, the six data points along a vertical axis with $i_{sd} = 8.04$ A correspond to the operational sequence described above. The last data point in this sequence, which corresponds with a torque setting of $T_e^* = 130$ Nm, was judicially chosen to coincide with the base operating point of the drive, which also intersects with the base speed MF ellipse as may be observed from Fig. 9.34. Note that during this first operating sequence, the flux linkage ψ_R and shaft torque T_e as found in the machine model match the chosen the reference values of may be deduced by running the simulation in question.

Also shown in Fig. 9.34 are a set of data points which are located on the MA circle and MTPF line. This set of data points can be found by maintaining the shaft torque reference at the value $T_e^* = 130$ Nm and varying the synchronous shaft speed over the range $n_s^* = 1672 \rightarrow 3072$ rpm in incremental steps of 200 rpm as discussed above. During this operating sequence the shaft torque and rotor flux linkage will no longer be equal to the reference values, given that field weakening is active. Note that the supply voltage u_s is kept within the specified maximum voltage value of the drive, as may be observed by running the simulation with the chosen operating sequence.

9.7.5.2 Part II

The second part of the tutorial considers drive operation under partial load conditions, where the torque reference is held at $T_e^* = 70$ Nm, and reference flux linkage is set to $\psi_R = 1.83$ Vs. Under these conditions the synchronous shaft speed is varied over the range $n_s^* = 1672 \rightarrow 3072$ rpm in incremental steps of 200 rpm which leads to a set of data points shown in Fig. 9.35. The set of asterisks represent the endpoint of the vector $\vec{i}_s^{dq}$ for the chosen operating sequence.

Also shown in Fig. 9.35 are the MA circle and MF ellipses which correspond to a synchronous shaft speed of $n_s = 1672$ rpm and $n_s = 3072$ rpm, respectively. Observation of the results shows that the set of data point which are not located on the MTPF line correspond to operation under constant torque conditions as discussed in Sect. 9.3. Observation of the simulation with said operating sequence should confirm that the shaft torque is help at the reference value as required. Once the MTPF line is reached field weakening continues where both flux weakening and the shaft torque reduce as the synchronous speed is increased.

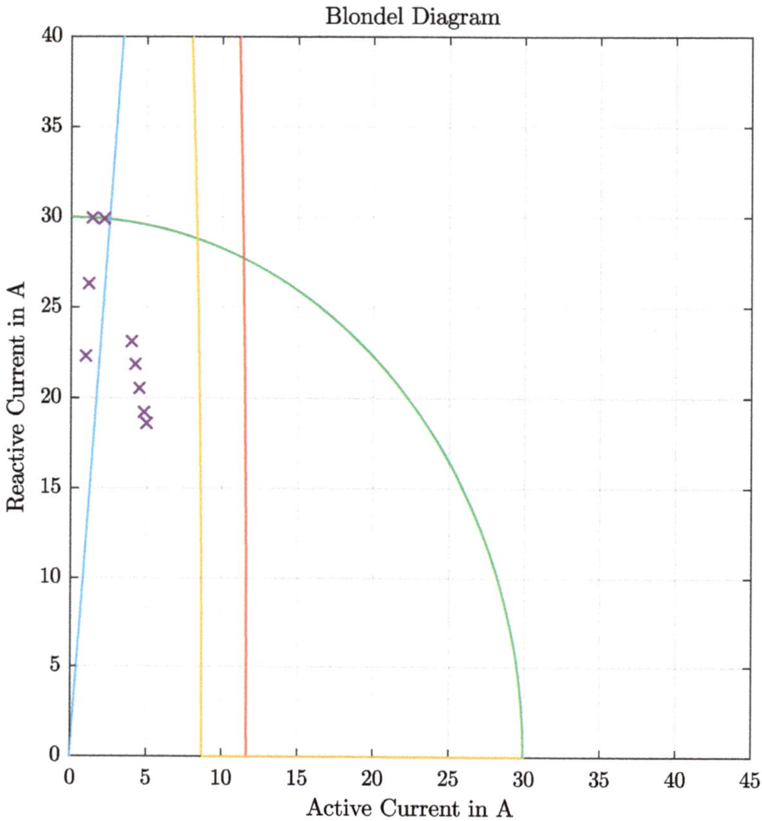

Fig. 9.35 Simulation results for induction machine drive with rotor flux oriented UFO controller and field weakening controller

References

1. Alakula M, Peterson B, Valis J (1992) Damping of oscillations in induction machines. In: PESC'92 record. 23rd annual IEEE power electronics specialists conference, 1992, vol 1, pp 133–138. https://doi.org/10.1109/PESC.1992.254702
2. Blaschke F (1972) The principle of field orientation as applied to the new transvektor closed-loop control system for rotating-field machines. Siemens Rev 39(5):217–219
3. De Doncker R (1991) Parameter sensitivity of indirect universal field oriented controllers. In: PESC'91 record. 22nd annual IEEE power electronics specialists conference, 1991, pp 605–612. https://doi.org/10.1109/PESC.1991.162737
4. De Doncker R, Novotny D (1994) The universal field oriented controller. IEEE Trans Ind Appl 30(1):92–100. https://doi.org/10.1109/28.273626
5. De Doncker R, Profumo F (1989) The universal field oriented controller applied to tapped stator windings induction motors. In: PESC'89 record. 20th annual IEEE power electronics specialists conference, 1989, vol 2, pp 1031–1036. https://doi.org/10.1109/PESC.1989.48592
6. Habetler T, Profumo F, Griva G, Pastorelli M, Bettini A (1998) Stator resistance tuning in a stator-flux field-oriented drive using an instantaneous hybrid flux estimator. IEEE Trans Pow Electr 13(1):125–133. https://doi.org/10.1109/63.654966

7. Hasse K (1969) Zur dynamik drehzahlgeregelter antriebe mit stromrichtergespeisten asynchron-kurzschlussläufermaschinen. PhD Thesis, TH Darmstadt
8. Hughes A (2006) Electric motors and drives: fundamentals, types and applications, 3rd edn. Newnes, Oxford
9. Lipo T, Krause P (1969) Stability analysis of a rectifier-inverter induction motor drive. IEEE Trans Pow Apparatus Syst 88(1):55–66. https://doi.org/10.1109/TPAS.1969.292338
10. Profumo F, Tenconi A, De Doncker R (1991) The universal field oriented (UFO) controller applied to wide speed range induction motor drives. In: PESC'91 record. 22nd annual ieee power electronics specialists conference, 1991, pp 681–686. https://doi.org/10.1109/PESC.1991.162749
11. Profumo F, Griva G, Pastorelli M, Moreira J, De Doncker R (1994) Universal field oriented controller based on air gap flux sensing via third harmonic stator voltage. IEEE Trans Ind Appl 30(2):448–455. https://doi.org/10.1109/28.287510
12. Qi F, Scharfenstein D, Schubert M, Doncker RWD (2017) Precise field oriented torque control of induction machines using thermal model based resistance adaption. In: 2017 IEEE 12th international conference on power electronics and drive systems (PEDS), pp 1055–1061. https://doi.org/10.1109/PEDS.2017.8289166
13. Walcarius H, Vandenput A, Jordan H, Geysen W (1978) Stability analysis of oscillating induction machines. In: International conference on electrical machines, 1978, 11–13 September, Brussels
14. Xu X, De Doncker R, Novotny D (1988) A stator flux oriented induction machine drive. In: PESC'88 record 19th annual IEEE power electronics specialists conference, 1988, vol 2, pp 870–876. https://doi.org/10.1109/PESC.1988.18219
15. Xu X, de Doncker R, Novotny D (1988) Stator flux orientation control of induction machines in the field weakening region. In: Conference record of the 1988 IEEE industry applications society annual meeting, vol 1, pp 437–443. https://doi.org/10.1109/IAS.1988.25097

Chapter 10
Switched Reluctance Drive Systems

The term *switched reluctance* (SR) may to the uninitiated reader convey the notion that the reluctance of the machine is *switched*. In reality the magnetic reluctance of the machine is rotor angle dependent and the term *switched* refers to the electronic commutation of the electrical phases by means of a power electronic converter. Torque production based purely on variation of the magnetic reluctance is well established and the first patent based on this approach stems from 1839 [24]. Despite being one of the oldest known machine concepts, it has not been able to maintain its hierarchical position in comparison to machines which utilize the Lorentz force as a basis for torque production. Since the development of power electronics, there is no readily identifiable single reason for this sequence of events, but it is perhaps useful to consider some of the factors which may well have facilitated this state of affairs.

From a learning perspective, the process of familiarization of, for example, a vector controlled induction machine drive concept starts from basic modeling principles, leading to a model based control concept. This model based approach to develop torque controllers (field-oriented control) may in turn be followed by a familiarization process that encompasses some form of simulation and practical application. To facilitate this learning process, the reader may acquire an "off-the-shelf" drive concept, with a machine which is well defined in terms of its model and parameters. If we attempt to emulate this learning process for a switched reluctance drive, the process is less well defined simply because this type machine is more difficult to model. Furthermore, the torque production concept of the switched reluctance machine (SRM) is considerably different to the Lorentz force based machine. The latter may for many readers have a degree of familiarity given the fact that the basic principles are often taught at an early stage of education. The learning curve to mastering SR technology is steep and comparatively unfamiliar given the need to understand a machine concept in which non-linear effects, i.e., magnetic saturation, play a key role. The inevitable non-linear modeling techniques required to handle saturation are not standard and machine characteristics are not

R. W. De Doncker et al., *Advanced Electrical Drives*, Power Systems,
https://doi.org/10.1007/978-3-030-48977-9_10

readily quantified by few parameters. Today, there are commercial tools available for modeling and designing switched reluctance machines, given the need to optimize the drive with respect to the application.

Notwithstanding the above, SR drive technology is increasingly seen as a cost saving alternative to conventional Lorentz force based machines. One of the key reasons for this is the simple doubly salient machine concept, devoid of permanent magnets, or any form of winding or squirrel cage on the rotor. Consequently, the rotor inertia can be kept low (if needed) and high speed operation, which is instrumental for realizing high power densities, is not inhibited by the presence of permanent magnets or rotor windings. The latter is also beneficial in terms of improving overall machine efficiency, i.e., no rotor losses due to the absence of rotor windings. However, this perceived advantage is offset by the need for a field which rotates faster than the rotor itself, which increases the magnetic stator losses in the machine. The fact that the SRM uses low cost concentrated stator winding with short end turns and high fill factors reduces stator ohmic losses. Furthermore, torque production is not current polarity dependent which provides an additional degree of freedom with respect to the converter topology. In this context, the conventional converter concept (full-bridge converter) of two series connected switches across the supply may be abandoned in favor of a short circuit proof asymmetric half-bridge converter topology which improves reliability [22].

Given the above, the reader may well wonder why the SR drive is not universally used today, given its obvious production and commercial benefits. The answer lies with the pulsed nature of torque production, which is accompanied by large radial mechanical forces, which may contribute to acoustical noise emissions and these may in turn be exacerbated by the type of current control technique used [11]. Note, that permanent magnet synchronous machines with concentrated stator windings exhibit similar elevated noise problems. As with Lorentz force based drives, the use of computer control and low cost DSPs recently provides the opportunity to improve drive performance [3]. For the SR drive the benefits are to reduce torque ripple and eliminating the need for a position sensor by implementing some form of position sensorless control [7]. In the face of equal converter and controller costs the benefits of a very simple rugged machine concept may serve to push SR drive development to greater heights in the future.

The number of SR drive manufactures is small in comparison with, for example, induction machine drive manufacturers. The manufactures are usually tailored to specific applications. It is instructive to consider applications where the merits of SR drives have warranted their use in comparison to other drive concepts. Its perceived ability to operate at very high speeds has seen its application in the textile processing industry (operating speeds of 100,000 rpm) [1, 14], aerospace industry (150 kW, aircraft turbine starter) [19], and trains (40 kW, high-performance compressors for air conditioning in ICE 3 high speed trains). Its inherently low cost construction benefit has been used to advantage in hand-held power tools [20], household appliances (mixer/kneading machines, vacuum cleaners with power range 0.5–2 kW), sliding doors, and electrical vehicles (traction drives for cars 50 kW range and scooters [10, 21]). Its benefit in terms of robustness has led to applications in aerospace other than the one indicated above and in the mining industry (e.g., in high-performance explosion proof drives with power levels up to 400 kW).

10.1 Basic Machine Concepts

The machine is characterized by discrete stator and rotor poles, which is commonly referred to as a doubly salient structure [2, 12, 13]. Figure 10.1 shows a typical machine configuration. The example three-phase machine ($N_{\mathrm{ph}} = 3$) has six stator teeth $N_{\mathrm{s}} = 6$ and four rotor teeth $N_{\mathrm{r}} = 4$, a configuration which is known as a 6/4 structure. The windings are typically positioned around each stator pole, which is why this winding arrangement is referred to as *concentrated* or *short pitched*. From a magnetic perspective, each phase of the 6/4 machine has two *magnetic* poles per phase, i.e., one pole pair ($p = 1$), given that diametrically opposed coils are electrically connected in series or in parallel. For example, phase A is made up of the coils located on teeth A and A'. If this phase is excited, a magnetic flux path is formed by the two stator teeth A–A', the rotor and stator yoke (see Fig. 10.1b).

Observing one phase of Fig. 10.1, it can be seen that two equilibrium positions of the rotor exist. The rotor position shown in Fig. 10.1a is called the unaligned position with respect to phase A–A'. The rotor angle at this position θ is defined as θ^{u}. The position of smallest magnetic reluctance is called the aligned position θ^{a}, as is shown in Fig. 10.1b. The width of the rotor and stator poles are defined as β_{s} and β_{r}. The interpolar arcs of stator and rotor (τ_{sp} and τ_{rp}), which represent the angle between two adjacent rotor or stator teeth, shown in Fig. 10.1b, are determined by

$$\tau_{\mathrm{sp}} = \frac{2\pi \text{ rad}}{N_{\mathrm{s}}}, \tau_{\mathrm{rp}} = \frac{2\pi \text{ rad}}{N_{\mathrm{r}}}. \tag{10.1}$$

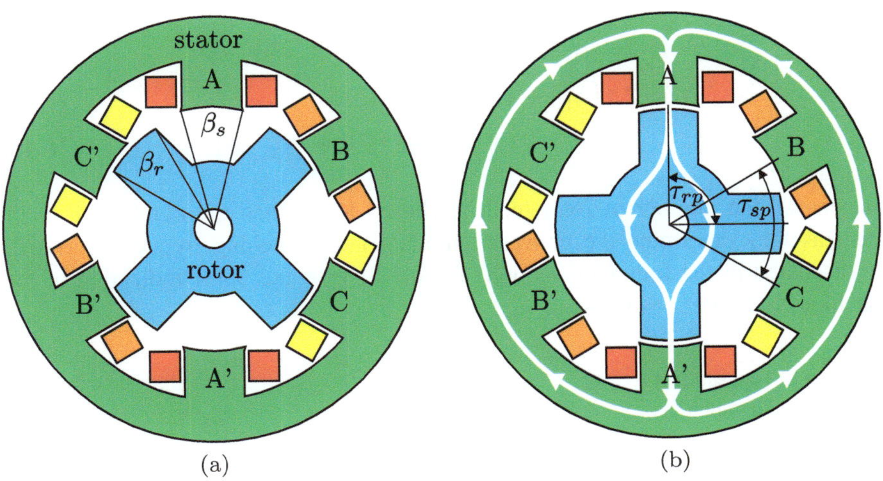

Fig. 10.1 Radial flux switched reluctance machine examples, showing a $N_{\mathrm{ph}} = 3$ phase machine with a 6/4 stator/rotor configuration. (**a**) Unaligned position (phase A). (**b**) Aligned position (phase A)

Definition of Electrical Angle
The position of the rotor is directly correlated with the mechanical angle θ_{m}. One period describes one revolution of the rotor. To become independent of the machine configuration it is reasonable to specify the rotor position of switched reluctance machines in electrical degrees. The electrical angle is defined by the periodicity of the machine. For example, in Fig. 10.1 the rotor repeats every 90° mechanical (τ_{rp}). Hence, this is defined as one electrical period of 360°. The relationship between electrical and mechanical angle is given in Eq. (10.2).

$$\theta_{\mathrm{e}} = N_{\mathrm{r}}\theta_{\mathrm{m}}. \tag{10.2}$$

Note that the switched reluctance machine illustrated in Fig. 10.1 is *radial flux* oriented, given that the flux which crosses the air-gap is predominantly in the radial direction. Most machines in use are of this type, given the fact that this approach allows the machine to be manufactured by stacked laminations, as used in, for example, induction and synchronous machines.

Possible Machine Configurations
The relationship between the number of stator teeth N_{s}, magnetic pole pair number p, and number of electrical phases N_{ph} for switched reluctance machines may be written as

$$N_{\mathrm{s}} = 2pN_{\mathrm{ph}}. \tag{10.3}$$

To avoid equilibrium positions in which the SRM cannot provide torque, the number of rotor teeth N_{r} must be different from N_{s}. Typically, N_{r} is calculated using

$$N_{\mathrm{r}} = 2p\left(N_{\mathrm{ph}} - 1\right), \quad N_{\mathrm{r}} < N_{\mathrm{s}}, \quad \text{for } N_{\mathrm{ph}} > 1 \tag{10.4a}$$

$$N_{\mathrm{r}} = 2p\left(N_{\mathrm{ph}} + 1\right), \quad N_{\mathrm{r}} > N_{\mathrm{s}}, \quad \text{for } N_{\mathrm{ph}} > 1 \tag{10.4b}$$

$$N_{\mathrm{r}} = 2p \quad \text{for } N_{\mathrm{ph}} = 1. \tag{10.4c}$$

In general, the number of rotor teeth is chosen lower than the number of stator teeth as they determine the converter fundamental switching frequency, as will be discussed in the following section. Table 10.1 provides some indication of the machine configurations in use.

Table 10.1 Typical SRM machine configuration

N_{ph}	1		2		3			4		5	
N_{s}	2	4	4	8	6	12	18	8	16	10	20
N_{r}	2	4	2	4	4	8	12	6	12	8	16
p	1	2	1	2	1	2	3	1	2	1	2

10.2 Operating Principles

The basic operating principles of the switched reluctance machine are discussed on the basis of a single-phase machine. A single-phase model is representative because even in multi-phase SRMs mutual coupling between electrical phases can be neglected. Consequently, the development of generic models for this single-phase machine is directly applicable to multi-phase concepts.

10.2.1 Single-Phase Motor Concept

The machine under consideration is a 2/2 configuration, i.e., two stator and two rotor teeth as shown in Fig. 10.2. In the given example the rotor is displaced by an angle θ_m from the stator teeth. An angle dependent current source $i(\theta_m)$ as shown in Fig. 10.2 is connected to the N-turn phase winding, which consists of two concentrically wound coils located on each of the two stator teeth.

The inter-pole arc τ_{rp} is in this example equal to 180° mechanical. This angle is also equal to one electric period of the two-pole rotor.

If the rotor is displaced by an angle $\theta_m = \tau_{rp}$, a single torque pulse will result, provided the appropriate phase excitation conditions are met. The number of torque pulses, defined by the variable N_{pu} for this machine equals two. For a multi-phase machine N_{pu} is given as

$$N_{pu} = N_r N_{ph} = 2pN_{ph}\left(N_{ph} - 1\right). \tag{10.5}$$

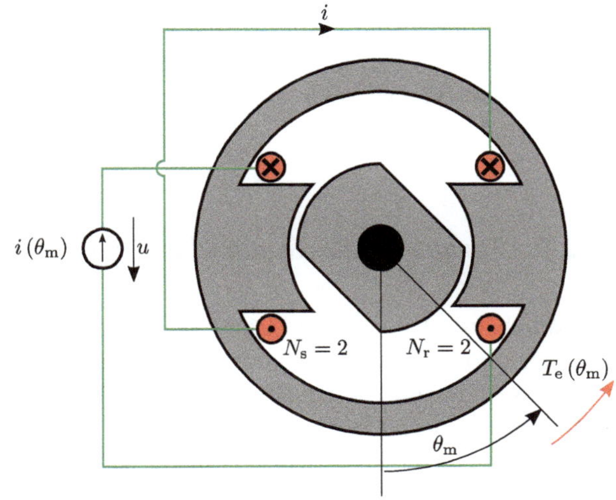

Fig. 10.2 2/2 single-phase switched reluctance motor

A mechanical shaft speed of ω_m (rad/s) is assumed, which in turn is linked to the fundamental electrical switching frequency f_e (Hz) of the excitation source. The electrical frequency may be written as

$$f_e = \frac{\omega_m}{\tau_{rp}} = N_r n_m \tag{10.6}$$

where n_m (rps) represents the shaft speed. Note that the electrical frequency must be equal to the number of torque pulses given the aim of controlling the excitation source in such a manner that the largest possible number of single torque pulses is realized during a rotor displacement interval. Furthermore, Eq. (10.6) shows that the speed of the stepwise moving magnetic field $\omega_e = 2\pi f_e$ is higher by a factor equal to the number of rotor teeth N_r than the shaft speed ω_m. The concept of a machine with a rotating magnetic field that is higher than the shaft speed may be compared in mechanical terms to the planetary gearing. This torque multiplication process, which is referred to as the *vernier* principle in electrical machines, leads to increased switching losses in the converter and core losses in the SR machine, given the need for a higher electrical frequency (see Eq. (10.6)). However, the presence of this vernier principle in SR machines is fundamental to its ability to produce a torque that is an average similar, if not higher, than that of an induction machine of the same frame size. In terms of choosing a higher or lower number of rotor teeth with respect to the stator teeth number, it is prudent to choose the latter, given that this leads to comparatively lower core and switching losses.

The terminal voltage equation for this machine is according to Fig. 10.2 and Kirchhoff's voltage law of the form

$$u(i, \theta_m) = R\,i\,(\theta_m) + \frac{d\psi\,(i, \theta_m)}{dt}, \tag{10.7}$$

where R represents the phase coil resistance and $\psi(i, \theta_m)$ the flux linkage depending on phase current and rotor angle, otherwise referred to as the *magnetization characteristics* of the machine.

10.2.2 Torque Production and Energy Conversion Principles

The simplicity of the doubly salient machine structure (see Fig. 10.2) may give the impression that understanding the nature of torque production and energy conversion principles is equally simple. Unfortunately, this is not the case as will become apparent in this section. Initially, the single-phase machine model is examined to ascertain the energy flows that are present between the supply source, magnetic energy in the air-gap and energy supplied to the shaft. In this process, the role of magnetic saturation is duly explored. Iron losses and copper losses will inevitably affect the overall machine performance but do not significantly affect the torque and energy conversion principles. Hence, these are ignored in this analysis. Under these circumstances, the terminal voltage Eq. (10.7) reduces to the following form:

$$u(i, \theta_m) = \frac{d\psi(i, \theta_m)}{dt}. \tag{10.8}$$

Energy Balance Equation

A suitable starting point for this analysis is the overall incremental energy balance equation of the machine. The latter builds on the understanding that the incremental input energy $dW_{in}(i, \theta_m)$ must be equal to the sum of the incremental magnetic energy in the air-gap $dW_f(i, \theta_m)$ and incremental energy linked to the shaft $T_e(i, \theta_m)\, d\theta_m$. Consequently, the following energy balance equation (in difference form) holds:

$$dW_{in}(i, \theta_m) = dW_f(i, \theta_m) + T_e(i, \theta_m)\, d\theta_m. \tag{10.9}$$

The incremental electrical input energy $dW_{in}(i, \theta_m)$ can also be written in terms of the terminal quantities, as shown in Eq. (10.10), where $u(i, \theta_m)$ is defined by Eq. (10.8). The reader is reminded of the fact that constant shaft speed is assumed, hence the variable time t may also be expressed in terms of the rotor angle $t = \theta_m/\omega_m$. The three key variables for this type of analysis are the flux linkage, current, and rotor angle. However, not all three variables may be chosen freely. Hence, it is a matter of selecting two of the three as so-called *independent* variables. For this analysis it is prudent to assign the rotor angle θ_m and current i as independent variables. Under these conditions Eq. (10.8) may also be written as shown in Eq. (10.11).

$$dW_{in}(i, \theta_m) = i\,(t)\,u\,(t)\;dt \tag{10.10}$$

$$u(i, \theta_m) = \left[\frac{\partial\psi(i, \theta_m)}{\partial i}\right]_{\theta_m=\text{const}} \frac{di}{dt} + \omega_m \left[\frac{\partial\psi(i, \theta_m)}{\partial\theta_m}\right]_{i=\text{const}} \tag{10.11}$$

where $\omega_m = \frac{d\theta_m}{dt}$ represents the shaft speed in rad/s. The partial derivative as shown in Eq. (10.12a) represents the flux-linkage derivative with respect to the current and is known as the incremental inductance, while the second term in Eq. (10.11) represents the so-called *motional EMF*.

$$l(i, \theta_m) = \left[\frac{\partial\psi(i, \theta_m)}{\partial i}\right]_{\theta_m=\text{const}} \tag{10.12a}$$

$$u_e(i, \theta_m) = \omega_m \left[\frac{\partial\psi(i, \theta_m)}{\partial\theta_m}\right]_{i=\text{const}}. \tag{10.12b}$$

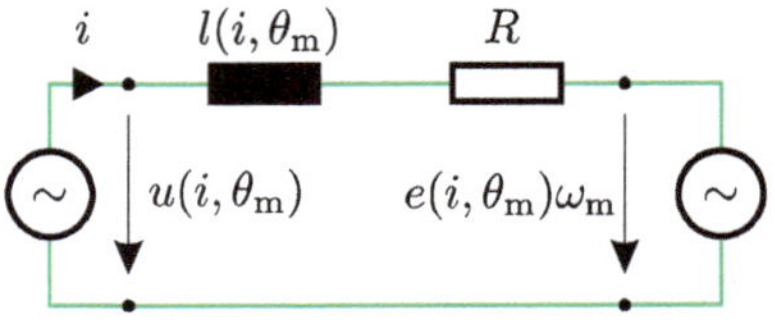

Fig. 10.3 Single-phase equivalent circuit of a switched reluctance machine

In practice, it is useful to introduce a so-called normalized *EMF variable* $e = u_e(i,\theta_m)/\omega_m$. Substitution of Eq. (10.12) into Eq. (10.11) leads to the terminal equation in terms of the chosen independent variables i and θ_m.

$$u(i,\theta_m) = \underbrace{l(i,\theta_m)\frac{di}{dt}}_{u_l} + \underbrace{e(i,\theta_m)\omega_m}_{u_e} . \tag{10.13}$$

Equation (10.13) shows that the per phase equivalent circuit of an SR machine can be represented by a series circuit (see Fig. 10.3), which consists of an inductance, a stator winding resistance (which can be easily added to this model), and a voltage source proportional to ω_m. The significance of this model representation should not be lost as it shows that the switched reluctance circuit model is similar to that of conventional motors.

The nature of the energy conversion process is however significantly different given that the flux linkage/current characteristics are rotor angle dependent and in most cases highly non-linear. In this context the energy conversion process is often referred to as a *parametric* type of energy conversion. Further insight into the energy flow can be obtained by defining the incremental input energy dW_{in} in terms of the energy linked with the circuit components which leads to

$$dW_{in}(i,\theta_m) = i\, l(i,\theta_m)\, di + i\, e(i,\theta_m)\, d\theta_m. \tag{10.14}$$

General Torque Equation

Substitution of Eq. (10.14) into Eq. (10.9) leads to an expression for the torque $T_e(i,\theta_m)$ in the event that the incremental field energy component dW_f present in Eq. (10.9) is expressed in terms of its partial derivatives, namely

$$dW_f(i,\theta_m) = \left[\frac{\partial W_f(i,\theta_m)}{\partial \theta_m}\right]_{i=\text{const}} d\theta_m + \left[\frac{\partial W_f(i,\theta_m)}{\partial i}\right]_{\theta_m=\text{const}} di. \tag{10.15}$$

Substitution of Eq. (10.15) into Eq. (10.9) and equating the latter with Eq. (10.14) in terms of the derivatives di and $d\theta_m$ lead to

$$W_f(i,\theta_m) = \left[\int_0^i i\, l(i,\theta_m)\, di\right]_{\theta_m=\text{const}} \tag{10.16a}$$

$$T_{\rm e}(i,\theta_{\rm m}) = i\ e(i,\theta_{\rm m}) - \left[\frac{\partial W_{\rm f}(i,\theta_{\rm m})}{\partial\theta_{\rm m}}\right]_{i={\rm const}} . \tag{10.16b}$$

Equation (10.16) may also be written in an alternative form by making use of $l(i,\theta_{\rm m}) = [\partial\psi(i,\theta_{\rm m})/\partial i]_{\theta_{\rm m}={\rm const}}$, $e(i,\theta_{\rm m}) = [\partial\psi(i,\theta_{\rm m})/\partial\theta_{\rm m}]_{i={\rm const}}$ which after some manipulation gives

$$W_{\rm f}(i,\theta_{\rm m}) = \left[\int_0^{\psi} i(\psi,\theta_{\rm m})\,{\rm d}\psi\right]_{\theta_{\rm m}={\rm const}} \tag{10.17a}$$

$$T_{\rm e}(i,\theta_{\rm m}) = \left[\frac{\partial}{\partial\theta_{\rm m}}\left(i\psi - W_{\rm f}(i,\theta_{\rm m})\right)\right]_{i={\rm const}} . \tag{10.17b}$$

The term $(i\psi - W_{\rm f}(i,\theta_{\rm m}))$ is known and defined in literature as the *co-energy* $W_{\rm f}'$, which may for all ψ and $\theta_{\rm m}$ be written as

$$W_{\rm f}'(i,\theta_{\rm m}) = \left[\int_0^{i} \psi(i,\theta_{\rm m})\,{\rm d}i\right]_{\theta_{\rm m}={\rm const}} . \tag{10.18}$$

Introduction of the term $W_{\rm f}'$ in Eq. (10.17b) allows the latter to be written as

$$T_{\rm e}(i,\theta_{\rm m}) = \left[\frac{\partial W_{\rm f}'(i,\theta_{\rm m})}{\partial\theta_{\rm m}}\right]_{i={\rm const}} . \tag{10.19}$$

A further observation of Eq. (10.17b) demonstrates that the largest achievable shaft torque level, defined as $\tilde{T}_{\rm e}$, can be realized in case $W_{\rm f} \to 0$. Under these conditions the torque is given as

$$\tilde{T}_{\rm e}(i,\theta_{\rm m}) = \left[\frac{\partial\,(i\psi)}{\partial\theta_{\rm m}}\right]_{i={\rm const}} . \tag{10.20}$$

Influence of Linear and Non-linear Inductance

The process of designing a machine with a torque output which approaches the theoretical upper limit value $\tilde{T}_{\rm e}$ is primarily concerned with the shape of the flux/linkage/current characteristics. This statement is demonstrated with the aid of Fig. 10.4, which shows an incremental rotor angle change of the rotor magnetization curves (under constant current conditions) for the linear and non-linear case.

For the linear case, as shown in Fig. 10.4a, the output torque is equal to half the maximum value $\tilde{T}_{\rm e}$. The reason for this is that the output energy $T_{\rm e}\,{\rm d}\theta_{\rm m}$, is tied to the incremental co-energy ${\rm d}W_{\rm f}'$, which corresponds to area $0 - B - E - 0$ in Fig. 10.4a. Note that in the linear case considered here, the magnetic energy $W_{\rm f}$ is equal to the co-energy $W_{\rm f}'$. If we consider the flux linkage current curve $\psi\,(i,\theta_{\rm m})$, then the co-energy and magnetic energy are represented by the areas $0 - A - B - 0$ and $0 - B - C - 0$, respectively. The energy which corresponds to $\tilde{T}_{\rm e}\,{\rm d}\theta_{\rm m}$ is given

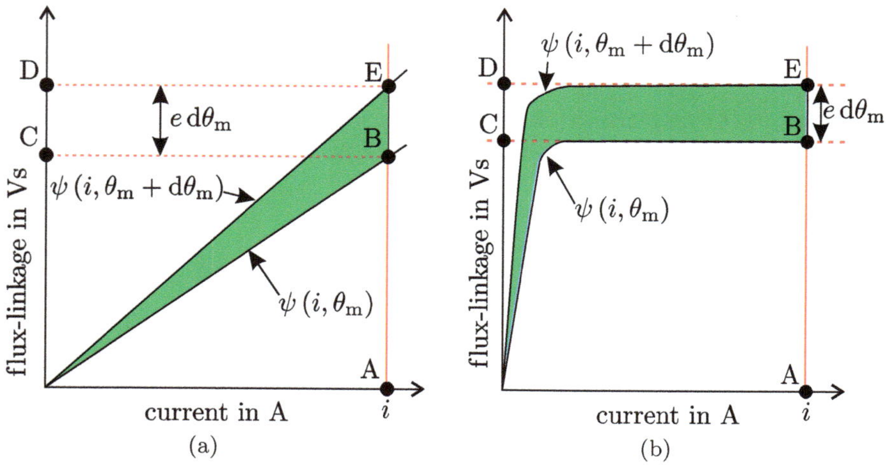

Fig. 10.4 Magnetization curves for the linear and non-linear case. (**a**) Linear case: $T_e = \frac{1}{2}\tilde{T}_e$. (**b**) Non-linear case: $T_e \approx \tilde{T}_e$

by area $B - C - D - E$, which is also equal to the energy $i e\, d\theta_m$ delivered by the supply source connected to the machine. Under constant current conditions, as assumed here, the energy $\tilde{T}_e\, d\theta_m$ represents the total amount of energy supplied to the machine and the area $B - C - D - E$ is double the area $0 - B - E - 0$ linked to the co-energy change.

If the machine is designed with a high magnetic saturation level, as shown (unrealistically high) in Fig. 10.4, it may be observed that the energy and co-energy levels linked to for example the flux linkage current curve $\psi(i, \theta_m)$ are very much different. Under these conditions the incremental co-energy as shown by the *green* area in Fig. 10.4 approaches the maximum output energy level $\tilde{T}_e\, d\theta_m$ which corresponds to area $B - C - D - E$. It is therefore not surprising to learn that switched reluctance machines are usually designed to operate with a substantial level of magnetic saturation (in the stator and rotor teeth of the machine). This means that the magnetization curves will be highly non-linear, which in turn has important implications for the simulation models tied to this type of machine.

Energy Flow of SR Motor

The understanding of the machine is enhanced by considering the energy flows which occur between the supply, load, and magnetic energy reservoir W_f as shown in Fig. 10.5.

Readily apparent in Fig. 10.5 are the incremental energy flow arrows that are linked with the torque variables $\tilde{T}_e$ and T_e. The incremental energy flow arrows tied to the magnetic energy reservoir W_f can be found by making use of Eq. (10.17) and the realization that the current supplied to the machine will be a function of the rotor angle, hence $i(\theta_m)$, $di = {}^{\partial i}/_{\partial \theta_m} d\theta_m$. Given the above, the energy flows tied to W_f may be written as

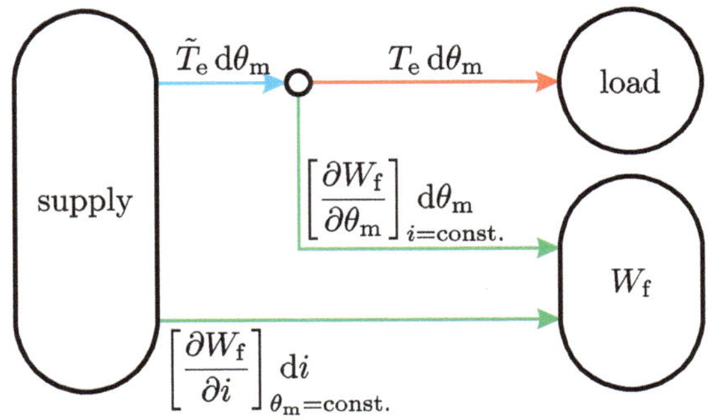

Fig. 10.5 Energy flow in SR machine

$$\left[\frac{\partial W_f}{\partial i}\right]_{\theta_m=\text{const}} di = i\,l\,(i,\ \theta_m)\,\frac{\partial i}{\partial \theta_m} d\theta_m \tag{10.21a}$$

$$\left[\frac{\partial W_f}{\partial \theta_m}\right]_{i=\text{const}} d\theta_m = \left(\tilde{T}_e - T_e\right) d\theta_m . \tag{10.21b}$$

If we consider the energy flow linked to Fig. 10.4, for the trajectory $B \rightarrow E$, we note that the contribution $[\partial W_f/\partial i]_{\theta_m=\text{const}} di$ will be zero, given that the current is constant under these conditions. Furthermore, the amount of energy delivered to the load and magnetic energy supply will be equal in the linear case. This situation is however drastically changed for the non-linear case, in which most of the energy from the supply is transferred to the load.

10.2.3 Single-Phase Switched Reluctance Machine: A Linear Example

To show the energy flow and torque production principles of a switched reluctance machine, it is helpful at this stage to consider a simple example of a single-phase switched reluctance motor. In this context, the term *linear* denotes that the machine is not operated in saturation. This leads to the simplification that the inductance only depends on rotor position so that the incremental inductance $l(i, \theta_e) = L(\theta_e)$. The flux linkage may therefore be written as

$$\psi = L\,(\theta_e)\,i\ . \tag{10.22}$$

The torque of a linear machine may be found using Eq. (10.22) with Eqs. (10.18) and (10.19) which leads to the following expression:

$$T_e(\theta_m) = \frac{1}{2} i^2 \frac{dL(\theta_m)}{d\theta_m}. \tag{10.23}$$

Due to the periodicity of the phase inductance of a switched reluctance machine, it is sufficient to pursue the analysis for one electrical period. Therefore, the rotor position is measured in electrical radians. In Eq. (10.23), torque is proportional to the derivative of inductance over the mechanical rotor position. Substituting the mechanical rotor position by its electrical equivalent leads, together with Eq. (10.2), to the following expression for the (per phase) torque:

$$T_e(\theta_e) = \frac{1}{2} N_r i^2 \frac{dL(\theta_e)}{d\theta_e}. \tag{10.24}$$

An important observation to be made from Eqs. (10.23) and (10.24) is the fact that the phase current appears in quadratic form, which implies that its polarity cannot influence the direction in which torque is produced. However, it is possible to produce negative torque by an excitation of the phase in the range of decreasing inductance (negative derivative).

Basically, two main operating modes for the switched reluctance machine can be identified depending on the speed of the machine. In the following, typical phase excitation patterns are examined in detail for operation at low and at high speed.

Phase Excitation at Low Speed

At low speed, the excitation pattern of one phase can be roughly divided into three steps:

1. Initial magnetization at maximum positive voltage
2. Torque production at constant current
3. De-magnetization at maximum negative voltage

Hence, it is convenient to use a piecewise linear model of the phase inductance, where the slope of the inductance is constant within each of the three ranges.

Figure 10.6 shows in its uppermost diagram the piecewise linear model of one phase inductance. In the range of $\tau_u/2$ to each side of the unaligned position, the inductance is constant at its unaligned value $L = L_u$. In the aligned position, the inductance also does not experience any changes as long as the rotor is not displaced further than $\tau_a/2$ to either side. In between, the inductance changes linearly at a constant slope according to $dL/d\theta_e = \text{const}$. From Eq. (10.24), it becomes obvious that torque can only be produced in the range of changing inductance. As a consequence, one aim during low speed operation is to have the active phase fully excited with the desired current right before the inductance starts to increase, as can

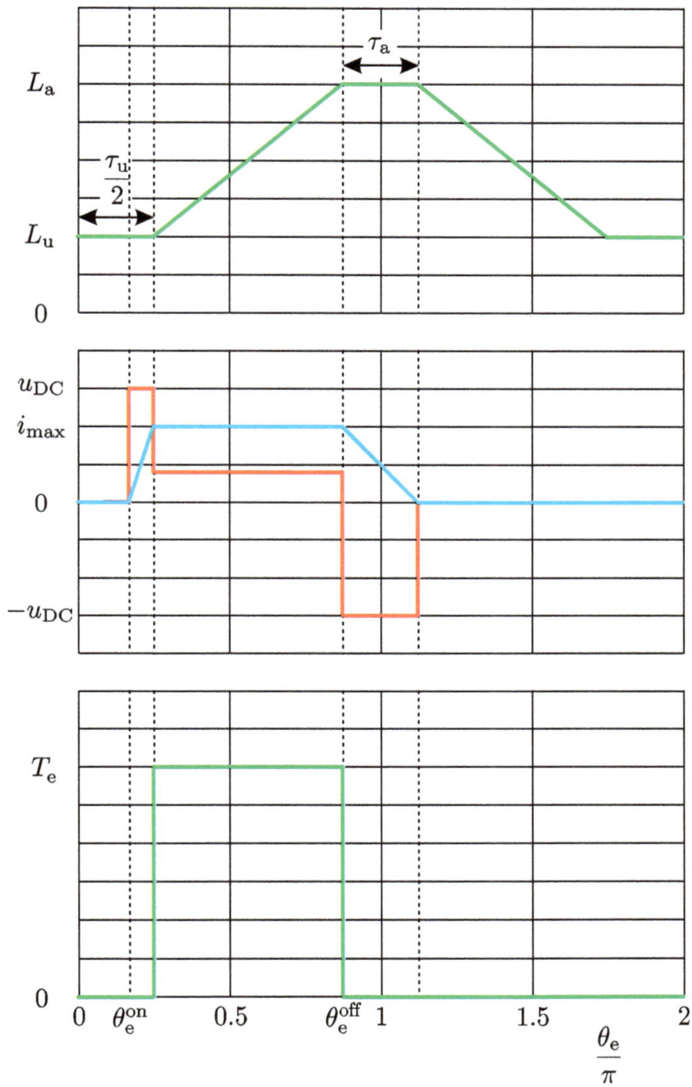

Fig. 10.6 Piecewise linear inductance, phase voltage (red), current (blue), and resulting phase torque over one electrical period at constant speed

be seen in the second graph of Fig. 10.6. This initial excitation can be carried out fast as L is low and the supply is at maximum dc-link voltage.

Once the current has reached the desired value and the inductance starts to increase, torque is produced according to Eq. (10.24). During this time, the phase voltage (blue curve in second diagram of Fig. 10.6) is adjusted to maintain a constant current waveform. Note that the flux linkage increases linearly according to Eq. (10.8). At the end of the region of increasing inductance, the phase is turned off (demagnetized) at maximum negative dc-link voltage. Due to the significantly larger inductance in the aligned position, the de-magnetization phase takes longer than the magnetization phase.

The resulting torque is shown in the third graph of Fig. 10.6. An important observation to be made in the torque waveform is that no torque is produced in the constant inductance region although the phase is already partially excited. As a result, the produced torque has, in contrast to the current waveform, a rectangular shape.

Although no torque is produced within the regions of constant inductance, which implies that no energy is supplied to the mechanical load, magnetic energy is fed into the phase during magnetization or fed back to the source during de-magnetization. During magnetization, Eq. (10.17a) together with Eq. (10.22) can be rewritten as

$$W_{\mathrm{f}}(i, \theta_{\mathrm{e}}) = \frac{1}{2} L(\theta_{\mathrm{e}})\, i^2 \,. \tag{10.25}$$

The waveform of the magnetic energy can be examined as the blue plot in Fig. 10.7. During magnetization and de-magnetization, the magnetic energy changes proportionally to the square of current. In the period of constant current, it rises linearly with the phase inductance. The green curve shows the mechanical energy supplied to the load which is equal to the integral of torque over the rotor position. It should be noted that the total change of mechanical energy (denoted as ΔW_{m}) is equal to the change of magnetic energy during the constant current period. The energy flows related to W_{f} reservoir can be calculated according to

$$\left[\frac{\partial W_{\mathrm{f}}}{\partial i}\right]_{\theta_{\mathrm{e}}=\mathrm{const}} \frac{\mathrm{d}i}{\mathrm{d}\theta_{\mathrm{e}}} = L(\theta_{\mathrm{e}})\, i \frac{\mathrm{d}i}{\mathrm{d}\theta_{\mathrm{e}}} \tag{10.26a}$$

$$\left[\frac{\partial W_{\mathrm{f}}}{\partial \theta_{\mathrm{e}}}\right]_{i=\mathrm{const}} = \frac{1}{2} i^2 \frac{\mathrm{d}L(\theta_{\mathrm{e}})}{\mathrm{d}\theta_{\mathrm{e}}} . \tag{10.26b}$$

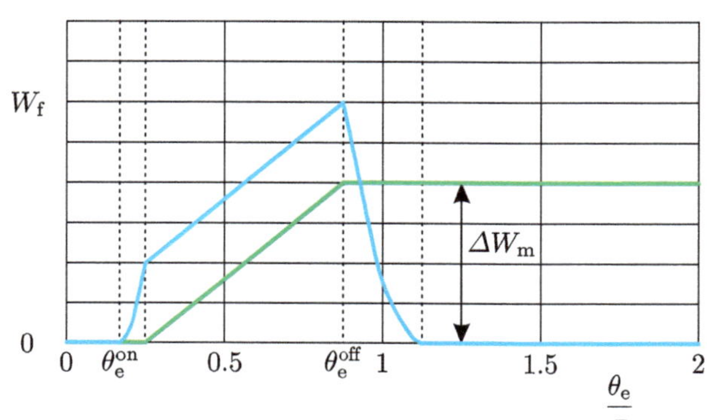

Fig. 10.7 Magnetic (blue) and mechanical (green) energy over one electrical period

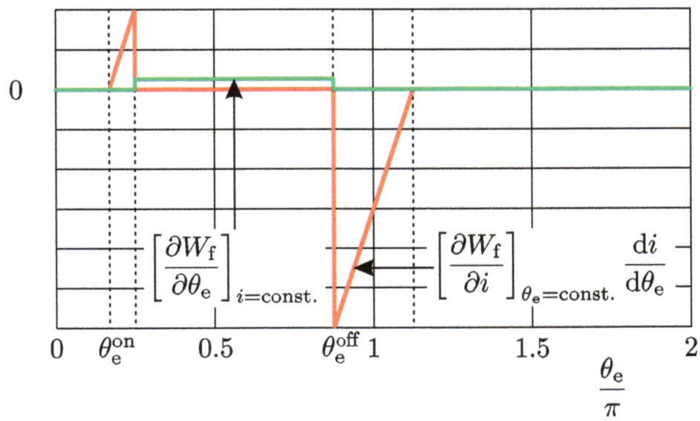

Fig. 10.8 Energy flows related to W_f reservoir

Fig. 10.9 Half cycle flux-linkage current diagram

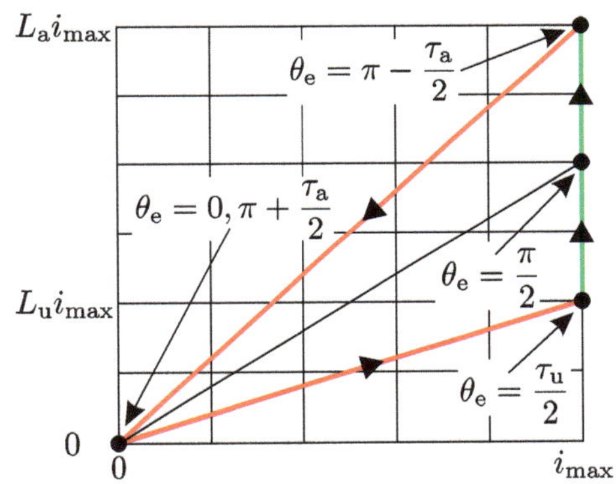

The waveforms of both are plotted in Fig. 10.8. Note, that the partial derivative of W_f over rotor position (green curve) has the same shape as the torque waveform. Another important observation to be made is that the magnetic energy recovered at the end of excitation is substantially larger than the magnetic energy provided by the supply during the magnetization phase.

For the analysis of switched reluctance drives it is helpful to consider the flux linkage/current locus which appears in the flux linkage/current diagram during an active rotor cycle. The locus which appears with the present choice of current excitation is displayed in Fig. 10.9. The two red trajectories in the diagram correspond to the magnetization (0 to $\tau_u/2$) and de-magnetization ($\pi - \tau_a/2$ to $\pi + \tau_a/2$) processes in the regions of constant inductance. The green leg reflects the constant current region, where the inductance rises linearly. The area, which is surrounded by the entire trajectory is equal to the mechanical energy delivered to the shaft.

Phase Excitation at High Speed (Single Pulse Mode)

At higher speeds, the maximum achievable slope of the stator current over rotor position decreases due to the induced voltage. As a consequence, when a certain speed is exceeded, the nominal phase current cannot be established within the constant inductance range anymore. The same applies for the de-magnetization at the end of the active cycle. In general, the time-spans of magnetization and de-magnetization overlap with the time-spans of changing inductance, a condition, which can be avoided at low speed operation.

Hence, at high speed it is more instructive to use a continuously differentiable inductance model to describe the mechanisms of energy conversion analytically. Such a model is represented in Eq. (10.27) and in the inductance plot of Fig. 10.10 over one electrical period.

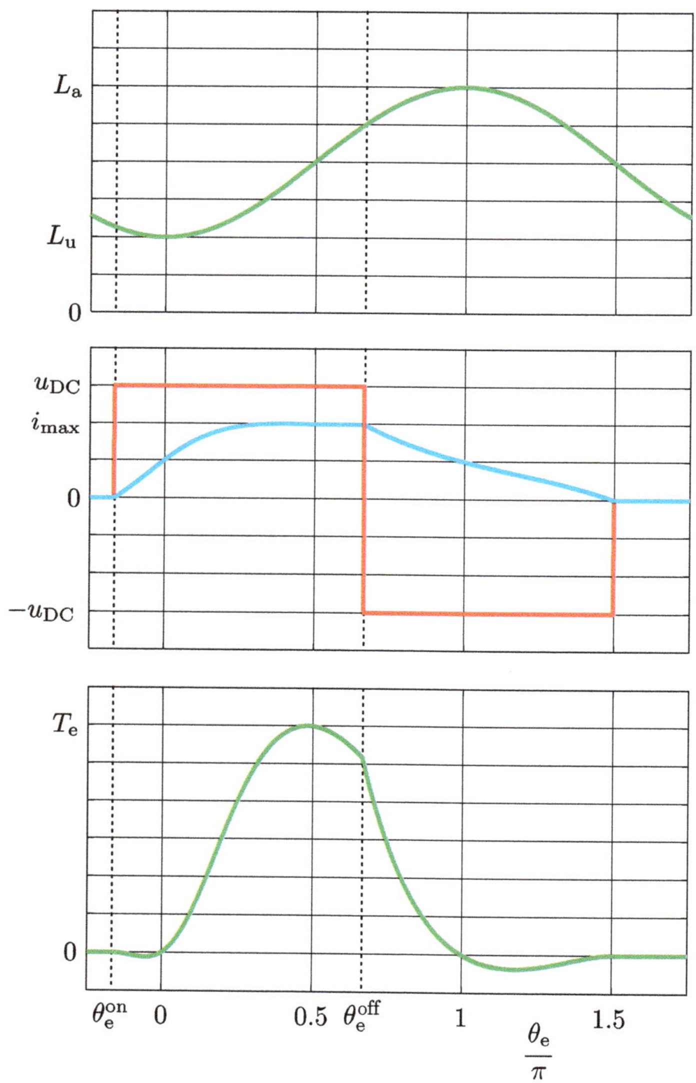

Fig. 10.10 Sinusoidal inductance, phase voltage (red), current (blue), and resulting phase torque over one electrical period at constant speed

$$L(\theta_e) = L_o\left((1+\kappa) + (1-\kappa)\cos(\theta_e)\right). \tag{10.27}$$

This inductance profile can be recognized as a harmonic approximation of the piecewise linear model presented at the beginning of this section given that $2L_o = L_u$ and $\kappa = L_a/L_u$. For a machine with two rotor teeth ($N_r = 2$), the given inductance profile repeats twice per mechanical revolution.

In single pulse operation, the positive dc-link voltage (red curve in second plot of Fig. 10.10) is applied for the entire first half of the excitation cycle, so that the flux linkage inside the phase increases linearly according to Eq. (10.22). During the second half of the active cycle, the negative dc-link voltage is applied to the phase leading to a linearly decreasing flux linkage. The flux-linkage profile during the entire excitation process is mathematically expressed as follows:

$$\psi(\theta_e) = \begin{cases} \psi\left(\theta_e^{on}\right) + \frac{u_{DC}}{\omega_e}\left(\theta_e - \theta_e^{on}\right) & \theta_e^{on} \leq \theta_e \leq \theta_e^{off} \\ \psi\left(\theta_e^{off}\right) - \frac{u_{DC}}{\omega_e}\left(\theta_e - \theta_e^{off}\right) & \theta_e^{off} < \theta_e. \end{cases} \tag{10.28}$$

Given a completely demagnetized phase at the beginning of the excitation, $\psi(\theta_e^{on}) = 0$, the current can be calculated using Eq. (10.22) with Eqs. (10.27) and (10.28):

$$i(\theta_e) = \begin{cases} \frac{u_{DC}}{\omega_e L_o} \frac{\theta_e - \theta_e^{on}}{1+\kappa+(1-\kappa)\cos(\theta_e)} & \theta_e^{on} \leq \theta_e \leq \theta_e^{off} \\ i\left(\theta_e^{off}\right) - \frac{u_{DC}}{\omega_e L_o} \frac{\theta_e - \theta_e^{off}}{1+\kappa+(1-\kappa)\cos(\theta_e)} & \theta_e^{off} < \theta_e. \end{cases} \tag{10.29}$$

The waveform of the phase current can be observed as the blue curve in the second plot of Fig. 10.10. It should be noted that the excitation starts in a region, where the slope of the inductance is negative but still rather small. This is usually done to ensure a fully excited phase right before the inductance rises again. However, a small amount of negative torque is produced during this initial magnetization, which can be seen in the third plot of Fig. 10.10.

The torque waveform for the complete active cycle can be calculated by inserting Eqs. (10.27) and (10.29) into Eq. (10.24):

$$T_e(\theta_e) = \frac{1}{2} N_r L_o (\kappa - 1) \sin(\theta_e)\, i^2(\theta_e) \tag{10.30}$$

and is a function of the choice of the turn-on angle θ_e^{on} and the turn-off angle θ_e^{off}. It becomes obvious that a non-zero average torque can be produced with the chosen excitation profile, although small amounts of negative torque are produced at the beginning and the end of the active cycle.

Figure 10.11 shows the magnetic energy W_f inside the phase (blue) and the mechanical energy W_m (green) supplied to the load, which is simply the integral of torque. The magnetic energy is recovered at the end of excitation and does therefore not contribute to torque production. However, this energy exchange causes losses inside a non-ideal drive configuration (ohmic losses inside machine, converter losses

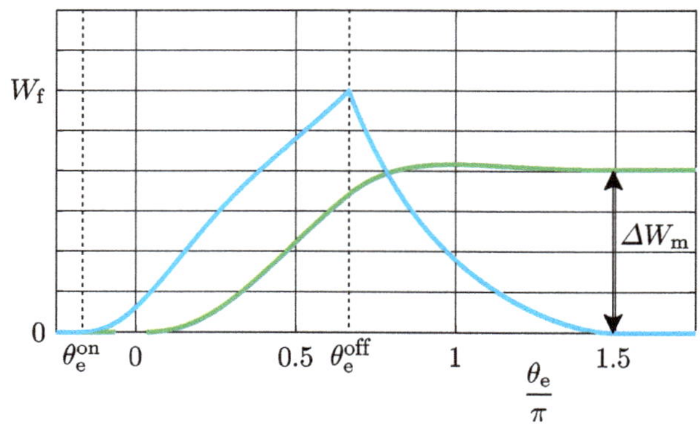

Fig. 10.11 Magnetic (blue) and mechanical (green) energy over one electrical period

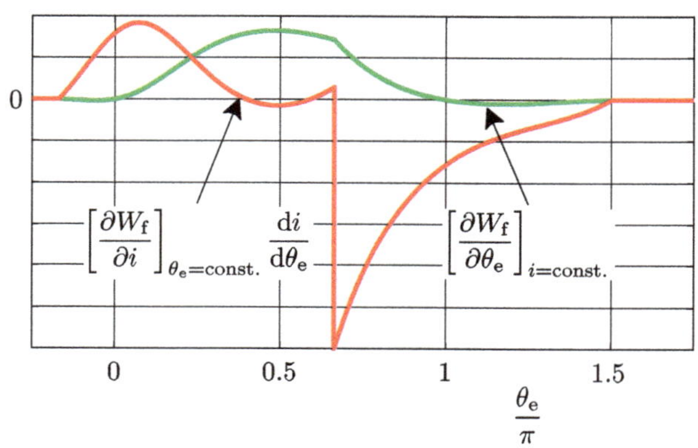

Fig. 10.12 Energy flows related to W_f reservoir

due to voltage drops across power switches) and should hence be minimized. A well-suited measure for the relative amount of magnetic energy is the so-called energy conversion factor EC, which is defined as

$$EC = \frac{\textit{energy delivered to the load}}{\textit{total energy supplied by the source}}. \tag{10.31}$$

In Eq. (10.31), the *total energy supplied by the source* is to be interpreted as the sum of the maximum magnetic energy and the mechanical *energy delivered to the load* after one active cycle.

Figure 10.12 denotes the energy flows related to the W_f reservoir according to Eqs. (10.26a) and (10.26b). The magnetic energy supplied at the start of the interval is lower than the amount recovered at the end. The reason for this is that for the linear case equal amounts of energy are supplied to the load and the magnetic energy reservoir. The amount of energy recovered at the end of the interval is therefore the sum of the magnetic energy supplied at the beginning of the active cycle plus the

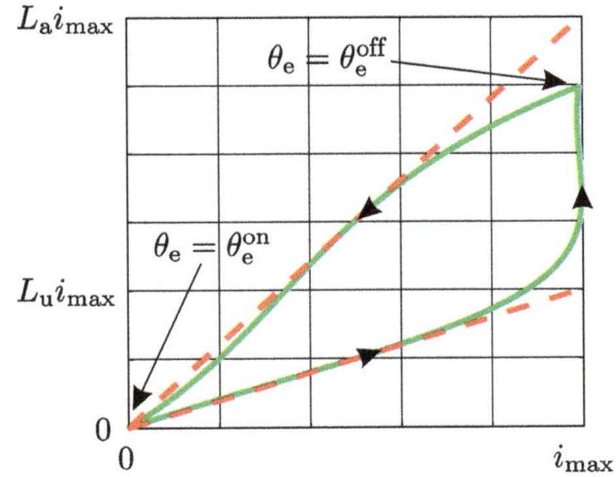

Fig. 10.13 Half cycle flux-linkage current diagram

amount diverted from the node (see Fig. 10.5) to the magnetic reservoir. The sum of these two contributions is the total energy supplied to the motor by the supply, as was mentioned previously.

The flux linkage/current locus for the chosen current waveform is displayed in Fig. 10.13. In contrast to the low speed case given in Fig. 10.9, the traces of magnetization and de-magnetization cannot be identified as straight lines anymore. Instead, a more smooth transition between magnetization and torque production takes place.

10.2.4 Switched Reluctance Modeling Concepts

The development of generic single SR models is readily undertaken with the aid of the theory given in Sect. 10.2.2. Initially, current based models are considered, which assumes that the current as function of the rotor angle is given. In the second part of this chapter voltage based models are discussed which are mostly used because these may be directly coupled to the converter structure, which are typically voltage source converters. The models considered will assume the presence of a phase coil resistance R.

Current based generic models require access to a current versus angle relationship $i(\theta_m)$.The output variables for this model should be the torque T_e and phase voltage u. The fundamental modeling component of the SR machine are the magnetization characteristics $\psi(i, \theta_m)$ which are represented by a single module, with inputs i, θ_m and output ψ. A similar module is also required for the torque, hence its output is the variable T_e, which is used in conjunction with the mechanical load equation

$$J\frac{d\omega_m}{dt} = T_e(i,\ \theta_m) - T_l \tag{10.32a}$$

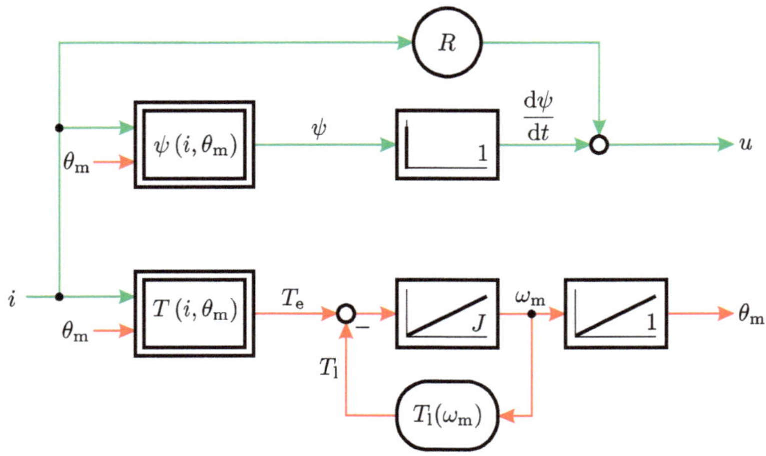

Fig. 10.14 Model of SRM with current excitation. Magnetization curves are used to generate the ψ and T_e tables as function of current and position

$$\omega_m = \frac{d\theta_m}{dt} \tag{10.32b}$$

where J and T_l represent the combined inertia and the load torque of the load and the machine respectively. The so-called static torque/current/angle characteristics are generated with the aid of the magnetization curves, using the theoretical approach given in Sect. 10.2.2. Use of terminal equations (10.7) and (10.32) leads to the generic model shown in Fig. 10.14.

Also shown in Fig. 10.14 is a module $T_l(\omega_m)$ which represents a user defined relationship between load torque and shaft speed. The reader is advised to consider the tutorials at the end of this chapter which demonstrate the use of all the generic models presented in this section. In Fig. 10.14, the phase voltage is derived by differentiation of the flux. However, an alternative approach may also be taken by using partial derivatives, as shown in the previous section, where an alternative terminal equation representation (10.13) was developed. The generic model as given in Fig. 10.15 uses the incremental inductance and normalized *EMF* curves, which are represented by non-linear modules with inputs i and θ_m.

The torque module and generic modules linked with Eq. (10.32a) are not shown in Fig. 10.15 given that these are identical to those shown in Fig. 10.14. The advantage of the first model (Fig. 10.14) lies with its simplicity, given that it only requires access to the magnetization and static torque curves. The second model (Fig. 10.15) generates the voltages u_e and u_l, which are useful in terms of enhancing the overall understanding of the machine. Furthermore, access to the voltage u_e is beneficial when considering sensorless SR drive applications. However, the model requires access to the partial derivatives of the magnetization curves. These must in turn be provided as non-linear modules, with inputs i, θ_m. Voltage source models

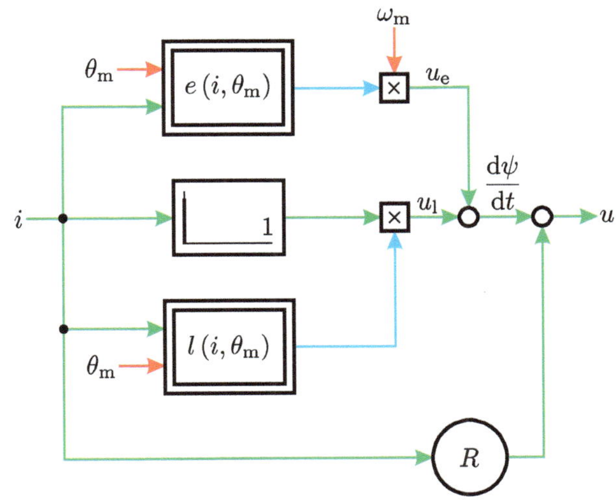

Fig. 10.15 Current fed model of SRM using incremental inductance and *EMF* curves

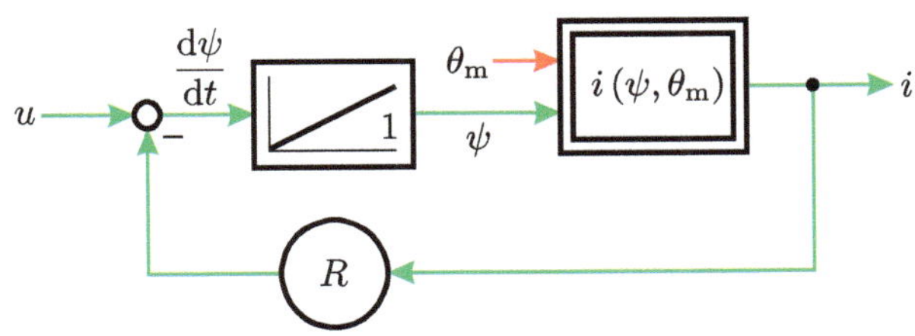

Fig. 10.16 Voltage fed model of SRM, using inverse magnetization curves

are commonly used for simulation purposes because most converter topologies are voltage source converters that control directly the phase voltage. The generic model as given in Fig. 10.16 uses the terminal voltage Eq. (10.7), which must be rewritten as an explicit function of the current. The torque module and generic modules linked with Eq. (10.32a) are not shown in the voltage based model, given that these remain unchanged.

The generic model according to Fig. 10.16 requires access to the *inverse* magnetization curves, which implies that the corresponding non-linear module has as output the current i and inputs the variables ψ and θ_m. The model in question is the causal counterpart of the model given in Fig. 10.14. The final model in this section is the causal counterpart of the current source model according to Fig. 10.15. This model, as given in Fig. 10.17, requires access to the incremental inductance and *EMF* curves.

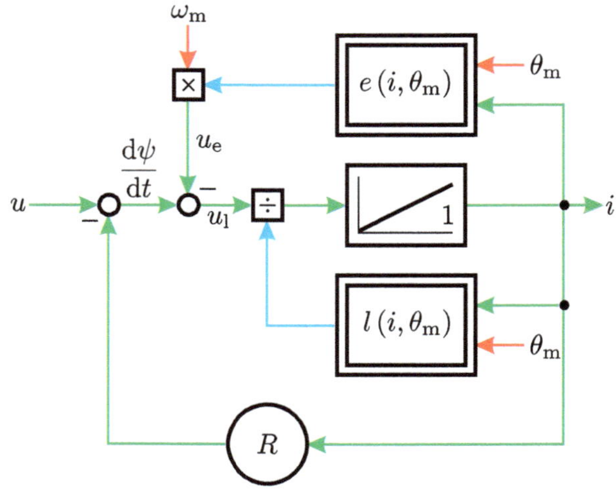

Fig. 10.17 Voltage fed SR model using incremental inductance and *EMF* curves

10.2.5 Representation of the Magnetization Characteristics

Central to the modeling process is the ability to represent the flux linkage depending on current and angle $\psi(i, \theta_e)$ and their derivatives as defined by Eq. (10.12). Due to the periodicity of the flux linkage, the angle will be defined in electrical degrees as described in Eq. (10.2). Figure 10.18 shows a typical set of magnetization curves which are represented in two different ways. Those given in Fig. 10.18a represent the flux linkage versus current relationship, whereby the rotor angle is held constant. These curves are defined over a current interval $0 \leq i \leq i_{\max}$.

As mentioned earlier, the magnetic reluctance of the circuit is at its minimum value for the aligned rotor position, hence magnetic saturation effects are at their highest level, which is reflected in the corresponding flux-linkage current curve $\psi(i, \theta_e = 180°)$ shown in Fig. 10.18a.

If on the other hand, the rotor is set to its unaligned position ($\theta_e = 0°$) magnetic saturation effects can be ignored, in which case the flux linkage $\psi(i, \theta_e = 0°)$ is a linear function of current, as may be observed from Fig. 10.18a.

It is also instructive to consider an alternative representation of the magnetization curves as given in Fig. 10.18b. This set of curves may be derived by using a conformal mapping technique in which, for example, the points $\psi(i = i^B, \theta_e = \theta_e^j)$ with parameter set $\theta_e^j = 0°, \theta_e^1, \theta_e^2, 180°$ are plotted in a flux linkage versus rotor angle diagram. The set of magnetization curves shown in Fig. 10.18a is in fact a subset that can be derived from a comprehensive set of flux linkage versus current curves, as will be discussed in Sect. 10.5.

Note that the flux linkage/angle characteristics $\psi(i = \text{const}, \theta_e)$ are periodic functions, which implies that the derivative $d\psi/d\theta_e$ at the aligned and unaligned rotor position must be zero. A further observation of Fig. 10.18a shows that the derivative of these characteristics is precisely the incremental inductance $l(i, \theta_e)$, which may in turn be expressed in terms of the variables i and θ_e. The partial derivative of the flux

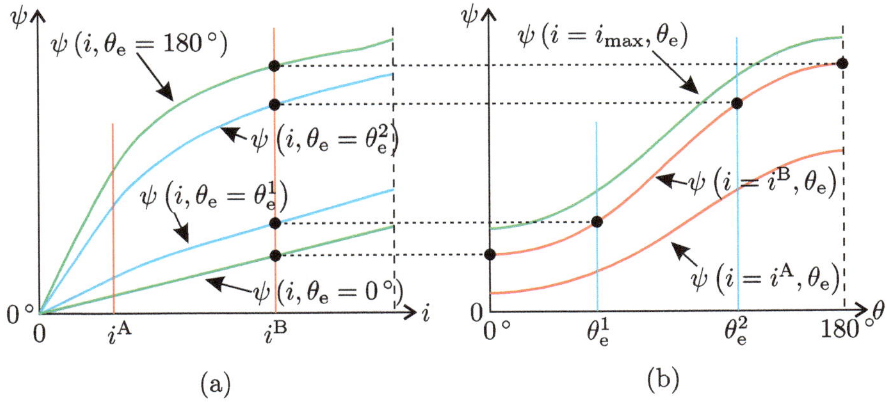

Fig. 10.18 Typical set of magnetization characteristics

with respect to the rotor angle, with i = const, referred to as the normalized *EMF* curves, $e(i, \theta_e)$ are precisely the derivatives of the flux linkage/angle characteristics as given in Fig. 10.18b.

10.2.6 Converter and Control Concepts

A commonly used switched reluctance drive configuration shown in Fig. 10.19, utilizes two switches per phase and this topology is referred to as an *asymmetrical half-bridge* converter. The drive has four switching states, which offer three basic modes of operation that will be discussed below with the aid of Table 10.2, Figs. 10.5, and 10.19.

- Mode 1, magnetization state: both switches S_b, S_t are closed, while the diodes D_b, D_t remain non-conducting. The phase voltage u is equal to the supply voltage u_{DC} and the supply current i_{supply} is equal to the phase current i. Energy from the supply source is transferred to the load and to the magnetic energy reservoir W_f.
- Mode 2, freewheeling state: switch S_b is closed and S_t is open. The diode D_b is conducting and D_t is non-conducting (or vice versa, i.e., there are two freewheeling states). The phase voltage u is equal to zero and the supply current i_{supply} is also zero. The supply source is disconnected from the machine, hence energy is transferred from the magnetic energy reservoir to the load.
- Mode 3, de-magnetization state: both switches S_b and S_t are open, while the diodes D_b and D_t are conducting. The phase voltage u is equal to $-u_{DC}$ and the supply current i_{supply} is equal to $-i$. This mode of operation can only persist for as long as the phase current is greater or equal to zero. Energy from the magnetic reservoir is transferred to the load as well as the supply.

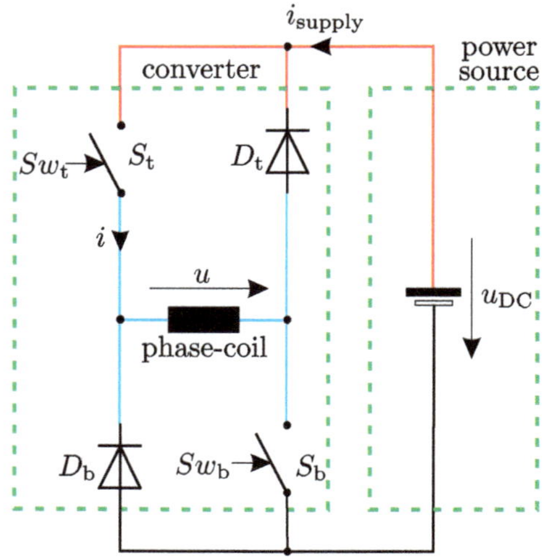

Fig. 10.19 SR drive with asymmetric half-bridge converter topology

Table 10.2 Switching states and resulting phase voltage of asymmetrical half-bridge (1: conducting, 0: non-conducting device)

S_t	S_b	D_t	D_b	u
1	1	0	0	u_{DC}
1	0	1	0	0
0	1	0	1	0
0	0	1	1	$-u_{DC}$ for $i > 0$

The active modes of operation are governed by a controller module which controls the two switches by way of the logic switch signals Sw_t and Sw_b. A logic *1* corresponds to a closed state for the corresponding switch. The inputs to the controller module are the reference i^* and measured phase currents as well as a logic *phase-active* signal, which is high during the period where the phase should be active. The latter is controlled by a commutator module, to be discussed at a later stage.

An example of a phase control module is given in Fig. 10.20, which utilizes *hysteresis* current control. Basically, the controller aims to keep the measured phase current within the boundaries $i^* \pm {}^{\Delta i^*}/2$, where Δi^* represents a user defined current tolerance band. The current controller controls the top converter switch S_t when the phase-active signal is set to 1. Comparator A is part of the actual hysteresis controller and its output is either 1 or -1, depending on the polarity of the error signal ε. The second comparator B is used to convert the bipolar output from the first comparator to a logic signal.

Note that the hysteresis current controller may be replaced by an alternative control concept where the top switch is connected to a PWM generator. In this case, the duty cycle δ for the top switch is the control variable, which implies that the average phase voltage during the phase-active period is equal to δu_{DC}. At the end of the phase-active period both switches are opened, in which case the phase voltage is equal to $u = -u_{DC}$ for the time interval in which the phase current is greater than zero. This type of control has a constant PWM switching frequency, which can be advantageous in terms of acoustic noise. A hysteresis type controller exhibits a

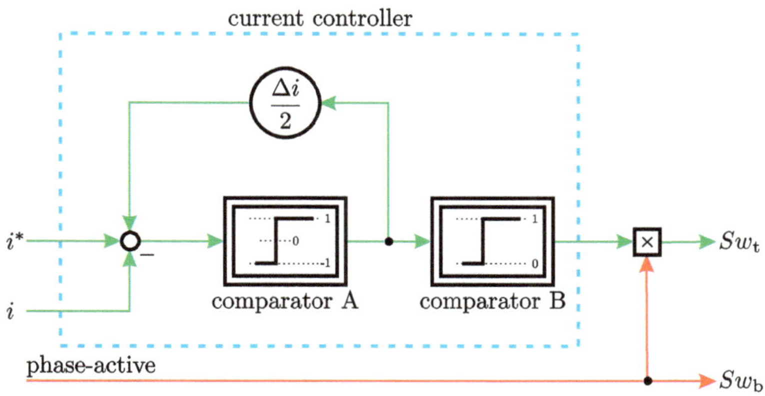

Fig. 10.20 Phase control module

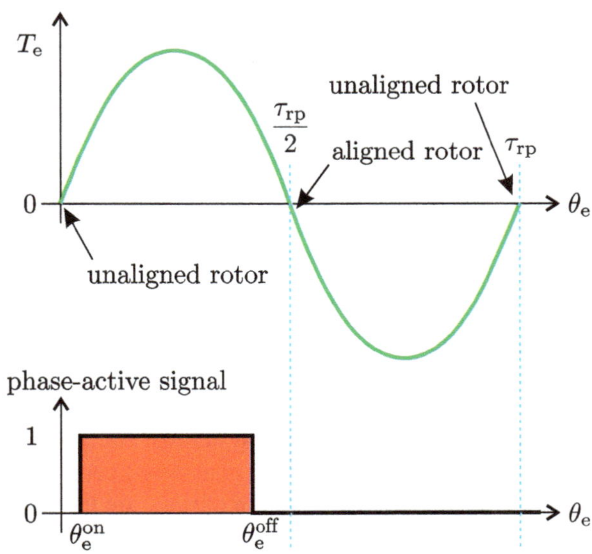

Fig. 10.21 Definition of *turn-on* θ_e^{on} and *turn-off* θ_e^{off} angles

variable switching frequency, which yields a decidedly different acoustical signature in comparison with drives that utilize a PWM based control method. The generation of the *phase-active* signal remains to be discussed and this may be done with the aid of Fig. 10.21.

This figure shows a typical stator torque versus shaft angle characteristic of a machine without saturation and inductance variation according to Eq. (10.27), in which case the torque equations (10.24) and (10.30) apply. The beginning and end shaft positions of the phase-active signal are defined with respect to the rotor position. In this context, the angles θ_e^{on} and θ_e^{off} are introduced as may be observed from Fig. 10.21. In this figure, an example is given which shows that the phase-active signal is set to *1* when the rotor angle θ_e is between the *turn-on* and *turn-off* angles ($\theta_e^{on} \leq \theta_e \leq \theta_e^{off}$).

If the phase excitation current resembles a square wave phase current, then the *turn-on* and *turn-off* angles should be set to 0° and 180°, respectively, to maximize the average torque per phase. However, the control angles will need to be adapted with speed in a practical drive to counter the effects of a finite rise and fall time of the current at the beginning and end of the phase-active period, as will become apparent in Sect. 10.4.1.

10.2.7 Example of Low and High Speed Drive Operation

The interaction between the converter and machine is very much dependent on the type of converter topology in use, the control methods deployed and the operating speed. Some insight into the behavior of the basic drive by way of typical waveforms for different speeds is given in this subsection. The fact that a single-phase based drive model is used does not affect the operational variable speed drive aspects considered here. The reason for this is that the phases are magnetically and electrically decoupled, hence a single-phase representation is sufficient. At a later stage, a more detailed analysis of drive behavior will be considered. The distinction between *low* and *high* speed operation is defined relative to the *base* operating speed, which is the speed ω_b at which the drive is able to operate at its rated current level and rated voltage. A detailed analysis with regards to calculating the base speed, is given in Sect. 10.4.1. An asymmetrical half-bridge converter topology (see Fig. 10.19) is assumed here.

Low Speed Operation

For low speed operation, some type of phase current control is required because the induced voltage levels in the machine are low in comparison with the supply voltage u_{DC}. In the previous section, hysteresis and PWM type control methods were discussed and these will also be considered here. The key variables (as defined in Figs. 10.14 and 10.17) as function of time, are the phase voltage u, induced voltage u_e, phase current i, and (per phase) torque T_e. An example of the waveforms which appear in the drive when operating at low speed with a hysteresis type controller is given in Fig. 10.22. These results were obtained with the tutorial given at the end of this chapter (see Sect. 10.6.2).

In this example, the *turn-on* angle is set to zero, hence the phase-active period starts when the rotor is at its unaligned position. The phase-active interval is terminated when the aligned rotor position at 180° is reached. The induced voltage waveform $u_e(t)$, as shown in Fig. 10.22a, underlines the presence of the unaligned/aligned rotor positions, given that u_e must then be zero. At the beginning of the phase-active interval both switches of the converter are closed (mode 1 operation), which causes the phase current $i(t)$ to increase because the phase voltage equals u_{DC} as shown in Fig. 10.22a. A hysteresis controller (see Fig. 10.20) is assumed at present with a reference current of $i^* = 10\,\text{A}$ and error current value of $\Delta i = 2\,\text{A}$, which means that the converter switches to mode 2 (freewheeling) operation (top switch open) as soon as the instantaneous current reaches $i^* + {}^{\Delta i}/_{2}$. For the remaining part of the phase-active cycle, the converter switches between

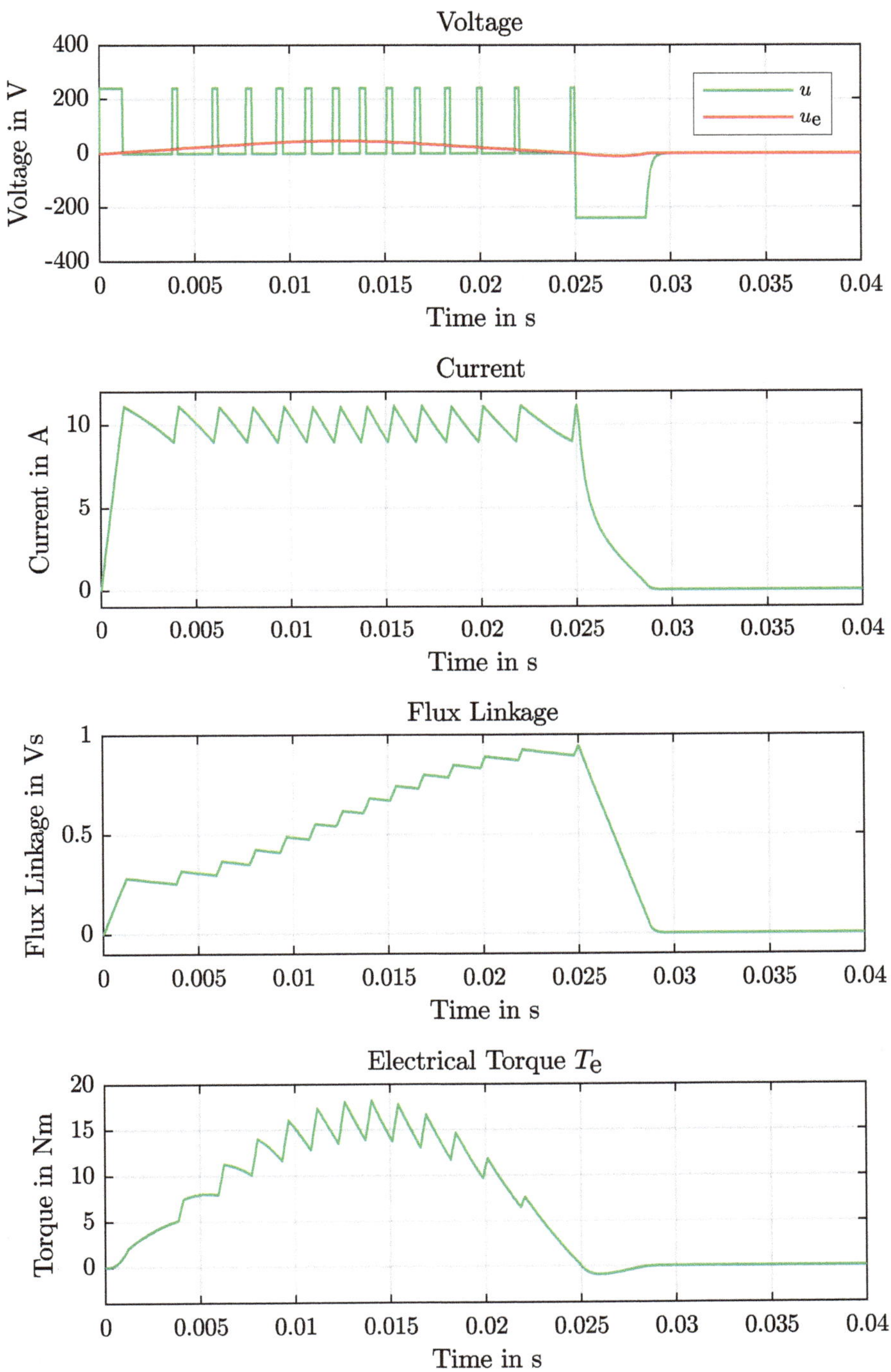

Fig. 10.22 SR drive with hysteresis controller: low speed operation, $n_{\mathrm{m}} = 200\,\mathrm{rpm}$, $\theta_{\mathrm{e}}^{\mathrm{on}} = 0°$, $\theta_{\mathrm{e}}^{\mathrm{off}} = 180°$

mode 1 and 2 in an effort to maintain the instantaneous current between the $i^* \pm \Delta i/2$ as may be observed in Fig. 10.22b. At the end of the excitation interval, both converter switches are opened (mode 3 converter operation) which causes the current to reduce to zero (de-magnetization phase). Also shown in Fig. 10.22 is the corresponding torque $T_e(t)$ waveform, which in this case has a small negative component caused by the fact that a non-zero phase current is still present after the rotor has passed the aligned position. The remaining waveform to be discussed for this drive configuration is the flux-linkage waveform (Fig. 10.22c), which basically represents the integral of the phase voltage, under the assumption that the voltage across the coil resistance is small in comparison with the supply voltage.

The second set of waveforms as given in Fig. 10.23 represent drive operation with the same converter/supply configuration and shaft speed as used to generate Fig. 10.23. However, in this example the hysteresis controller has been replaced by a PWM based controller with a switching frequency of 500 Hz. This implies that the top switch controls the average voltage across the phase coil (when active). Under these condition the average voltage is simply Δu_{DC}. The switching frequency has been purposely chosen low (factor ten lower than normally in use), to better visualize this operating mode. A duty cycle value of $\delta = 0.13$ is used, while the control angles were set to $\theta_e^{on} = -42°$ and $\theta_e^{off} = 138°$.

A comparison between Figs. 10.22 and 10.23 shows that these two modes of operation are significantly different. The basic difference between the two is that hysteresis type controllers aim to control the current during the phase-active period. This is in contrast to the PWM type controller which controls the average phase voltage (which is equal to δu_{DC}) during the phase-active interval. The latter is reflected by the nature of the flux linkage waveform (Fig. 10.23) during the phase-active interval, which must in this case correspond to a linear function, with a gradient that is proportional to the average phase voltage. The phase current waveform follows from the magnetization characteristics with the flux linkage and rotor angle as input variables, together with the chosen control angles. An observation of Fig. 10.23 shows that the induced voltage crosses the time axis at $t = 25$ ms when the rotor is in its unaligned position. Furthermore, the phase-active interval starts before the unaligned position ($\theta_e^{on} \leq 0°$), which leads to the slightly negative torque after the phase is turned on.

High Speed Operation

As the shaft speed is increased, the induced voltage increases which reduces the current slope and extends the rise time. Consequently, the average phase voltage during the phase-active interval must be increased. For drives which utilize a hysteresis type controller this occurs *automatically* because the top inverter switch will remain closed if the phase current is unable to reach the reference current value. The duty cycle of the PWM based controller must increase with speed, which inevitably leads to the identical situation, namely that the top switch remains on continuously during the phase excitation interval.

An example of drive operation at a speed higher than the base speed ω_b is given in Fig. 10.24. These characteristics resemble those shown for the PWM based current

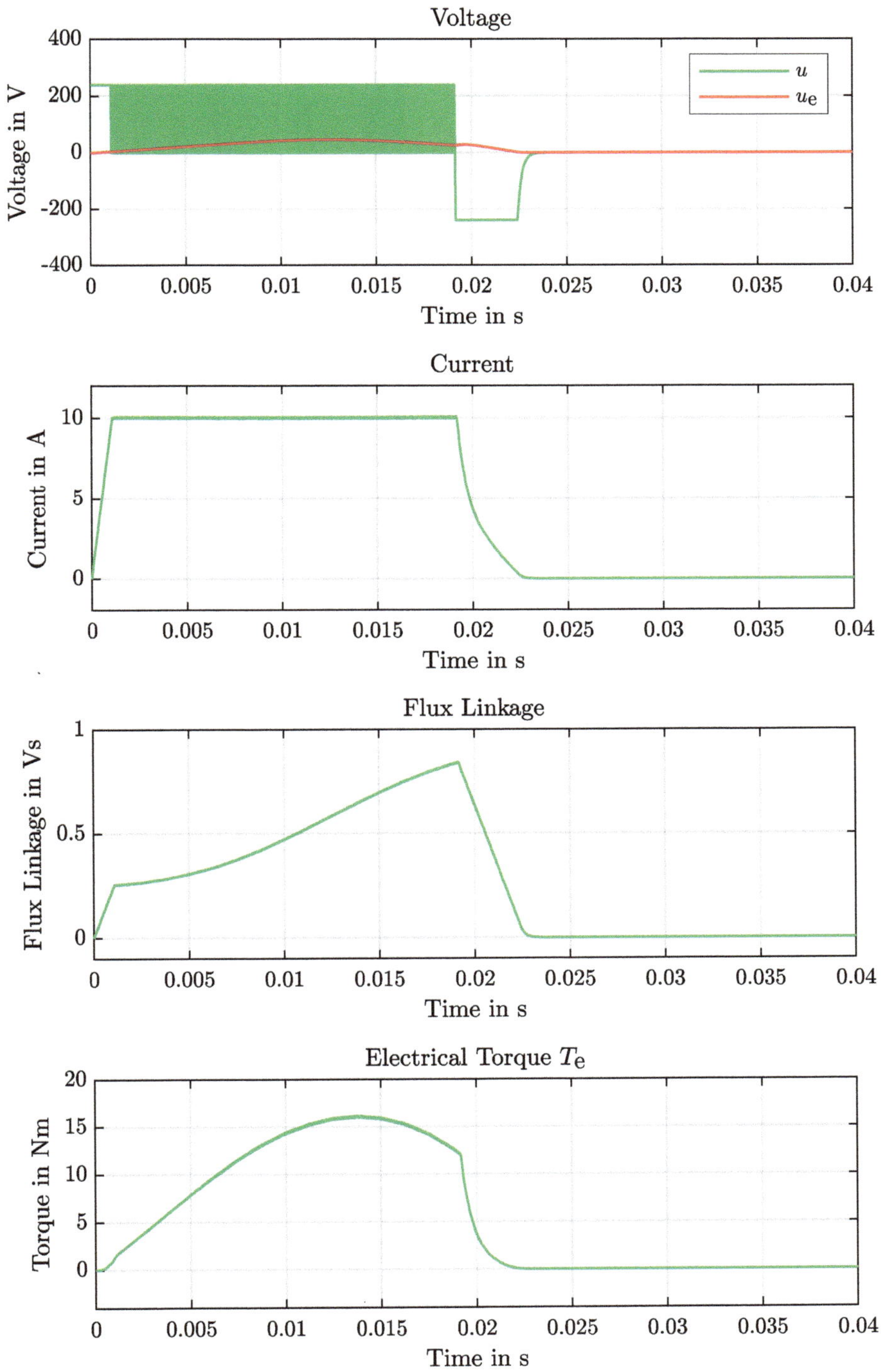

Fig. 10.23 SR drive with PWM controller: low speed operation, $n_m = 200\,\text{rpm}$, $\theta_e^{on} = -42°$, $\theta_e^{off} = 138°$

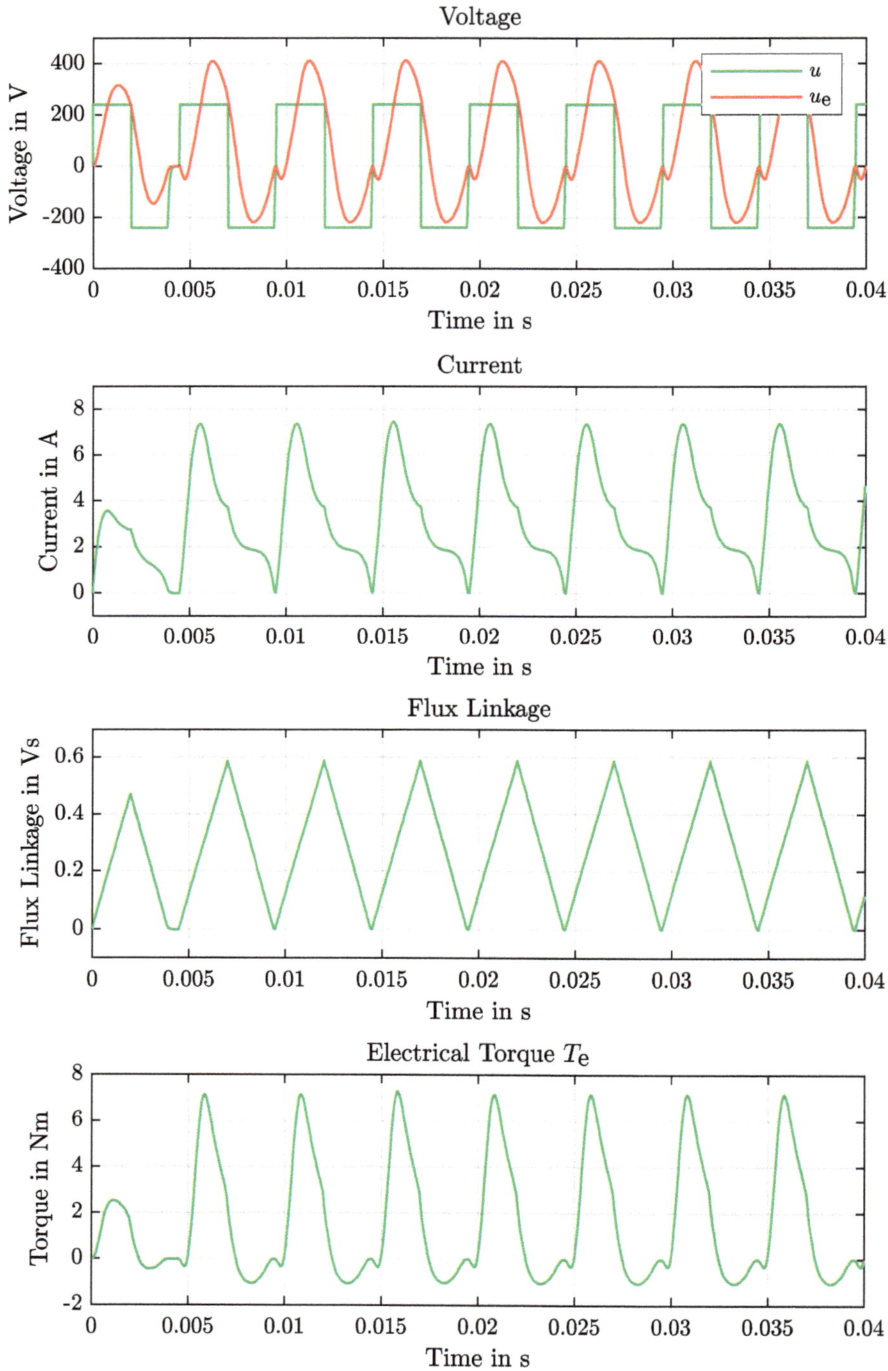

Fig. 10.24 SR drive: high speed operation, $n_{\mathrm{m}} = 2000\,\mathrm{rpm}$, $\theta_{\mathrm{on}} = -36°$, $\theta_{\mathrm{off}} = 144°$

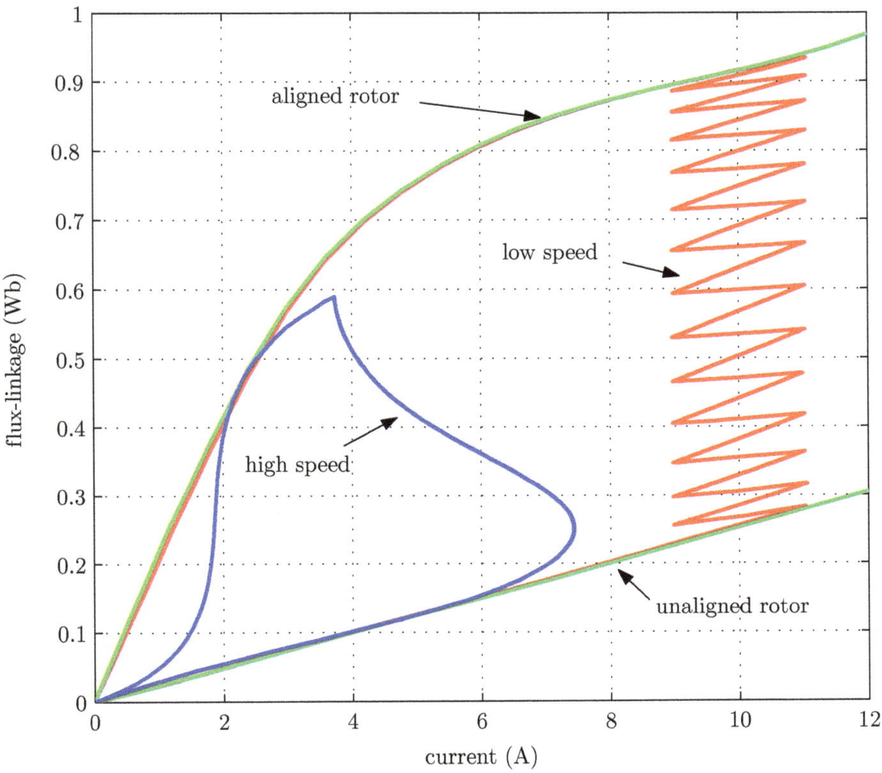

Fig. 10.25 Typical low and high speed flux linkage/current trajectories of SRM

controller (see Fig. 10.23). The notable difference is that the duty cycle for high speed operation is set to unity value. This is directly reflected in the gradient of the flux linkage versus time waveform during the phase-active period, which is now equal (in absolute terms) to that present during the de-magnetization interval. An observation of the induced voltage waveform in Fig. 10.24 confirms that its magnitude is considerably larger than found for low speed operation. The zero EMF cross-over time points indicate the unaligned and aligned rotor positions, respectively. The control angles were set to a value of $\theta_{\mathrm{e}}^{\mathrm{on}} = -36°$ and $\theta_{\mathrm{e}}^{\mathrm{off}} = 144°$.

It is instructive to carefully examine the current as function of the flux linkage for the two examples given in Figs. 10.23 and 10.24, respectively. If the plot is executed over one electrical period (which in mechanical degrees corresponds to τ_{rp}), as is the case here, a flux linkage/current locus will result of which the enclosed area is proportional to the average torque produced by the SRM, as was discussed in Sects. 10.2.2 and 10.2.3. The low and high speed flux linkage/current trajectories as given in Fig. 10.25 are complemented by the aligned and unaligned flux linkage/current characteristics.

An observation of Fig. 10.25 confirms that the low speed flux linkage/current trajectory encompasses the maximum available area, within the available magnetization characteristics for the chosen reference current value. The high speed average torque is substantially lower as may also be deduced from the size of the enclosed surface area generated by the corresponding flux linkage/current locus.

10.3 Multi-Phase Machines

In the previous section, the single-phase switched reluctance machine was discussed in considerable detail. Such an approach is warranted because the machine phases are electrically and magnetically independent. However, in a practical machine, design criteria, such as maximizing the torque per volume ratio, the need for a self-starting capability, and minimizing torque ripple, favor a multi-phase approach. The total shaft torque T_{e} of the machine is equal to the instantaneous phase contributions namely

$$T_{\mathrm{e}} = \sum_{\mathrm{j}=1}^{\mathrm{j}=N_{\mathrm{ph}}} T_{\mathrm{ej}} \tag{10.33}$$

where N_{ph} represents the number of electrical phases and T_{ej} the instantaneous torque per phase. In the multi-phase machine, the individual torque versus angle waveforms are displaced with respect to each other by a so-called *step angle* τ_{step} defined as

$$\tau_{\mathrm{step}} = \frac{\tau_{\mathrm{rp}}}{N_{\mathrm{ph}}}. \tag{10.34}$$

Figure 10.26 shows the torque per phase waveforms and corresponding phase-active signals for a two and three-phase machine. Also shown in this diagram are the unaligned rotor positions $\theta_{\mathrm{e}}^{\mathrm{u}}$ for each of the phases. In this example, a square wave current pulse is assumed which corresponds to the length of the phase-active signal. The *turn-on* and *turn-off* angles are set to 0° and 180°, respectively. Furthermore, a linear machine model is assumed which implies that the torque per phase is defined by Eq. (10.24).

The self-starting capability, i.e., the ability to produce a non-zero unidirectional torque T_{e} (as defined by Eq. (10.33)) at any rotor position, is present for the three-phase machine example shown in Fig. 10.26.

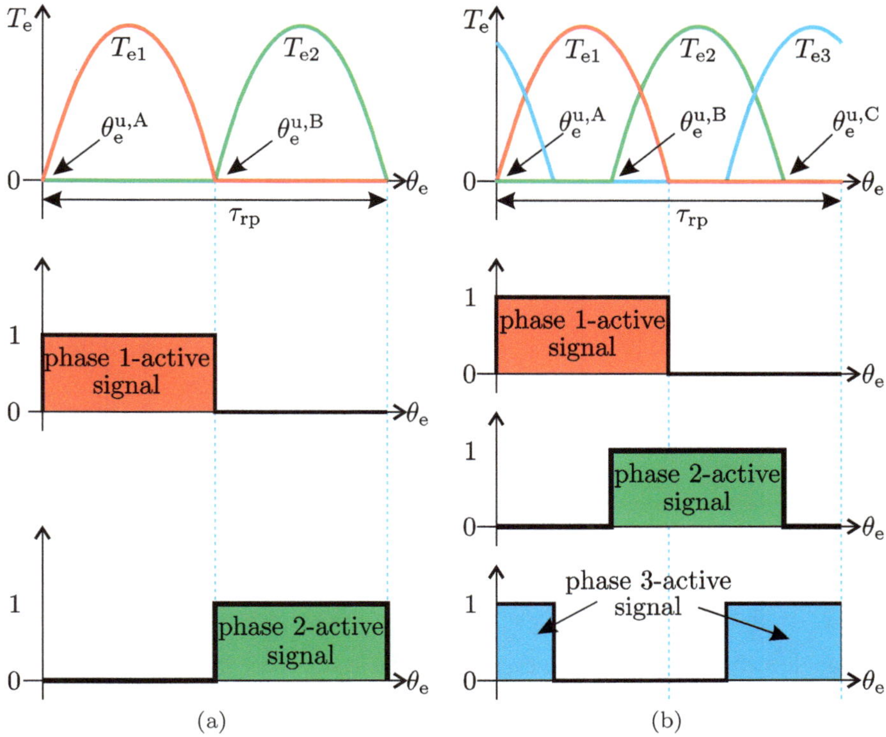

Fig. 10.26 Multi-phase operation. (**a**) Two phases, $N_{ph=2}$. (**b**) Three phases, $N_{ph=3}$

10.3.1 Converter Concepts

The asymmetric bridge concept as shown in Fig. 10.19 remains the preferred configuration for SR drives, given its simplicity and effectiveness in terms of energy recovery during the de-magnetization phase (see Sect. 10.2.6). However, other concepts are in use and two such examples are shown in Fig. 10.27. The dissipative (discharge) based model, shown in Fig. 10.27a, derives its name from the fact that the de-magnetization energy which is released after the phase is de-activated, is dissipated in an external resistance R. The duration of this phase may be shortened by inserting a Zener diode (anode connected to the positive supply source) between the external resistance and positive supply terminal. The use of the Zener diode increases the negative voltage across the phase during the time interval where the switch is opened and the diode D is conducting, which shortens the de-magnetization interval.

The second configuration, as shown in Fig. 10.27b requires an additional *fly-back* coil which is bifilar wound with the phase winding. During the de-magnetization interval the energy is returned to the supply, in a manner which is identical to that found in fly-back type converters. The advantage of this converter lies with the use

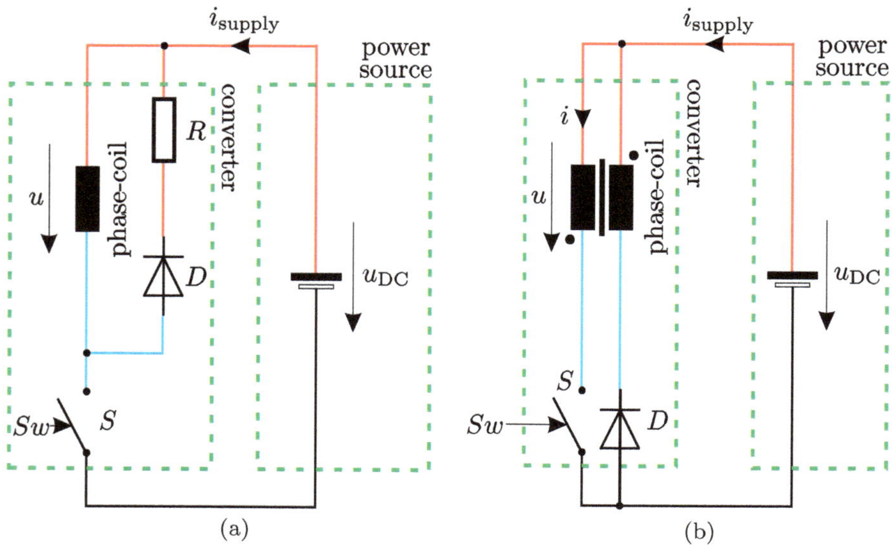

Fig. 10.27 Single switch per phase converter concepts. (**a**) Dissipative based. (**b**) Fly-back based

of a single switch per phase. However, the available slot space in the machine which houses the phase must be shared with the fly-back coil. The multi-phase converter provides an opportunity for reducing the overall number of semiconductors. The configuration shown in Fig. 10.28 utilizes a common top switch S_t for the arbitrarily chosen three-phase machine configuration. The bottom phase switches S_b are controlled by their respective phase-active signals (see Fig. 10.26b). If a phase-active signal is at a logic level *1*, the corresponding selection switch is closed.

The top switch may be connected to a PWM modulator, which in effect controls the duty cycle δ. The corresponding average voltage on the common phase connection with respect to the zero voltage node of the supply is in this case equal to δu_{DC}. Controlling the duty cycle provides a means of controlling the phase current amplitude for all phases. However, no individual phase current control is possible. Furthermore, the voltage across the average phase during the de-magnetization phase is equal to $(1 - \delta)u_{DC}$, which implies that the de-magnetization process will become less effective as the duty cycle is increased. In the extreme case, i.e., with unity duty cycle (switch S_t closed continuously), the de-magnetization energy cannot be adequately removed, hence the drive becomes inoperative.

Furthermore, a capacitive based energy recovery concept is considered. Drives which utilize this concept make use of one or more capacitors to store the magnetic energy which must be recovered during the de-magnetization interval of each phase. The diode-capacitor combination as shown in Fig. 10.29 is an example of an approach which utilizes this type of energy recovery concept. The diode/capacitor combination may be positioned between the power source and a converter of the type given by, for example, Fig. 10.19 or Fig. 10.28.

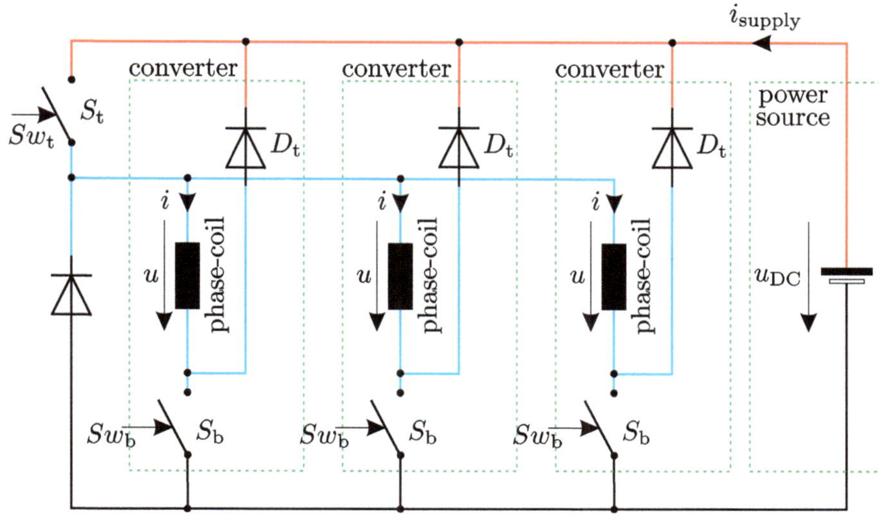

Fig. 10.28 Three-phase drive concept using a four-switch converter

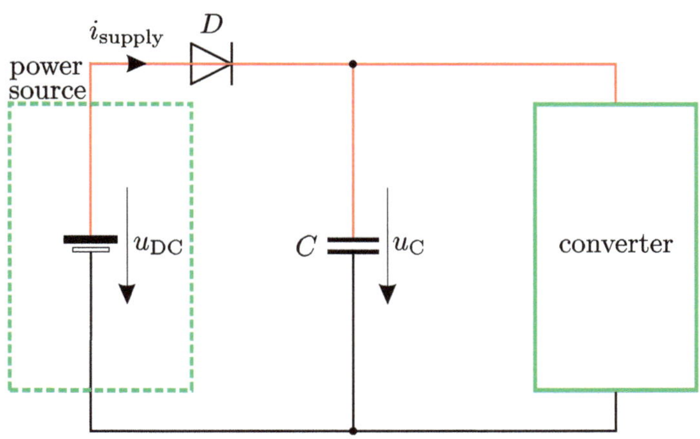

Fig. 10.29 Capacitive based energy recovery

The purpose of the diode D is to ensure that the energy which is normally returned to the supply is now delivered to a relatively small capacitor C, which is initially charged to the supply voltage u_{DC}. During the de-magnetization interval the capacitor voltage u_{C} will increase by an amount which is governed by the amount of energy to be stored and capacitor size. This effect can be used to reduce the demagnetizing interval (in motoring mode), when the phase inductance is high. At the start of the phase excitation interval the current rise time will be initially shorter than that achievable without the diode/capacitor combination, which leads to higher torque levels in the drive and a wider speed range [15].

10.4 Control of Switched Reluctance Drives

The drive performance of an SR drive cannot be considered without taking into account the converter topology in use, the nature of current control, and choice of control angles. It is therefore instructive to provide further insight in terms of how drive performance is affected by these drive variables. In this context, the drive characteristics are initially discussed with the aid of so-called *SR control* diagrams. In the sequel to this section, a modern control strategy known as *Direct Instantaneous Torque Control* (DITC) is considered [8, 9]. The reader is urged to closely examine and run the simulations at the end of this chapter to fully appreciate the concepts discussed in this section.

10.4.1 Drive Characteristics and Operating Range

The output power versus shaft speed characteristic of the SR drive is normally used as the main performance indicator. A typical example of such a characteristic is given in Fig. 10.30 for the non-linear SR motor prototype used in this chapter (see Sect. 10.5). An asymmetrical half-bridge converter topology, as discussed in Sect. 10.2.6, is assumed to be connected to the machine. A DC supply source

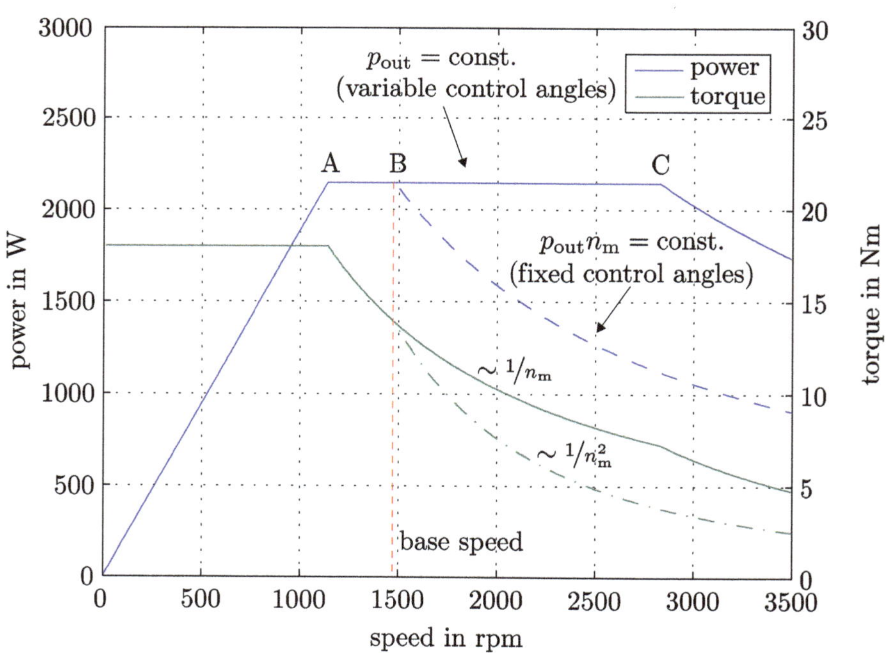

Fig. 10.30 Operation limits of a switched reluctance machine

of magnitude, $u_{\mathrm{DC}} = 240\,\mathrm{V}$ is connected to the converter. In terms of current control, two control strategies are considered. The first is hysteresis control, where the hysteresis current band Δi (see Fig. 10.20) is assumed to be infinitely small, which in effect implies that the current ripple associated with this type of control is ignored. The second control strategy adapts the average voltage during the phase-active interval to control the current. A sufficiently high PWM switching frequency, with duty cycle δ is assumed with the same objective as above, namely to reduce the currently ripple caused by the phase voltage switching functions present during the phase-active cycle. The current reference i^* value for the controller is taken to be 10.2 A, which is a convenient value as will become apparent shortly.

The performance characteristic as shown in Fig. 10.30 is represented in a *stylized* form, which implies that variations normally present with non-ideal controllers are not shown. Also the choice of control angles $\theta_{\mathrm{e}}^{\mathrm{on}}$ and $\theta_{\mathrm{e}}^{\mathrm{off}}$ will significantly influence the actual performance curve, which can be understood with the aid of the simulation given in the tutorial section.

An observation of Fig. 10.30 shows that operation is distinctly different for shaft speeds above or below the base operating speed $n_{\mathrm{m}}^{\mathrm{b}}$. Its numerical value for the current choice of drive variables is equal to $n_{\mathrm{m}}^{\mathrm{b}} = 1475\,\mathrm{rpm}$ as will be shown in Sect. 10.4.2. In the operating region 0-A (see Fig. 10.30), constant torque operation is possible, which implies that the ratio of output power to shaft speed is controlled by the average torque produced by the machine. The latter is in turn controlled by the choice of reference current and control angles. The highest current reference value is governed by the thermal limitations of the machine. Typically, phase current densities in the order of $10\,\mathrm{A/mm^2}$ are permissible without deploying forced cooling techniques in the machine. However, specific design and construction techniques can significantly increase the permissible current density.

As the SR drive operation approaches point A, the output power level reaches its maximum value which in turn defines the maximum supply power requirement for the drive. At point A, the supply source must at least deliver the required output power which is equal to the product of the DC supply voltage and average supply current. In addition, the supply must cover the magnetic core, converter switching, and mechanical losses. Operating region A–B (see Fig. 10.30) is a transitional region and the actual power versus speed curve is again strongly influenced by the choice of drive parameters. For operation above the base speed, two distinct modes of operation are found. The first utilizes a fixed set of control angles, which results in a typical output power curve that is approximately inversely proportional to the shaft speed. High speed operation in this region is referred to as the *natural* performance trajectory, for reasons to be discussed in Sect. 10.4.2. If the control angles are judicially chosen to maximize the output power for each speed, operation along the trajectory B–C is possible, where the output power level remains approximately constant. However, as speed is increased beyond point C, drive operation reverts to its *natural* mode, i.e., further control angle variation is unwarranted for reasons to be discussed below. Consequently, the operating range of an SR drive can be similar to that of an induction machine.

10.4.2 Drive Operational Aspects

It is apparent that the choice of control angles significantly affects drive performance. To be able to comprehend drive behavior as typified by Fig. 10.30, it is instructive to introduce a so-called *SR control* diagram. Basically, the diagram is a flux linkage versus angle diagram, as shown in Fig. 10.18, in which the flux linkage versus angle locus for drive operation over one period τ_{rp} is shown. Prior to discussing the use of this type of control diagram it is instructive to reconsider the flux-linkage time derivative $\frac{d\psi}{dt}$ in terms of its partial derivative with respect to the rotor angle. The latter may be readily undertaken with the aid of Eq. (10.7) and partial differentiation of the flux with respect to angle, which gives

$$\frac{d\psi}{d\theta_m} \simeq \frac{u}{\omega_m} \tag{10.35}$$

where it is assumed that we can ignore the resistance R in Eq. (10.7). Furthermore, a constant shaft speed $\omega_m = \frac{d\theta_m}{dt}$ is assumed. Equation (10.35) is significant because it defines the gradient of the flux linkage/angle trajectory within the SR control diagram. During operation in the phase-active interval, the gradient will be equal to $\delta u_{DC}/\omega_m$ if a PWM controller is used in the low speed region. If a hysteresis type controller is used the gradient will be equal to u_e/ω_m, where u_e represents the induced voltage (see Fig. 10.17). A constant phase current equal to the reference value can be maintained in the phase-active interval provided the condition

$$u_e \leq u_{DC} \tag{10.36}$$

is met. If this condition is exceeded the gradient is defined by the term u_{DC}/ω_m, which also corresponds to the flux linkage/angle gradient value that is possible for high speed operation. During the de-magnetization interval, the phase voltage is equal to $-u_{DC}$ for the asymmetrical bridge topology assumed here, in which case the gradient is defined as $-u_{DC}/\omega_m$.

Control Diagram at Low Speeds

The first SR control diagram to be considered is related to operation in the low speed region of the performance curve (see Fig. 10.30) for the specific case where the control angles θ_e^{on} and θ_e^{off} are 0° and 180°, respectively. Furthermore, a hysteresis type current controller is assumed. The analysis of the drive in its present form is instructive because it provides insight with regard to the need for adapting the control angles, as will become apparent shortly. The SR control diagram, given in Fig. 10.31, shows the flux linkage/current versus angle curves for the motor prototype described in Sect. 10.5 for a set of discrete current values in the range $0 \rightarrow 12$ A. The machine in question has six rotor teeth, hence one electrical period is equal to $\tau_{rp} = 60\,°$mech. If we initially ignore the voltage limitations imposed on the current controller, the latter would be able to deliver a rectangular current pulse with a magnitude equal to the reference value $i^* = 10.2$ A during the

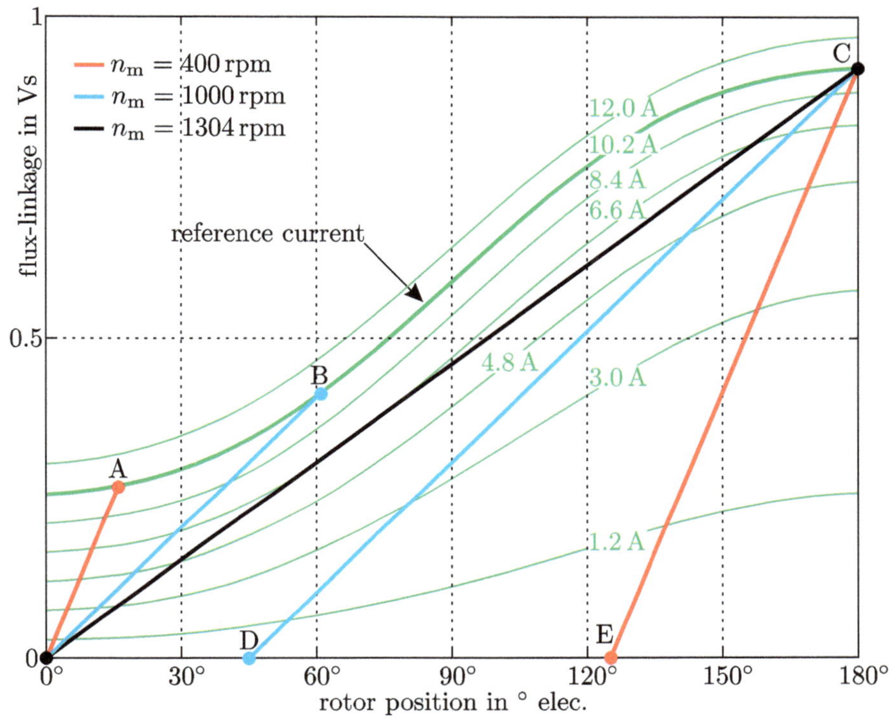

Fig. 10.31 Drive operation, with hysteresis controller

phase-active interval. Under these conditions, the flux/angle trajectory in the control diagram would begin at coordinates (0°, 0 Vs) and proceed vertically until the flux linkage/angle curve $\psi(i = 10.2\,\text{A}, \theta_e)$ is reached. Thereafter, the locus proceeds along this curve until the aligned position, point C, in Fig. 10.31 is reached. At that point of operation, the phase-active interval is terminated and the de-magnetization sequence begins which will in this case result in a trajectory that is along the vertical axis from point C → (180°, 0) in the control diagram. The instantaneous torque follows from the torque versus angle curves, which are given in Fig. 10.45. Operation along the line $\psi(i = 10.2\,\text{A}, \theta_e)$ for $\theta_e = 0° \rightarrow 180°$ (see Fig. 10.30) is reflected in Fig. 10.45 by considering the cross-sectional view with a constant current of 10.2 A. Hence, the torque per phase is zero at the unaligned and aligned angles and maximum in the vicinity of $\theta_e = 90°$.

In reality, the voltage limitations imposed on the current controller must be taken into account as will become apparent in the following example for drive operation at $n_m = 400$ rpm and given supply voltage value u_{DC}. At the start of the phase-active interval, the hysteresis controller will switch to the phase voltage $u = u_{DC}$ until current $i = i^*$ is reached. In the control diagram, this part of the locus corresponds to a trajectory from coordinate (0°, 0 Vs) to point A. The gradient (in absolute terms) of this part of the trajectory is given by Eq. (10.35), with $u = u_{DC} = 240$ V and

$\omega_m = {}^{2\pi 400}/_{60}$ Hz. Once point A is reached, the hysteresis controller will maintain the current at its 10.2 A reference value, which leads to an operating trajectory from point A to C along the line $\psi(i = 10.2\,\text{A}, \theta_e)$. At point C the de-magnetization sequence starts and the phase voltage is set to $u = -u_{DC}$. The gradient of the locus for this part of the trajectory is negative compared to what is shown at the start of the interval. Furthermore, the expectation would be that the locus endpoint would lie behind the aligned rotor position (180°). However, the flux linkage/angle characteristics are symmetrical with respect to the vertical axis at $\theta_e = 0°$ and $\theta_e = 180°$. Consequently, this part of locus can also be plotted within the boundaries of the control diagram, with the understanding that the instantaneous torque values are taken to be negative. Hence, the line C to E represents the locus which corresponds to the de-magnetization phase of drive operation. Note that this trajectory must be parallel with the line (0°, 0 Vs)–A in Fig. 10.31 given that the speed and supply voltage are identical.

Control Diagram at High Speeds

The process described above is repeated for the higher shaft speed of $n_m = 1000$ rpm which implies that the constant voltage part of the locus will have a gradient which is factor ${}^{400}/_{1000}$ lower with respect to the previous case. Consequently, the first part of the locus starts at coordinate (0°, 0 Vs) and proceeds to point B in Fig. 10.31. Hence, the period of constant current operation is shortened given that it extends from point B to C. Similarly, the de-magnetization part of the locus is now represented by the section C–D which implies that a larger (in comparison to the previous case) negative torque component will be present. Note also that the interval D–(0°, 0 Vs) along the rotor axis represents the interval of the drive cycle τ_{rp} in which the phase current, flux, and torque are zero. As the speed is increased, the hysteresis controller becomes progressively ineffective in terms of its ability to maintain the phase current at the set reference value. The limit speed ω_m^{limit} as shown in Fig. 10.31 corresponds to the shaft speed at which the interval D–(0°, 0 Vs) becomes zero and the locus follows trajectory (0°, 0 Vs)–C–(0°, 0 Vs). In this case the current waveform is identical during the phase-active and non-phase-active part of the drive cycle. The torque waveforms are also identical in absolute terms, but positive during trajectory (0°, 0 Vs)–C and negative for the period C–(0°, 0 Vs). The limiting speed ω_m^{limit}, at which the drive can operate with *turn-on* and *turn-off* angles at 0° and 180° is therefore given as

$$\omega_m^{limit} = \frac{u_{DC}}{\psi\,(i = i^*,\ \theta_e = 0)} \frac{1}{2N_r} \tag{10.37}$$

where $\psi(i = i^*, \theta_e = 180°)$ represent the flux value equal to 0.92 Wb at point C. With the given choice of drive variables the limiting speed according to Eq. (10.37) is equal to $n_m^{limit} = 1304$ rpm. This speed value is below the base speed value of $n_m^b = 1478$ rpm (to be calculated shortly), which underlines the need to adapt the control angle values.

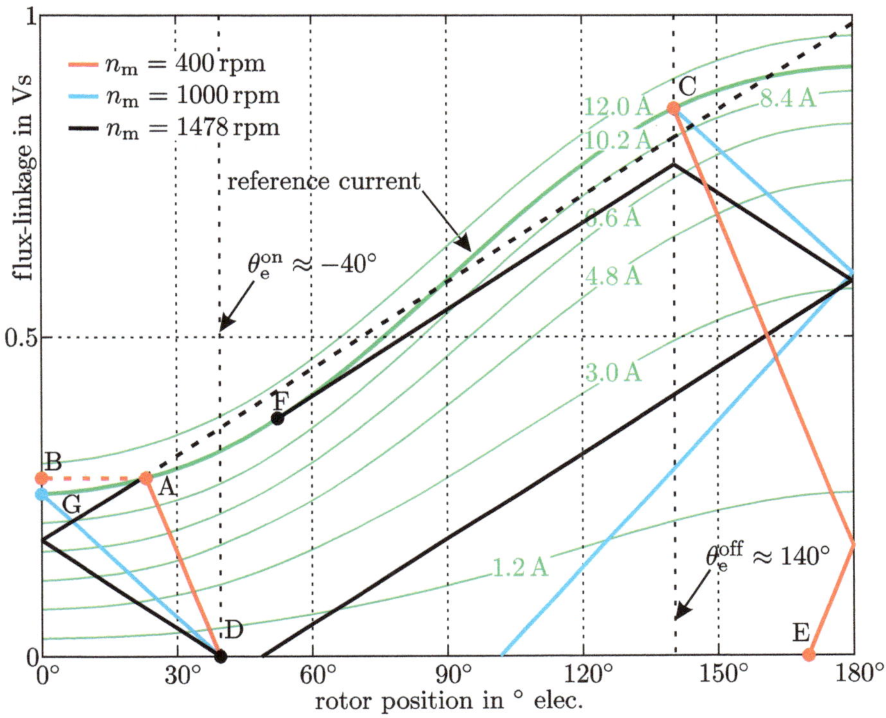

Fig. 10.32 Drive operation, hysteresis controller

Control Diagram for Advanced Turn-on and Turn-off Angles

The second application of the SR control diagram relates to operation with a hysteresis type current controller for three operating speeds, of which two n_m = 400 rpm, n_m = 1000 rpm are chosen in the low speed drive area. The third $n_m^b = 1478$ rpm is at the so-called base operating speed of the drive. In this example, a fixed *turn-on* angle of $\theta_e^{on} \approx -40°$ and *turn-off* angle of $\theta_e^{off} \approx 140°$ is assumed for reasons to be discussed at a later stage.

The operating trajectory for the speed n_m = 400 rpm is initiated when the rotor position is equal to $\theta_e = \theta_e^{on}$, which implies that the phase-active interval starts 40° BEFORE the unaligned position. However, as mentioned earlier, the flux linkage/angle curves are symmetrical with respect to the vertical axis $\theta_e = 0°$, hence the first part of the locus may also be plotted with a starting point at $\theta_e = \theta_e^u - \theta_e^{on}$ (see Fig. 10.32). The gradient of the locus from D to A is again defined by Eq. (10.35). At point A the reference current is reached and the hysteresis controller continues to maintain the current at the 10.2 A reference value. However, the next part of the trajectory is from A to B which is *above* the chosen reference value. The reason for this is that the current controller in its present form only controls the phase voltage using the top switch of the asymmetrical converter (bottom switch is normally closed during the phase-active period). This implies that a flux reduction cannot be initiated by the controller. An alternative chopping technique is to allow the current controller to control both switches during the phase-active interval, in which case the trajectory would be along the reference current line from point A to G. Controlling both switches during the phase-active period increases the switching frequency, which is not beneficial in terms of converter switching losses. The normal approach is to only allow freewheeling states during this phase, which is the approach taken in this book. The trajectory which is taken at the beginning of the phase-active interval is therefore D–A–B, where the instantaneous torque will be negative. Once the unaligned position (point B) is reached, the locus will again return to point A and from there onwards along the curve $\psi(i = i^*, \theta_e)$ until point C is reached which is the end of the phase-active interval, i.e., $\theta_e = \theta_e^{off}$. During the de-magnetization interval, the locus follows a path which is parallel to the trajectory A–D (shaft speed and supply voltage magnitude remain unchanged) until the aligned rotor position is reached. The trajectory endpoint E is in reality above 180°, but it can also be shown within the control diagram (as is the case here), given the symmetry of the flux linkage/angle curves to be mirrored with respect to the vertical axis at $\theta_e = 180°$. It is however important to realize that the instantaneous torque during the trajectory from the aligned rotor position to point E will be negative.

An increase in speed from $n_m = 400$ rpm to $n_m = 1000$ rpm leads to a locus which is similar to that described above. The control angles remain unchanged, but the gradient of the initial part of the trajectory D–G is reduced by a factor $^{400}/_{1000}$ as may be observed from Fig. 10.32. The fact that the rotor angle for point G coincides with the unaligned rotor position is purely coincidental. The hysteresis current controller maintains the current for the remaining part of the phase-active interval, which leads to trajectory G–C along the curve $\psi(i = i^*, \theta_e)$. The de-magnetization trajectory is similar to that described above, but the gradient will be lower (and parallel to locus D–G), as is apparent from Fig. 10.32. For this drive speed, the trajectory reaches the aligned position at a higher flux-linkage value (in comparison to the previous case) and then terminates approx. in the vicinity of the coordinates (100°, 0 Vs) in the control diagram.

As the operating speed is increased further, a situation will arise where for a substantial part of the phase-active interval current control cannot be maintained because the induced voltage u_e is greater than the available supply voltage u_{DC}. The trajectory with shaft speed $n_m = 1478$ rpm is precisely such an example as may be observed from Fig. 10.32. The control angles remain unchanged, hence the locus starting point is again at coordinates (40°, 0 Vs) of the control diagram. During the first part of the phase-active period, the drive is operating under voltage control until the reference current curve is reached at point A. From that point onwards the trajectory is along the curve $\psi(i = i^*, \theta_e)$ until point F is reached, where the induced voltage becomes equal to the supply voltage $u_e = u_{DC}$. From point F onwards the trajectory proceeds along a trajectory with a gradient defined by Eq. (10.35). For this particular operating speed, known as the *base speed* ω_m^b, the gradient of the trajectory can be represented by a linear function. This linear function is of the form

$$\psi_e = k_\psi \theta_e + \psi_e^o \tag{10.38}$$

for $0 \leq \theta_e \leq 180°$. The base speed may then be found by making use of Eqs. (10.35) and (10.38), which gives

$$\omega_m^b = \frac{u_{DC}}{k_\psi}. \tag{10.39}$$

In the given example, k_ψ is equal to 1.55 and ψ_e^o to 0.02. This leads to a base speed of $n_m^b = 1478$ rpm. When the drive is operating at base speed, the gradient of the locus trajectory (after point F) is equal to the gradient of Eq. (10.38) when hysteresis control is no longer possible. However, the *turn-on* angle must be chosen in such a manner that the flux linkage ψ_e^o at 0° is reached. The required *base turn-on* angle at the operating speed which satisfies this condition is given as

$$\theta_e^{on,b} = \frac{\psi_e^{min}}{k_\psi} \tag{10.40}$$

with $\psi_e^{min} = \psi_e(\theta_e = 0°)$. The choice of *turn-on* angle is such that the first part of the locus from D to the unaligned angle (with gradient $u_{DC}/\omega_m^b = k_\psi$) intersects the linear function at $\theta_e = 0°$. With the current choice of parameters, the *base turn-on* angle is equal to $\theta_e^{on,b} \approx -40°$. The remaining control variable to be defined is the *turn-off* angle. Observation of Fig. 10.32 (for the base speed locus) shows that its value must be smaller or equal to $\theta_e^{off,b} = \theta_e^{on,b} + 180°$ to maintain a non-conduction interval which is greater or equal to zero. As a guideline, the *base turn-off* angle is set to 140°.

Control Diagram for Pulse Width Modulation with Constant Duty Cycle

The use of pulse width modulation for controlling the average phase voltage during the phase-active interval is considered here with the aid of the SR control diagram given in Fig. 10.33. As with the previous case, three operating speeds are considered, which are precisely those used for the hysteresis controller. For this type of control the (constant) duty cycle δ must be a function of speed. The control angles are set to their base values for reasons to be discussed. In terms of identifying a suitable duty cycle it is beneficial to use the linear function as defined by Eq. (10.38). This linear function represents the best possible fit to the flux linkage/angle curve for the chosen reference current $i^* = 10.2$ A. Consequently, it is prudent to choose the duty cycle and *turn-on* angle in such a manner that the PWM control trajectory during a substantial part of the phase-active is coincidental with this linear function. This control strategy in effect states that the gradient of the locus should be set equal to the gradient k_ψ of the linear function $\psi_e(\theta_e)$ which leads to a duty cycle value of

$$\delta = \frac{k_\psi \omega_m}{u_{DC}}. \tag{10.41}$$

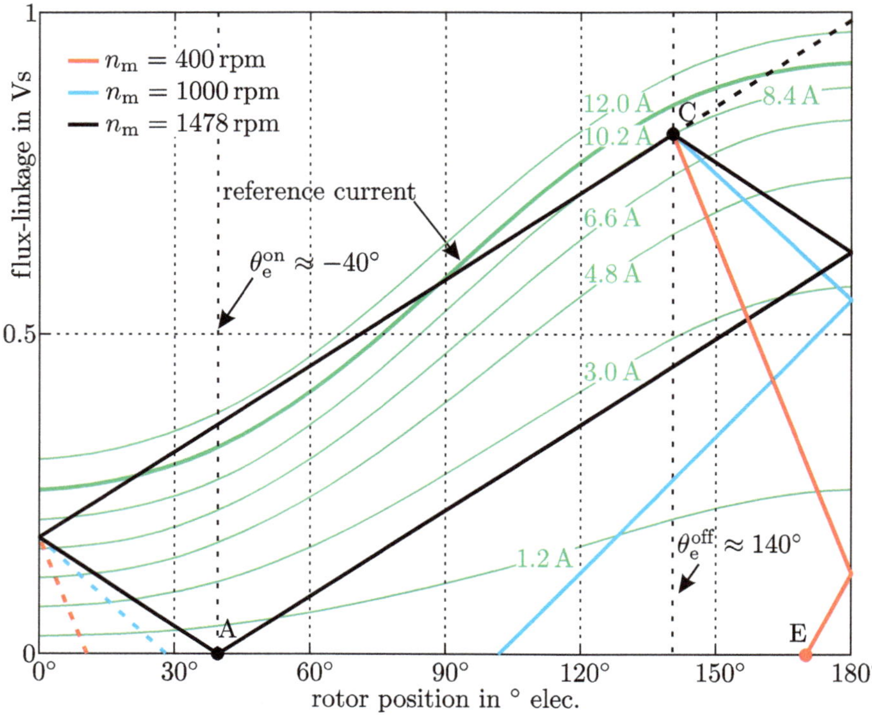

Fig. 10.33 Drive operation, PWM control

As speed increases, the duty cycle must increase; however, the gradient of the trajectory during the phase-active period remains constant as may be observed from Fig. 10.33. The speed at which the duty cycle becomes one is exactly the base speed as defined by Eq. (10.39).

In this case, the trajectory starts at point A and follows the linear function $\psi_e(\theta_e)$ at the unaligned position. The gradient of this part of the trajectory is maintained at the value k_ψ. Furthermore, the *turn-on* angle required for the PWM controller is precisely the base value given the fact that the gradient and endpoint of this part of the trajectory are equal to k_ψ and ψ_e^{min}, respectively. This means that the *turn-on* angle follows from Eq. (10.40). The second part of the trajectory is along the linear function from $(0°,\ \psi_e^{min})$ to point C, at which the rotor angle is equal to the *turn-off* angle value.

The de-magnetization part of the locus is a function of speed given that the phase voltage during this part of the cycle is set to $u = -u_{DC}$. This implies that the gradient of the locus is governed by Eq. (10.35) in exactly the same manner as described for the hysteresis type current controller. This is perhaps not surprising given that this mode of operation is the same for both control strategies. Consequently, this part of the locus is identical for both control strategies, as may be observed by comparing Figs. 10.33 and 10.32, respectively. Prior to discussing the use of SR control diagram in the high speed drive region, it is instructive to consider a minor change to the PWM control strategy. The standard approach activates the PWM generator at the start of the phase-active period. An alternative would be to compare the measured current with the reference current at the beginning of the active cycle and switch on the PWM generator once the condition $i = i^*$ is met. Such a change requires a change to the *turn-on* angle given that the gradient of this first part of the trajectory is then defined by the ratio of the supply voltage and shaft speed. Figure 10.33 (dashed blue and red lines) shows how this control strategy affects the first part of the locus. The positive effect is that the negative torque contribution, which is inherent for this part of the trajectory, is reduced. However, as speed is increased, its effectiveness is reduced.

High Speed Operating Range of SR Machine

This section focuses on the high speed operating range of the drive (see Fig. 10.30). The control strategy with constant control angles is discussed with the aid of the SR control diagram shown in Fig. 10.34. In this example, the operating trajectory for three shaft speeds $n_\mathrm{m} = 1500$, 2000 rpm and $n_\mathrm{m} = 3000$ rpm is shown, with control angles $\theta_\mathrm{e}^\mathrm{on} = -36°$ and $\theta_\mathrm{e}^\mathrm{off} = 144°$, which are in the vicinity of the base speed values. At the beginning of the phase-active interval, the trajectory starting point is

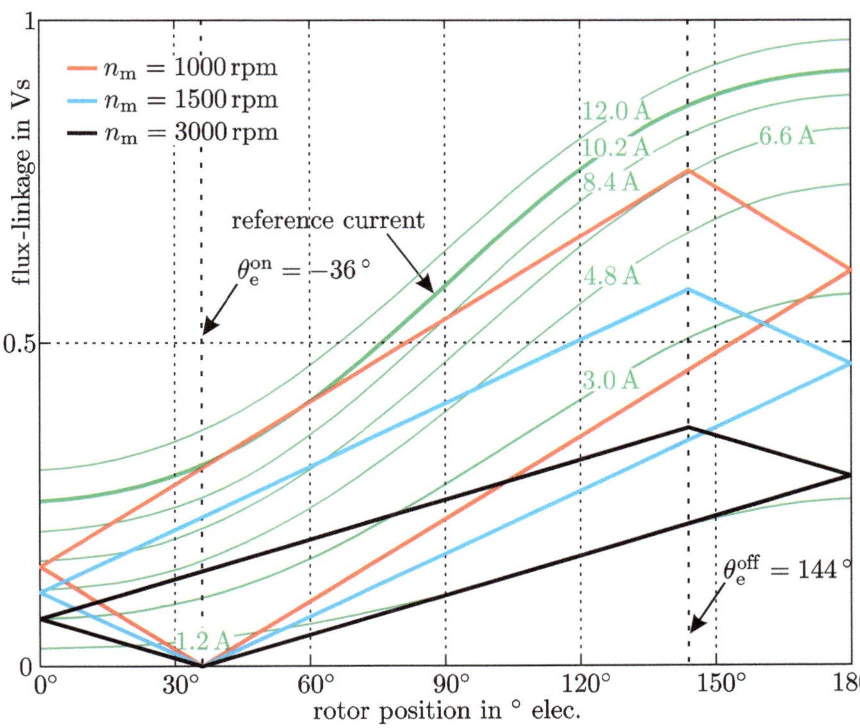

Fig. 10.34 High speed operation with fixed control angles

at coordinates (θ_e^{on}, 0 Vs) and proceeds to the unaligned position, where the gradient of the trajectory is defined by the ratio of the supply voltage and shaft speed (see Eq. (10.35)). The trajectory from the unaligned position to the end of the interval remains unchanged in terms of the absolute gradient value. At the position θ_e^{off}, the end of the phase-active interval is reached which means that the gradient of the locus is reversed (phase voltage switches from u_{DC} to $-u_{DC}$). The de-magnetization trajectory is similar to that shown in the low speed SR control diagrams. However, it is noted that the complete locus will be trapezoidal, because the absolute value of the trajectory gradient before and during the de-magnetization phase remains unchanged. If the *turn-off* angle is decreased, an opening in the trapezoid will appear on the rotor axis, which signifies the presence of a zero conduction interval of operation over the period τ_{rp}. As speed is increased the gradient of the trajectory reduces, as may be observed from Fig. 10.34.

The operating trajectories for the three shaft speed values, as given in Fig. 10.34 may also be shown in a conventional flux linkage versus current diagram as given in Fig. 10.35. The locus for each speed encompasses an area which is directly proportional to the average torque per phase, cycle τ_{rp}. Consequently, an indication of the change in average torque with increasing speed may be found by observing the

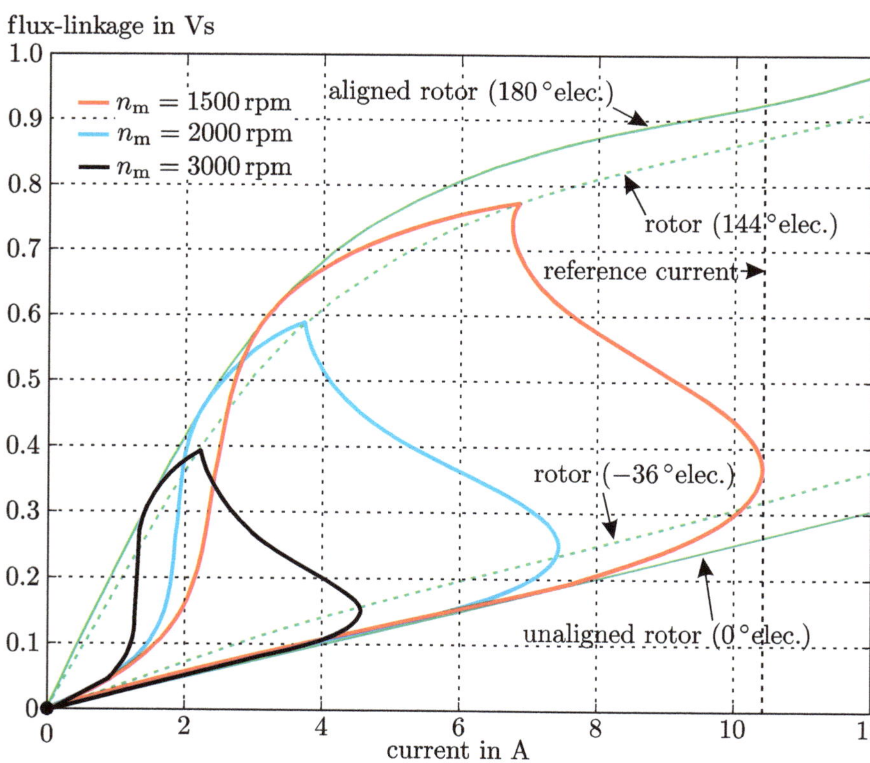

Fig. 10.35 Flux linkage/current locus for high speed operation

change of area enclosed by the locus. Observation of Fig. 10.35 shows that doubling the speed (from $n_m = 1500\,\text{rpm}$ to $n_m = 3000\,\text{rpm}$) leads to an approximate fourfold reduction in the average torque. Hence, the corresponding output power level is approximately halved, as may also be observed from the *fixed angle* high speed operating trajectory shown in Fig. 10.30.

In the sequel to this section, the SR control diagram is used as a means of determining how the control angles may be varied to maintain an approximately constant output power level in the high speed operating range, i.e., operation along trajectory B–C in Fig. 10.30. The trajectory for $n_m = 1500\,\text{rpm}$, as illustrated in Fig. 10.36, is shown for a set of control variables $\theta_e^{on} = -36°$ and $\theta_e^{off} = 144°$ which is close to the base speed values given earlier. This is perhaps not surprising given that the trajectory under consideration is only slightly above the base speed $n_m = 1478\,\text{rpm}$. In terms of maximizing the average torque, it is important to realize that only the top two trajectories of the trapezoidal locus yield a positive torque contribution. The two positive torque parts of the locus are those which correspond to rotor displacement from the unaligned $\theta_e^u = 0°$ to the aligned $\theta_e^a = 180°$ position.

As the shaft speed is increased, the gradient of the locus reduces, as was shown earlier (see Fig. 10.34), which means that the positive torque trajectories of the locus encounter lower currents which in turn leads to a lower average torque. The average torque can be increased if the *turn-on* angle is advanced in such a manner, that the first part of the positive torque producing locus (from the unaligned position to the phase-active *turn-off* angle) coincides for at least one rotor angle with the reference flux linkage/angle curve with $i = 10.2\,\text{A}$. This control strategy as depicted in Fig. 10.36 provides a guideline for a suitable choice of *turn-on* angle.

The *turn-off* angle is chosen to satisfy the equation $\theta_e^{off} \leq \theta_e^{on} + 180°$ at the given speed to avoid an operating mode where the flux linkage is greater than zero for the entire period τ_{rp}.

For the test prototype motor under consideration here, the control angles for the operating speeds $n_m = 2000, 3000$ were set to $\theta_e^{on} = -60°, -108°$ and $\theta_e^{off} = 120°, 72°$ respectively. With this choice of *turn-on* and *turn-off* angles, the output power can be maintained at an approximately constant level as shown in Fig. 10.30. As speed is increased even further, the gradient of the locus is also reduced. At a given operating speed, which corresponds to point C in Fig. 10.30, further variation in the control angles become ineffective given that no part of the positive torque producing locus coincides with the flux linkage/angle curve $\psi(i = i^*, \theta_e)$. From this speed onwards the control angles are kept constant in which case the output power decreases in a manner as described previously.

10.4.3 Direct Instantaneous Torque Control (DITC)

In the previous part of this chapter, current and PWM voltage (flux) based control techniques were discussed. The disadvantage of these methods is that the shaft torque of the machine will contain a large torque ripple, as was shown in Fig. 10.26b

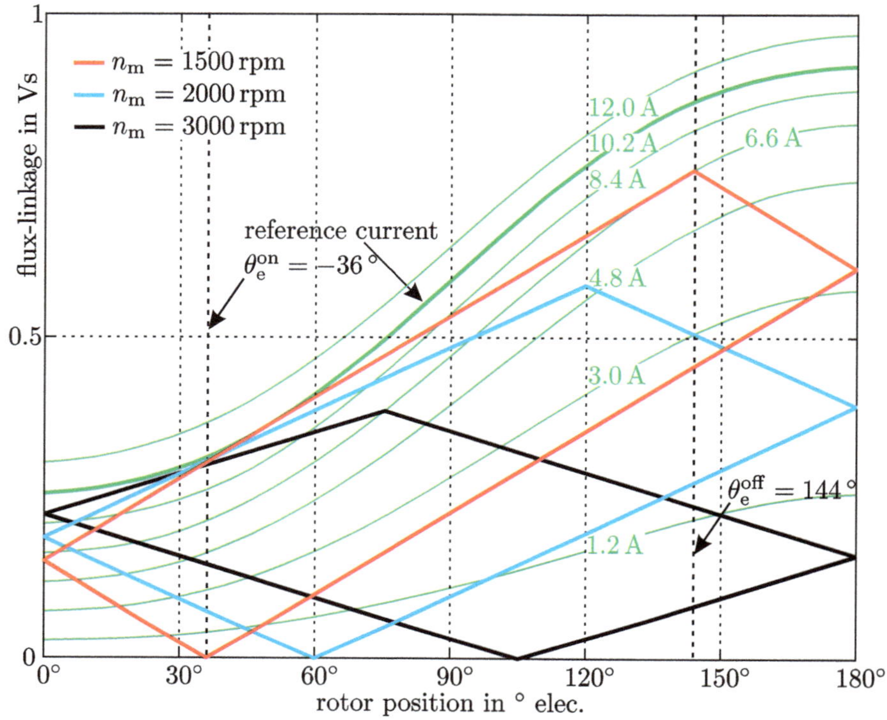

Fig. 10.36 High speed operation, variable control angles

for the linear SR model. In reality, saturation will exacerbate the problem further, which means that mechanical or electronic counter measures are usually taken to realize a smoother torque. A variety of mechanical techniques are available, such as for example, rotor skewing and pole shaping, which unfortunately compromise the overall torque producing capability of the machine. A range of electronic methods are also available which are basically concerned with profiling the phase current or flux-linkage waveform in such a manner as to minimize the torque ripple component present in the shaft torque.

This section considers an alternative approach, referred to as *direct instantaneous torque control* (DITC) [9], which controls the converter switches in such a manner as to ensure that the estimated shaft torque $\tilde{T}_e$ is held at approximately (within user specified limits) the reference torque level T_e^*. The operating principles of DITC will be discussed with the aid of Fig. 10.37, which makes use of the four-phase 8/6 non-linear test machine (see Sect. 10.5).

The machine, as illustrated in Fig. 10.37, is connected to a converter module, which may, for example, be formed by four asymmetric half-bridge phase units as shown in Fig. 10.19. The four signal pairs Sw_t and Sw_b, which control the top and bottom phase switches, respectively, are represented by the switch array variable $[Sw]$. A control module, as given in Fig. 10.37 generates the required converter

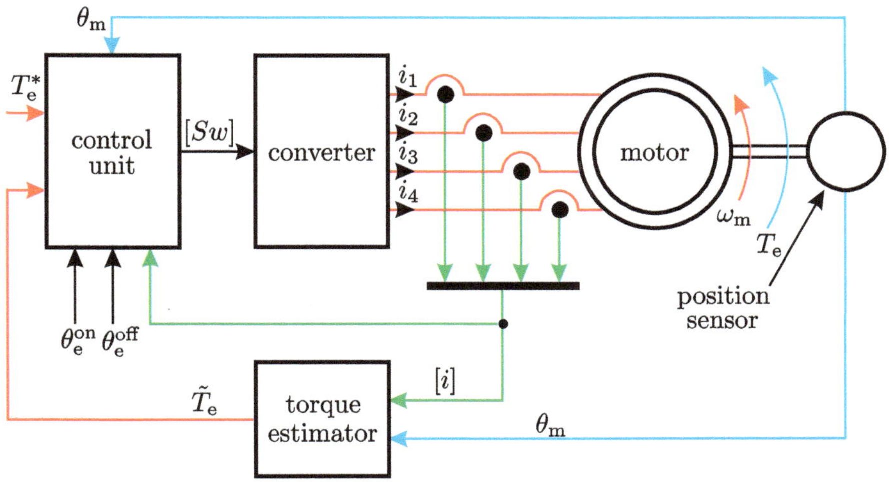

Fig. 10.37 Basic block diagram of a direct instantaneous torque controlled (DITC) SRM drive

switch signals on the basis of the input variables T_e^* and $\tilde{T}_e$ and control angles θ_e^{on} and θ_e^{off}. In addition, this module requires access to the shaft angle θ_m, which is provided by the position sensor attached to the shaft. Furthermore, the measured phase currents i_1, i_2, i_3, i_4, as represented by the matrix variable $[i]$, are also fed into the control unit.

The torque estimator calculates the instantaneous shaft torque with the aid of the per phase static torque versus current/angle curves of the machine in question in accordance with the approach shown in Fig. 10.14. These torque per phase contributions are then added with the aid of Eq. (10.33) which yields the required shaft torque estimate.

In the generic representation of the controller, as given in Fig. 10.38, three torque hysteresis comparators are used, which provide three logic outputs S_t^{comp}, S_c^{comp}, and S_b^{comp}, respectively [8]. The switching algorithm for these comparators is of the form

$$\text{if } \tilde{T}_e > \left(T_e^* + \Delta T_t^*\right) + \frac{\Delta T^*}{2} \quad \text{comparator output } S_t^{comp} = 0 \tag{10.42a}$$

$$\text{if } \tilde{T}_e < \left(T_e^* + \Delta T_t^*\right) - \frac{\Delta T^*}{2} \quad \text{comparator output } S_t^{comp} = 1 \tag{10.42b}$$

$$\text{if } \quad \tilde{T}_e > T_e^* + \frac{\Delta T^*}{2} \quad \text{comparator output } S_c^{comp} = 0 \tag{10.42c}$$

$$\text{if } \quad \tilde{T}_e < T_e^* - \frac{\Delta T^*}{2} \quad \text{comparator output } S_c^{comp} = 1 \tag{10.42d}$$

$$\text{if } \tilde{T}_e > \left(T_e^* - \Delta T_b^*\right) + \frac{\Delta T^*}{2} \quad \text{comparator output } S_b^{comp} = 0 \tag{10.42e}$$

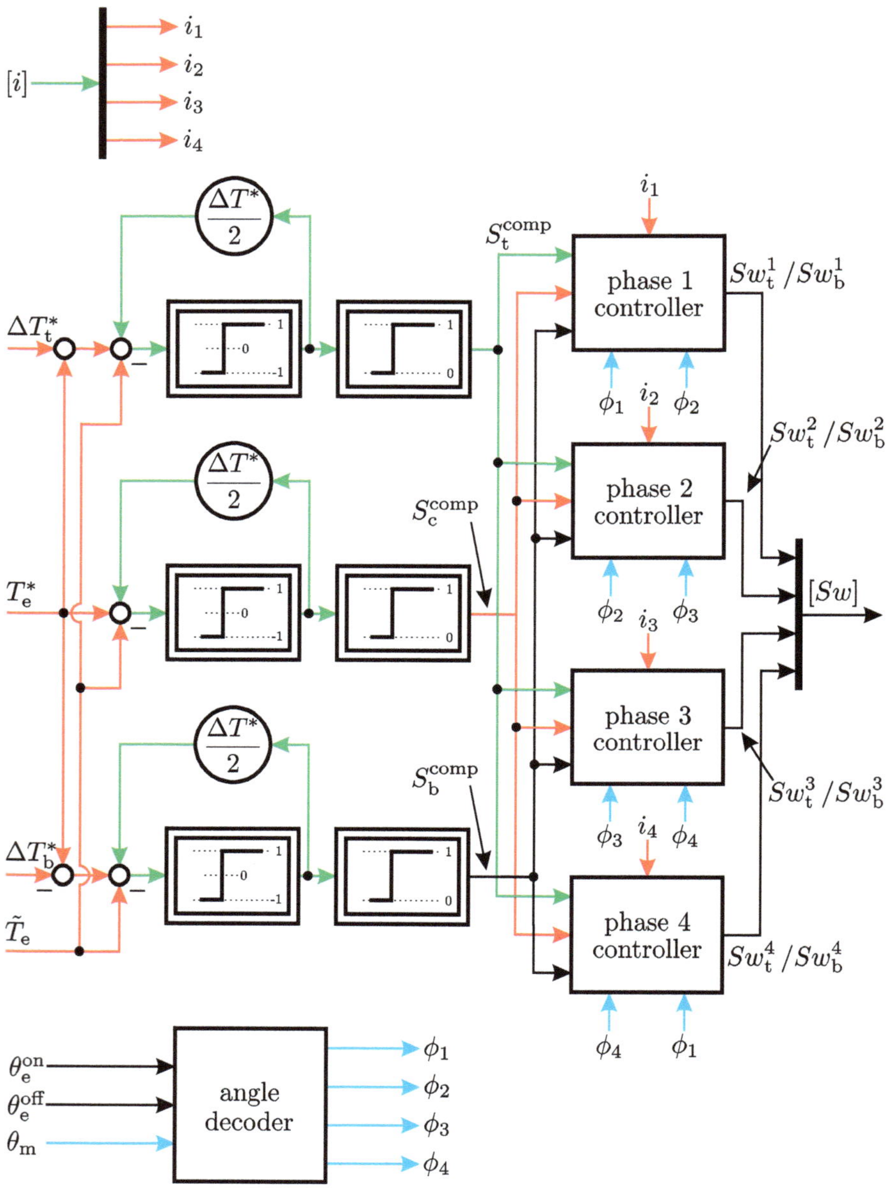

Fig. 10.38 DITC controller block diagram, showing torque comparators and phase controllers (for a four-phase SRM)

$$\text{if } \tilde{T}_{\rm e} < \left(T_{\rm e}^{*} - \Delta T_{\rm b}^{*}\right) - \frac{\Delta T^{*}}{2} \quad \text{comparator output } S_{\rm t}^{\rm comp} = 1 \qquad (10.42\text{f})$$

where $\Delta T_{\rm t}^{*}$, $\Delta T_{\rm b}^{*}$, and ΔT^{*} are user defined values which determine the error band of the torque controller as may be observed from the example shown in Fig. 10.39.

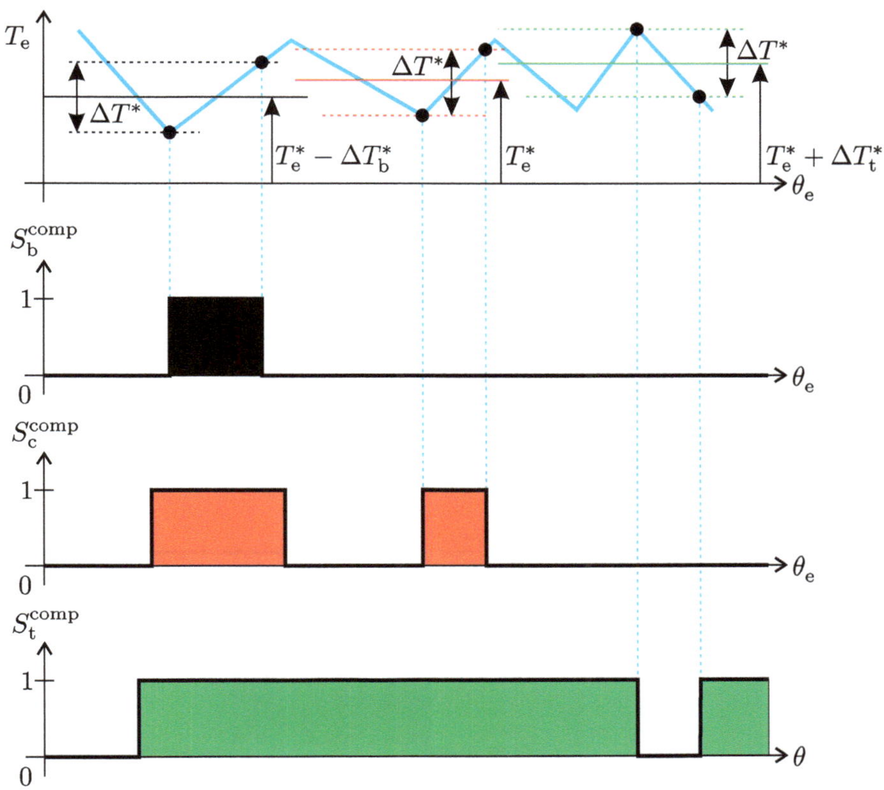

Fig. 10.39 DITC comparator signals

In this example, an arbitrary torque versus angle function $T_e(\theta_e)$ is shown together with the reference torque value T_e^* and set of incremental reference torque settings $\Delta T_t^* = \Delta T_b^*$, ΔT^*. The logic outputs from the three comparators, as shown in Fig. 10.39, is governed by the switching algorithm according to equation set (10.42).

Central to the DITC approach is the way in which the three comparator signals are used by the phase controller modules of Fig. 10.38. The comparator signals S_t^{comp}, S_c^{comp}, S_b^{comp} are connected to all the phase controller modules. However, at any one time (or rotor angle) a phase controller module uses either S_c^{comp} or a combination of S_t^{comp}, S_b^{comp} as inputs. This choice of signals is realized by sub-dividing each phase-active interval into a so-called *incoming phase* (IPI) and *outgoing phase* (OPI) interval as shown in Fig. 10.40. This figure shows an idealized static torque/angle curve for phase 1, together with the phase-active intervals for phases 1 and 2, namely ϕ_1, ϕ_2. For a four-phase machine $\tau_{rp} = 60°$mech. and the phase-active intervals are equal to 30 °mech. and displaced by 15 °mech., with respect to each other. The incoming phase 1 interval represents the first part of the phase 1 active interval, where there is NO overlap with the next phase ϕ_2. Likewise, the outgoing phase 1 interval is that part of the phase 1 active interval which overlaps

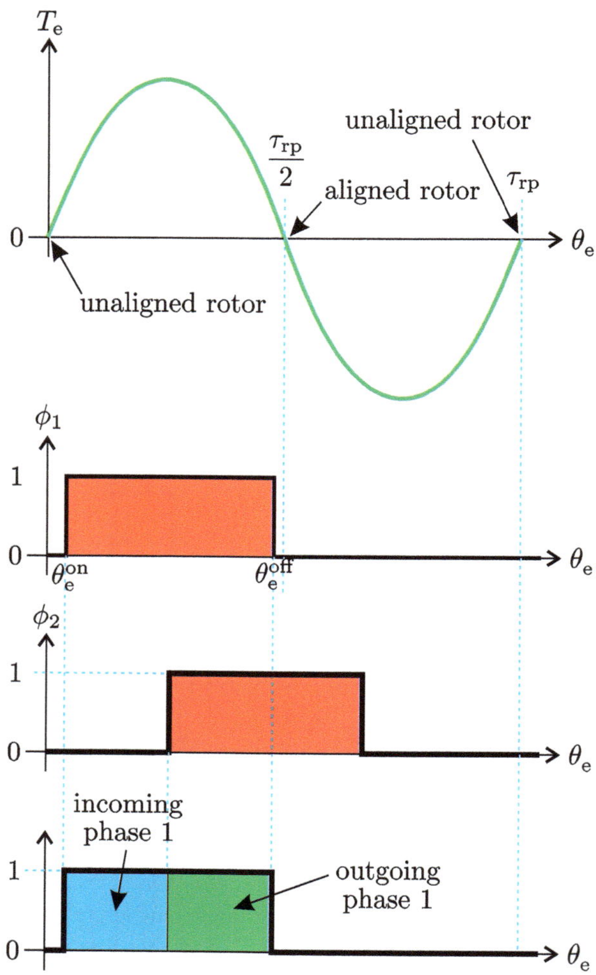

Fig. 10.40 DITC control signals

with phase 2, as may be observed from Fig. 10.38. For the four-phase machine there will be at any rotor angle an active ICI and OPI interval and these are calculated with the aid of the four phase-active signals which in turn are generated by the angle decoder module. For example, during the ICI phase 2 interval, an OPI interval for phase 1 will be present.

The basic DITC control strategy is based on allocating the primary task of producing the required reference torque value with the incoming phase. The designated outgoing phase is then used to augment the torque production for the duration of the corresponding OPI interval. To implement this control strategy the converter switching signals are directly generated from the three comparator signals according to the algorithm given in Table 10.3.

The algorithm basically states that the top converter switch of the incoming phase is controlled by the comparator with output S_c^{comp}. This implies that the

Table 10.3 Phase controller switching algorithm

	Incoming phase	Outgoing phase
Sw_t	S_c^{comp}	S_b^{comp}
Sw_b	Phase-active signal	S_t^{comp} × phase-active signal

incoming phase controller aims to keep the estimated torque between the limits $T_e^* \pm \Delta T^*/2$. At the same time, the outgoing controller will make use of the *outer* comparators with outputs S_b^{comp} and S_t^{comp} and this controller will activate the top switch if the estimated torque (from the torque observer) falls below the level $T_e^* - \Delta T_b^* - \Delta T^*/2$. If the outgoing controller detects an estimated torque in excess of level $T_e^* + \Delta T_t^* + \Delta T^*/2$, both switches of the outgoing controller will be opened to de-magnetize that particular phase. Finally, the measured phase current values are also connected to the respective phase controller modules. These current signals are used by an internal current hysteresis module which compares the measured current against the specified maximum value. If the latter is met or exceeded, the converter top switch will be modulated irrespective of Table 10.3 to insure that the current cannot exceed the maximum value.

A DITC tutorial is given in the tutorial section at the end of the chapter and an example of the performance, which can be achieved, is given in Fig. 10.41. In this case, the torque reference value is set to $T_e^* = 10\,\text{Nm}$ and the incremental torque reference value were taken to be $\Delta T_t^* = \Delta T_b^* = 0.05\,\text{Nm}$. A 240 V converter supply is assumed with a maximum phase current rating of 11.8 A and the *turn-on* and *turn-off* angles were set to $\theta_e^{on} = -30°$ and $\theta_e^{off} = 150°$.

Shown in Fig. 10.41 are two phase currents, the phase voltage and resultant shaft torque with the 8/6 machine operating at 1000 rpm. It is instructive to compare the DITC based drive with a *standard* hysteresis based drive operating under the same load and speed conditions. The reference current is set to 6.45 A to realize the same average shaft torque as developed in the DITC based drive. The control angles and the supply voltage remain unchanged when compared to the previous case. An observation of the results as given in Fig. 10.42 demonstrates that the current hysteresis based drive has a considerably higher torque ripple component in the shaft torque, when compared to the DITC based drive.

However, the price for this performance lies with the need for a higher dynamic current control range, as may be observed by comparing the phase currents for the two drives. As the shaft torque is increased, the DITC drive peak currents will be limited by the protective current controller, and this in turn will impede the operation of the DITC controller. Furthermore, both controller methods do not operate with constant switching frequencies in the converter, which impedes realization optimized control (to minimize acoustic emissions or to maximize efficiency). Using predictive control principles it is possible to implement DITC with constant switching frequencies. We refer to specialized literature, such as [4, 5].

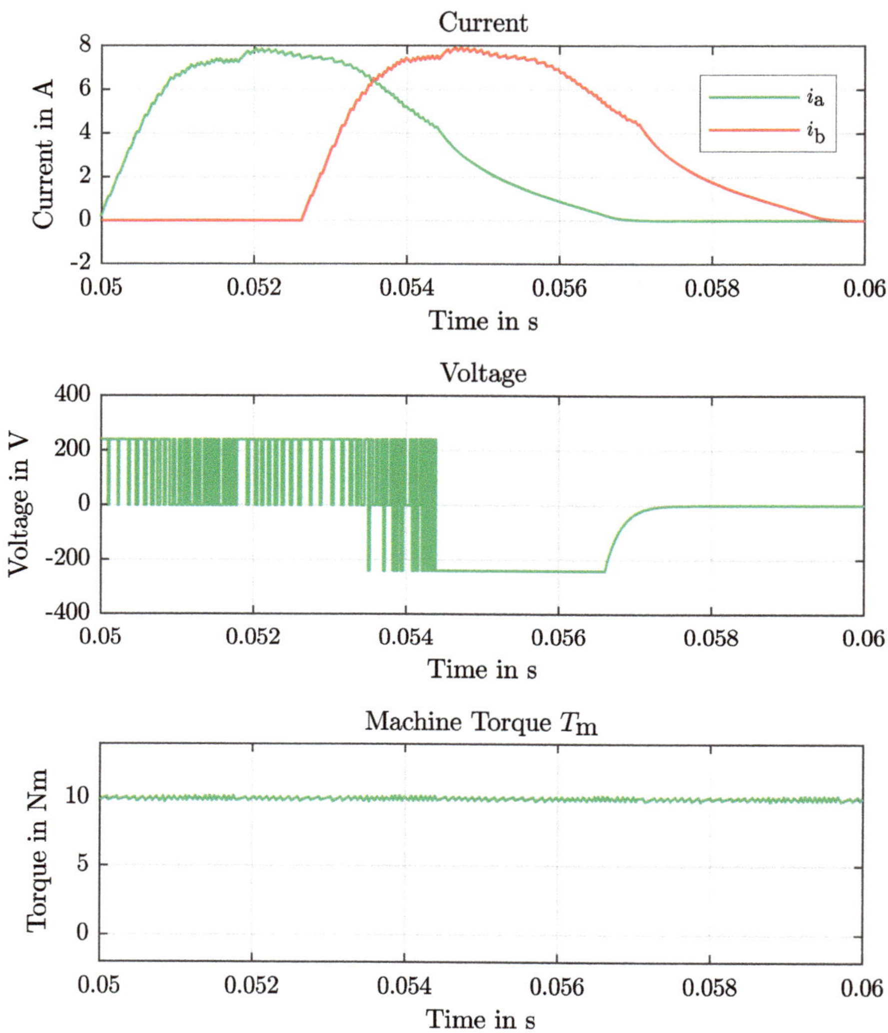

Fig. 10.41 Waveforms of DITC based SR drive, showing low torque ripple

10.5 Switched Reluctance Demonstration Machine

In various sections of this chapter, including the tutorials, numerical examples are used to demonstrate specific aspects of SR drive behavior. For this purpose, the 8/6 SR axial flux (SRAF) machine [18], [16] as shown in Fig. 10.43 is used. The magnetization characteristics of the machine were derived via a set of *locked rotor* measurements, whereby a phase voltage pulse is applied and the transient current response is measured. Use of Eq. (10.7) leads to the representation of the flux linkage/current curve (for a given rotor angle θ_k) in the form of a set of data

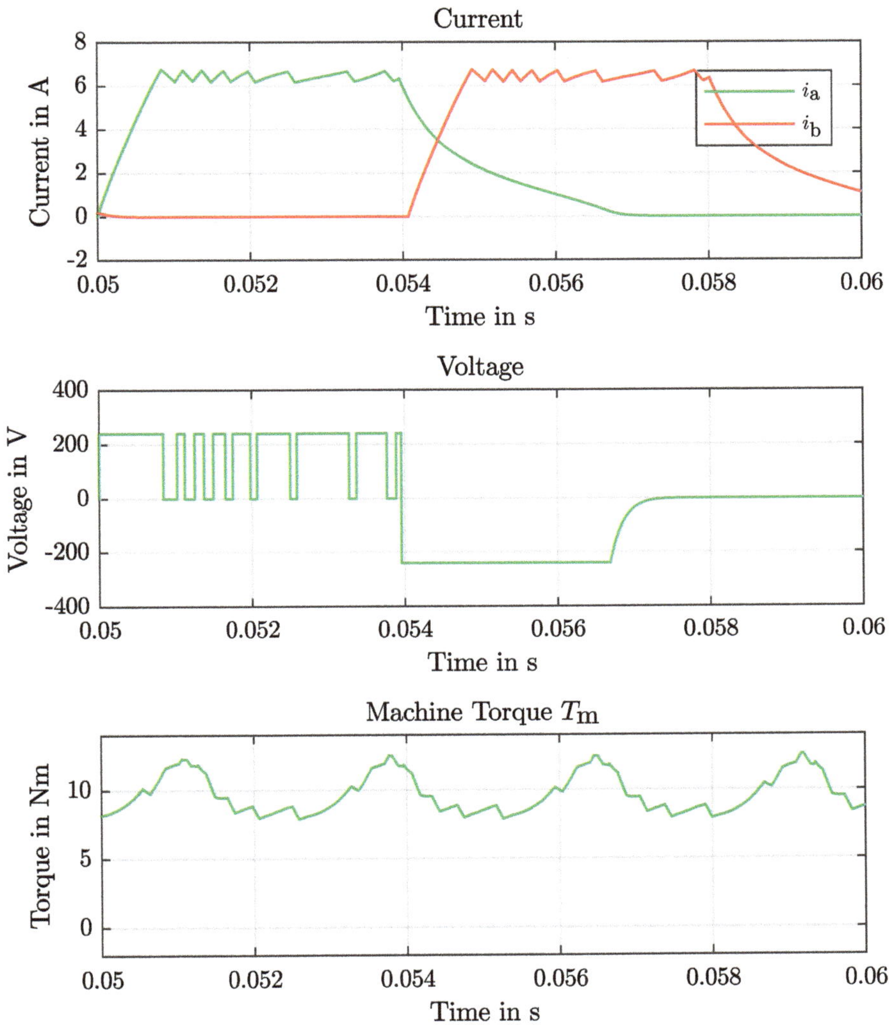

Fig. 10.42 Waveforms of current hysteresis controlled SR drive

points $\Psi_j\left(i_j, \theta_k\right)$, $j = 1 \ldots N_i$, where N_i represents the number of measured (or computed) flux linkage/current data pairs. This measurement process must be repeated for a set $k = 1 \ldots N_\theta$ rotor angles starting from the (in this example) unaligned $\theta_e = 0°$; $k = 1$ to the aligned $\theta_e = \theta_{max}$; $k = N_\theta$ position. Typical values for N_θ and N_i are 11 and 200, respectively, and the actual values used are governed by the degree of saturation, i.e., the shape of the magnetization characteristics. Considerable efforts have been undertaken by researchers on the topic of developing a suitable mathematical representation of these characteristics [18] and to automate these measurements [6]. A relatively simple approach to solving this problem

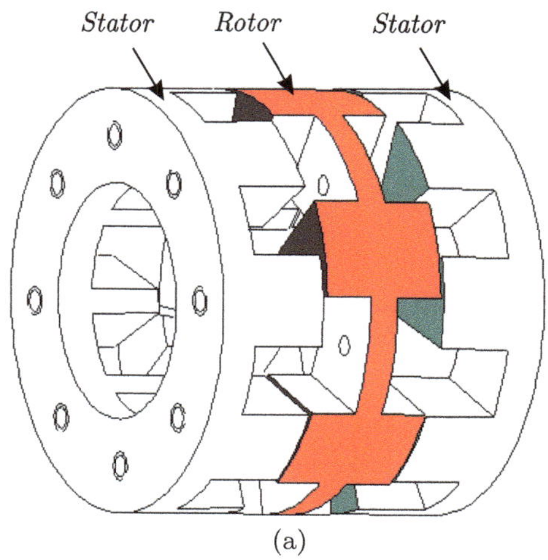

Fig. 10.43 Examplary 8/6 Switched Reluctance Axial Flux (SRAF) [18]. (**a**) Stator and rotor module

is to use a two-dimensional lookup table directly based on a set of data points $\Psi_{\mathrm{j}}\left(i_{\mathrm{j}},\theta_{\mathrm{k}}\right)$, $\mathrm{j} = 1\ldots N_i$ derived via measurements [6], finite element analysis or analytical flux path techniques [23]. In the given example, the lookup table is obtained via measurements.

For some models (see for example Fig. 10.15), derivatives of the magnetization curves are needed. Note also that the torque characteristics can be either measured (lookup tables can be used to represent the data) using locked rotor techniques or derived from the magnetization curves, in which case further processing involving differentiation and integration (see Eqs. (10.18) and (10.19)) is needed. If further processing of the magnetization curves is required, i.e., to obtain derivatives $\partial\psi/\partial i$, $\partial\psi/\partial\theta_{\mathrm{e}}$, and torque T_{e}, then the use of cubic *B-splines* is particularly effective [17]. The reason for this is that so-called *B-splines* are continuous in terms of the function itself and its derivatives. Furthermore, the latter is readily available, which is beneficial given the need to find derivative functions, as mentioned above.

This approach has been used here to represent the characteristics of the machine, in the form of a set of *spline coefficients*. All characteristics in this book are characterized by the magnetization data which is represented in terms of a set of bi-variate splines which in turn are defined by a set of current $[0\ i_1^{\mathrm{s}}\ i_2^{\mathrm{s}}\ i_{\mathrm{max}}]$ and angle knots $[0\ \theta_1^{\mathrm{s}}\ \theta_2^{\mathrm{s}}\ \theta_{\mathrm{max}}]$ [17] as shown in Fig. 10.18. The values assigned to these knots and the phase resistance value are given in Table 10.4.

Application of the spline curve fitting techniques, as mentioned above, leads to the characteristics shown in Fig. 10.18. A set of spline based routines [17] can be used to generate a bi-variate cubic spline representation of the magnetization curves $\psi(i,\theta_{\mathrm{e}})$ (see Fig. 10.44) and *inverse* function $i(\psi,\theta_{\mathrm{e}})$. In addition, the static torque/current/angle curves $T(i,\theta_{\mathrm{e}})$ (see Fig. 10.45), incremental inductance curves $l(i,\theta_{\mathrm{e}})$ (see Fig. 10.46) and normalized *EMF* versus current/angle curves: $e(i,\theta_{\mathrm{e}})$ (see Fig. 10.47).

Table 10.4 SR parameters

Parameters	Value
Current knot 1 (i_1^s)	3 A
Current knot 2 (i_2^s)	5 A
Maximum current $(i_{\max})$	12 A
Angle knot 1 (θ_1^s)	10 °mech.
Angle knot 2 (θ_2^s)	20 °mech.
Unaligned angle $(\theta_{\max})$	30 A
Phase resistance (R)	1.1 Ω

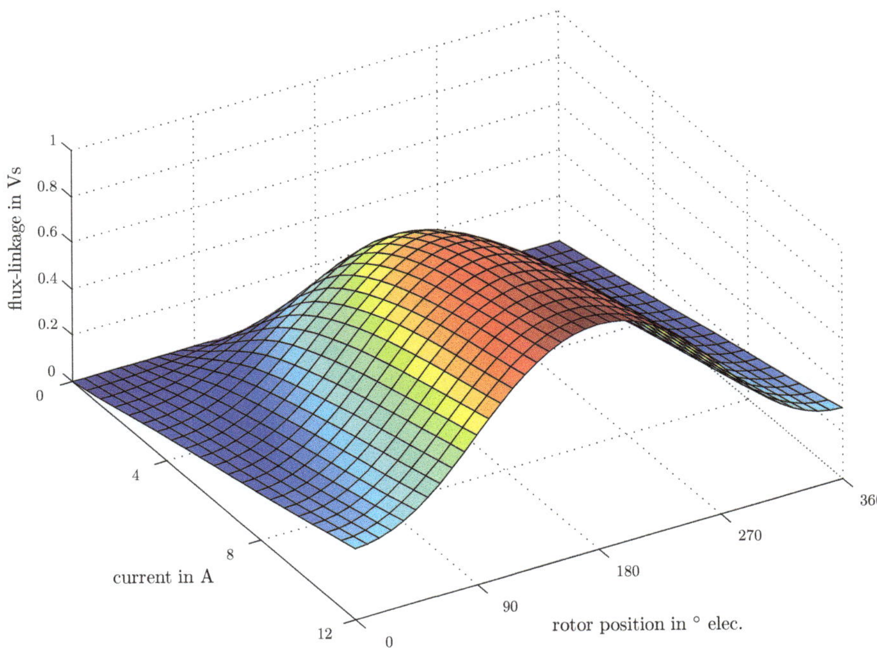

Fig. 10.44 Flux-linkage curves: $\psi(i, \theta_e)$

10.6 Tutorials

10.6.1 Tutorial 1: Analysis of a Linear SR Machine, with Current Excitation

This tutorial considers a simplified SR machine, based in part on the generic models shown in Figs. 10.14 and 10.15, respectively. A linear 8/6 machine representation is assumed, which is according to the model discussed in Sect. 10.5. In this case, a single-phase (phase a) is connected to a current source, while the shaft speed is set to a constant value of 1000 rpm. The simulation makes use of lookup tables to represent the machine. The aim of the tutorial is to allow the reader to examine

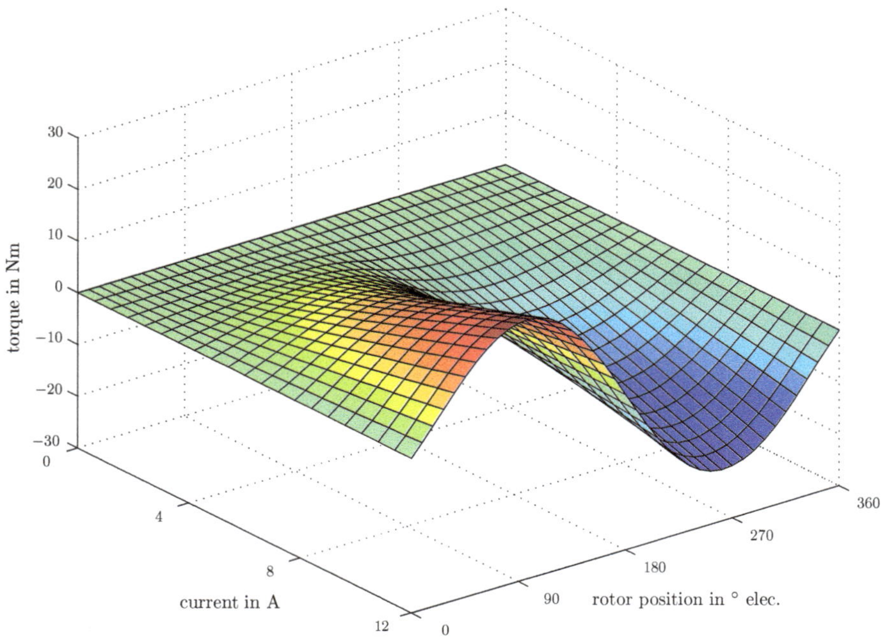

Fig. 10.45 Torque versus current/angle curves: $T(i, \theta_e)$

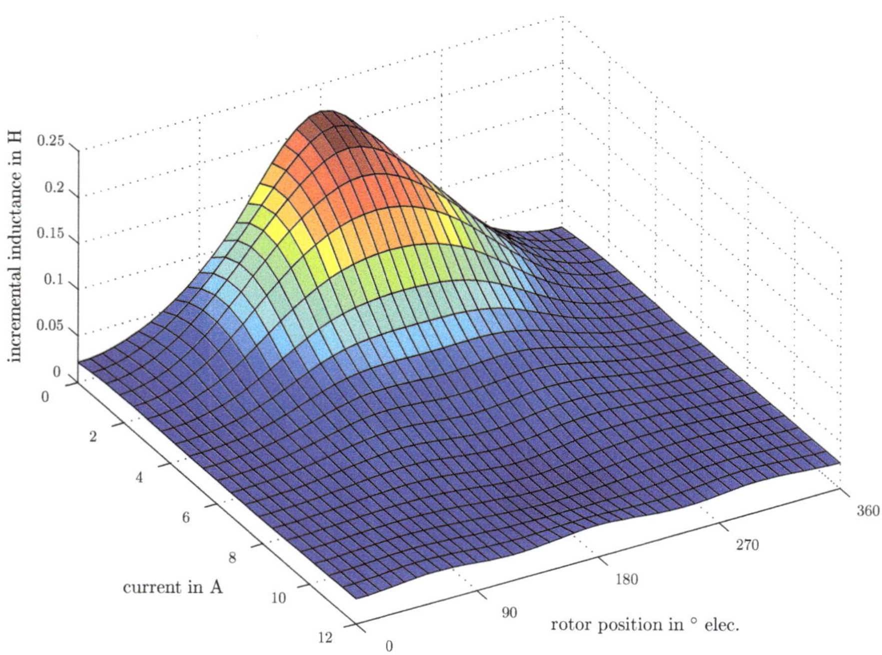

Fig. 10.46 Incremental inductance versus current/angle curves: $l(i, \theta_e)$

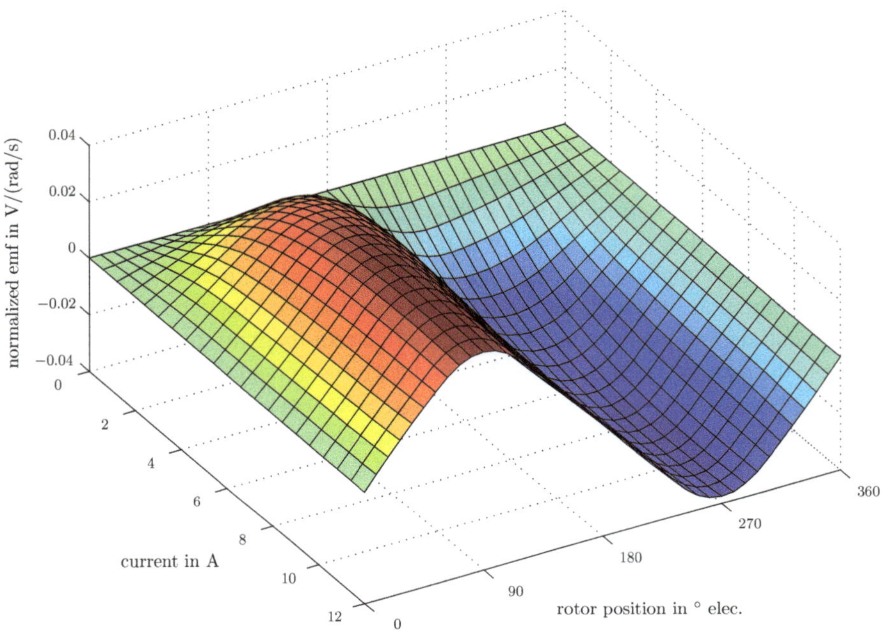

Fig. 10.47 Normalized *EMF* versus current/angle curves: $e(i, \theta_e)$

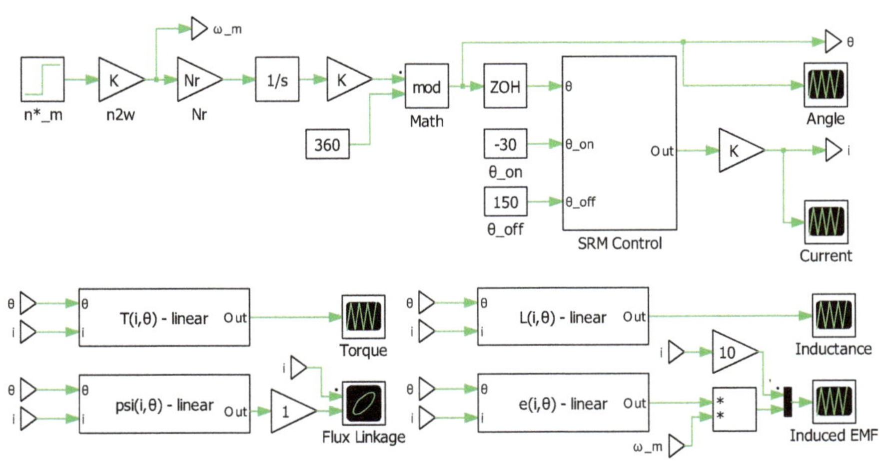

Fig. 10.48 Simulation with a linear SRM model, using current excitation

the time dependent waveforms which represent the instantaneous torque $T_e(t)$, flux linkage $\psi(t)$, incremental inductance $l(t)$, induced voltage u_e, and input phase current $i(t)$ of phase a under constant speed operation. Furthermore, the tutorial allows the reader to examine the effect of changing the *turn-on* θ_{on} and *turn-off* θ_{off} control angles (see Fig. 10.21) on these waveforms.

The simulation model *SR machine*, given in Fig. 10.48, shows the four generic modules which contain the machine characteristics in the form of lookup tables. Inputs to these modules are the phase current i and rotor phase angle θ. The latter variable is a phase related rotor angle, which is derived from the shaft angle θ_m. The rotor phase angle θ is defined with respect to the unaligned position of the phase in question. The rotor phase angle variable and the control angles θ_{on} and θ_{off} are used by the *SR controller* module to generate the phase-active signal Sw. The latter is multiplied by a gain factor (6 in this case) in order to derive a signal which represents the ideal phase a current i waveform with a magnitude of 6 A.

The simulation results given in Figs. 10.49 and 10.50 show the torque, induced voltage and the inductance, and flux linkage / current diagram, respectively. In the example the *turn-on* angle is set to $-30°$, and the turn-off angle to 150°. This implies that the phase current is set to the value of 6 A, $-30°$ before the unaligned position (for phase a) is reached and switched off when the rotor reaches 150° of the phase in question. If the *turn-on* angle is set to zero, the torque waveform T_e will be positive only. The flux linkage/current trajectory as shown in Fig. 10.50 should be similar, in terms of shape, to Fig. 10.9. The angle values will be different, because the latter figure refers to a 2/2 configuration.

10.6.2 Tutorial 2: Nonlinear SR Machine, with Voltage Excitation and Hysteresis Current Controller

A single-phase representation of an 8/6 nonlinear SR machine is considered, with the characteristics as outlined in Sect. 10.5. The machine in question is connected to an asymmetrical half-bridge converter, as in Fig. 10.19, which in turn is fed by a DC supply $u_{DC} = 240$ V. Furthermore, a hysteresis type controller, as shown in Fig. 10.20, is used with a reference current setting of $i^* = 10$ A and a current error setting of $\Delta i = 0.4$ A. As with the previous case, the rotor shaft is taken to rotate at a constant speed of $n_m = 1000$ rpm. In addition, the control angles are adjustable to allow the user to examine their effect on the simulation results. A generic model of the machine as shown in Fig. 10.17 is used for this tutorial.

The simulation given in Fig. 10.51 shows the generic model representation of the machine, which is connected to the converter. Four modules, i.e., T, ψ, l, and e, with input variables i and θ contain the lookup tables required to build the nonlinear model of the machine, with characteristics as shown in Sect. 10.5. A *circuit* model representation is used for the converter, which in turn is linked with the machine via a *single-phase interface module* which provides the voltage excitation for the machine. The phase current variable is returned to the circuit model via the same *single-phase interface module* placed between the two ideal IGBT based converter switches. The simulation modules used to generate the phase-active signal

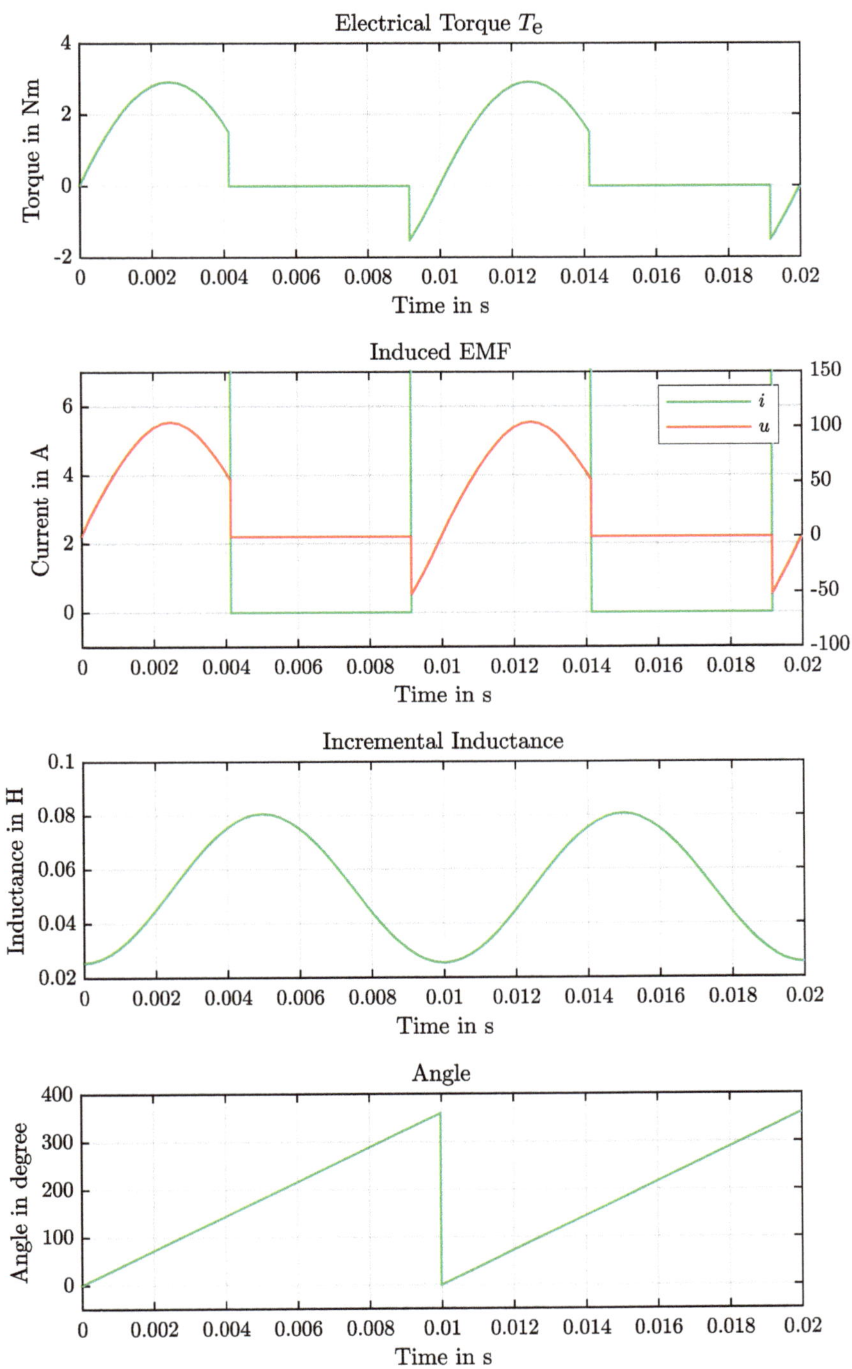

Fig. 10.49 Simulation results for a linear SRM model, with current excitation, $n_\mathrm{m} = 1000\,\mathrm{rpm}$ and $\theta_\mathrm{on} = -30°$, $\theta_\mathrm{off} = 150°$

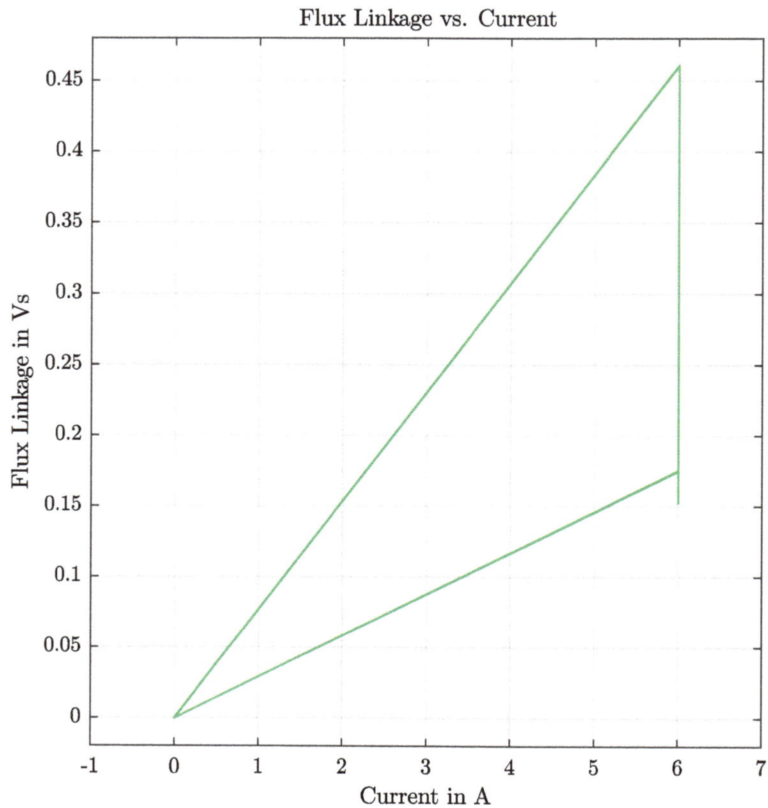

Fig. 10.50 Simulation results for a linear SRM model with current excitation, flux linkage/current diagram with $n_{\mathrm{m}} = 1000\,\mathrm{rpm}$ and $\theta_{\mathrm{on}} = -30°$, $\theta_{\mathrm{off}} = 150°$

Sw remains unchanged with respect to the previous tutorial. An adjustable current reference value, set to 10 A, is also shown in Fig. 10.51. This variable, together with the phase current variable i, form the input to the hysteresis controller. The waveforms generated by the simulation are plotted in Figs. 10.52 and 10.53. The effects of a switching voltage source with a hysteresis control on the phase current can be observed by comparing Figs. 10.50 and 10.53.

10.6.3 Tutorial 3: Nonlinear SR Machine, with Voltage Excitation and PWM Controller

An alternative control approach to the hysteresis current control is a PWM based method mentioned in Sect. 10.2.6. In this case the duty cycle is simply held constant. However, this parameter can also be controlled by a PI controller with input the

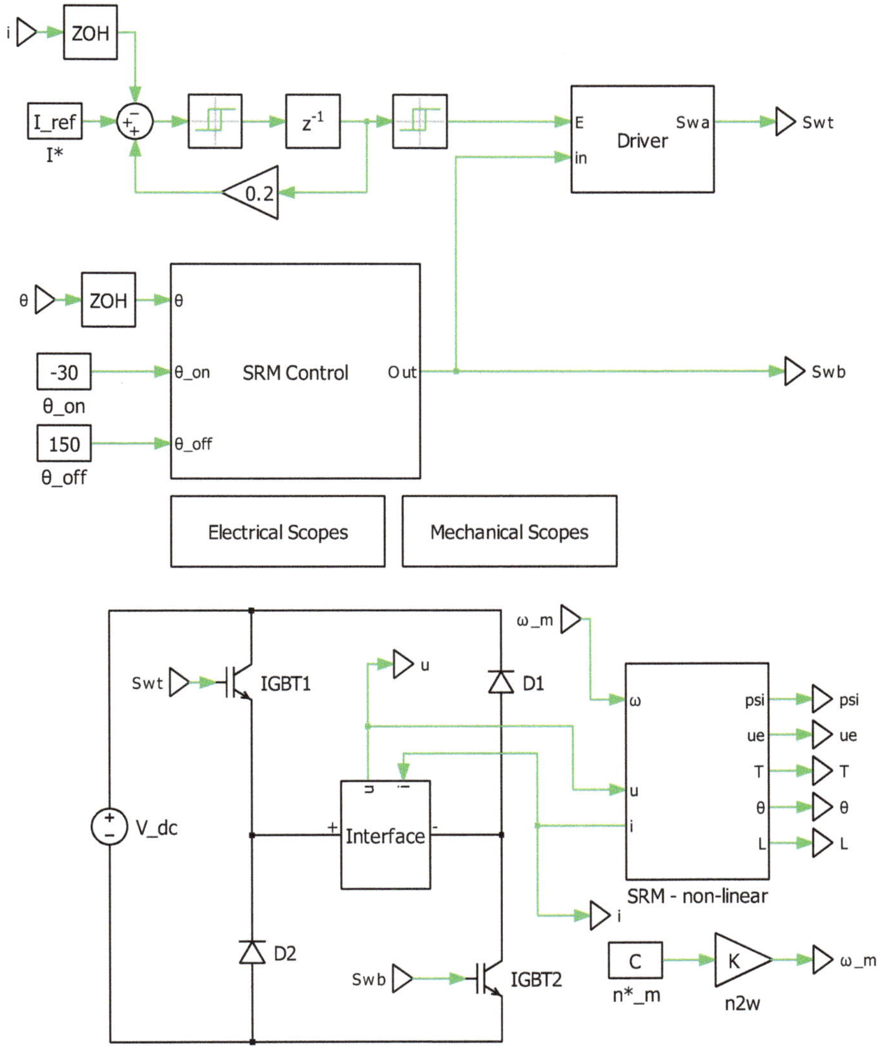

Fig. 10.51 Simulation model of a nonlinear SRM, with voltage excitation and hysteresis current controller

current error signal (i^*-i_{phase}). It is left to the user to implement this approach. The object of this tutorial is to modify the simulation discussed in the previous tutorial by replacing the hysteresis controller with a PWM based controller. The machine model, machine characteristics, and drive operating conditions remain unchanged compared to the previous case. An example of the simulation, as given in Fig. 10.54 shows the revised PWM controller with the duty cycle as input variable.

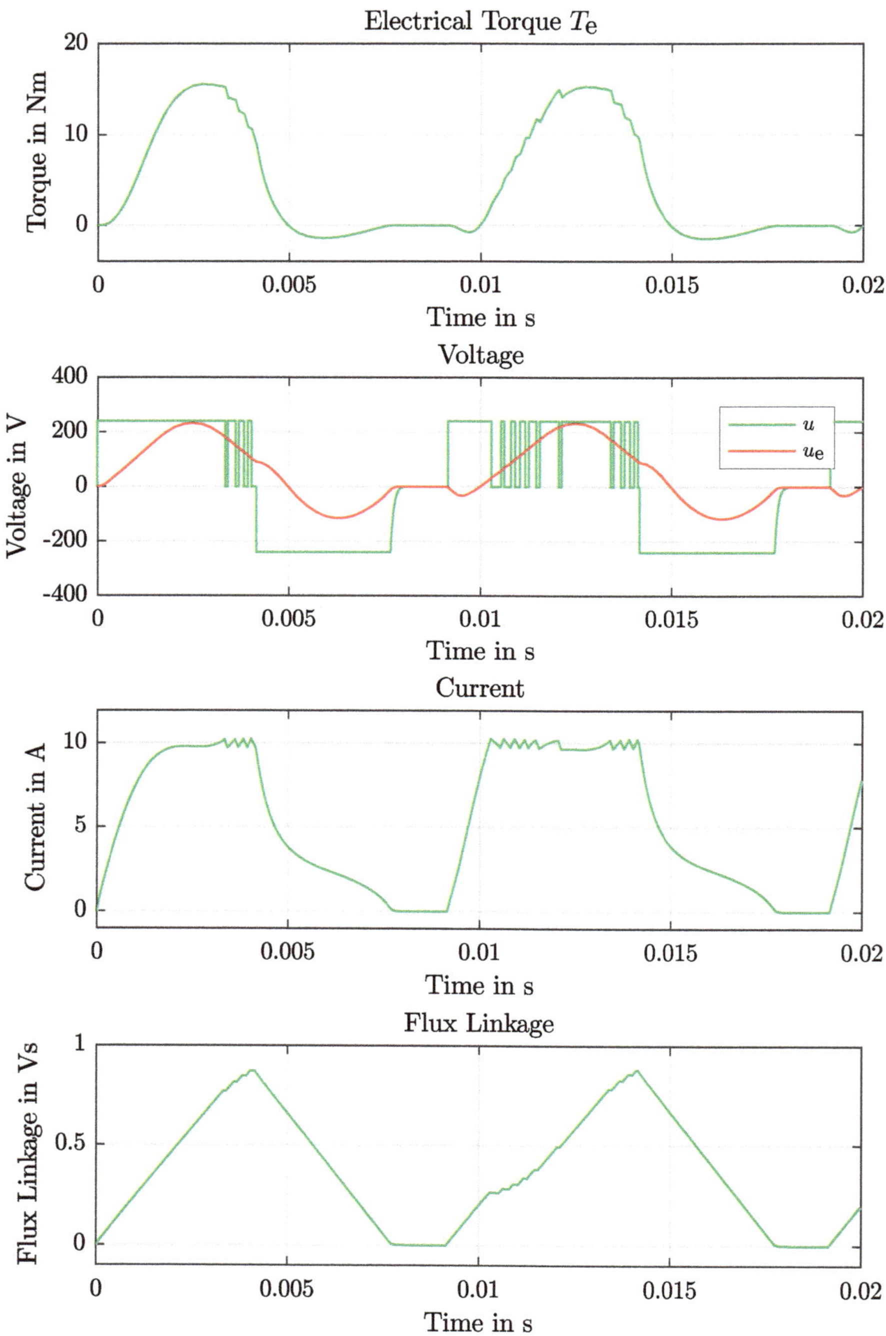

Fig. 10.52 Simulation results of a nonlinear SRM, with voltage excitation and hysteresis current controller, $n_m = 1000\,\text{rpm}$, $\theta_{on} = -30°$, $\theta_{off} = 150°$

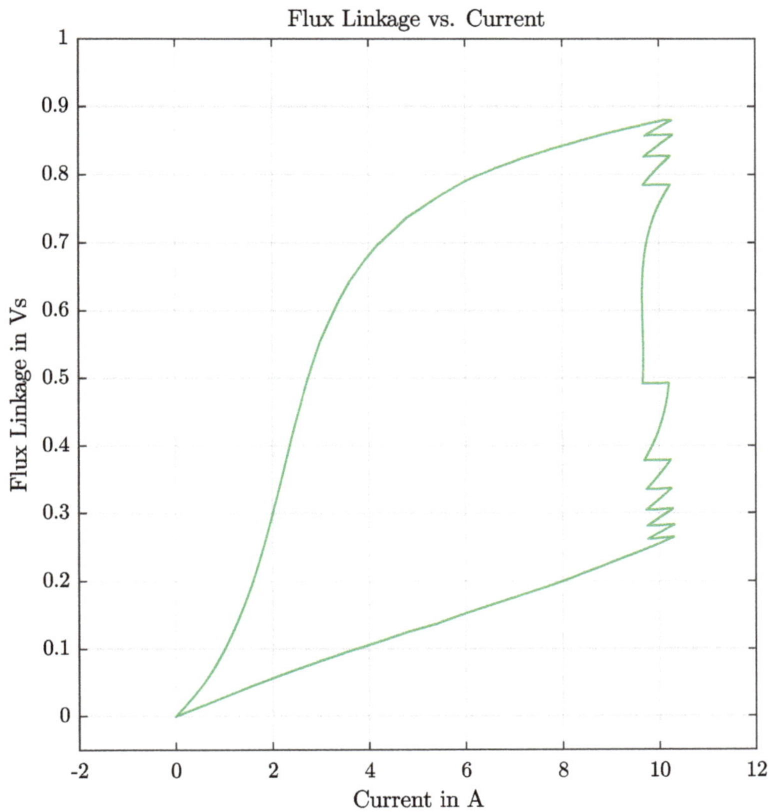

Fig. 10.53 Simulation results of a nonlinear SRM, with voltage excitation and hysteresis current controller, flux linkage/current diagram with $n_{\mathrm{m}} = 1000\,\mathrm{rpm}$ and $\theta_{\mathrm{on}} = -30°$, $\theta_{\mathrm{off}} = 150°$

In the simulation, the PWM switching frequency has been purposely set low to $f_{\mathrm{PWM}} = 2000\,\mathrm{Hz}$ to better visualize the impact on the waveforms with this type of control strategy. The maximum phase current which appears in the simulation should be limited by the user to a value below the maximum current value of 12 A by a prudent choice of the duty cycle. Otherwise, the lookup tables used for the representation of the machine characteristics may yield incorrect results. For the selected shaft speed of $n_{\mathrm{m}} = 1000\,\mathrm{rpm}$, a duty cycle value of $\delta = 0.4$ is used. The output is represented with the same set of scopes used in the previous tutorial, as given in Figs. 10.55 and 10.56.

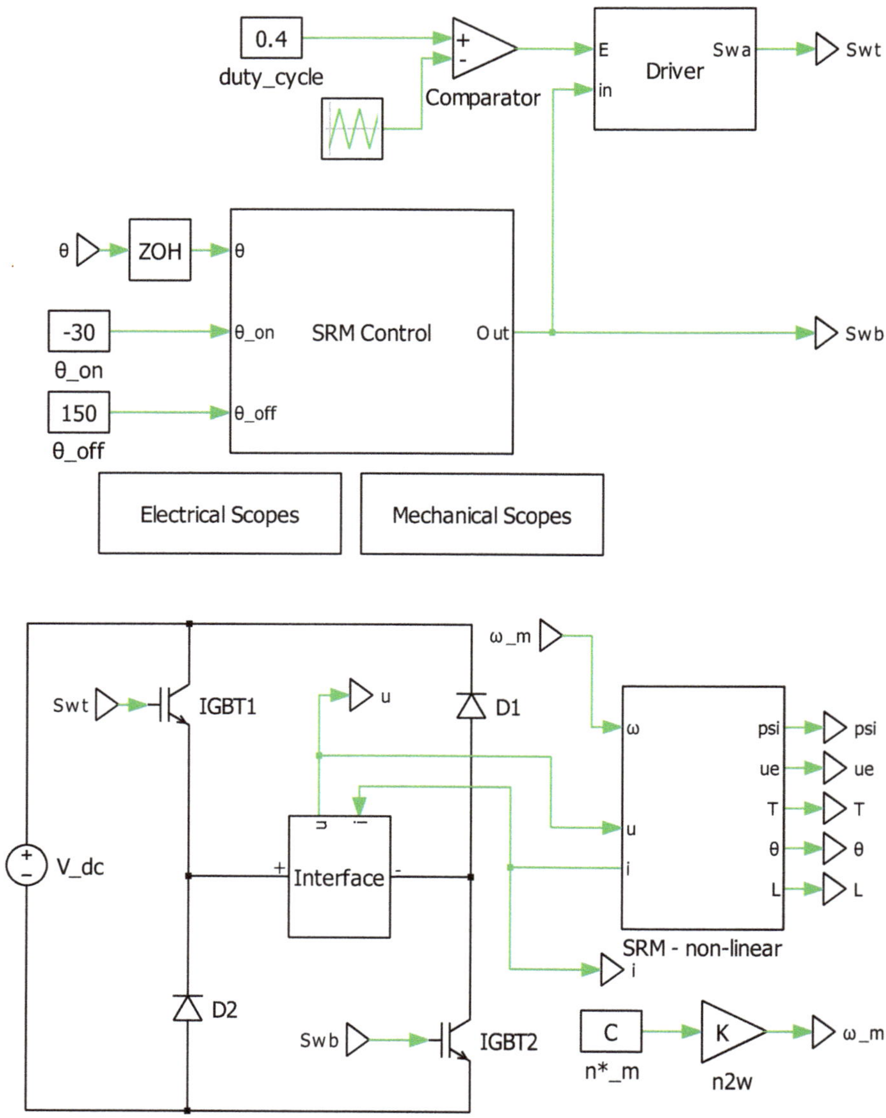

Fig. 10.54 Simulation model of a nonlinear SRM, with voltage excitation and PWM controller

10.6.4 Tutorial 4: Four-Phase Nonlinear SR Model, with Voltage Excitation and Hysteresis Control

The tutorial according to Sect. 10.6.2 was concerned with a single-phase representation of an 8/6 SR drive operating with hysteresis type current control. This tutorial considers an extension of this simulation model to encompass all four phases of the

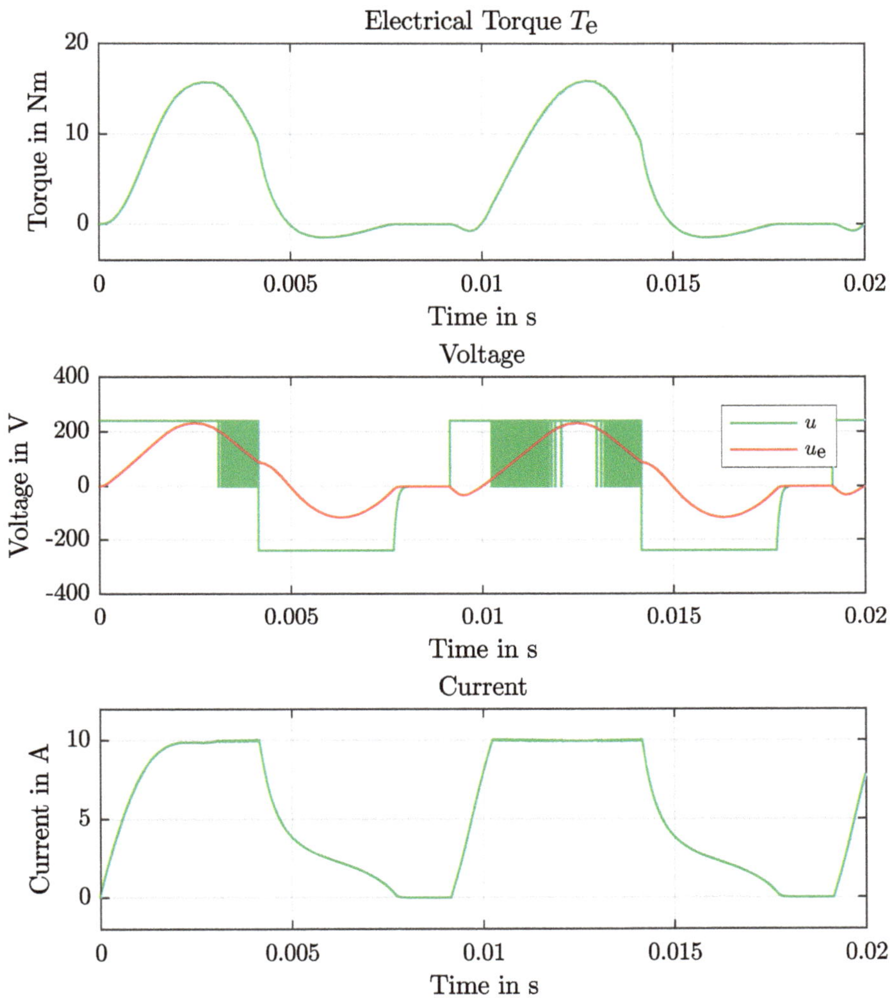

Fig. 10.55 Simulation results of a nonlinear SRM, with voltage excitation and PWM controller

drive. Subsequently, the resultant shaft torque may be examined and furthermore the simulation may be extended to accommodate, the combined load/machine inertia, set to $J = 0.001\,\text{kgm}^2$. A quadratic load torque speed characteristic is assumed which must be set to provide a load torque of 100 Nm at $n_m = 3000\,\text{rpm}$. The simulation should be able to accommodate drive operation under constant and variable speed operation.

An example of a simulation model as given in Fig. 10.57 shows the complete dynamic SR drive structure. Immediately observable are the four-phase asymmetric half-bridge converter, a four-phase SR machine (each phase is modeled separately), and four control modules, which include the control of the angle and the hysteresis

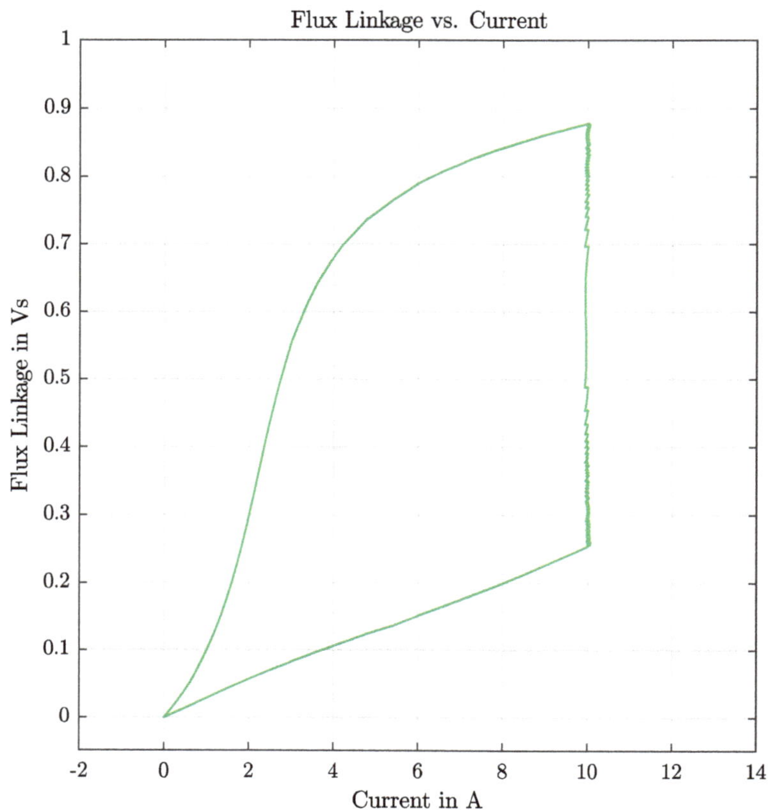

Fig. 10.56 Simulation results of a nonlinear SRM, with voltage excitation and with voltage excitation and PWM controller, flux linkage/current diagram

controller, similar to Sect. 10.6.2. The machine module also represents the model discussed in Sect. 10.6.2. The torque outputs of these modules are added, which gives the total machine torque T_{m}. The total torque is fed to a mechanical load, to calculate the speed and the position of the rotor. This block uses an integrator block with the gain $1/J$, to simulate the inertia of the rotor.

The first set of results, as given in Fig. 10.58, shows machine start-up without load. For this example, the current reference setting and control angles were set to $i^* = 6.45$ A and $\theta_{\mathrm{on}} = 0°$, $\theta_{\mathrm{off}} = 132°$, respectively. A total simulation time of 60 ms was chosen, which allows the machine to accelerate from zero speed to approximately 3000 rpm.

The second set of results, as given in Fig. 10.59, shows operation under quadratic load. In this case, the shaft speed quickly settles at about 100 rad/s, while the current reference setting and control angles were set to $i^* = 6.45$ A and $\theta_{\mathrm{on}} = 0°$, $\theta_{\mathrm{off}} = 132°$, respectively. Note that operation under constant speed is convenient for optimizing drive performance, for example in terms of selecting the required control angles which yield the highest average torque.

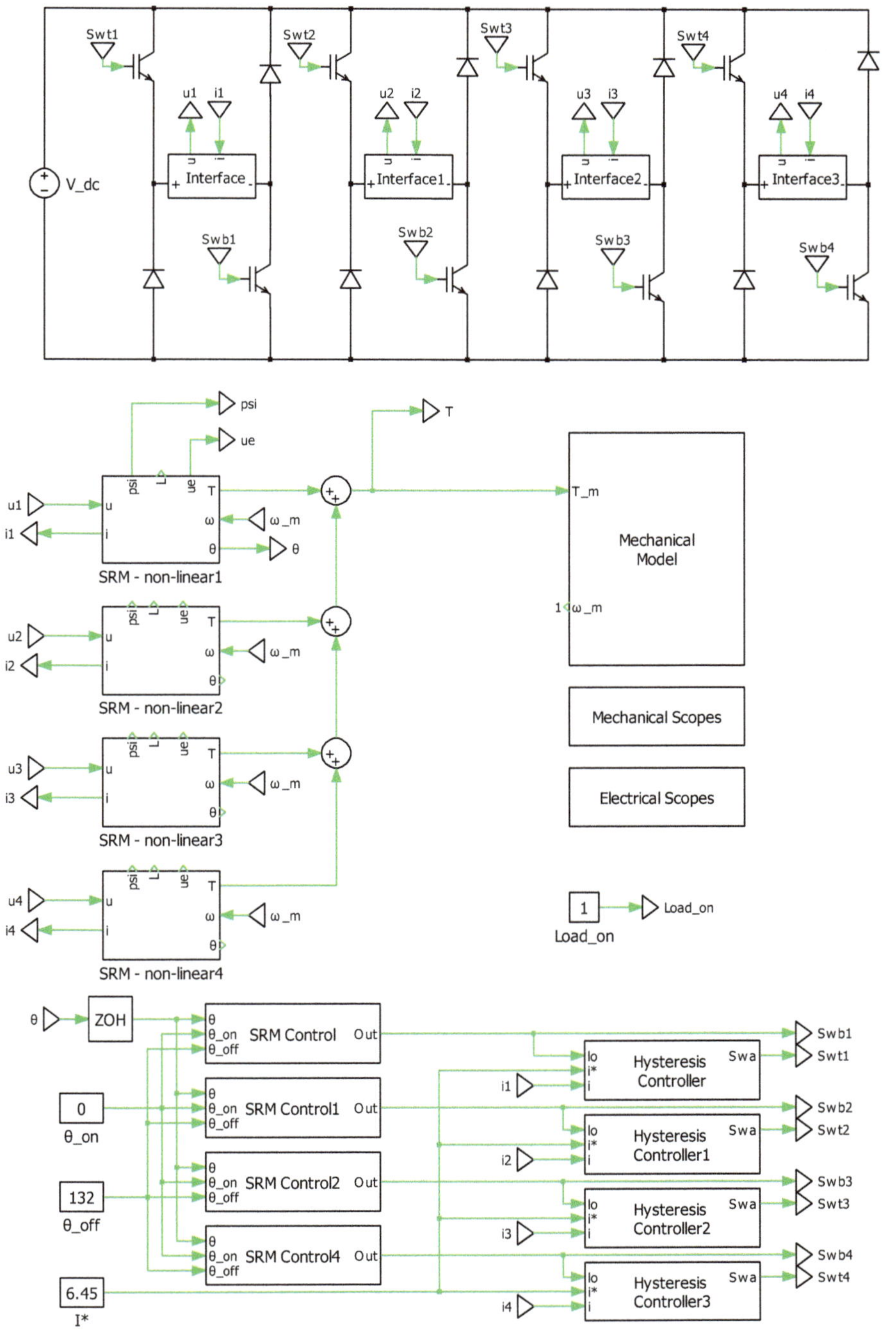

Fig. 10.57 Simulation model of a four-phase nonlinear SRM, with voltage excitation and hysteresis current controller

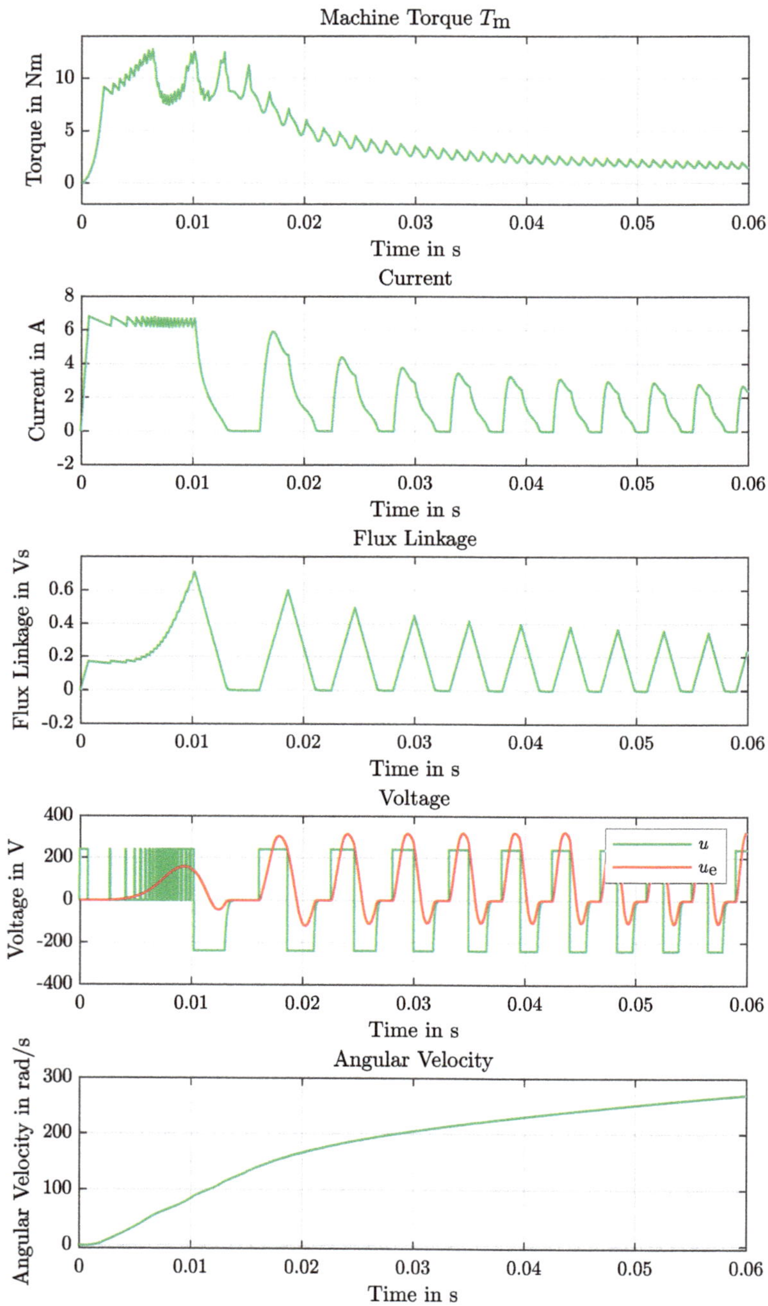

Fig. 10.58 Simulation results of the start-up of a nonlinear four-phase SRM, with voltage excitation and hysteresis controller, no-load, $i^* = 6.45$ A, $\theta_{on} = 0°$, $\theta_{off} = 132°$

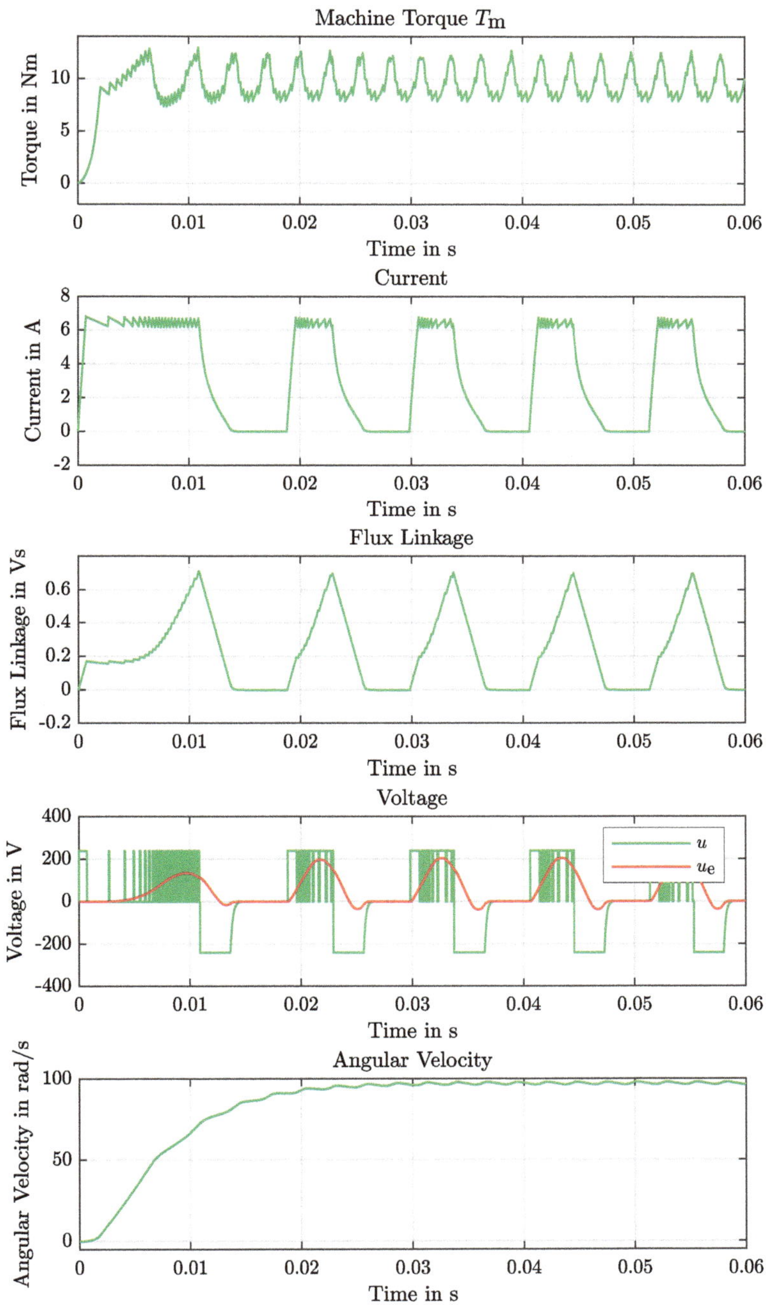

Fig. 10.59 Simulation results of the start-up of a nonlinear four-phase SRM, with voltage excitation and hysteresis controller, quadratic load, $i^* = 6.45$ A, $\theta_{on} = 0°$, $\theta_{off} = 132°$

An interesting exercise, left to the reader, is to adapt this simulation model to a drive configuration shown (for a three-phase machine) in Fig. 10.28. In this case, the single top switch operation may be equally well modeled by modulating all four top switches simultaneously (and continuously, i.e., independent of the phase-active interval signal), with the aid of a PWM generator as used in the tutorial given in Sect. 10.6.3. Of interest is to consider the impact on the simulation waveforms as a result of varying the duty cycle as well as drive speed.

10.6.5 Tutorial 5: Four-Phase Nonlinear SR Model, with Voltage Excitation and Direct Instantaneous Torque Control (DITC)

The final tutorial in this chapter considers the use of a DITC type control method for an SR drive as discussed in Sect. 10.4.3. A suitable starting point for this exercise is the previous tutorial, which deploys a set of hysteresis type current controller together with the phase-active signals to manipulate the converter switches. For this tutorial, a DITC type control structure as given in Fig. 10.38 is to be examined. For this purpose, the simulation model according to Fig. 10.57 is to be modified to accommodate a DITC control module which uses the measured torque T_{m} (in this case taken directly from the SR machine model), the reference torque T_{e}^{*}, the four *measured* phase currents i_1, i_2, i_3, i_4, and four phase-active signals $g_1^{\mathrm{hi,lo}}$, $g_2^{\mathrm{hi,lo}}$, $g_3^{\mathrm{hi,lo}}$, $g_4^{\mathrm{hi,lo}}$.

The simulation model as given in Fig. 10.60 shows a possible implementation of the proposed tutorial. Readily apparent in the Fig. 10.60 is the presence of a DITC controller unit with the required inputs. The module itself has a control structure, which is in accordance with the model given in Fig. 10.38. In this simulation the control angles θ_{on} and θ_{off} may be chosen by the user. In industrial DITC SR drive applications lookup tables are used to generate the optimum set of angles at each shaft speed. Both converter switches of each of the four phases are directly controlled by the DITC module, as may be observed from Fig. 10.60.

In this example the torque reference value is set to $T_{\mathrm{e}}^{*} = 10\,\mathrm{Nm}$, which is also used to generate the numerical results shown in Sect. 10.4.3. A set of scopes identical to those used in the previous tutorial are used to evaluate the simulated waveforms. An example of the results, obtained with the DITC controller as given in Fig. 10.61, shows a start-up sequence under no-load. Both control angles have been purposely set to $\theta_{\mathrm{on}} = -20°$ and $\theta_{\mathrm{off}} = 132°$.

Observation of the results shown in Fig. 10.61 shows that the controller is able to maintain the torque at the reference value below the base speed of the drive, which is 1468 rpm for this drive. As speed increases, the back-EMF increases (as may be observed from SCOPE3, trace u_{e}) which implies that the current levels reduce. Operation above the base speed is for both drives (hysteresis and DITC based) identical given that both use the same control angle settings.

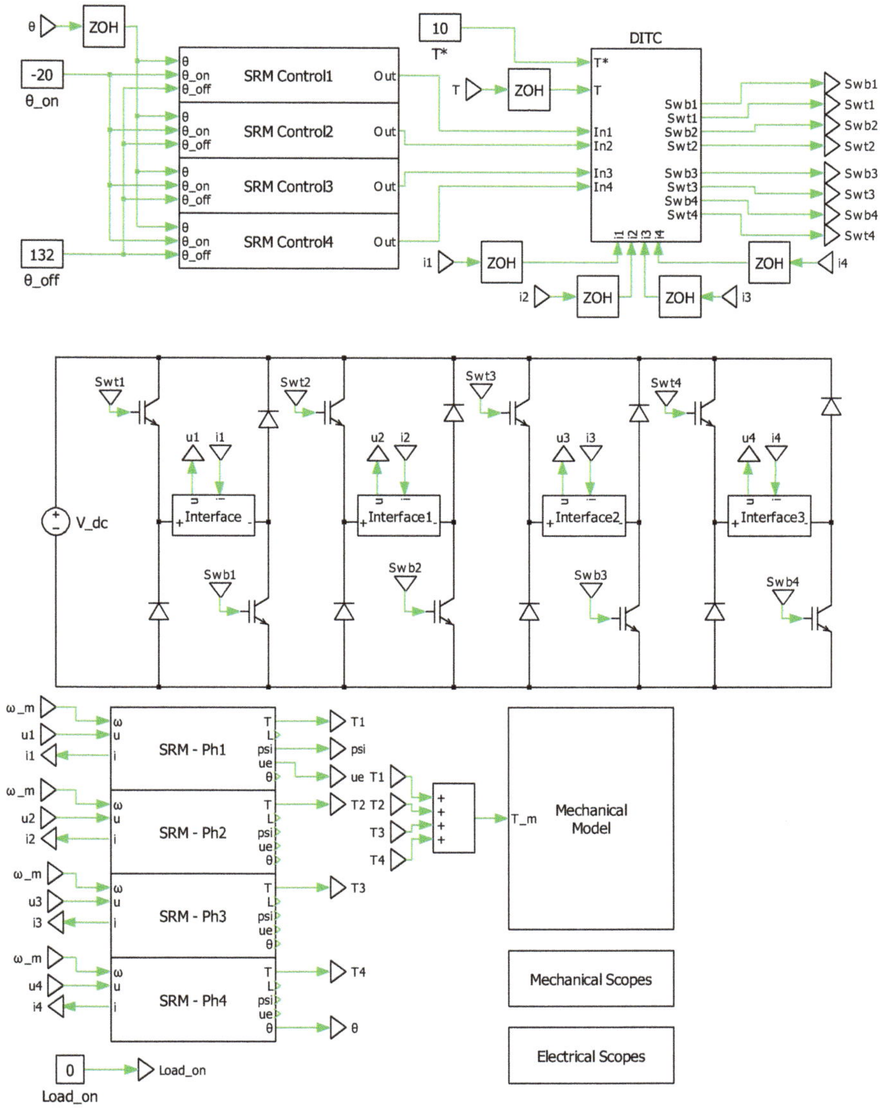

Fig. 10.60 Simulation model of a four-phase nonlinear SRM, with voltage excitation and DITC

The second set of results, illustrated in Fig. 10.62, shows operation under quadratic load, resulting in an almost constant speed. Observation of the results according to Fig. 10.62 shows that the DITC controller is able to maintain a virtually constant shaft torque at the chosen speed. This is markedly different compared to the results obtained with the hysteresis type control under similar drive conditions as may be observed by comparing Figs. 10.62 and 10.59.

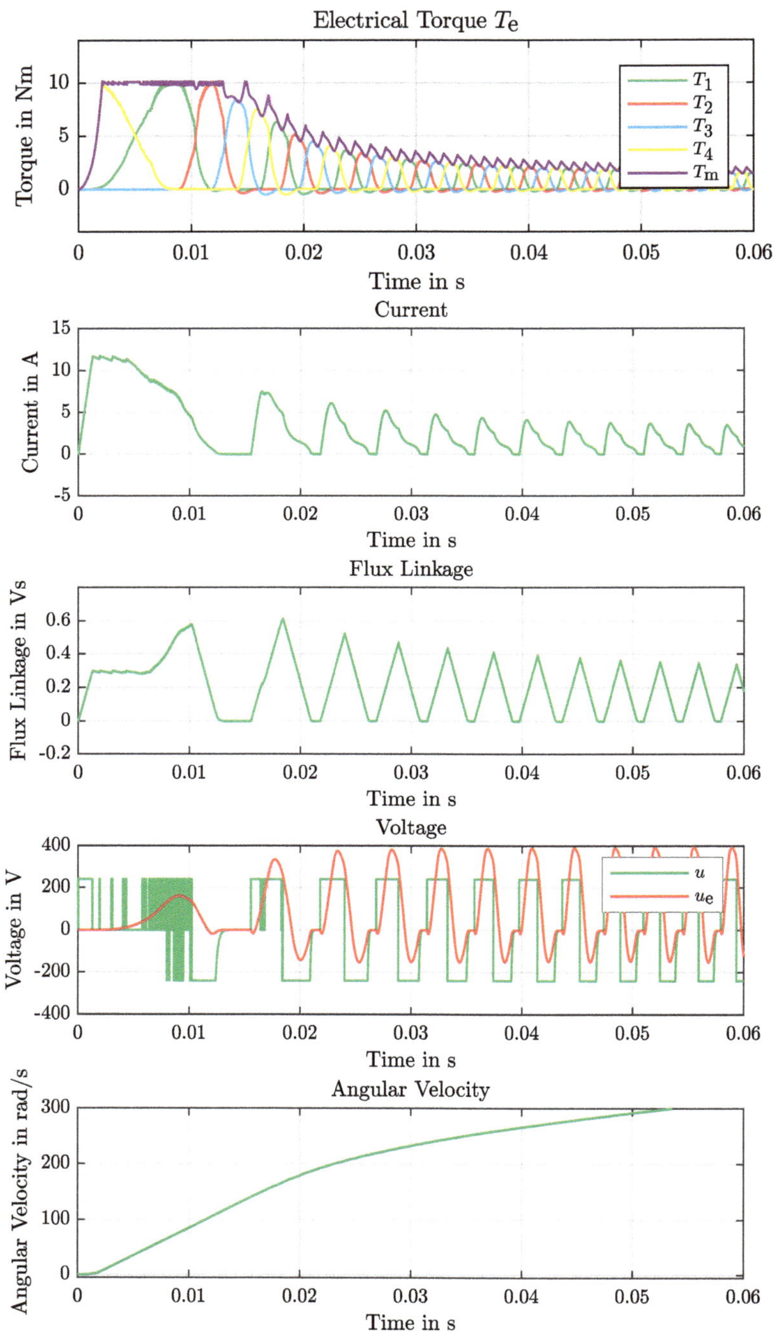

Fig. 10.61 Simulation results of a nonlinear four-phase SRM, with voltage excitation and DITC based controller, no-load, $i^* = 6.45$ A, $\theta_{on} = 0°$, $\theta_{off} = 132°$

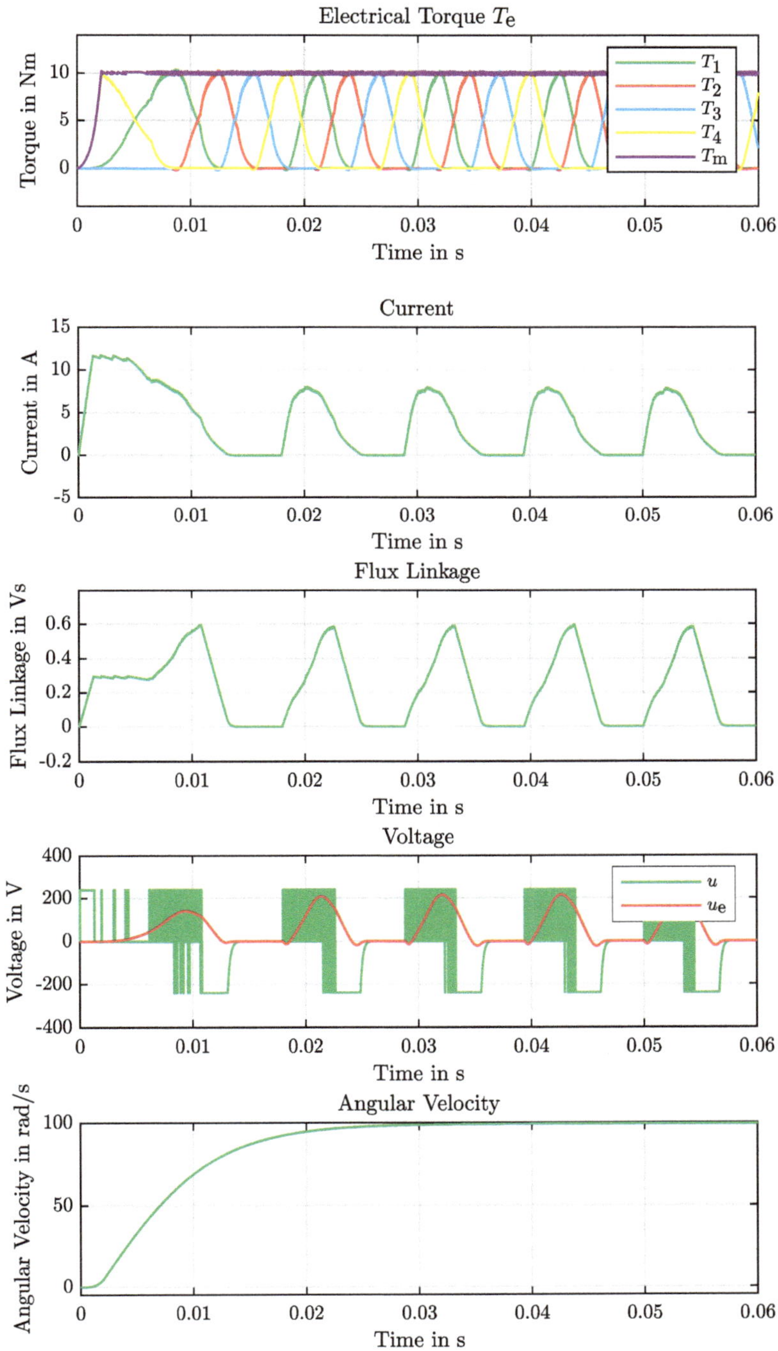

Fig. 10.62 Simulation results of a nonlinear four-phase SRM, with voltage excitation and DITC based controller, $T_e^* = 10.0\,\text{Nm}$, quadratic load of 10 Nm at 100 rad/s

A negative turn-on angle also allows the controller to get rid of the dips in the total torque. The reader is encouraged to explore how the turn-on and turn-off angles affect the output phase torques and the total torque.

References

1. Backhaus K, Link L, Reinert J (1995) Investigations on a high-speed switched reluctance drive incorporating amorphous iron. In: European conference on power electronics and applications, vol 1, pp 1.460–1.464
2. Bausch H, Rieke R (1978) Performance of thyristor-fed electric car reluctance machines. In: Proceedings of the international converence on electrical machines, Brussels, pp E4/2–1–E4/2–10
3. De Doncker R (2003) Twenty years of digital signal processing in power electronics and drives. In: ECON'03. 29th annual conference of the IEEE industrial electronics society, 2003, vol 1, pp 957–960. https://doi.org/10.1109/IECON.2003.1280112
4. Fuengwarodsakul N, Bauer S, Krane J, Dick C, De Doncker R (2005) Sensorless direct instantaneous torque control for switched reluctance machines. In: 2005 European conference on power electronics and applications, pp 10 pp. - P.7. https://doi.org/10.1109/EPE.2005.219423
5. Fuengwarodsakul N, Menne M, Inderka R, De Doncker R (2005) High-dynamic four-quadrant switched reluctance drive based on ditc. IEEE Trans Ind Appl 41(5):1232–1242. https://doi.org/10.1109/TIA.2005.853381
6. Fuengwarodsakul NH, Bauer SE, De Doncker RW (2006) Characteristic measurement system for automotive class switched reluctance machines. Eur Pow Electr Drives (EPE) J 16(3):10 pp - P.10
7. Gallegos-Lopez G, Kjaer P, Miller T (1999) High-grade position estimation for SRM drives using flux linkage/current correction model. IEEE Trans Ind Appl 35(4):859–869. Also in Industry Applications Conference, 1998. Thirty–Third IAS Annual Meeting. The 1998 IEEE. https://doi.org/10.1109/28.777195
8. Inderka R, De Doncker R (2003) DITC-direct instantaneous torque control of switched reluctance drives. IEEE Trans Ind Appl 39(4):1046–1051. Also in Industry Applications Conference, 2002. Conference Record of the 37th IAS Annual Meeting. https://doi.org/10.1109/TIA.2003.814578
9. Inderka R, De Doncker R (2003) High-dynamic direct average torque control for switched reluctance drives. IEEE Trans Ind Appl 39(4):1040–1045. Also in Industry Applications Conference, 2001. Conference Record of the 2001 IEEE Thirty–Sixth IAS Annual Meeting. https://doi.org/10.1109/TIA.2003.814579
10. Inderka R, Menne M, De Doncker R (2002) Control of switched reluctance drives for electric vehicle applications. IEEE Trans Ind Electr 49(1):48–53. https://doi.org/10.1109/41.982247
11. Kasper K, Bosing M, De Doncker R, Fingerhuth S, Vorlander M (2007) Noise radiation of switched reluctance drives. In: PEDS'07. 7th international conference on power electronics and drive systems, 2007, pp 967–973. https://doi.org/10.1109/PEDS.2007.4487821
12. Lawrenson P, Stephenson J, Fulton N, Blenkinsop P, Corda J (1980) Variable-speed switched reluctance motors. IEE Proc B Electr Pow Appl 127(4):253–265. https://doi.org/10.1049/ip-b:19800034
13. Miller T (1993) Switched reluctance motors and their control, illustrated edition. Oxford University Press, Oxford
14. Morel L, Fayard H, Vives Fos H, Galindo A, Abba G (2000) Study of ultra high speed switched reluctance motor drive. In: Conference record of the 2000 IEEE industry applications conference, 2000, vol 1, pp 87–92. https://doi.org/10.1109/IAS.2000.881030

15. Pulle DWJ (1982) Prediction and analysis of variable reluctance stepmotor-drive systems. PhD Thesis, University of Leeds, Leeds
16. Pulle D (1989) Switched reluctance motor. In: Australian patent AU4005589
17. Pulle DWJ (1991) New database for switched reluctance drive simulation. In: IEE Proceedings B electric power applications, vol 138, pp 331–337
18. Pulle DWJ (1991) A novel axial flux switched reluctance motor for variable speed drive operations. In: Australian conference on industrial drives, Townsville
19. Radun A (1992) High-power density switched reluctance motor drive for aerospace applications. IEEE Trans Ind Appl 28(1):113–119. https://doi.org/10.1109/28.120219
20. Reinert J, Schroder S (2002) Power-factor correction for switched reluctance drives. IEEE Trans Ind Electr 49(1):54–57. https://doi.org/10.1109/41.982248
21. Reinert J, Inderka R, Menne M, De Doncker R (1998) Optimizing performance in switched reluctance drives. In: APEC'98 Thirteenth annual applied power electronics conference and exposition, 1998, vol 2, pp 765–770. https://doi.org/10.1109/APEC.1998.653984
22. Stephens CM (1989) Fault detection and management system for fault tolerant switched reluctance motor drives. In: Conference record of the IEEE industry applications society annual meeting, vol 1, pp 574–578. https://doi.org/10.1109/IAS.1989.96707
23. Stephenson JM, Corda J (1979) Computation of torque and current in doubly salient reluctance motors form nonlinear magnetisation data. IEE Proc Part B 126:5
24. Taylor W (1839). Obtaining motive power, Patent No. 8255

Index

R. W. De Doncker et al., *Advanced Electrical Drives*, Power Systems,
https://doi.org/10.1007/978-3-030-48977-9

www.ingramcontent.com/pod-product-compliance
Ingram Content Group UK Ltd.
Pitfield, Milton Keynes, MK11 3LW, UK
UKHW021315130926
15127UKWH00002B/11

9783030489762

1 1
1 5
2 1